D0026166

Introduction to
Holography

Introduction to
Holography

Vincent Toal

CRC Press
Taylor & Francis Group
Boca Raton   London   New York

CRC Press is an imprint of the
Taylor & Francis Group, an **Informa** business

The front cover image was prepared by Dr. Izabela Naydenova using the technique of dye deposition holography, which is described in Chapter 17.

CRC Press
Taylor & Francis Group
6000 Broken Sound Parkway NW, Suite 300
Boca Raton, FL 33487-2742

© 2012 by Taylor & Francis Group, LLC
CRC Press is an imprint of Taylor & Francis Group, an Informa business

No claim to original U.S. Government works

Printed in the United States of America on acid-free paper
Version Date: 20110815

International Standard Book Number: 978-1-4398-1868-8 (Hardback)

This book contains information obtained from authentic and highly regarded sources. Reasonable efforts have been made to publish reliable data and information, but the author and publisher cannot assume responsibility for the validity of all materials or the consequences of their use. The authors and publishers have attempted to trace the copyright holders of all material reproduced in this publication and apologize to copyright holders if permission to publish in this form has not been obtained. If any copyright material has not been acknowledged please write and let us know so we may rectify in any future reprint.

Except as permitted under U.S. Copyright Law, no part of this book may be reprinted, reproduced, transmitted, or utilized in any form by any electronic, mechanical, or other means, now known or hereafter invented, including photocopying, microfilming, and recording, or in any information storage or retrieval system, without written permission from the publishers.

For permission to photocopy or use material electronically from this work, please access www.copyright.com (http://www.copyright.com/) or contact the Copyright Clearance Center, Inc. (CCC), 222 Rosewood Drive, Danvers, MA 01923, 978-750-8400. CCC is a not-for-profit organization that provides licenses and registration for a variety of users. For organizations that have been granted a photocopy license by the CCC, a separate system of payment has been arranged.

Trademark Notice: Product or corporate names may be trademarks or registered trademarks, and are used only for identification and explanation without intent to infringe.

Library of Congress Cataloging-in-Publication Data

Toal, Vincent.
   Introduction to holography / Vincent Toal.
      p. cm.
   ISBN 978-1-4398-1868-8 (hardcover : alk. paper)
   1. Holography. I. Title.

QC449.T63 2011
621.36'75--dc22   2011011414

Visit the Taylor & Francis Web site at
http://www.taylorandfrancis.com

and the CRC Press Web site at
http://www.crcpress.com

For
Catherine, Helen, and Phoebe
and for Emilia
and in memory of my parents

* * * * *

"Explanations come to an end somewhere."

L. Wittgenstein

CONTENTS

Preface

The author, as a boy of nine, saw his first black and white photograph slowly appear in a developing dish and the image from his first hologram a number of decades later. The two experiences evoked much the same sense of wonder and amazement.

Holography, since its invention by Nobel laureate Denis Gabor in 1948 and the advent of the laser in the late 1950s, has evolved into a mature technology with a wide range of applications beginning with the three-dimensional images of astounding realism produced by Emett Leith and Juris Upatneiks in the mid-1960s. It is perhaps not surprising that semiologist Umberto Eco devoted some thought to holographic images and their significance in his essay "Travels in Hyperreality."

The first holograms could only be viewed with laser light, but 3-D holographic images, which could be viewed in white light, soon followed, with the inventions by Yuri Denisyuk of reflection holography and by Stephen Benton of rainbow holography. It is interesting to note that the Denisyuk technique is very closely related to the century-old process developed by Nobel laureate Gabriel Lippmann for color photography, whose present-day success is a result of the drive toward higher resolution silver halide emulsions needed for full color holography.

These developments have made holography almost all pervasive with its use in advertising and on credit cards, theater tickets, passports, driving licenses, and packaging. At the same time, artists have exploited holographic techniques in order to create very striking and original art objects, despite the fact that a hologram invariably has to be created from an object that is real and concrete, or at least exists as computed data.

Conventional photographic and video techniques have been combined with holography with increasing success both for production of extremely large holograms and of moving images, and the goal of video holography appears to be drawing close to reality.

The 1970s saw holography striking out in a number of different directions. The possibilities inherent in the production of conjugate waves led to a number of interesting applications including diffraction-limited imaging and the descrambling of light wavefronts. Evanescent wave holography was also actively researched and exploited for the study of biomaterials.

The realization that the usually unwanted fringe patterns appearing on a holographically reconstructed image were an indication of submicrometer movement, led to the development of holographic interferometry, one of the most sensitive and powerful tools now available for nondestructive evaluation and testing, along with the closely related technique of electronic speckle pattern interferometry.

The fact that many optical devices have holographic equivalents has stimulated holographic optical element design for a wide range of applications from simple optical components and optical circuitry to display screens and solar energy devices.

The vast data capacity inherent in volume, as opposed to surface, storage media, has led to holographic data storage with its added advantages of associative recall and mathematical operations. A range of data multiplexing techniques and systems have been explored including phase-coded systems with no moving parts.

Methods for overcoming the resolution barrier associated with CCD cameras, as well as powerful computational hardware and software, enabled digital holography and holographic microscopy to develop rapidly as research tools.

Computer-generated holography has resulted in the creation and display of images, as well as the subtle control of light for manipulation of submicrometer-sized objects using holographic optical tweezers. These have provided researchers with new tools for study of phenomena at the subcellular level and for nanoscale assembly and manufacturing and the dynamic control of light itself for driving micromachines.

Holographic light-in-flight has enabled us to confirm with unprecedented clarity how light propagates through space and how it is reflected, transmitted, and focused. Holographic first-arriving light also enables us to image through tissue including living tissue.

Low-cost holographic sensing techniques have also been developed and look set to make a valuable contribution to health care and environmental monitoring and protection.

The history of holography over the past 60 years has been characterized by extraordinary ingenuity in overcoming both real and apparent obstacles, combined with the ability to exploit innovations unrelated to optics. Holography has enabled us to develop new insights into the nature of light and to exploit it in new and exciting ways.

This book is written for advanced undergraduate and graduate students including those contemplating a research career as well as for more experienced scientists and engineers considering the application of holographic techniques to the solution of specific problems in widely diverse fields. It deals with the state of development of various holographic applications, including more recent ones. Numerical problems are provided at the end of each chapter. Some are designed to supplement the material presented in the main text. No more mathematical skill is assumed than strictly necessary.

Part I reviews some essential optics. Part II is concerned with basic holographic principles, beginning with an introduction in Chapter 3. The theory of thick holograms is discussed in Chapter 4, which concludes with a rather less demanding and more insightful path to some of the more important results for which I and, I am sure, many others are grateful to the late J. E. Ludman. Part III deals with holography in practice beginning in Chapter 5 with essential conditions for successful holography, including the various lasers in common use for holography. Recording materials are described in Chapter 6 and their important holographic characteristics in the following chapter.

Part IV, the remainder of the book, deals with the applications of holography under various headings including imaging (Chapters 8 and 9), holographic interferometry (Chapter 10), holographic optical elements (Chapter 11), and data storage (Chapter 12). Discussions of digital (Chapter 13) and computer-generated holography (Chapter 14) then follow. Light-in-flight and first-arriving light techniques and their applications are considered in Chapter 15. Polarization holography is treated separately in the following chapter and finally we look at holography for sensing applications.

In general, the tools of the trade are introduced as and when required. It is hoped that this approach will enable the reader to learn more about optics and optical devices as part of the exploration of holography.

A Nobel laureate has suggested that a textbook is perhaps better written by someone afraid to make mistakes. The author can therefore only apologize for the mistakes that have inevitably arisen from his temerity in attempting the task.

ACKNOWLEDGMENTS

Dr. Suzanne Martin, manager of the Center for Industrial and Engineering Optics (IEO), who has been my coworker for almost two decades and a true friend. I hope that her vision of what a research center should be will continue long into the future. Her assistance and advice have always been of enormous help. Suzanne read parts of the manuscript and made useful comments.

Dr. Izabela Naydenova, lecturer in the School of Physics at Dublin Institute of Technology (DIT), who read the manuscript and gave helpful advice. I thank her for her invaluable contribution to the work of IEO and for interesting discussions about holography and many other subjects. I thank Izabela also for allowing me to use results of her work on holographic sensors.

Past research students and staff at IEO who have helped in many ways.

Julie de Foubert, Tina Hayes, David Casey, and other library staff at DIT. I doubt if there is a better library service anywhere or a more helpful staff.

Dr. Yan Li at the General Engineering Research Institute, Liverpool John Moore's University for allowing me to make use of her doctoral thesis on digital holography.

James Callis at the School of Physics, DIT, for frequent advice and practical help with computers and software.

Dr. Dana Mackey at the School of Mathematical Sciences, DIT.

Colleagues at the School of Physics, DIT.

Luna Han at Taylor & Francis for advice and guidance.

AUTHOR

Vincent Toal is director of the Center for Industrial and Engineering Optics at the Dublin Institute of Technology (DIT). He gained the BSc degree in physics and mathematics from the National University of Ireland, an MSc in optoelectronics at Queen's University, Belfast, and a PhD in electronic engineering at the University of Surrey. He is a fellow of the Institute of Physics. Toal was formerly head of the School of Physics at DIT and has taught optics for two decades. He has worked on research in holography and related topics for most of his professional life and has written or cowritten a large number of research papers.

I

Optics

CHAPTER 1

Light, Waves, and Rays

1.1 INTRODUCTION

We need a clear and thorough understanding of the nature of light in order to be able to control and use it in holography and its applications.

1.2 DESCRIPTION OF LIGHT WAVES

James Clerk Maxwell discovered that light consists of electromagnetic waves by manipulating the four equations that bear his name that describe the behavior of the electric field, **E**, and the magnetic field, **H**.

The four equations are

$$\mathbf{\nabla} \times \mathbf{E} \left(= \frac{\partial E_z}{\partial y} - \frac{\partial E_y}{\partial z}, \frac{\partial E_x}{\partial z} - \frac{\partial E_z}{\partial x}, \frac{\partial E_y}{\partial x} - \frac{\partial E_x}{\partial y} \right) = -\mu_0 \mu_r \frac{\partial \mathbf{H}}{\partial t} \tag{1.1}$$

$$\mathbf{\nabla} \times \mathbf{H} = \varepsilon_0 \varepsilon_r \frac{\partial \mathbf{E}}{\partial t} + \sigma \mathbf{E} \tag{1.2}$$

$$\mathbf{\nabla} \cdot \mathbf{D} = \frac{\partial E_x}{\partial x} + \frac{\partial E_y}{\partial y} + \frac{\partial E_z}{\partial z} = 0 \quad \text{(assuming no volume charges)} \tag{1.3}$$

$$\mathbf{\nabla} \cdot \mathbf{H} = 0 \tag{1.4}$$

where μ_r is the relative permeability (=1 in a medium which is not magnetic), μ_0 is the permeability of a vacuum, ε_0 its permittivity, and ε_r, σ are, respectively, the relative permittivity and the conductivity of the medium in which the fields are present. $\mathbf{D} = \varepsilon_0 \varepsilon_r \mathbf{E}$ is the electric displacement. We use heavy type to denote vector quantities.

From Equations 1.1 and 1.2,

$$\mathbf{\nabla} \times \mathbf{\nabla} \times \mathbf{E} = -\mu_0 \frac{\partial (\mathbf{\nabla} \times \mathbf{H})}{\partial t} = -\mu_0 \frac{\partial \left(\varepsilon_0 \varepsilon_r \dfrac{\partial \mathbf{E}}{\partial t} + \sigma \mathbf{E} \right)}{\partial t} = -\mu_0 \varepsilon_0 \varepsilon_r \frac{\partial^2 \mathbf{E}}{\partial t^2} \tag{1.5}$$

since $\sigma = 0$.

$$\nabla \times \nabla \times \mathbf{E} = \nabla(\nabla \cdot \mathbf{E}) - \nabla^2 \mathbf{E} \qquad (1.6)$$

From Equation 1.3,

$$\nabla \cdot \mathbf{D} = 0 = \nabla \cdot (\varepsilon_0 \varepsilon_r \mathbf{E}) = \varepsilon_0 \nabla \cdot \nabla(\varepsilon_r) + \varepsilon_0 \varepsilon \nabla \cdot \mathbf{E} \qquad (1.7)$$

In a vacuum ε_r (=1) has zero gradient so that $\nabla \cdot \mathbf{E} = 0$ and, from Equation 1.6,

$$\nabla \times \nabla \times \mathbf{E} = -\nabla^2 \mathbf{E}$$

And from Equation 1.5,

$$\nabla^2 \mathbf{E} = \mu_0 \varepsilon_0 \frac{\partial^2 \mathbf{E}}{\partial t^2} \qquad (1.8)$$

which can be rewritten

$$\nabla^2 \mathbf{E} = \frac{1}{c^2} \frac{\partial^2 \mathbf{E}}{\partial t^2} \qquad (1.9)$$

Equation 1.9 is a classical wave equation and the velocity, c, of the wave is $(\mu_0 \varepsilon_0)^{-1/2}$. Inserting the values of μ_0($4\pi \times 10^{-7}$H·m^{-1}) and ε_0(8.85×10^{-12}F·m^{-1}), we find c is 3.0×10^8 ms^{-1}, which is the velocity of light in a vacuum. The logical deduction is that light itself must be an electromagnetic wave. The direction and amplitude of the electric field may vary. The speed of light depends on the medium in which it is traveling and is reduced from its vacuum value by $1/(\mu_r^{1/2} \varepsilon_r^{1/2})$whose value is characteristic of the medium. The factor $(\mu_r \varepsilon_r)^{1/2}$ is called the *refractive index* and is 1.33 for water and about 1.5 for many ordinary types of glass. The number of complete oscillations of the wave per second is the frequency, ν, which is also the number of crests or maxima of the wave that pass any point in 1 second and $1/\nu = T$, the period of the wave. The distance between crests is the wavelength, λ, so that the speed of light, c, is given by

$$c = \lambda \nu$$

The wavelength of visible light, ranging from about 380 to about 760 nm, is extremely small (a fraction of a micrometer), so its frequency of around 10^{15} Hz makes it nearly impossible for us to measure the instantaneous value of the electric field directly. We can only measure the intensity or irradiance. The fact that visible light has such a small wavelength is of great importance for scientists and engineers as well as optical metrologists and holographers. Figure 1.1 shows a "snapshot" of a light wave at a single instant in time.

The simplest source of light that one can imagine is an ideal point source radiating in all directions into a space, in which the refractive index is the same everywhere. Suppose the source is monochromatic; that is, only one unique wavelength is radiated. We assume the sinusoidal behavior of the electric field is continuous without interruptions or sudden changes in field strength. If we *could* measure the instantaneous electric field at points equidistant from the source, the readings would be perfectly synchronous, meaning that

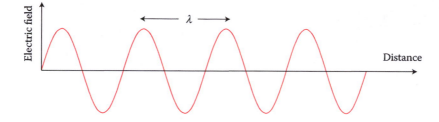

FIGURE 1.1 Light wave.

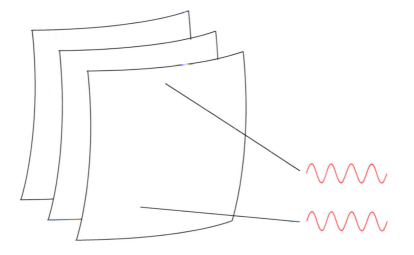

FIGURE 1.2 Spherical wavefronts; the electric field oscillates in perfect synchronism at all points on the wavefront.

the readings would both be maximum at the same instant and minimum a half period of oscillation later. In other words, the electric fields at these points are in phase with one another. A surface on which the electric field oscillates in perfect synchronism, or in phase, at all points is called a wavefront, and, in the case of a point source of monochromatic light, the wavefronts are spherical (Figure 1.2).

Light sources are not usually points, so wavefronts are rarely spherical and can be quite convoluted in shape.

1.3 SPATIAL FREQUENCY

Spatial frequency is a very important concept that we can use to arrive at a basic understanding of how holography works. Just as a wave has a specific *temporal frequency*, υ, which is the number of times per second that the electric field oscillates between maximum and minimum and back to maximum again, it also has a *spatial frequency*, which is the number of such oscillations that occur within a unit of distance, say 1 m. However, unlike the temporal frequency, the spatial frequency of a wave does not have a single value as we shall see. Let us imagine *plane* wavefronts, at a given instant in time, spaced exactly one wavelength, λ, apart as shown in Figure 1.3. We could choose to have them spaced

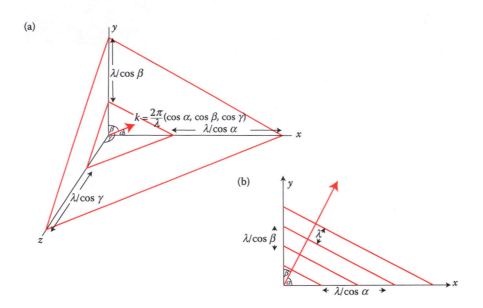

FIGURE 1.3 (a) Plane wavefronts and (b) their projections on the *x,y* plane.

apart by any fixed distance but λ is the most useful choice. The normal to the wavefronts makes angles of α, β, and γ, respectively, with the *x*, *y*, and *z* orthogonal axes. The distances between neighboring wavefronts measured along the axes are $\lambda/\cos\alpha$, $\lambda/\cos\beta$, and $\lambda/\cos\gamma$. The reciprocals of these distances, $\cos\alpha/\lambda$, $\cos\beta/\lambda$, $\cos\gamma/\lambda$, are the numbers of cycles of the wave that are measured per unit length along each of the axes and are called the *spatial frequencies*. The spatial frequency of a light wave obviously depends on the direction in which it is measured. We also define *angular spatial frequencies*

$$k_x = 2\pi\cos\alpha/\lambda, \; k_y = 2\pi\cos\beta/\lambda, \quad \text{and} \quad k_z = 2\pi\cos\gamma/\lambda.$$

1.4 THE EQUATION OF A PLANE WAVE

Now we are in a position to set up an equation that completely describes a plane wave in space and time. Suppose for simplicity the normal to the wavefront makes angle α with the *x*-axis and angles of $\pi/2$ with each of the other axes. This means that the spatial frequencies measured in those latter directions are both zero and

$$\mathbf{E}(x,y,z,t) = \mathbf{E}_0(t)\cos(k_x x - \omega t)$$

where $\omega = 2\pi v = 2\pi/T$. If the wave is traveling along the *x*-axis, α is zero and we have

$$\mathbf{E}(x,t) = \mathbf{E}_0(t)\cos\left(\frac{2\pi x}{\lambda} - \omega t\right)$$

By plotting $\mathbf{E}(x,t)$ at times $t = 0$ and $T/4$, we can easily show that this wave travels in the $+x$ direction.

If the wave is traveling in a direction that makes angles α, β, and γ respectively with the x, y, and z axes, we have

$$\mathbf{E}(x,y,z,t) = \mathbf{E}_0(t)\cos(k_x x + k_y y + k_z z - \omega t + \phi) \tag{1.10}$$

(An arbitrary phase ϕ is included so that, at $t = 0$ and $\mathbf{r} = (0,0,0)$, $\mathbf{E}(0,0,0,0) = \mathbf{E}_0\cos\phi$ although we can simply move the origin of coordinates so that $\phi = 0$). The expression $\mathbf{E}_0(t)$ implies that the *electric field can change direction with time*. For example, the electric field vector can continuously rotate around the direction of wave propagation while varying sinusoidally in magnitude between maximum and minimum values with frequency ν. The light is then *elliptically polarized*. If $\mathbf{E}_0(t)$ does not rotate, but is always in the same plane, the light is *plane polarized*, which we take to be the case here and we can drop the dependence on t. We can write $\mathbf{k}$ as the vector with components (k_x, k_y, k_z), known as the wave vector and the vector, $\mathbf{r}$, with components (x,y,z) so that finally Equation 1.10 can be written as

$$\mathbf{E}(\mathbf{r},t) = \mathbf{E}_0\cos(\mathbf{k}\cdot\mathbf{r} - \omega t + \phi) \tag{1.11}$$

The term $\mathbf{k}\cdot\mathbf{r} - \omega t + \phi$ is the phase. We have used the cosine and *assumed* that $\mathbf{E}$ is $\mathbf{E}_0\cos\phi$ at $t = 0$ and position $(0,0,0)$. We could equally well have used the sine form in Equation 1.11. In fact, it is often convenient, especially when we have to deal with more than one wave, to use the complex exponential form for each one and write Equation 1.11 as

$$\mathbf{E}(\mathbf{r},t) = \text{Re}\,\mathbf{E}_0\exp j(\mathbf{k}\cdot\mathbf{r} - \omega t + \phi) \tag{1.12}$$

and, in practice, we often use the expression $\mathbf{E}_0\exp j(\mathbf{k}\cdot\mathbf{r} - \omega t + \phi)$, which we call the *complex amplitude,* because it is easier to add exponential functions together than trigonometric ones.

In Section 1.2, it is pointed out that we cannot measure the instantaneous value of the electric field because it changes far too quickly. We can only measure the time average of the rate at which electromagnetic energy flows across unit area normal to the direction of flow. The *instantaneous* rate of energy flow across unit area is given by the vector cross product, $\mathbf{S}$, of the electric field vector, $\mathbf{E}$, and the magnetic field vector, $\mathbf{H}$, that is

$$\mathbf{S} = E_y H_z - E_z H_y,\ E_z H_x - E_x H_z,\ E_x H_y - E_y H_x = \mathbf{E}\times\mathbf{H} \tag{1.13}$$

$\mathbf{S}$ is called the Poynting vector and, since $\mathbf{E}$ and $\mathbf{H}$ are mutually perpendicular, its magnitude, S, is EH. The time average of S is given by (see problem 23)

$$\langle S\rangle = \langle EH\rangle = \left\langle \varepsilon_0 c E^2\right\rangle \tag{1.14}$$

where the angle brackets denote time average.

We can find $\langle S\rangle$ for a plane wave.

$$\langle S\rangle = \left\langle \varepsilon_0 c E^2\right\rangle = \varepsilon_0 c\left\langle \frac{E_0^2}{2}(1+\cos[2(\mathbf{k}\cdot\mathbf{r} - \omega t)])\right\rangle$$

The average of the cosine term is zero and we have

$$\langle S \rangle = \varepsilon_0 c \frac{E_0^2}{2} \tag{1.15}$$

In other words, the rate of flow of electromagnetic energy through unit area normal to the direction of energy flow, that is, the intensity or irradiance, is proportional to the square of its electric field amplitude.

Alternatively, the intensity, I, is proportional to $(1/2)\tilde{E}\tilde{E}^*$ where $\tilde{E}^*$ is the complex conjugate of $\tilde{E}$.

If we focus the beam from a laser whose total output power is 1 W into a spot of diameter 50 μm, the electric field at the focus, E_0, is

$$E_0 = \sqrt{2S/(\varepsilon_0 c)} = 2.5 \text{kVm}^{-1}.$$

A similarly focused laser of 10 kW power could easily induce an electrical spark in air.

1.5 NONPLANAR WAVEFRONTS

Plane wavefronts are a special case of wavefront shape. We make liberal use of them because any wavefront, no matter how convoluted its shape, can be regarded as made up of a large number of *plane* wavefronts, each tangential to the actual wavefront and all propagating in different directions (Figure 1.4). This statement is the basis of what is known as Fourier Optics. When we look at its implications more closely, a very wide range of interesting applications emerges, some of which we look at later.

The statement is mathematically the same as saying an electrical signal of arbitrary shape can be decomposed by Fourier analysis into a set of pure sinusoidal electrical signals of various amplitudes and with various phase differences between them. The difference

FIGURE 1.4   A wavefront of arbitrary shape consists of a large set of plane wave segments.

here is that optical Fourier components always involve *two independent spatial* frequencies as opposed to just one *temporal* frequency.

1.6 GEOMETRICAL OPTICS

Geometrical optics considers light in the form of rays traveling in straight lines, which are the *normals* to the *wavefronts*. The rays obey fairly simple laws at the boundaries between media of differing refractive index, enabling us to calculate or specify the important properties of many optical components used in holography. Examples are the focal length of a lens or of a spherical mirror. All of the laws governing reflection and refraction of light are derived from Fermat's principle, which states that the precise *route* which a beam of light follows in traveling between two points is such that the time taken is a maximum, constant, or minimum. For example, when light is reflected at a plane mirror, the journey time is minimum.

At first sight, the geometrical optics approach may seem a confusing departure from the concept of a light wave, but the two are compatible, as we shall see.

Imagine plane waves passing from air, whose refractive index is almost that of a vacuum (i.e., 1.0) into a sphere made of glass whose radius is R_1 and whose refractive index is n. In Figure 1.5, the distance, x, traveled by the part of the wavefront at P to reach the glass, is given by

$$x \cong \frac{y^2}{2R_1}.$$

Note that we are using the well-known geometrical theorem of intersecting chords of a circle. Strictly, we should write $x(2R_1 - x) = y^2$, but x^2 is small compared with other terms,

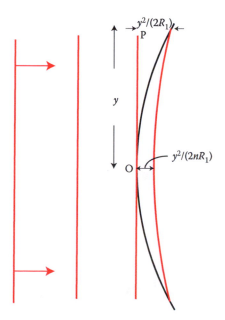

FIGURE 1.5 Plane wavefronts traversing a spherical glass surface.

if R_1 is large, and is ignored here. (Its exclusion leads to *aberrations* in optical systems involving spherical surfaces.)

In the same time, the part of the wavefront at O on the axis travels x/n, where n is the refractive index of the glass, because the velocity is reduced by the factor n. The wavefront inside the glass has radius of curvature, R', given by

$$\left(x - \frac{x}{n}\right)2R' \cong y^2 \cong 2xR_1$$

and therefore

$$\frac{n}{R'} = \frac{n-1}{R_1} \tag{1.16}$$

1.6.1 THE THIN LENS

We now introduce a second surface of radius $-R_2$ to form a *thin lens*, which is a lens *whose thickness at its axis is zero*. Note the negative sign here indicates that this surface has its center of curvature to its left, whereas the first surface has its center of curvature to its right. The total path traveled by the part of the wavefront at O to reach the second surface (Figure 1.6) is then

$$\frac{y^2}{2R_1} + \frac{y^2}{2R_2}.$$

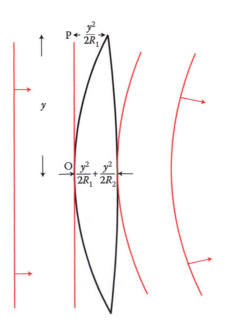

FIGURE 1.6 The thin lens.

Meanwhile, the part at P travels

$$n\left(\frac{y^2}{2R_1}+\frac{y^2}{2R_2}\right)$$

and the radius of curvature, f, of the spherical wavefront emerging from the lens is given by

$$\frac{y^2}{2f}\cong(n-1)\left(\frac{y^2}{2R_1}+\frac{y^2}{2R_2}\right)$$

$$\frac{1}{f}=(n-1)\left(\frac{1}{R_1}+\frac{1}{R_2}\right) \tag{1.17}$$

Equation 1.17 is the lensmaker's formula for the focal length of a lens. It tells us that a plane wavefront passing through a lens along its axis ultimately comes to a point on the axis called the focus located a distance f from the lens. There are two foci one on each side. The distance f is called the focal length. A plane containing the focus and perpendicular to the lens axis is called the *focal plane*. It is the plane in which infinitely distant objects are imaged since light from such objects takes the form of plane waves. Another way of saying this is that the focal plane is *conjugate* to an infinitely distant plane.

We now consider a *ray* of light, normal to the wavefront and therefore parallel to the lens axis at a distance y as shown in Figure 1.7. This ray must go through the focus F′ when it passes through the lens because that is the point to which the wavefront converges. Also, the two lens surfaces are effectively parallel in the region close to the axis if the surfaces have large radii of curvature and a ray directed at this region must travel in practically the same direction after passing through the lens. These two simple rules of geometrical optics for lenses enable us to construct ray diagrams that tell us the magnification effect of a lens, and, if we know the location of an object, where its image will be. From Figure 1.7,

$$\text{magnification} = \frac{\text{image height}}{\text{object height}} = \frac{\text{IK}}{\text{OQ}} = \frac{\text{PI}}{\text{OH}} = \frac{v}{-u} = \frac{v-f}{f} \tag{1.18}$$

Rearranging (1.18) gives

$$\frac{1}{v}+\frac{1}{u}=\frac{1}{f} \tag{1.19}$$

We have taken u and v both to be positive. Some textbooks use a Cartesian system of coordinates with the lens center as its origin so that the value of u in Figure 1.7 is *negative* and Equation 1.19 is altered accordingly. As an exercise, consider the nature of the image in cases in which the object is positioned (1) to the left of the first (left hand) focus more than two focal lengths away from the lens, (2) two focal lengths left of the lens, (3) at the first focus, and (4) between the first focus and the lens. Determine whether the image is upright or inverted and magnified or reduced in size.

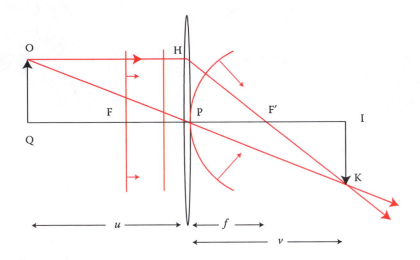

FIGURE 1.7 Wavefronts and ray paths through a thin lens.

We can differentiate Equation 1.19 with respect to object distance u to obtain the *longitudinal magnification*

$$\frac{\mathrm{d}v}{\mathrm{d}u} = -\frac{v^2}{u^2} \tag{1.20}$$

1.6.2 SPHERICAL MIRROR

Similar reasoning to that which led to Equation 1.16 enables us to find the focal length of a spherical mirror. However, if we simply replace n in Equation 1.16 by -1 to make the surface of radius $-R_1$ a reflecting one, we find that the radius of curvature of the wavefront, R', is simply $-R_1/2$. So the wavefront converges to a focus at this distance from the mirror (Figure 1.8) and the focal length of the mirror as $-R_1/2$.

1.6.3 REFRACTION AND REFLECTION

At the beginning of this chapter, it was mentioned that the velocity of light depends on the material through which it travels being a maximum, c, in a vacuum and c/n in a material of refractive index, n.

Let us now consider a plane wave that crosses the plane boundary between two materials of differing refractive index. The angle between the wave vector, or ray, and the normal at the boundary is θ_i, called the angle of incidence. Referring to Figure 1.9, we see that the part of the plane wave reaching the boundary last travels a distance BD in a time BDn_1/c. In the same time, the light that reached the boundary first must travel a distance (c/n_2) $(BDn_1/c) = BDn_1/n_2$. So we draw a circle of this radius and then a tangent from D to this circle, touching it at C. This tangent is the plane wavefront in material 2, and the first arriving part of the wave travels a distance AC in material 2. Hence,

$$AC = BDn_1/n_2 \quad \text{so} \quad BD/AC = n_2/n_1$$

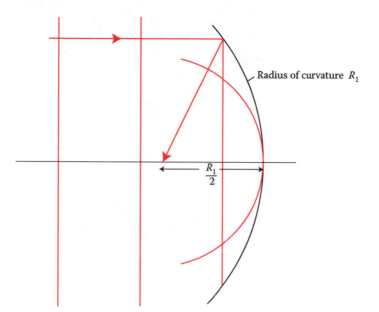

FIGURE 1.8 Spherical mirror.

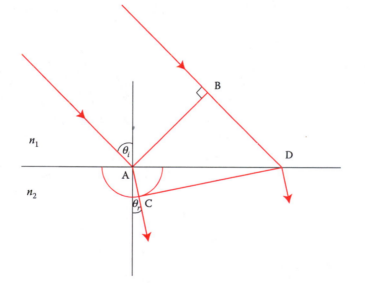

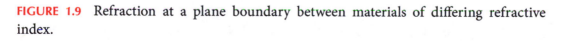

FIGURE 1.9 Refraction at a plane boundary between materials of differing refractive index.

We also draw the rays from A and D and note that the light has changed direction in material 2.

$$\sin\theta_i = \frac{BD}{AD}, \quad \sin\theta_r = \frac{AC}{AD}$$

$$\frac{\sin\theta_i}{\sin\theta_r} = \frac{n_2}{n_1} \tag{1.21}$$

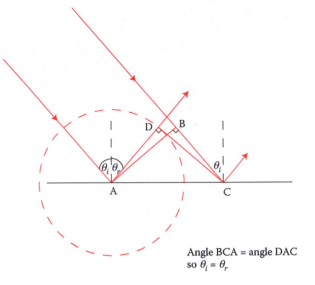

Angle BCA = angle DAC
so $\theta_i = \theta_r$

FIGURE 1.10 Reflection at a plane boundary.

Equation 1.21, which is Snell's law of refraction, enables us to trace the paths of light rays across the boundaries between materials of differing refractive indices and is the basis of a number of powerful optical system design software packages such as OSLO and ZEMAX.

To see what happens when the surface is simply a flat mirror n_2 is simply replaced by $-n_1$ we see that $\theta_r = -\theta_i$, and so the angle of reflection is equal but opposite in sign to the angle of incidence. Figure 1.10 shows how to prove this law of reflection from basic principles.

A plane wave approaches a flat mirror at angle of incidence θ_i, defined as the angle between the wave vector, or ray, and the surface normal. By the time the part of the wavefront at B reaches the surface at point C, the part at A will have traveled a distance equal to BC after reflection. So to construct the reflected wavefront, we draw a circle centered on A and radius AD = BC and draw a tangent CD to this circle. The triangles ADC and ABC are clearly congruent, so angles BCA and DAC are equal; that is the *angle of incidence is equal to the angle of reflection*, the angle between the reflected ray and the surface normal.

1.7 REFLECTION, REFRACTION, AND THE FRESNEL EQUATIONS

In this section, we use the laws of electromagnetism to obtain Fresnel's equations, which play an important role in optics and holography as they can be used to calculate the intensity reflectivity and transmissivity of optical components. On the way to obtaining the equations, we shall incidentally prove again the laws of reflection and refraction.

1.7.1 REFLECTION AND REFRACTION

In this discussion, we assume a locally planar boundary, parallel to the x,y plane at $z = z_1$ between media of refractive indices n_i and n_t respectively (Figure 1.11) where subscripts

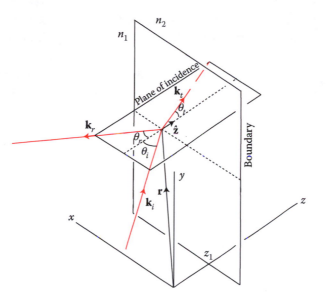

FIGURE 1.11 Reflection and transmission at a plane boundary between media of differing refractive index.

i and t refer to the *incident* medium in which light is traveling before reaching the boundary and the *transmission* medium beyond the boundary. We assume the incident light in medium i is in the form of a plane monochromatic wave described by the equation

$$\mathbf{E}_i(\mathbf{r},t) = \mathbf{E}_{0i}\cos(\mathbf{k}_i \cdot \mathbf{r} - \omega_i t)$$

Here, $\mathbf{E}_{0i}$ is independent of time and of spatial coordinates and the wave is therefore plane polarized.

The reflected and transmitted waves are

$$\mathbf{E}_r(\mathbf{r},t) = \mathbf{E}_{0r}\cos(\mathbf{k}_r \cdot \mathbf{r} - \omega_r t + \phi_r) \quad \text{and} \quad \mathbf{E}_t(\mathbf{r},t) = \mathbf{E}_{0t}\cos(\mathbf{k}_t \cdot \mathbf{r} - \omega_t t + \phi_t)$$

where ϕ_r and ϕ_t are respectively, phase changes in the reflected and transmitted waves relative to the incident wave. Now, we apply the condition that the total electric fields parallel to the boundary, immediately on either side of it, are the same. Violation of this condition would require the local divergence of the electric field, $\mathbf{\nabla} \cdot \mathbf{E}$, to be infinite, requiring an infinitely large density of electrical charge. This condition implies

$$\hat{\mathbf{z}} \times \mathbf{E}_i + \hat{\mathbf{z}} \times \mathbf{E}_r = \hat{\mathbf{z}} \times \mathbf{E}_t \tag{1.22}$$

$\hat{\mathbf{z}}$ being the unit vector in the direction of the z-axis. The vectors in Equation 1.22 are all normal to $\hat{\mathbf{z}}$ and are therefore parallel to the boundary.

Equation 1.22 is quite general so $\mathbf{E}_i, \mathbf{E}_r$, and $\mathbf{E}_t$ must all depend on t in the same way and therefore

$$\omega_i = \omega_r = \omega_t \tag{1.23}$$

which we would expect, as the incident light energy causes the bound charges in medium 2 to oscillate and emit light of the same frequency, whether transmitted into medium 2 or reflected.

Also, $(\mathbf{k}_i - \mathbf{k}_r) \cdot \mathbf{r} = \phi_r$ which is the equation of the boundary plane so $\hat{\mathbf{z}} \times (\mathbf{k}_i - \mathbf{k}_r) = 0$, $\mathbf{k}_i, \mathbf{k}_r$, and $\hat{\mathbf{z}}$ are coplanar, and

$$\mathbf{k}_i \sin\theta_i = \mathbf{k}_r \sin\theta_r.$$

Since the incident and reflected waves are in the same medium, $\mathbf{k}_i = \mathbf{k}_r$ and

$$\theta_i = \theta_r$$

Using the simple wave approach in Section 1.6.3, we already established this last result, now further validated using electromagnetic theory.

In the plane of incidence and since $(\mathbf{k}_i - \mathbf{k}_t) \cdot \mathbf{r} = \varepsilon_r$, $\hat{\mathbf{z}} \times (\mathbf{k}_i - \mathbf{k}_t) = 0$ so that $\mathbf{k}_i, \mathbf{k}_t$, and $\hat{\mathbf{z}}$ are coplanar and

$$\mathbf{k}_i \sin\theta_i = \mathbf{k}_t \sin\theta_t$$

$$\frac{\sin\theta_i}{\sin\theta_t} = \frac{\lambda_i}{\lambda_t} = \frac{\lambda_i \omega_i}{\lambda_t \omega_t} = \frac{c_i}{c_t} = \frac{c/n_i}{c/n_t} = \frac{n_t}{n_i}$$

which is Snell's law.

1.7.2 THE FRESNEL EQUATIONS

We now derive the equations that tell us how much light is actually transmitted and reflected at the boundary between media of differing refractive indices. These equations will provide important results that are useful in the practice of holography. We will deal with the important cases when the incident electric field is (1) perpendicular and (2) parallel to the plane of incidence.

1.7.2.1 Electric Field Perpendicular to the Plane of Incidence

Referring to Figure 1.12a, we first note that the total electric field component *parallel* to the boundary has to be the same immediately on either side of the boundary at all times and at every point, as noted in the previous section, so

$$\mathbf{E}_{0i} + \mathbf{E}_{0r} = \mathbf{E}_{0t} \tag{1.24}$$

The total component of $\mathbf{H}$ *parallel* to the boundary also has to be continuous across it when the materials are both dielectric, which is normally the case when dealing with optical components; that is they are nonmetallic and therefore electrically nonconducting. Otherwise $\nabla \cdot \mathbf{H}$ is infinite, which in turn implies an infinitely large electrical current density. Therefore

$$-\mathbf{H}_{0i} \cos\theta_i + \mathbf{H}_{0r} \cos\theta_r = -\mathbf{H}_{0t} \cos\theta_t \tag{1.25}$$

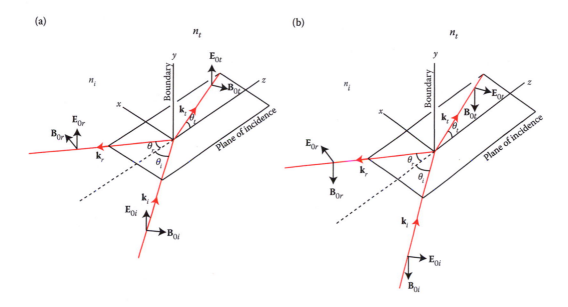

FIGURE 1.12 Reflection and transmission at a plane boundary between media of differing refractive index: (a) electric field perpendicular to plane of incidence and (b) electric field normal to plane of incidence.

As $\theta_i = \theta_r$, $\mu_i = \mu_r = \mu_t = \mu_0$ since the relative permeabilities of dielectrics are all 1.0 and (see problem 23) H $= \varepsilon_0 c$E

$$-\mathbf{E}_{0i}n_i\cos\theta_i + \mathbf{E}_{0r}n_r\cos\theta_r = -\mathbf{E}_{0t}n_t\cos\theta_t \tag{1.26}$$

Using Equation 1.24 and that $n_i = n_r$

$$-\mathbf{E}_{0i}n_i\cos\theta_i + \mathbf{E}_{0r}n_i\cos\theta_r = -(\mathbf{E}_{0i} + \mathbf{E}_{0r})n_t\cos\theta_t \tag{1.27}$$

Using Snell's law

$$\left.\frac{\mathbf{E}_{0r}}{\mathbf{E}_{0i}}\right|_{\perp} = \frac{n_i\cos\theta_i - n_t\cos\theta_t}{n_i\cos\theta_i + n_t\cos\theta_t} = \frac{\sin(\theta_t - \theta_i)}{\sin(\theta_t + \theta_i)} \tag{1.28}$$

From Equations 1.24 and 1.28 $(\mathbf{E}_{0t}/\mathbf{E}_{0i}) = 1 + (\mathbf{E}_{0r}/\mathbf{E}_{0i}) = 1 + (n_i\cos\theta_i - n_t\cos\theta_t)/(n_i\cos\theta_i + n_t\cos\theta_t) = (2n_i\cos\theta_i)/(n_i\cos\theta_i + n_t\cos\theta_t)$

Again using Snell's law $(\mathbf{E}_{0t}/\mathbf{E}_{0i})|_{\perp} = (2\cos\theta_i)(\cos\theta_i + (n_t/n_i)\cos\theta_t)$

$$\left.\frac{\mathbf{E}_{0t}}{\mathbf{E}_{0i}}\right|_{\perp} = \frac{2\sin\theta_t\cos\theta_i}{\sin(\theta_t + \theta_i)} \tag{1.29}$$

Note that

$$\left.\frac{\mathbf{E}_{0r}}{\mathbf{E}_{0i}}\right|_{\perp} + \left.\frac{\mathbf{E}_{0t}}{\mathbf{E}_{0i}}\right|_{\perp} = \frac{\sin(\theta_t - \theta_i) + 2\sin\theta_t \cos\theta_i}{\sin(\theta_t + \theta_i)} = 1.$$

We have assigned all field directions in Figure 1.12 following the directions indicated by the Poynting vector equation (Equation 1.13) and indicated that $\mathbf{E}_{0r}$ is in the direction of $+y$. We did not know in advance whether or not this is correct, and in fact if $n_2 > n_1$ $\mathbf{E}_{0r}$ should be in the $-y$ direction because, according to Equation 1.28, $\mathbf{E}_{0r}/\mathbf{E}_{0i}$ is negative, since $\theta_t > \theta_i$. It follows that when light in a dielectric is reflected at the boundary with a dielectric of higher refractive index, the reflected light is 180° out of phase with the incident light.

1.7.2.2 Electric Field Parallel to the Plane of Incidence

Figure 1.12b is identical to Figure 1.12a except that now the electric field is parallel to the plane of incidence and we have assigned the **B** field directions accordingly. Applying continuity rules, that the component of **E** parallel to the boundary is continuous across it

$$\mathbf{E}_{0i}\cos\theta_i - \mathbf{E}_{0r}\cos\theta_r = \mathbf{E}_{0t}\cos\theta_t \tag{1.30}$$

and that **H** parallel to the boundary is continuous across it

$$\mathrm{H}_{0i} + \mathrm{H}_{0r} = \mathrm{H}_{0t}$$

that is (problem 1.23), $\hspace{6cm}$ (1.31)

$$(\mathrm{E}_{0i} + \mathrm{E}_{0r})\,n_i = \mathrm{E}_{0t}n_t$$

Inserting $\mathbf{E}_{0t}$ from Equation 1.30

$$\left(\mathbf{E}_{0i} + \mathbf{E}_{0r}\right)n_i = \mathbf{E}_{0t}\,n_t = \left(\mathbf{E}_{0i} - \mathbf{E}_{0r}\right)\frac{\cos\theta_i}{\cos\theta_t}$$

$$\left(1 + \frac{\mathbf{E}_{0r}}{\mathbf{E}_{0i}}\right)n_i = \left(1 - \frac{\mathbf{E}_{0r}}{\mathbf{E}_{0i}}\right)\frac{\cos\theta_i}{\cos\theta_t}n_t$$

$$\frac{\mathbf{E}_{0r}}{\mathbf{E}_{0i}} = \frac{n_t\cos\theta_i - n_i\cos\theta_t}{n_t\cos\theta_i + n_i\cos\theta_t} \tag{1.32}$$

$$\frac{\mathbf{E}_{0r}}{\mathbf{E}_{0i}} = \frac{n_t\cos\theta_i - n_i\cos\theta_t}{n_t\cos\theta_i + n_i\cos\theta_t} = \frac{\cos\theta_i - \dfrac{n_i}{n_t}\cos\theta_t}{\cos\theta_i + \dfrac{n_i}{n_t}\cos\theta_t} = \frac{\sin\theta_i\cos\theta_i - \sin\theta_t\cos\theta_t}{\sin\theta_i\cos\theta_i + \sin\theta_t\cos\theta_t}$$

$$= \frac{\sin(2\theta_i) - \sin(2\theta_t)}{\sin(2\theta_i) + \sin(2\theta_t)} = \frac{\cos(\theta_i + \theta_t)\sin(\theta_i - \theta_t)}{\sin(\theta_i + \theta_t)\cos(\theta_i - \theta_t)}$$

$$\left.\frac{\mathbf{E}_{0r}}{\mathbf{E}_{0i}}\right|_{\|} = \frac{\tan(\theta_i - \theta_t)}{\tan(\theta_i + \theta_t)} \tag{1.33}$$

An interesting result is obtained in the special case when $\theta_i + \theta_t = 90°$ and *no light is reflected*. A simple substitution using Snell's law shows that this occurs when $\theta_i = \sin^{-1}(n_t/n_i)$. The particular value of θ_i here is known as the Brewster angle and light reflected at this angle is always plane polarized perpendicular to the plane of incidence.

From Equations 1.30 and 1.32

$$\left.\frac{\mathbf{E}_{0t}}{\mathbf{E}_{0i}}\right|_{\parallel} = \left(\frac{2n_i\cos\theta_i}{n_i\cos\theta_t + n_t\cos\theta_i}\right) \tag{1.34}$$

$$\frac{\mathbf{E}_{0t}}{\mathbf{E}_{0i}} = \left(\frac{2n_i\cos\theta_i}{n_i\cos\theta_t + n_t\cos\theta_i}\right) = \frac{2\cos\theta_i}{\cos\theta_t + \dfrac{n_t}{n_i}\cos\theta_i} = \frac{2\cos\theta_t\cos\theta_i}{\sin\theta_t\cos\theta_t + \sin\theta_i\cos\theta_i}$$

$$\left.\frac{\mathbf{E}_{0t}}{\mathbf{E}_{0i}}\right|_{\parallel} = \frac{2\cos\theta_i\sin\theta_t}{\sin\left(\theta_i + \theta_t\right)\cos\left(\theta_i - \theta_t\right)} \tag{1.35}$$

1.7.2.3 Antireflection Coatings

The four Fresnel equations (1.28), (1.29), (1.33), and (1.35) again are

$$\left.\frac{\mathbf{E}_{0r}}{\mathbf{E}_{0i}}\right|_{\perp} = \frac{\sin\left(\theta_t - \theta_i\right)}{\sin\left(\theta_t + \theta_i\right)} \tag{1.28}$$

$$\left.\frac{\mathbf{E}_{0t}}{\mathbf{E}_{0i}}\right|_{\perp} = \frac{2\cos\theta_i\sin\theta_t}{\sin\left(\theta_t + \theta_i\right)} \tag{1.29}$$

$$\left.\frac{\mathbf{E}_{0r}}{\mathbf{E}_{0i}}\right|_{\parallel} = \frac{\tan\left(\theta_i - \theta_t\right)}{\tan\left(\theta_i + \theta_t\right)} \tag{1.33}$$

$$\left.\frac{\mathbf{E}_{0t}}{\mathbf{E}_{0i}}\right|_{\parallel} = \frac{2\cos\theta_i\sin\theta_t}{\sin\left(\theta_i + \theta_t\right)\cos\left(\theta_i - \theta_t\right)} \tag{1.35}$$

Figure 1.13 shows the amplitude reflectivities and transmissivities plotted against θ_i for $n_i = 1$ and $n_t = 1.5$.

Let us first examine what happens at normal incidence. Equation 1.28 shows immediately that

$$\left.\frac{\mathbf{E}_{0r}}{\mathbf{E}_{0i}}\right|_{\perp\theta_i=0} = \frac{n_i - n_t}{n_i + n_t} \tag{1.36}$$

Equation 1.33 produces the same result as the small angle approximation allows replacement of tangents with sines.

$$\left.\frac{\mathbf{E}_{0r}}{\mathbf{E}_{0i}}\right|_{\parallel\theta_i=0} = \frac{\tan\left(\theta_i - \theta_t\right)}{\tan\left(\theta_i + \theta_t\right)} = \frac{\sin\left(\theta_i - \theta_t\right)}{\sin\left(\theta_i + \theta_t\right)} = \frac{\sin\theta_i\cos\theta_t - \cos\theta_i\sin\theta_t}{\sin\theta_i\cos\theta_t + \cos\theta_i\sin\theta_t} = \frac{n_i\cos\theta_t - n_i\cos\theta_i}{n_i\cos\theta_t + n_i\cos\theta_i} = \frac{n_t - n_i}{n_t + n_i}$$

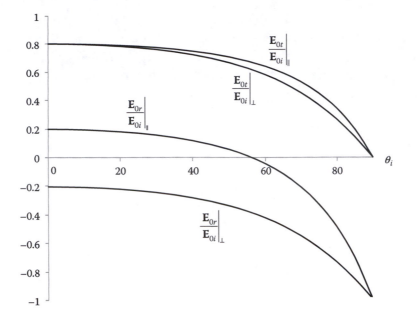

FIGURE 1.13 Amplitude reflection and transmission coefficients ($n_i = 1$, $n_t = 1.5$).

which is −0.2, in the case of the boundary between air ($n_i = 1.0$) and ordinary glass ($n_t = 1.5$). The negative sign indicates that *the reflected light is 180° out of phase with the incident light*. The *intensity* reflectivity is 0.04. Thus, at least a 4% of the light incident at every air/glass boundary is reflected.

Note here that, although there is no negative sign in front of the expression $(n_t - n_i)/(n_t + n_i)$, Figure 1.12b shows that the components of the incident and reflected electric fields parallel to the boundary are exactly 180° out of phase so that in this case, we made the correct choices of *relative* directions of the incident and reflected electric fields.

The reader might like to calculate the total loss in a simple camera lens consisting of three separated components, assuming normal incidence at each of the six boundaries. It is these reflection losses that give rise to the flare effect in images produced by a lens because light multiply reflected at the internal surfaces of a lens system cannot be focused at the same image point as the light that is transmitted on first arrival at every surface. In holographic recording systems, which almost invariably require the use of lasers, the laser light is reduced in intensity on crossing each such boundary, ultimately limiting the beam intensity available for holographic recording. Also, light reflected back toward a laser source can re enter the laser cavity and seriously affect the stability of the laser beam.

These problems are alleviated by covering every surface with a very thin antireflection coating, which has a refractive index intermediate between those of the dielectrics, air and glass, on either side of the boundary. The thickness of a single layer coating is such that light reflected from its front surface interferes destructively with light reflected from its back surface in the same direction. From Figure 1.14, the optical path difference, opd, between light reflected at the surfaces of such a layer of thickness d is $2n_t d\cos\theta t$. At normal incidence, assuming $n_i < n_t < n_s$, where n_s is the refractive index of the optical component,

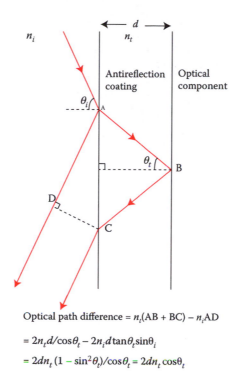

Optical path difference = $n_t(AB + BC) - n_iAD$

$= 2n_td/\cos\theta_t - 2n_id\tan\theta_t\sin\theta_i$

$= 2dn_t(1 - \sin^2\theta_t)/\cos\theta_t = 2dn_t\cos\theta_t$

FIGURE 1.14 Antireflection coating.

there are phase changes of 180° on reflection at the two boundaries so that, if $2n_td = \lambda/2$, that is, $d = \lambda/(4n_td)$, the two reflected light beams will destructively interfere with each other. The effect is that the reflectivity is significantly reduced. Further reductions can be obtained using multiple layers.

In practice, optical components, including lenses, beam splitters, and polarizing components, can be coated to reduce the reflection losses at a specified wavelength, or range of wavelengths, and most reputable optical component manufacturers offer this facility.

1.7.2.4 Total Internal Reflection and Evanescent Waves

Suppose $n_i > n_t$ then $\theta_t > \theta_i$ and from Equation 1.28 the ratio $\mathbf{E}_{0r}/\mathbf{E}_{0i}\big|_\perp$ is always positive and there is no phase difference between incident and reflected light. As θ_i increases, θ_t eventually reaches 90° and $\mathbf{E}_{0r}/\mathbf{E}_{0i}\big|_\perp = 1$ and all the light is reflected. Equation 1.33 produces the same result.

Now consider the transmitted wave in the case where $n_i > n_t$ and $\theta_t > \theta_i$.

$$\mathbf{E}_t = \mathbf{E}_{0t}\exp[j(\mathbf{k}_t \cdot \mathbf{r} - \omega t)] \tag{1.37}$$

In Figure 1.12, $\mathbf{k}_t = k_t(\sin\theta_t, 0, \cos\theta_t)$.

$$k_t\cos\theta_t = k_t\sqrt{1-\sin^2\theta_t} = \pm jk_t\left(\frac{n_i^2}{n_t^2}\sin^2\theta_i - 1\right)^{\frac{1}{2}}$$

$$\mathbf{E}_t = \mathbf{E}_{0t} \exp\left[j\left(k_t x \sin\theta_t + jk_t \left(\frac{n_i^2}{n_t^2} \sin^2\theta_i - 1 \right)^{\frac{1}{2}} - \omega t \right) \right]$$

We have used

$$k_t \cos\theta_t = +jk_t \left(\frac{n_i^2}{n_t^2} \sin^2\theta_i - 1 \right)^{\frac{1}{2}}$$

as use of the negative root would lead to an exponentially increasing amplitude.

$$\mathbf{E}_t = \mathbf{E}_{0t} \exp\left[-k_t \left(\frac{n_i^2}{n_t^2} \sin^2\theta_i - 1 \right)^{\frac{1}{2}} z \right] \exp j(k_t x \sin\theta_t - \omega t) \tag{1.38}$$

Equation 1.38 shows that light propagates *along* the interface (in the x direction), but its amplitude decays rapidly with z. Waves of this type are called *evanescent waves* and they can be utilized in *evanescent wave holography* as we shall see in Chapter 9.

We can rewrite Equation 1.38 as

$$\mathbf{E}_t = \mathbf{E}_{0t} \exp\left(\frac{-z}{z_0} \right) \exp j(k_t x \sin\theta_t - \omega t) \tag{1.39}$$

where

$$z_0 = \frac{1}{k_t} \left(\frac{n_i^2}{n_t^2} \sin^2\theta_i - 1 \right)^{-\frac{1}{2}}$$

is the distance in the z direction over which the amplitude decays by a factor $1/e$ and is called the *skin depth*. Substitution of typical values for n_i, n_t, k_t, and θ_i will show that z_0 is of the order of a wavelength. If another dielectric material is brought into proximity with the first one, so that the gap between them is z_0 or less, total internal reflection does not take place and the light can cross the gap and propagate in the second dielectric, a phenomenon called *frustrated total internal reflection*.

1.7.2.5 Intensity Reflection and Transmission Ratios

Returning again to the Fresnel equations, we need to calculate the intensity reflectivity and transmissivity. Imagine a beam of light with circular cross-section incident at angle θ_i to the surface normal. It illuminates an elliptical area, a, on the boundary so that the cross-sectional area of the incident (and reflected) beams must be $a\cos\theta_i$ while that of the transmitted beam must be $a\cos\theta_t$. Say the intensity of the incident beam, that is, the power (or energy per second) crossing unit area *normal* to the beam is I_i, then the total power in the incident beam is $I_i a\cos\theta_i$. Similarly, the reflected beam has total power $I_r a\cos\theta_r$ and the transmitted beam has power $I_t a\cos\theta_t$.

We now define the *reflectance, R,* as the ratio of reflected power to incident power.

$$R = \frac{I_r a \cos\theta_r}{I_i a \cos\theta_i} = \frac{I_r}{I_i} \tag{1.40}$$

and the *transmittance, T,* as the ratio of transmitted power to incident power

$$T = \frac{I_t a \cos\theta_t}{I_i a \cos\theta_i} = \frac{I_t \cos\theta_t}{I_i \cos\theta_i} \tag{1.41}$$

From Equation 1.15, $I = \langle S \rangle = \varepsilon_0 c E_0^2 / 2$, which is rewritten for a dielectric of refractive index n and permittivity ε as $I = \varepsilon c E_0^2 / 2$.

So R and T now become

$$R = \frac{\varepsilon_r c_r E_{0r}^2}{\varepsilon_i c_i E_{0i}^2} = \left(\frac{E_{0r}}{E_{0i}}\right)^2 \tag{1.42}$$

$$T = \frac{I_t \cos\theta_t}{I_i \cos\theta_i} = \frac{\varepsilon_t c_t \dfrac{E_{0t}^2}{2} \cos\theta_t}{\varepsilon_i c_i \dfrac{E_{0i}^2}{2} \cos\theta_i} = \frac{n_t^2 n_i \cos\theta_t}{n_i^2 n_t \cos\theta_i} \left(\frac{E_{0t}}{E_{0i}}\right)^2 = \frac{n_t \cos\theta_t}{n_i \cos\theta_i} \left(\frac{E_{0t}}{E_{0i}}\right)^2.$$

Since $\varepsilon = n^2$ in a dielectric (see Section 1.1)

$$T = \frac{n_t \cos\theta_t}{n_i \cos\theta_i} \left(\frac{E_{0t}}{E_{0i}}\right)^2 \tag{1.43}$$

We now have

$$R_\perp = \left|\frac{\mathbf{E}_{0r}}{\mathbf{E}_{0i}}\right|_\perp^2, \quad R_\parallel = \left|\frac{\mathbf{E}_{0r}}{\mathbf{E}_{0i}}\right|^2 \tag{1.44}$$

$$T_\perp = \frac{n_t \cos\theta_t}{n_i \cos\theta_i} \left|\frac{\mathbf{E}_{0t}}{\mathbf{E}_{0i}}\right|_\perp^2 \quad T_\parallel = \frac{n_t \cos\theta_t}{n_i \cos\theta_i} \left|\frac{\mathbf{E}_{0t}}{\mathbf{E}_{0i}}\right|^2 \tag{1.45}$$

Figure 1.15 shows the intensity reflectivities and transmissivities plotted against θ_i for $n_i = 1$ and $n_t = 1.5$. These graphs are useful for calculating reflection losses in holographic recording systems.

1.8 INTRODUCTION TO SPATIAL FILTERING

All optical components such as lenses, mirrors, and apertures influence the light passing through, or reflected from them, in a way that is analogous to the effects of passive electrical networks on electrical signals. In other words, just as an electronic filter can alter the temporal frequency content of an electrical signal, so also can a spatial filter consisting of lenses, mirrors, apertures, or simply materials having various refractive indices, alter the

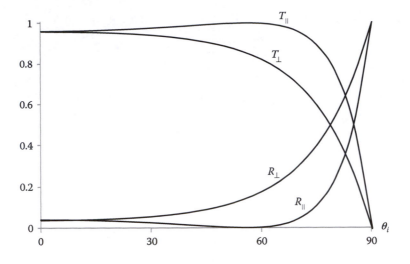

FIGURE 1.15 Intensity reflectivities and transmissivities ($n_i = 1$, $n_t = 1.5$) versus angle of incidence.

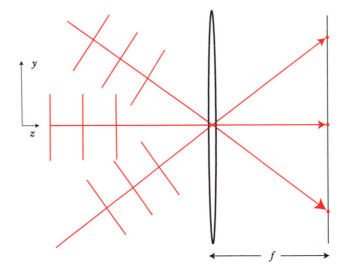

FIGURE 1.16 One-to-one correspondence between spatial frequencies and points in the focal plane of a lens.

spatial frequency content of a light wavefront. The important difference is that here we are dealing with signals that vary in three spatial dimensions rather than just the single one of time. A simple example illustrates the process of spatial filtering. Consider a number of sets of plane waves of the same wavelength, all traveling in different directions relative to the optical axis of a thin lens (Figure 1.16). For simplicity, we assume the propagation vectors are all parallel to the y,z plane. Each converges to a point in the focal plane of the lens. To locate the point, we use the fact that a ray normal to the wavefronts and aimed at the center of the lens is undeviated and must cross the focal plane at the point. Then, the wavefronts themselves must converge to the same point on passing through the lens. In Figure 1.16, we see that in the focal plane, there is a single point that corresponds to each set of wavefronts. Each set has a single spatial frequency with which it is associated. The higher the spatial frequency, the further is the convergence point from the optical axis.

By placing an aperture in the focal plane of the lens, we can prevent the light associated with some spatial frequencies from proceeding any further. If the aperture is located on the lens axis, it acts as a low-pass spatial filter. If the aperture is replaced by an obstacle, we have a high-pass filter. An annular aperture acts as a bandpass filter.

1.8.1 PHASE CONTRAST IMAGING

An interesting example of spatial filtering is one in which the *phase* of the light in some spatial frequencies is altered with respect to the phases of others. It has very important practical applications enabling high-contrast imaging of almost transparent materials, such as biological samples. These samples are normally stained in order to obtain high-contrast images and the staining usually results in cessation of biological processes of interest. Consider a plane wave $E_0 \cos(kz - \omega t)$ propagating along the optical (z) axis of a lens. On passing through the sample, only a phase retardation $\phi(x', y')$ occurs and we have in the focal (x', y') plane

$$E_0 \cos[kz - \omega t + \phi(x', y')] = E_0 \cos(kz - \omega t) \cos\phi(x', y') - E_0 \sin(kz - \omega t) \sin\phi(x', y')$$
$$\cong E_0 \cos(kz - \omega t) - E_0 \sin(kz - \omega t)\phi(x', y')$$

if the retardation is small.

We see that the light now consists of the original illuminating part,

$$E_0 \cos(kz - \omega t)$$

and a wave

$$-E_0 \sin(kz - \omega t)\phi(x', y')$$

which is not zero at points that are displaced from the lens axis. In the plane (x', y'), we introduce a circularly symmetrical transparent plate whose thickness close to the optical axis is different from the thickness elsewhere (Figure 1.17).

The light close to the axis passes through the thicker inner region and the remaining light passes through the thinner outer region. The thickness difference is such that a phase lag of $\pi/2$ between the two is introduced and we now have a light wave

$$E_0 \cos(kz - \omega t) - E_0 \sin(kz - \omega t - \pi/2)\phi(x', y')$$
$$= E_0 \left[1 + \phi(x', y')\right] \cos(kz - \omega t)$$

Thus, the phase-modulated light has been converted to amplitude, and therefore intensity-modulated light. In the final image, the thickness and refractive index variations in the sample are converted into contrast variations. This version is known as positive phase contrast, regions of the sample where the phase retardation is greater appearing brighter due

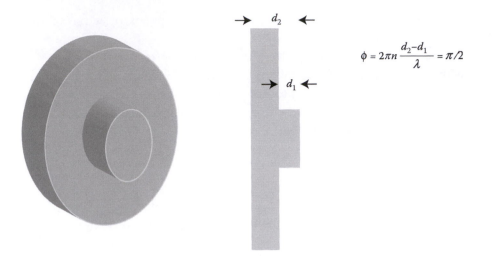

$$\phi = 2\pi n \frac{d_2 - d_1}{\lambda} = \pi/2$$

FIGURE 1.17 Phase contrast plate.

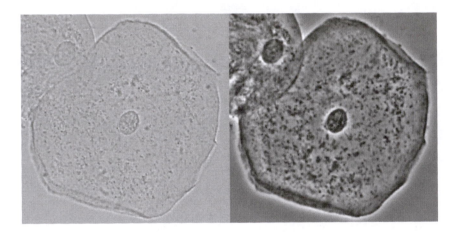

FIGURE 1.18 Unstained epithelial cell in normal view (left) and in phase contrast (right). (Courtesy of Gregor Overney.)

to increased amplitude. Conversely, a plate that is thinner near the axis would produce a light wave

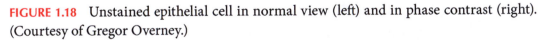

$$E_0 \cos(kz - \omega t) - E_0 \sin(kz - \omega t + \pi/2)\phi(x', y')$$
$$= -E_0 \left[1 - \phi(x', y') \right] \sin(kz - \omega t)$$

and here greater values of ϕ appear darker due to reduced amplitude. In the 1930s, Zernike developed this technique, for which he received the Nobel Prize in Physics in 1953.

A good example of the use of the technique is shown in Figure 1.18. Epithelial cells are seen in normal illumination on the left while the result of applying a phase-only spatial filter is seen on the right.

1.9 SUMMARY

This chapter begins with a description of light waves and explains the concepts of refractive index and of spatial frequency. We derive the general equation of a plane wave and explain the meaning of intensity. We use simple arguments to derive the basic properties of thin lenses and spherical mirrors. Following derivation of the Fresnel equations, we use them to confirm laws of reflection and refraction as well as to determine reflection and transmission coefficients. We also introduce spatial filtering, with a discussion of phase contrast microscopy as a useful example.

PROBLEMS

1. Prove Equation 1.6.
2. What is the velocity of light (a) in water of refractive index 1.33 and (b) in glass of index 1.55?
3. Visible light ranges in wavelength from about 380 to about 760 nm. What is its frequency range?
4. A plane wave, wavelength 500 nm, propagates in the direction of the vector (1,2,3). What are its spatial frequencies measured along the x, y, and z axes?
5. In Chapter 2, it is shown that a collimated light beam of wavelength λ filling a lens of diameter D is focused into a circular spot approximately $2.44\lambda f/D$ in width where f is the focal length of the lens. Assume the diameter of the lens is 10 mm and its focal length is 20 mm. What laser power would be needed to cause a spark at the focus of the lens if the breakdown electric field of dry air is about 30 kV per cm?
6. Using an argument similar to that which leads to Equation 1.16, show that the focal length of a spherical mirror of radius R' is $R/2$.
7. Cats' eyes are commonly used to delineate the central white line on single carriageway roads at night. They consist of spheres of glass (refractive index 1.5) in tough rubber mounts set in the roadway. Trace a typical ray of light from a car headlamp through a cat's eye to show how it works. (The cat's eye principle has much in common with red eye, the unflattering effect often seen in flash photographs.)
8. A point source of light is at a distance l from a glass sphere of radius R'. Find a relationship between l and l', the distance between the image of the point and the surface. (*Hint:* Use the method given in Section 1.6 to find the radius of curvature of the wavefront after it has passed through the surface and the fact that its radius of curvature on reaching the surface is l.)
9. One way of measuring the focal length is the displacement method. An estimate of the focal length f is found first by forming an image of a fairly distant object such as a lighting fixture in the ceiling. If an illuminated transparency as object and a white screen are then set up a distance $D > 4f$, apart, images can be obtained with the lens in *two* positions distance d apart. Show how the method works and find an expression for the focal length.
10. The lensmaker's formula assumes that the lens is of negligible thickness. Find an expression for the distance of the second focus of a lens of *thickness t* from its second spherical surface of radius R_2. The first surface has radius R_1.
11. Many lenses consist of more than one component. Consider one having two thin components of focal lengths f_1 and f_2 distance d apart. Find the locations of its focal points.
12. Laser beams are usually expanded (or reduced) in diameter by means of a telescope consisting of two lenses separated by a distance equal to the algebraic sum of their focal lengths, one of the lenses being of longer focal length than the other. Trace a beam of collimated

light through such a combination to see how it works in the case of a telescope in which the shorter focal length is (a) positive and (b) negative.

13. Find the position of the image of a point, which is located 100 cm from a spherical convex mirror of radius 50 cm. An object of 1 cm high is located at the same position. What is the size of its image?

14. A small lamp is located at the bottom of a swimming pool 2 m in depth. What is the diameter of the illuminated surface of the water of refractive index 1.33? What is the apparent depth of the lamp in the water? A collimated beam of laser light 5 mm in diameter is incident at a flat glass surface at angle of incidence 45°. What is the diameter of the beam inside the glass?

15. Triangular prisms are used as means of spatially separating the different wavelengths in a beam of light because the refractive index is wavelength dependent. Show that a beam of light traversing such a prism of refractive index n is deviated by an angle δ where

$$\delta = \theta_{i1} + \theta_{r2} - A$$

where θ_{i1} is the angle of incidence at the point of entry, θ_{r2} the angle at which the light emerges, and A is the apex angle.

Then, show that the minimum angle of deviation is given by

$$\delta_{\min} = (n-1)A.$$

Explain why at minimum deviation, the passage of the light beam must *logically* be symmetrical, that is $\theta_{i1} = \theta_{r2}$.

16. A collimated light beam is incident at angle θ on one face of a triangular prism with apex angle A. Find an expression for the ratio of the diameters of the output and input beams in terms of θ and A.

17. Recent research into invisibility uses photonic bandgap materials (Section 11.11), which bend light waves tightly around an object but allow the wavefronts to recover their original shape downstream of the object. In *The Invisible Man* by H. G. Wells, the protagonist renders himself invisible by changing his refractive index to that of air. Why would this make him invisible? Comment on the feasibility of actually becoming invisible by the method described in the novel.

18. A lens made from glass of refractive index 1.53 is antireflection coated with magnesium fluoride (MgF_2) of refractive index 1.38, for use with a laser of wavelength 633 nm. What is the coating thickness?

19. Diamond has a refractive index of 2.417 for visible light and is considered to be among the highest valued gems. Explain why. (Rutile, TiO_2, has a refractive index of 2.9 but it is rather brittle).

20. Rhinestone is essentially glass which is cut so that it has many surfaces which are then *reflection* coated. What should be the thickness of a MgF_2 coating?

21. The figure below shows a method of modulating the intensity of a laser using a piezoelectric transducer (PZT). Explain how the device works.

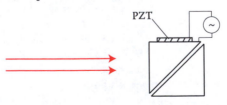

22. One form of protective eyewear for use in sunlight is made from sheets of polyvinyl alcohol (PVA) uniformly strained to align the long chain polymer molecules parallel to one another. Iodine, which has low affinity for its rather large number of electrons, is then bound to the PVA so loosely bound iodine electrons can freely oscillate along the

molecular backbone and thus absorb an oscillating electric field. In what direction (vertical or horizontal) would you normally expect the strain to be applied in such sunglasses and why?

23. A collimated linearly polarized beam of light is incident at a flat glass surface (refractive index 1.53) at 45° to the normal and with its plane of polarization at 30° to the plane of incidence. Find the ratio of the transmitted and reflected light intensities.

24. Using Equation 1.1 prove by a simple example that $H = \varepsilon_0 cE$ in a vacuum.

CHAPTER 2

Physical Optics

2.1 INTRODUCTION

Physical optics is the study of the behavior of light. When light encounters an aperture or obstacle or a change in refractive index, its direction of propagation and its phase are both altered. This effect, known as *diffraction*, is an important aspect of physical optics. Another important behavior that light exhibits is *interference*, which happens when light waves encounter one another usually having traveled rather different routes. A third and extremely important property of light is its *coherence*, and a fourth is its state of *polarization*. All these aspects of light are of great importance for holography, and we will examine each of them.

2.2 DIFFRACTION

As already stated, diffraction happens when light waves encounter an aperture, an obstacle, or a change in refractive index. Light from a source encounters an aperture or an obstacle, which partially blocks the light, or there may be a collection of apertures or obstacles in the path of the light, or indeed the light may simply pass through a region where there are abrupt or gradual changes in the refractive index. The light is *diffracted* in all these cases. (We have already considered, in Section 1.6.3, the special case of an abrupt change of refractive index as light is refracted across a plane boundary between different media.) We assume that the light which is diffracted is originally emitted by a point source with complex amplitude

$$\tilde{E}(r,t) = \frac{E_0}{r} e^{j(kr - \omega t)}$$

To calculate the complex amplitude $\tilde{E}_\mathrm{P}$ at a point, P, in a plane beyond the region where the diffraction has occurred, we use Equation 2.1 whose derivation is given in Appendix A (Equation A.8).

$$\tilde{E}_\mathrm{P} = \frac{E_0 e^{j(kr - \omega t - (\pi/2))}}{\lambda r r'} \iint_S e^{jkr'} \, dS \qquad (2.1)$$

S is a surface, shown in Figure A.3, comprising a hemisphere of infinite radius and an infinite plane x,y, in which diffraction takes place, while r' is the distance from an element dS on the surface S to the point P. The value of the expression $e^{jkr'}$ is zero for elements on the hemisphere because they are so far from P, and we only have to carry out the integration over those regions in the x,y plane, which actually allow light to propagate beyond that plane.

2.2.1 THE DIFFRACTION GRATING

Figure 2.1 shows a diffraction grating consisting of parallel, long, narrow slits each of width, w, and spaced d apart in an opaque screen, in the plane x,y. We can consider this screen as part of the surface S, completely enclosing the plane x',y' on which we need to find the light intensity distribution. The only contributions to that disturbance are from light passing through the slits, and therefore, we only have to carry out the integration over the slit areas. The source is assumed to be far from the grating plane (x,y) so that r' is large enough for the wavefronts arriving at the grating can be considered plane. The fact that the slits are long allows us to assume cylindrical symmetry about the x-axis and ignore any x' dependence in the x',y' plane.

$$\frac{1}{r'} \cong \frac{1}{r'_0} \quad \text{and} \quad r' \cong r'_0 - y\sin\theta$$

with r'_0 the distance from the middle of the first slit to P and θ the angle with the horizontal axis at which the light is diffracted. These simplifications mean that from Equation 2.1, we have

$$\tilde{E}_P = \frac{E_0}{\lambda r r'_0} e^{j(kr - \pi/2 - \omega t)} \iint_{\text{diffracting plane}} e^{jk(r'_0 - y\sin\theta)} dS \tag{2.2}$$

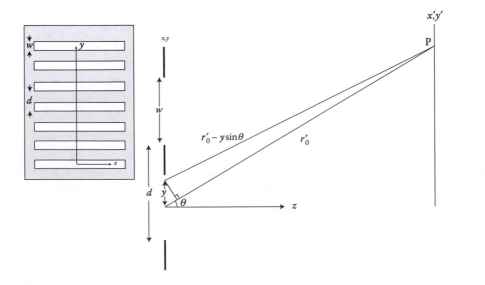

FIGURE 2.1 The diffraction grating.

Equation 2.2 is perfectly general as the nature and dimensions of the diffracting aperture(s) are not specified.

$$\tilde{E}_P = \frac{E_0 L}{\lambda r r_0'} e^{j\left[k(r+r_0')-\pi/2-\omega t\right]} \int_{\text{aperture plane}} e^{-jky\sin\theta} dy \tag{2.3}$$

where L is the length of each slit.

$$\tilde{E}_P = \frac{E_0 L}{\lambda r r_0'} e^{j\left[k(r+r_0')-\pi/2-\omega t\right]} \left\{ \begin{array}{l} \int_{-w/2}^{w/2} e^{-jky\sin\theta} dy + \int_{d-w/2}^{d+w/2} e^{-jky\sin\theta} dy + \int_{2d-w/2}^{2d+w/2} e^{-jky\sin\theta} dy \\ + \cdots + \int_{(N-1)d-w/2}^{(N-1)d+w/2} e^{-jky\sin\theta} dy \end{array} \right\} \tag{2.4}$$

with N, the number of slits.

$$\tilde{E}_P = \frac{E_0 L}{\lambda r' r_0} e^{j\left[k(r+r_0')-\pi/2-\omega t\right]} \left\{ \frac{e^{-jky\sin\theta}}{-jk\sin\theta}\Big|_{-w/2}^{w/2} + \frac{e^{-jky\sin\theta}}{-jk\sin\theta}\Big|_{d-w/2}^{d+w/2} + \frac{e^{-jky\sin\theta}}{-jk\sin\theta}\Big|_{2d-w/2}^{2d+w/2} + \cdots + \frac{e^{-jky\sin\theta}}{-jk\sin\theta}\Big|_{(N-1)d-w/2}^{(N-1)d+w/2} \right\}$$

$$= \frac{E_0 L}{\lambda r r_0'} e^{j\left[k(r+r_0')-\pi/2-\omega t\right]} \left\{ \frac{e^{-jky\sin\theta}}{-jk\sin\theta}\Big|_{-w/2}^{w/2} \left[1 + e^{-jkd\sin\theta} + e^{-2jkd\sin\theta} + \cdots + e^{-jk(N-1)d\sin\theta} \right] \right\}$$

$$= \frac{E_0 L}{\lambda r r_0'} e^{j\left[k(r+r_0')-\pi/2-\omega t\right]} \frac{e^{-jky\sin\theta}}{-jk\sin\theta}\Big|_{-w/2}^{w/2} \frac{e^{-Njkd\sin\theta}-1}{e^{-jkd\sin\theta}-1} = \frac{-E_0 L}{k\sin\theta\lambda r'} e^{jk(r'+r_0)} 2jk\frac{w}{2}\sin\theta \frac{e^{-jNkd\sin\theta}-1}{e^{-jkd\sin\theta}-1}$$

$$= \frac{-E_0 L \frac{w}{2}}{\lambda r r_0'} e^{j\left[k(r+r_0')-\pi/2-\omega t\right]} 2j \frac{\sin\left(k\frac{w}{2}\sin\theta\right)}{k\frac{w}{2}\sin\theta} \frac{e^{-jNkd\sin\theta}-1}{e^{-jkd\sin\theta}-1}$$

$$= \frac{-2E_0 L}{\lambda r r_0'} e^{j\left[k(r+r_0')+\pi/2-\omega t\right]} \frac{\sin\left(k\frac{w}{2}\sin\theta\right)}{k\frac{w}{2}\sin\theta} \frac{\sin(Nkd\sin\theta/2)}{\sin(kd\sin\theta/2)} e^{jk(N-1)\sin\theta/2}$$

$$= \frac{-E_0 L \frac{w}{2}}{\lambda r r_0'} e^{j\left[k(r+r_0')+\pi/2-\omega t\right]} \frac{\sin\left(k\frac{w}{2}\sin\theta\right)}{k\frac{w}{2}\sin\theta} \frac{-2\sin^2(Nkd\sin\theta/2) - j2\sin(Nkd\sin\theta/2)\cos(Nkd\sin\theta/2)}{-2\sin^2(kd\sin\theta/2) - j2\sin(kd\sin\theta/2)\cos(kd\sin\theta/2)}$$

$$= \frac{-2E_0 L}{\lambda r r_0'} \frac{\sin\left(k\frac{w}{2}\sin\theta\right)}{k\frac{w}{2}\sin\theta} \frac{\sin(Nkd\sin\theta/2)}{\sin(kd\sin\theta/2)} e^{j(k[r+r_0'+(N-1)\sin\theta/2]+\pi/2-\omega t)}$$

and finally intensity

$$I = \frac{1}{2} \left\langle \text{Re}^2 \left[\frac{-2E_0 L}{\lambda r r_0'} \frac{\sin\left(k\frac{w}{2}\sin\theta\right)}{k\frac{w}{2}\sin\theta} \frac{\sin(Nkd\sin\theta/2)}{\sin(kd\sin\theta/2)} e^{j(k[r+r_0'+(N-1)\sin\theta/2]+\pi/2-\omega t)} \right] \right\rangle$$

$$= \frac{1}{2} \left[\frac{-2E_0 L}{\lambda r r_0'} \frac{\sin\left(k\frac{w}{2}\sin\theta\right)}{k\frac{w}{2}\sin\theta} \frac{\sin(Nkd\sin\theta/2)}{\sin(kd\sin\theta/2)} e^{j(k[r+r_0'+(N-1)\sin\theta/2]+\pi/2-\omega t)} \right]$$

$$\times \left[\frac{-2E_0 L}{\lambda r r_0'} \frac{\sin\left(k\frac{w}{2}\sin\theta\right)}{k\frac{w}{2}\sin\theta} \frac{\sin(Nkd\sin\theta/2)}{\sin(kd\sin\theta/2)} e^{j(k[r+r_0'+(N-1)\sin\theta/2]+\pi/2-\omega t)} \right]^*$$

$$I_\theta = \frac{1}{2} \left(\frac{2E_0 L}{\lambda r' r_0} \right)^2 \frac{\sin^2\left(k\frac{w}{2}\sin\theta\right)}{\left(k\frac{w}{2}\sin\theta\right)^2} \frac{\sin^2(Nkd\sin\theta/2)}{\sin^2(kd\sin\theta/2)}$$

$$I_\theta = I_0 \frac{\sin^2\beta \sin^2(N\gamma)}{\beta^2 \sin^2\gamma} \tag{2.5}$$

where $\beta = kw\sin\theta/2$, $\gamma = kd\sin\theta/2$, $I_0 = (2E_0 L/\lambda r' r_0)^2$.

We now consider two special cases of the diffraction grating, the single slit and the double slit and then multiple slits.

2.2.1.1 Single Slits

$$I_\theta = I_0 \frac{\sin^2\beta}{\beta^2} \tag{2.6}$$

which has zeroes when $\beta = n\pi$, n an integer, that is, $w\sin\theta = n\lambda$. It is worth noting here that a single slit produces a diffraction pattern, which has the same mathematical form as the Fourier transform of an electrical rectangular pulse (Figure 2.2).

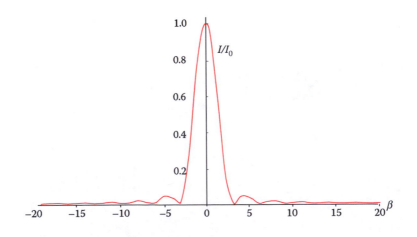

FIGURE 2.2 Intensity distribution due to a single slit.

2.2.1.2 Double Slits

$$I_\theta = I_0 \frac{\sin^2\beta \sin^2(2\gamma)}{\beta^2 \sin^2\gamma}$$

When the slits are narrow, then

$$\frac{\sin^2\beta}{\beta^2} = 1$$

$$I_\theta = I_0 \frac{\sin^2(2\gamma)}{\sin^2\gamma} = 4I_0 \cos^2\gamma$$

This intensity distribution is that obtained in the celebrated double slit experiment performed by Thomas Young to demonstrate the wave nature of light.

2.2.1.3 The Diffraction Grating (Multiple Slits)

Diffraction gratings are of great importance in optics particularly for applications in spectroscopy, which is the study of the relative intensities of electromagnetic waves emitted by light sources at different wavelengths. Spectroscopy is probably the single most important tool for scientific investigation available to us. It has provided us with almost all that we know about the structure of galaxies and stars as well as their movements relative to one another. Infrared, visible light, and ultraviolet light as well as millimeter and radio wavelength spectroscopy have all been brought to bear in our quest for an understanding of the universe. Every element and chemical compound has its own particular spectroscopic signature enabling us to determine both its presence and abundance.

Diffraction gratings and their properties are also of importance in holography for two reasons. One is that in holographic recording, we can consider all holographic recordings as consisting of sets of diffraction gratings with differing orientations and grating spacings. The other is that holographically recorded diffraction gratings offer significant advantages for spectroscopic instruments.

For all these reasons, it is worth examining the characteristics of diffraction gratings in some detail. We need only consider the expression $\sin^2(N\gamma)/\sin^2\gamma$ whose turning points can be found by differentiating with respect to γ and equating the result to zero.

There are three sets of solutions:

1. $\sin\gamma = 0$, $\gamma = n\pi$ with n an integer, known as the order of diffraction. To test whether these points produce maxima or minima of intensity, we alter γ by a small amount δ.

   $\sin^2 N(\gamma \pm \delta)/\sin^2(\gamma \pm \delta) = \{\sin(Nn\gamma \pm N\delta)/\sin(n\pi \pm \delta)\}^2 = \{[\sin(Nn\gamma)\cos(N\gamma) \pm \cos(Nn\gamma)\sin(N\delta)]/[\sin(n\gamma)\cos\delta \pm \cos(n\gamma)\sin\delta]\}^2$

   As $\delta \Rightarrow 0$, this last expression becomes $\{\sin(N\delta)/\sin\delta\}^2 = N^2$ so these are *maxima* and the intensities in directions given by $d\sin\theta = n\lambda$ are very large compared with the intensity due to a single slit.

2. $\sin(N\gamma) = 0$, $\gamma = n\pi/N$ except where n/N is an integer. These points are zeroes of intensity. The zeroes on either side of a principal maximum are $2\pi/N$ apart so that the base width of this peak is $2\pi/N$.
3. $\tan(N\gamma) = N\tan\gamma$. These produce maxima approximately midway between the minima of (2), that is, at $\gamma \cong (2n + 1)\pi/(2N)$, which are usually very small compared with the maxima in (1). Compare the intensities of the major peak at $\gamma = \pi$ and the minor peak immediately adjacent to it when there are five slits in the grating.

$$\sin^2(N\gamma)/\sin^2\gamma = 25 \quad \text{at} \quad \gamma = \pi$$

$$\sin^2(N\gamma)/\sin^2\gamma = 1/\sin^2(3\pi/10) = 1.53 \quad \text{at} \quad \gamma = \pi + 3\pi/50$$

Generally

$$\sin^2(N\gamma)/\sin^2\gamma = 1/\sin^2[(2n+1)\pi/(2N)] \cong 4N^2/[(2n+1)\pi]^2$$

so the ratio of the major peak intensity to an adjacent minor peak intensity is $(2n+1)^2\pi^2$.

At the major peak $\gamma = n\pi = kd\sin\gamma/2$

so

$$d\sin\theta = n\lambda \tag{2.7}$$

Equation 2.7 is the basis of grating spectroscopy and is known as the diffraction grating equation. If we measure the angle θ at which a particular peak is found and know the order n, we can find λ. Figure 2.3 shows the intensity distribution produced by a grating of seven identical slits with neighboring slits separated by three times the slit width. Notice that the $\mathrm{sinc}^2\beta$ term steadily attenuates the peaks due to the array of slits and that the peak at 3π is entirely suppressed because the $\mathrm{sinc}^2\beta$ term becomes zero at that point.

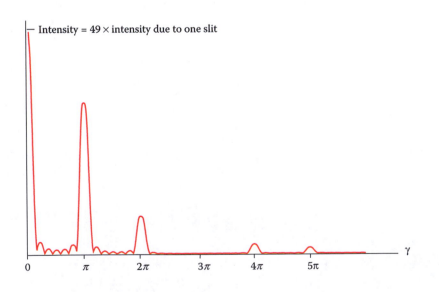

FIGURE 2.3 Intensity distribution from grating with seven slits with $d/w = 3$.

The base width of a major peak in the intensity pattern

$$\Delta\gamma = 2\pi/N = \Delta\,(kd\sin\theta/2) = kd\cos\theta\,\Delta\theta\,/2$$

This gives the angular width $\Delta\theta$ of a peak at its base as

$$\Delta\theta = 4\pi/(Nkd\cos\theta) = 2\lambda/(Nd\cos\theta)$$

From Equation 2.7, we obtain an expression for the dispersion

$$\frac{\partial\theta}{\partial\lambda} = \frac{n}{d\cos\theta}$$

2.2.1.4 Resolution and Resolving Power

We adopt the criterion that diffraction peaks associated with wavelengths, which are close to one another, are resolved if the peaks are separated at least by half the base width of either one as in Figure 2.4.

So the angular separation has to be at least

$$\Delta\theta_{min} = \lambda/(Nd\cos\theta)$$

giving a minimum resolvable wavelength separation of

$$\Delta\lambda_{min} = [\lambda/(Nd\cos\theta)]/(n/d\cos\theta)$$

The resolving power, R, is given by

$$R = \lambda/\Delta\lambda_{min} = nN,$$

which is maximum when θ is 90° and so $R = d\sin 90°\,N/\lambda = Nd/\lambda$.

The discussion so far has assumed that the incident light is normal to the plane of the grating but suppose that it arrives at an angle θ_i to the grating normal (Figure 2.5a).

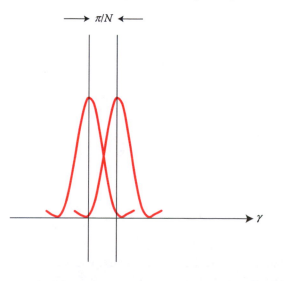

FIGURE 2.4 Minimum resolvable separation between diffraction pattern peaks.

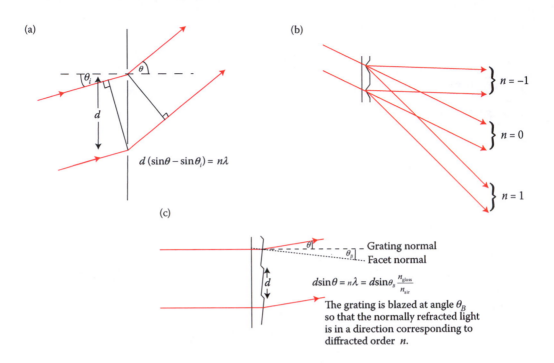

FIGURE 2.5 Blazed transmission grating.

We need not go through the whole analysis again but note that Equation 2.7 expresses the fact that, for an intense peak of diffraction to occur in direction θ (at P), the path difference between light waves arriving at P from neighboring slits must be a whole number of wavelengths. So from Figure 2.5a, we can see that the condition for an intense peak of diffraction should now be

$$d(\sin\theta - \sin\theta_i) = n\lambda \qquad (2.8)$$

In practice, diffraction gratings consist of parallel grooves in a usually flat glass plate (Figure 2.5b). The zero order of diffraction ($n = 0$, $\theta = \theta_i$) is the most intense, but we cannot obtain any useful information from it. This is a serious disadvantage, especially when we wish to study the spectra of weak light sources such as distant stars. To solve this problem, a technique known as blazing is used to shift the $\sin^2\beta$ function relative to the $\sin^2(N\gamma)/\sin^2\gamma$ term in the general expression for the grating intensity distribution. Gratings can be shaped so that the strongest diffracted beam of light is in a direction which is the normal direction of refraction (Figure 2.5c). Alternatively, a *reflection* grating can be made by cutting grooves in a reflecting metal such as aluminum coated onto a glass plate (Figure 2.6a) and Equation 2.8 applies.

Similarly, a reflection grating may be blazed to ensure that the wasteful zero-order intensity is negligible (Figure 2.6b). Gratings with as many as a thousand grooves per millimeter are common and with a width of say 10 cm have a resolution of 100,000 in the first order. Copies of transmission and reflection gratings can be made in gelatin at modest cost and are of adequate quality for many research purposes.

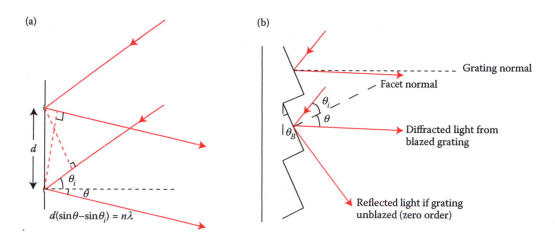

FIGURE 2.6 Blazed reflection grating: (a) oblique incidence, (b) normal transmission grating, (c) blazed transmission grating.

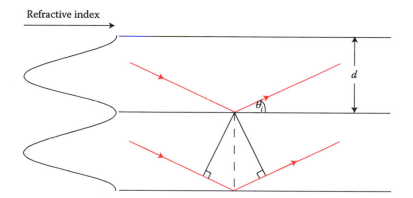

FIGURE 2.7 Volume diffraction grating.

Diffraction gratings of both reflection and transmission types can be fabricated using holographic methods, as we shall see in Chapter 11. Transmission holographic gratings are fundamentally different from the planar gratings we have considered so far, being of finite thickness. The performance of such holographic, volume, diffraction gratings can exceed those made by conventional methods, often at greatly reduced cost. They are usually of the form shown in Figure 2.7, consisting of a sinusoidal spatial variation of refractive index. Thick transmission gratings normally only produce one order of diffraction.

Incident light is partially transmitted and partially reflected as it encounters the changes in refractive index. Regions of constant refractive index are parallel planes and the condition for a maximum of diffracted light intensity is given by Equation 2.8, except that, additionally, $\theta = -\theta_i$ by the law of reflection and we have

$$2d \sin \theta_i = n\lambda \qquad (2.9)$$

Equation 2.9 is also applied to diffraction from atoms which lie in equally spaced planes in crystals. It is known as Bragg's equation, and it will reappear in Chapter 4 in which we deal with volume holography using coupled wave theory.

2.2.2 CIRCULAR APERTURE

In the case of a circular aperture (Figure 2.8), there is no cylindrical symmetry about the x-axis so that, in this calculation, we must take into account the variation in length of each strip of width dy in the aperture.

$$\tilde{E}_p = \frac{E_0}{rr_0'}e^{j(kr-\pi/2-\omega t)}\iint_{\text{aperture}} e^{jk(r_0'-y\sin\theta)}dS = \frac{E_0}{rr_0'}e^{j\left[k(r+r_0')-\pi/2-\omega t\right]}\iint_{\text{aperture}} e^{-jky\sin\theta}dS$$

$$= \frac{2E_0}{rr_0'}e^{j\left[k(r+r_0')-\pi/2-\omega t\right]}\int_{-a}^{a} e^{-jky\sin\theta}\sqrt{a^2-y^2}\,dy$$

where the aperture has radius a, dS has been replaced by a strip of width dy and length $2(a^2 - y^2)^{1/2}$ whose distance from the center of the aperture is y.

Let $u = y/a$ and $v = ka\sin\theta$

$$\tilde{E}_p = \frac{2E_0 a}{rr_0'}e^{j\left[k(r+r_0')-\pi/2-\omega t\right]}\int_{-1}^{1} e^{-juv}\sqrt{1-u^2}\,du = \frac{E_0 a}{rr_0'}e^{j\left[k(r+r_0')-\pi/2-\omega t\right]}$$

$$\left[\int_{-1}^{1}\left(\cos(uv)\sqrt{1-u^2}\right)du + j\int_{-1}^{1}\sin(uv)\sqrt{1-u^2}\,du\right]$$

$$= \frac{4jE_0 a}{rr_0'}e^{j\left[k(r+r_0')-\pi/2-\omega t\right]}\int_{0}^{1}\cos(uv)\sqrt{1-u^2}\,du$$

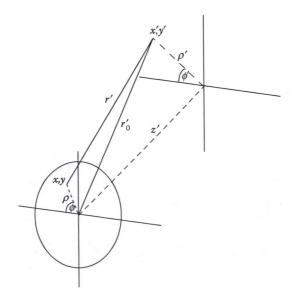

FIGURE 2.8 Circular aperture.

A first-order Bessel function is defined as

$$B_1(v) = \frac{2v}{\pi} \int_0^1 \cos(uv)\sqrt{1-u^2}\,du$$

$$\tilde{E}_P = \frac{2\pi j E_0 a}{rr_0'} e^{j\left[k(r+r_0')-\pi/2-\omega t\right]} \frac{B_1(v)}{v} = \frac{\pi E_0 a}{rr_0'} e^{j(k[r+r_0']-\omega t)} \frac{2B_1(v)}{v}$$

$$I_\theta = I_0 \left[\frac{2B_1(ka\sin\theta)}{ka\sin\theta}\right]^2 \qquad (2.10)$$

The intensity distribution given by Equation 2.10 is shown in Figure 2.9. It is maximum for $ka\sin\theta = 0$ and decreases to zero, for the first time, at $ka\sin\theta = 3.83$, that is when

$$\sin\theta = 0.61\lambda/a \qquad (2.11)$$

The pattern consists of a bright disc, known as the Airy disc, in which most of the light is concentrated, surrounded by concentric rings of alternating maximum and minimum intensity. If the focal length of the lens is f, the diameter of the disc is given by

$$2f\tan\theta \cong 1.22\lambda f/a, \quad \text{since } \theta \text{ is usually small.}$$

This is one of the most important results in optics. We have assumed that the amplitude of the light arriving at the circular aperture is the same everywhere in it. We have assumed that r is large so that the aperture is in fact illuminated by a plane wave. This can be ensured by using a point source of light located at the first focus of a lens that is coaxial with the aperture. We have also assumed the plane in which we measure I_θ is far from the aperture, effectively at infinity. Again, we could ensure this in practice by placing a second lens to the right of the aperture and coaxial with it so that the measurement plane is now the focal

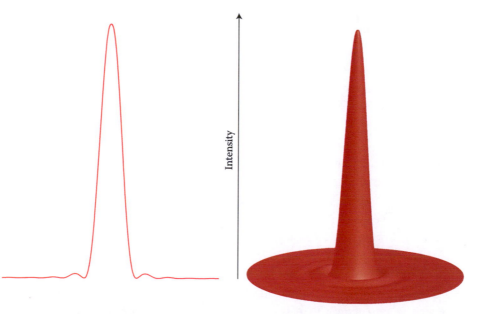

FIGURE 2.9 Diffraction pattern from a circular aperture.

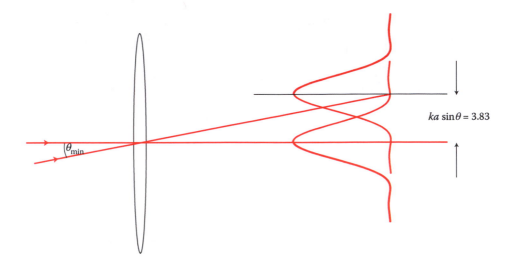

$ka \sin\theta = 3.83$

FIGURE 2.10 Minimum angular separation of distant point sources, which can be resolved by a lens.

plane of this lens. The conditions that the source and the measurement plane are infinitely far from the aperture or, equivalently, that the wavefronts arriving at and leaving the aperture are plane, are known as Fraunhofer diffraction conditions.

Now, suppose the second lens is replaced by the one whose diameter is that of the aperture, so that it fits *into* the aperture. Then, remove the aperture altogether. Light that formerly was incident outside the aperture area and so did not contribute to I_θ now bypasses the lens and still cannot contribute to I_θ. In other words, Equation 2.10 gives the intensity distribution of light in the focal plane of a lens due to a distant monochromatic source. This means that it is not physically possible to form a perfect image of a point. It will always have the form shown in Figure 2.9 at best. Now, we can easily see that there is a physical limit to how close together two points of light can be and yet still be distinguished, or resolved, as two separate points. We use as a criterion that they are resolved if the central maximum of the diffraction pattern for one lies directly over the first minimum of the other (Figure 2.10).

Since θ is very small we have from Equation 2.11,

$$\theta_{\min} = 0.61\lambda/a.$$

This result tells us that if we wish to improve the resolution of any optical instrument, we need to either increase its aperture radius, which is why astronomical telescopes are very large, or we must reduce the wavelength of the light used to illuminate what we wish to see, which is why we have ultraviolet and electron microscopes.

The spatial resolution limit d is simply

$$d = 1.22\lambda f / D \tag{2.12}$$

where f is the focal length of the lens and D its diameter. This spatial resolution limit leads to the concept of *depth of field*. By this, we mean that all object points within a certain range of distance from the lens are imaged with acceptable sharpness. The corresponding range

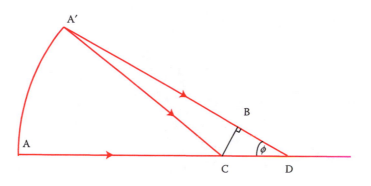

FIGURE 2.11 Depth of field.

of image distance is called the *depth of focus.* To determine the depth of field, we use the so-called Rayleigh criterion, that an image point is acceptably sharp if the wavefronts that form it do not deviate from sphericity by more than a quarter of a wavelength (Figure 2.11). In Figure 2.11, A and A′ are points in the middle and at one edge of a spherical wavefront emerging from a lens and converging to a diffraction limited image at C, meaning that the intensity of light in the image is given by Equation 2.10. Now consider the image formed on a screen at D some distance from C. If the image at D is to be just about acceptably sharp, it must be just at the limit of depth of focus from C and the path difference between light from A and A′ arriving at D, is $\lambda/8$. Then, if the radius of curvature of the wavefront is large

$$A'D - AD = (CD - BD) = CD(1 - \cos\phi) = \lambda/8 = 2CD\sin^2\left(\frac{\phi}{2}\right)$$

since $\phi = A\widehat{C}A' \cong A\widehat{D}A$. The depth of focus is 2CD since an acceptably sharp image can be obtained in either direction. Now, we set $\phi \cong D/2f$ with D the lens diameter and f its focal length.

The depth of focus is then

$$\frac{\lambda}{8(D/4f)^2} \cong \frac{2d^2}{\lambda} \tag{2.13}$$

where d is given by Equation 2.12.

The corresponding depth of field is obtained from Equation 1.20 and, for unity magnification, is therefore also given by Equation 2.13.

2.3 DIFFRACTION AND SPATIAL FOURIER TRANSFORMATION

We can show that Fraunhofer diffraction and spatial Fourier transformation are equivalent. Rewriting Equation 2.1, we have

$$\tilde{E}_P = \frac{-j}{\lambda}\iint_{\text{diffracting plane}} \frac{E_0 e^{j(kr-\omega t)}}{rr'}e^{jkr'}\,dS$$

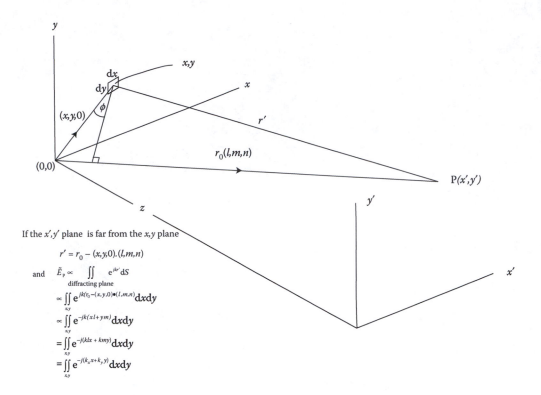

FIGURE 2.12 Equivalence of Fraunhofer diffraction and spatial Fourier transformation.

in which it is assumed that the light source illuminating the diffracting plane is described by the expression

$$\tilde{E}(r,t) = \frac{E_0}{r} e^{j(kr-\omega t)}$$

We can make Equation 2.1 more general by simply including the amplitude distribution in the x,y plane. Omitting the time dependence,

$$\tilde{E}_\mathrm{p} = \frac{-j}{\lambda} \iint_{x,y} \tilde{A}(x,y) \frac{e^{jkr'}}{r'} dxdy \qquad (2.14)$$

$\tilde{A}(x, y)$ may also be an object with complex transmittance $\tilde{A}(x, y)$ illuminated normal to the x,y plane by a plane wave of unit amplitude.

The complex amplitude, $\tilde{E}_\mathrm{p}$, at point P(x′,y′) can be written $\tilde{A}(x', y')$

$$\tilde{A}(x',y') = \frac{-j}{\lambda} \iint_{x,y} \tilde{A}(x,y) \frac{e^{jkr'}}{r'} dxdy = \frac{-j}{\lambda} \iint_{x,y} \tilde{A}(x,y) \frac{\exp\left\{ jk\left[(x'-x)^2 + (y'-y)^2 + z^2 \right]^{\frac{1}{2}} \right\}}{\left[(x'-x)^2 + (y'-y)^2 + z^2 \right]^{\frac{1}{2}}} dxdy$$

Since z is large compared with $x' - x$ and $y' - y$

$$\tilde{A}(x',y') \cong \frac{-j}{\lambda z} \iint_{x,y} \tilde{A}(x,y) \exp\left\{ jkz\left[1 + \frac{(x'-x)^2 + (y'-y)^2}{z^2} \right]^{\frac{1}{2}} \right\} dxdy,$$

Neglecting higher order terms, that is, $z^3 \gg [(x'-x)^2 + (y'-y)^2]^2/8$

$$\tilde{A}(x',y') \cong \frac{-je^{jkz}}{\lambda z} \iint_{x,y} \tilde{A}(x,y) \exp\left\{ \frac{jk}{2z}\left[(x'-x)^2 + (y'-y)^2 \right] \right\} dxdy \tag{2.15}$$

$$\tilde{A}(x',y') = \frac{-je^{jkz} e^{\left[jk/2z(x'^2 + y'^2) \right]}}{\lambda z} \iint_{x,y} \tilde{A}(x,y) e^{\left[-jk/z(xx'+yy') \right]} e^{\left[jk/2z(x^2+y^2) \right]} dxdy \tag{2.16}$$

Since z is large compared with x, y, x' and y', $\exp[jk/2z(x^2 + y^2)] \cong 1$ and $\exp[jk/2z(x'^2 + y'^2)] \cong 1$.

$$\tilde{A}(x',y') = \frac{-je^{jkz}}{\lambda z} \iint_{x,y} \tilde{A}(x,y) \exp\left[-\frac{jk}{z}(xx' + yy') \right] dxdy \tag{2.17}$$

The right-hand side of Equation 2.17 has the form of a two-dimensional (2-D) Fourier transform and describes the complex amplitude distribution in the x',y' plane, due to the amplitude distribution in the x,y plane.

Alternatively if, in Figure 2.12, the x,y and x',y' planes are sufficiently far apart, we can write

$$r' = r_0 - (x^2 + y^2)^{\frac{1}{2}} \sin\theta = r_0 - (x, y, 0) \bullet (l, m, n)$$

where $(x,y,0)$ is the vector from the origin of the x,y plane to the element $dxdy$ and (l,m,n) is the unit vector in the direction from the origin of the x,y plane to the point P in the x',y' plane. We can also approximate $1/r'$ by $1/r_0$ as we did for Equation 2.2 and Equation 2.14 becomes

$$\tilde{A}(x',y') = \frac{-j}{\lambda r_0} \iint_{x,y} \tilde{A}(x,y) \exp\left\{ jk\left[r_0 - (x, y, 0).(l, m, n) \right] \right\} dxdy$$

$$= \frac{-je^{jkr_0}}{\lambda r_0} \iint_{x,y} \tilde{A}(x,y) \exp\left[-jk(xl + ym) \right] dxdy = \frac{-je^{jkr_0}}{\lambda r_0} \iint_{x,y} \tilde{A}(x,y) \exp\left[-j(klx + kmy) \right] dxdy$$

kl and km are set equal to k_x, and k_y respectively, and we have

$$\tilde{A}(x',y') = \frac{-je^{jkr_0}}{\lambda r_0} \iint_{x,y} \tilde{A}(x,y) e^{-j(k_x x + k_y y)} dxdy \tag{2.18}$$

Equation 2.18 is identical in form to Equation 2.17, if r_0 is substituted for z, again having the form of a 2-D Fourier transform and describing the complex amplitude distribution in the x', y' plane due to the amplitude distribution in the x, y plane. Fraunhofer diffraction and Fourier transformation are equivalent, the latter being simply the mathematical expression of the former. Fraunhofer diffraction decomposes a convoluted wavefront into its constituent spatial frequencies, each corresponding to a single point in the x', y' plane. The spatial frequencies l/λ and m/λ are, respectively, the same as $x'/(\lambda z)$ and $y'/(\lambda z)$ in Equation 2.17.

In the derivation of Equation 2.18, we placed the x', y' plane far from the x, y plane as a necessary condition. Strictly, the planes should be infinitely far apart so that the line joining the element $dxdy$ to P is parallel to that joining the origin of the x, y plane to P. Of course, the insertion of a lens brings the x', y' plane into the focal plane of the lens.

In Fraunhofer diffraction, it is assumed that all the wavefronts contributing to the diffraction pattern in the x', y' plane are plane. If the x, y and x', y' planes are close together, then we cannot ignore the curvatures of wavefronts arriving at the x, y plane or diffracted in that plane.

2.4 PHASE EFFECT OF A THIN LENS

We now wish to calculate the effect of a lens on a complex amplitude distribution of light arriving at one of the lens surfaces.

In Figure 2.13, we consider the phase difference between light of wavelength, λ, passing through the lens at point (x_l, y_l) and light passing through at $(0,0)$. At $(0,0)$, the additional phase imposed on a light wave passing through the lens is

$$2\pi n D / \lambda$$

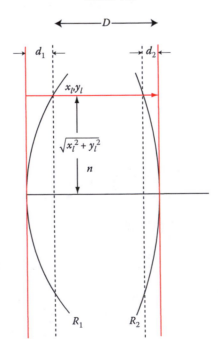

FIGURE 2.13 Phase effect of a lens.

where D is the thickness of the lens at its axis and n is its refractive index. At x_l, y_l the additional phase is

$$\frac{2\pi d_1}{\lambda} + \frac{2\pi n(D - d_1 - d_2)}{\lambda} + \frac{2\pi d_2}{\lambda}$$

which leads the phase at (0,0) by

$$\phi(x_l, y_l) = \frac{2\pi}{\lambda}(n-1)(d_1 + d_2) = \frac{2\pi}{\lambda}(n-1)(x_l^2 + y_l^2)\left(\frac{1}{2R_1} + \frac{1}{2R_2}\right)$$

$$\phi(x_l, y_l) = \frac{k}{2f}(x_l^2 + y_l^2) \tag{2.19}$$

using the lensmaker's formula (Equation 1.16). Here, we have again used the theorem of intersecting chords and assumed the radii of curvature of the lens R_1 and R_2 to be very large compared with the lens aperture (see Section 1.6.1). Equation 2.19 enables us to determine the amplitude in the plane tangential to the right-hand surface given the complex amplitude in the plane tangential to the left-hand surface. The phase at (x_l, y_l) *leads* that at (0,0) by which we mean that the oscillating electric field will reach a particular value at (0,0) *later* than at (x_l, y_l).

In Section 2.5, we demonstrate the ability of a lens to compute the spatial Fourier transform of the complex amplitude distribution of light. Some mathematics is involved including use of the convolution theorem and, although it is not too difficult, some readers may prefer to go straight on to Section 2.6 in which we demonstrate the Fourier property from what we already know about the *physical* properties of lenses.

2.5 FOURIER TRANSFORMATION BY A LENS

Suppose we place an object with complex transmittance $\tilde{A}(x, y)$ in the front focal plane of a lens (Figure 2.14) and illuminate it with a plane wave of unit amplitude normal to the x, y plane.

The complex amplitude at the lens plane x_l, y_l is given by Equation 2.15 with f substituted for z.

$$\tilde{A}(x_l, y_l) = \frac{-je^{jkf}}{\lambda f}\iint_{x,y}\tilde{A}(x, y)\exp\left\{\frac{jk}{2f}\left[(x_l - x)^2 + (y_l - y)^2\right]\right\}dxdy \tag{2.20}$$

The complex amplitude immediately following the lens is, from Equation 2.19, $\tilde{A}(x_l, y_l)\exp[-(jk/2f)(x_l^2 + y_l^2)]$ the negative sign in the exponent indicating the fact that the phase at (x_l, y_l) leads that at (0,0).

The complex amplitude in the back focal plane is given by Equation 2.16, with f again substituting for z.

$$\tilde{A}(x', y') = \frac{-je^{jkf}}{\lambda f}\exp\left[\frac{jk}{2f}(x'^2 + y'^2)\right]\iint_{x_l, y_l}\tilde{A}(x_l, y_l)\exp\left[-\frac{jk}{f}(x_l x' + y_l y')\right]dx_l dy_l \tag{2.21}$$

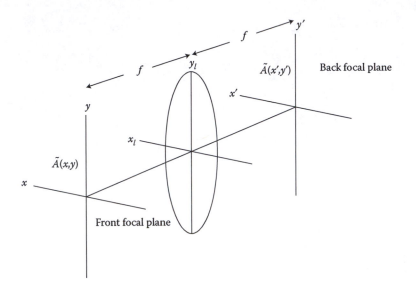

FIGURE 2.14 Spatial Fourier Transformation by a lens.

The expression

$$\frac{1}{\lambda f}\iint_{x_l,y_l} \tilde{A}(x_l,y_l)\exp\left[-\frac{jk}{f}(x_l x' + y_l y')\right]dx_l dy_l$$

is the spatial Fourier transformation of $\tilde{A}(x_l, y_l)$ which, in its turn, is a *convolution* (Equation 2.20) of $\tilde{A}(x, y)$ and $\exp[(jk/2f)(x_l^2 + y_l^2)]$. Applying the *convolution theorem* (see Appendix A.2), we have

$$\tilde{A}(x',y') = \frac{-e^{2jkf}}{\lambda^2 f^2}\exp\left[\frac{jk}{2f}(x'^2 + y'^2)\right]\mathrm{SFT}\{\tilde{A}(x,y)\}\mathrm{SFT}\left\{\exp\left[\frac{jk}{2f}(x_l^2 + y_l^2)\right]\right\}$$

where SFT means spatial fourier transform.

$$\tilde{A}(x',y') = \frac{-e^{2jkf}}{\lambda^2 f^2}\exp\left[\frac{jk}{2f}(x'^2 + y'^2)\right]\iint_{x,y}\tilde{A}(x,y)\exp\left[-\frac{jk}{f}(xx' + yy')\right]dxdy$$

$$\times \iint_{x_l,y_l}\exp\left[\frac{jk}{2f}(x_l^2 + y_l^2)\right]\exp\left[-\frac{jk}{f}(x_l x' + y_l y')\right]dx_l dy_l$$

$$= \frac{-e^{2jkf}}{\lambda^2 f^2}\iint_{x,y}\tilde{A}(x,y)\exp\left[-\frac{jk}{f}(xx' + yy')\right]dxdy$$

$$\times \iint_{x_l,y_l}\exp\left[\frac{jk}{2f}(x_l - x')^2 + (y_l - y')^2\right]dx_l dy_l$$

We usually can ignore the constant phase factor $-je^{2jkf}$. The second integral is evaluated using the identity $\int_{-\infty}^{\infty} \exp(jx^2)\,dx = \sqrt{(\pi/2)}(1+j)$ to give

$$\tilde{A}(x', y') = \frac{1}{\lambda f} \iint_{x,y} \tilde{A}(x, y)\exp\left[-\frac{jk}{f}(xx' + yy')\right]dxdy \qquad (2.22)$$

Equation 2.22 shows that the lens performs spatial Fourier transformation of the complex amplitude distribution of light in its front focal plane and the point x',y' in the back focal plane represents spatial frequency $x'/\lambda f, y'/\lambda f$. There are limitations, the most important one being that the result is not quite accurate since the lens itself is a low-pass spatial filter and high spatial frequency light bypasses it. Nonetheless, it is a very important result. A lens acting as an optical computer of Fourier transforms and operating at light speed is an extremely powerful tool with a range of holographic applications as we shall see.

The object transparency may also be placed anywhere in front of the lens. The Fourier transform appears again in the back focal plane although multiplied by a spherical phase factor. The advantage here is that the Fourier transform is more accurate if the lens diameter is larger than the transparency. The transparency may also be located between the lens and the back focal plane with a collimated coaxial light beam entering the lens in which case the Fourier transform appears in the back focal plane as before but its transverse spatial scale may be adjusted by moving the transparency along the lens axis.

2.6 FOURIER TRANSFORM PROPERTY OF A LENS—A PHYSICAL ARGUMENT

We saw from Equations 2.17 and 2.18 that the Fraunhofer diffraction pattern appearing in a plane is the spatial Fourier transform of the complex amplitude of the light distribution in a parallel plane some distance away. The condition is that the distance between the planes be large compared with other distances involved. In the statement immediately preceding Equation 2.17, this was expressed as

$$z \gg \frac{x^2 + y^2}{\lambda} \quad \text{and} \quad z \gg \frac{x'^2 + y'^2}{\lambda} \qquad (2.23)$$

This is the Fraunhofer or far-field condition. We can effectively satisfy this condition by placing a lens in the path of the light diffracted by the object. The plane x',y', ideally infinitely distant from the x,y plane, now has a *conjugate plane*, which is the back focal plane of the lens as we saw in Section 1.6.1. The front focal plane of the lens is also conjugate to one which is infinitely far away and an object placed in the front focal plane is effectively infinitely far away. Thus, when an object is placed in the front focal plane of the lens, its spatial Fourier transform appears in the back focal plane, with the Fraunhofer condition satisfied.

2.7 INTERFERENCE BY DIVISION OF AMPLITUDE

Interference which happens when light waves encounter one another is the physical phenomenon on which all holography depends. We have already encountered interference

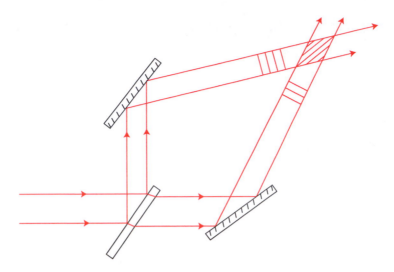

FIGURE 2.15 The Mach–Zehnder interferometer.

phenomena in Section 2.2 in which we considered the effects of apertures on light. The phenomenon of diffraction is often called *interference by division of wavefront*. As this name implies, the *wavefront* is divided up by virtue of its encounter with one or more apertures, which prevent some of the light from continuing on its way. Any system of apertures is often called a *wavefront splitting interferometer*. However, we are concerned here with a rather different method of producing interference effects, which is of great importance in holography and is known as *interference by division of amplitude*. The method involves splitting a beam of light into at least two beams, each of which has a reduced amplitude, then allowing the two separate beams to travel different routes in space and finally making them recombine. The splitting is easily accomplished by a partially reflecting and partially transmitting device such as a partially silvered mirror and the recombination by means of mirrors, as shown in Figure 2.15, which depicts a very useful device called a Mach–Zehnder interferometer. It is only one of many different forms of interferometer and is largely used in wind tunnel experiments to study airflow over scale models of aircraft and automobiles. Small variations in the refractive index of the air in the flow cause changes in the phase differences between the interfering waves and thus in the interference pattern that is observed.

As we shall see later, many practical holographic recording arrangements are essentially Mach–Zehnder interferometers.

When two plane waves interfere, the resultant intensity at any location, $\mathbf{r}$, at which interference occurs, is given by

$$
\begin{aligned}
I &= \left\langle \left[E_1 \cos(\mathbf{k}_1 \cdot \mathbf{r} - \omega t) + E_2 \cos(\mathbf{k}_2 \cdot \mathbf{r} - \omega t) \right]^2 \right\rangle \\
&= \frac{1}{2} E_1^2 + \frac{1}{2} E_2^2 + \left\langle 2 E_1 E_2 \cos(\mathbf{k}_1 \cdot \mathbf{r} - \omega t) \cos(\mathbf{k}_2 \cdot \mathbf{r} - \omega t) \right\rangle \\
&= I_1 + I_2 + 2\sqrt{I_1 I_2} \left\{ \left\langle \left[\cos(\mathbf{k}_1 + \mathbf{k}_2) \cdot \mathbf{r} - 2\omega t \right] \right\rangle + \cos(\mathbf{k}_1 - \mathbf{k}_2) \cdot \mathbf{r} \right\} \\
&= I_1 + I_2 + 2\sqrt{I_1 I_2} \cos(\mathbf{k}_1 - \mathbf{k}_2) \cdot \mathbf{r}
\end{aligned}
$$

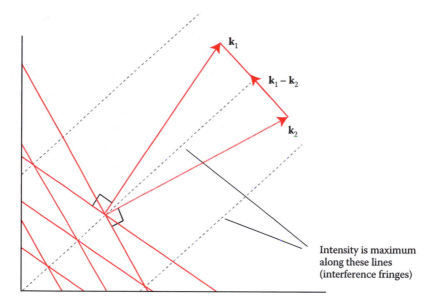

FIGURE 2.16 Two sets of wavefronts with different spatial frequencies overlap in space to produce an interference pattern.

where $\mathbf{k}_1$ and $\mathbf{k}_2$ are the wave vectors of the two sets of wavefronts. We have also assumed that the electric fields of the waves are parallel. The intensity is maximum for $(\mathbf{k}_1 - \mathbf{k}_2) \cdot \mathbf{r} = 2n\pi$ and minimum for $(\mathbf{k}_1 - \mathbf{k}_2) \cdot \mathbf{r} = (2n + 1)$, with n an integer. What we see in the space where the waves overlap is an interference pattern of light and dark bands or fringes. Their spatial frequency measured in any direction is simply the difference between the spatial frequencies of the two contributing sets of wavefronts (Figure 2.16). The visibility, or contrast, γ, of the fringes is

$$\gamma = \frac{I_{\max} - I_{\min}}{I_{\max} + I_{\min}} = \frac{2\sqrt{I_1 I_2}}{I_1 + I_2}$$

Note that the visibility is everywhere constant, but this result is based on the assumption that the light source is perfectly monochromatic and is essentially a point source. In other words, the source is *coherent*.

2.8 COHERENCE

Coherence is a subtle property of light, of which we are not normally aware but which is of enormous importance in holography. In order to understand coherence, we need to look at how light is produced in the first place.

2.8.1 PRODUCTION OF LIGHT

In all light sources, regardless of type, one fundamental physical process occurs repeatedly and that is the excitation of atoms, whether by heating, using an electric current, or by bombardment by electrons in an electric current. In the excitation process, the atoms gain energy only to lose it again by emitting electromagnetic radiation, often in the form of visible light. Broadband, or white light sources, emit light throughout the visible spectrum of color from red to violet. Normal household incandescent light bulbs, or tungsten lamps, emit infrared light as well. They are inefficient, producing heat too, and are being replaced by fluorescent lamps, which produce light of only certain colors, the remainder being added by coating a fluorescent material on the inside surface of the glass.

If you look at a mercury lamp through a spectroscope, you see pure violet, blue, green, and some yellow light, but not red. That is why, in a mercury lamp, we need to add a red fluorescing material to obtain white light. Some light sources such as sodium street lamps emit almost, but not quite, monochromatic light. That is why they appear almost pure yellow. Lamps such as sodium and mercury lamps are called spectral lamps because they emit light in narrow ranges of wavelength, unlike tungsten lamps. Laser light sources are in a different category as they emit light of just one color in an extremely narrow wavelength range, although some lasers can be tuned to different wavelengths.

2.8.2 THE BANDWIDTH OF LIGHT SOURCES

We will confine our attention to a single atom struck by an electron and undergoing excitation. This means that one of the atom's electrons changes from an orbital that confines it to regions close to the nucleus, to an orbital that confines it to regions further away implying that the atom changes to a state of higher energy. The atom later returns to its original state, releasing the energy gained in the excitation, as an electromagnetic wave, also called a *photon* reflecting the fact that it carries a discrete amount or *quantum* of energy. We normally use the term photon when referring to electromagnetic energy emitted (or absorbed) by a single atom in course of transition between energy states, as here, but we will stick to the more familiar idea of a wave for the purpose of the present discussion.

The wave is of finite duration, and this is critically important because it means that the wave cannot be purely monochromatic. We can show that this is the case by carrying out a temporal Fourier transformation of the function describing a light wave of finite duration. We will consider the consequences of the result for interference in general and holography in particular.

Suppose we have a wave $E(t) = E_0\cos(\omega_0 t)$ of duration δt, which begins at time $t = 0$. The frequency spectrum $f(\omega)$ is obtained by finding the Fourier transform of $E(t)$

$$f(\omega) = \int_0^{\delta t} E_0\cos(\omega_0 t)\cos(\omega t)\mathrm{d}t$$

which only has significant values in the frequency range between $(\omega - \omega_0)\delta t = +/- \pi$. That is, $\delta\omega\delta t = 2\pi$ where $\delta\omega = |\omega - \omega_0|$

$$\delta v \delta t = 1 \qquad (2.24)$$

This result, one of the most important in all of physics, simply tells us that a wave of duration δt is not perfectly monochromatic but has a spread of frequencies, δv, which in turn implies a wavelength spread of $\delta\lambda$ which is found as follows:

$$v = \frac{c}{\lambda}$$

$$\frac{\mathrm{d}v}{\mathrm{d}\lambda} = \frac{-c}{\lambda^2}$$

$$\delta\lambda = \frac{-\lambda^2 \delta v}{c} = \frac{-\lambda^2}{c\delta t}$$

and the length of the wavetrain, or coherence length, δl, is given by

$$\delta l = c\delta t = \frac{\lambda^2}{\delta\lambda}$$

For a mercury or sodium spectrum lamp, the coherence length is typically a few millimeters. It is because the coherence lengths of most light sources are so short that interference effects are not that easy to observe in everyday experience. Interference effects arising from white light can be obtained using a Michelson interferometer (Section 5.3). With some care and patience, they can even be observed by looking at the full moon through a double glazed window [1].

One could argue that interference effects could be observed easily even with short coherence length sources since, when the wavetrain splits into two parts, the one taking the *longer* route through the interferometer could interfere with part of a *later* wavetrain, which has taken the *shorter* route. This is certainly true, but the resulting intensity whether maximum or minimum intensity, or something in-between, would only persist for the duration of a single wavetrain that is, $\sim 3 \times 10^{-3} / 3 \times 10^8 = 0.01$ ns. This is because the phase relationship between *consecutively* emitted waves changes randomly as the atom is randomly bombarded by electrons.

The coherence length is shortened by the fact that even while atoms are in the process of emitting light waves, they are being struck by electrons and other atoms and the emission process is constantly interrupted. Thus, the wavetrain undergoes random phase changes during emission and δt is reduced. This leads to an increased wavelength spread, and therefore frequency spread and the effect is called collision broadening.

The picture is complicated still further when we turn our attention to the fact that the light from a single atom is not very bright. We will need the light from lots of them to serve any useful purpose, but in a spectrum lamp, the light-emitting atoms are all in continual random motion and the wavelength of the light emitted by a single atom depends on its velocity relative to the viewer. This results in a further increase in the spread of wavelengths emitted from a light source, known as Doppler broadening.

To measure the coherence length of a light source, we use it in an interferometer whose design does not matter, so long as we can vary and measure the path difference. When the path difference is zero, the interference fringes have maximum visibility, decreasing as the path difference increases, becoming zero when the path difference exceeds the coherence length. The change in path difference in the interferometer as the fringe pattern visibility goes from zero through maximum visibility and back to zero again is twice the coherence length.

2.8.3 SPATIAL COHERENCE

In addition to the problem of temporal coherence a further limitation is imposed by the *physical dimensions* of the light source, that is, by the fact that a source consists of vast numbers of atoms in different places all emitting light independently of one another. There is generally no fixed phase relationship between the light emitted by two different atoms. To understand the consequences of this, let us return to the interference pattern produced when light is diffracted by a double slit.

$$I_\theta = I_0 \frac{\sin^2(2\gamma)}{\sin^2\gamma} = 4I_0 \cos^2\gamma = 2I_0(1+\cos 2\gamma) = 2I_0\left[1+\cos(kd\sin\theta)\right]$$

$$I_{y'} \cong 2I_0\left[1+\cos\left(\frac{ky'd}{D'}\right)\right]$$

where $\sin\theta \cong y'/D'$ with D' the distance between the slits and the screen on which the intensity $I_{y'}$ is observed at point y'.

In obtaining this result and indeed all the results in Section 2.2, it is assumed that the diffracting aperture(s) are illuminated by a plane wavefront (effectively a very distant point source or a point source at the first focus of a lens).

Now, suppose that the double slit is in fact illuminated by an *extended* source (Figure 2.17) and all parts of the wavefront are not in phase with one another.

However, we can divide the source into small elements, within each of which all parts are in phase.

The optical path difference between paths from a point at y'' on the source through the two slits to a point y' on the screen is given by

$$\sqrt{\left(y'+\frac{d}{2}\right)^2 + D'^2} + \sqrt{\left(y''+\frac{d}{2}\right)^2 + D^2} - \sqrt{\left(y'-\frac{d}{2}\right)^2 + D'^2} - \sqrt{\left(y''-\frac{d}{2}\right)^2 + D^2}$$

$$= D'\left[1+\left(\frac{y'}{D'}+\frac{d}{2D'}\right)^2\right]^{\frac{1}{2}} + D\left[1+\left(\frac{y''}{D}+\frac{d}{2D}\right)^2\right]^{\frac{1}{2}} - D'\left[1+\left(\frac{y'}{D'}-\frac{d}{2D'}\right)^2\right]^{\frac{1}{2}}$$

$$- D\left[1+\left(\frac{y''}{D}-\frac{d}{2D}\right)^2\right]^{\frac{1}{2}}$$

$$\cong \frac{y'd}{D'} + \frac{y''d}{D}$$

since both y' and y'' are small compared with D, the distance between the source and the slits, and D'.

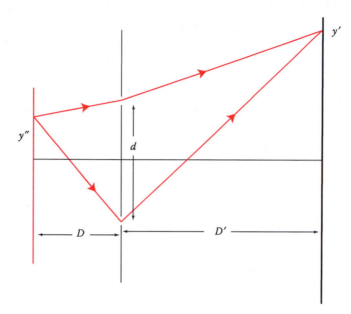

FIGURE 2.17 Effect of extended light source.

The intensity distribution, $dI_{y'}$ on the screen due to an element of width dy'' of the source is given by

$$dI_{y'} = 2I(y'')\left[1+\cos\left\{k\left(\frac{y''d}{D}+\frac{y'd}{D'}\right)\right\}\right]$$

and the total intensity

$$I_{y'} = \int_{\text{source}} 2I(y'')\left[1+\cos\left\{k\left(\frac{y''d}{D}+\frac{y'd}{D'}\right)\right\}\right]dy''$$

Suppose now the source intensity is spatially uniform, so that $I(y'')=I_0'$ for $-y_0''/2 \le y'' \le y_0''/2$ and zero elsewhere. That means that its form is the same as that of a rectangular aperture of width y_0''.

$$I_{y'} = \int_{-y_0''/2}^{y_0''/2} 2I_0'\left[1+\cos\left(\frac{ky''d}{D}+\frac{ky'd}{D'}\right)\right]dy''$$

$$= 2I_0'y_0''+2I_0'\frac{\sin\left(\dfrac{ky_0''d}{2D}+\dfrac{ky'd}{D'}\right)-\sin\left(-\dfrac{ky_0''d}{2D}+\dfrac{ky'd}{D'}\right)}{\left(\dfrac{kd}{D}\right)}$$

$$= 2I_0'y_0''+2I_0'y_0''\frac{\sin\left(\dfrac{ky_0''d}{2D}\right)}{\left(\dfrac{ky_0''d}{2D}\right)}\cos\left(\frac{ky'd}{D'}\right)$$

Setting $I_0' y'' = I_0$ the total intensity due to one of the slits

$$I_{y'} = 2I_0 + 2I_0 \frac{\sin(ky_0''d/2D)}{(ky_0''d/2D)} \cos\left(\frac{ky'd}{D'}\right) \tag{2.25}$$

Note that if the source is a point source, $\sin(ky_0''d/2D)/(ky_0''d/2D) = 1$ and we revert to the original form of intensity distribution

$$I_{y'} \cong 2I_0\left[1+\cos\left(\frac{ky'd}{D'}\right)\right]$$

The visibility or contrast of the fringe pattern is given by $\gamma = (I_{max} - I_{min})/(I_{max} + I_{min})$, which is simply $\sin(ky_0''d/2D)/(ky_0''d/2D)$. Thus, the effect of an extended light source is to reduce the visibility of the fringe pattern that is obtained. The expression $\sin(ky_0''d/2D)/(ky_0''d/2D)$ has the same form as the diffraction pattern that is obtained from a single slit.

Similarly, the visibility, or contrast, of the interference pattern produced by a double slit illuminated by a *circular* source of spatially uniform intensity has the form of the diffraction pattern produced by a circular aperture. These are examples of the consequences of the perfectly general VanCittert-Zernike theorem, one form of which states that the spatial coherence of a light source is given by the form of the far-field diffraction pattern of the source. It has already been noted that a single slit produces a diffraction pattern, which has the same mathematical form as the Fourier transform of a rectangular electrical pulse, which suggests that the mathematical form of a diffraction pattern is in fact the 2-D spatial Fourier Transform of the spatial distribution of light that produces the pattern. This leads us to restate VanCittert -Zernike theorem as: the spatial coherence of a light source is given by the form of the Fourier transform of the source intensity distribution. We therefore use the visibility of the fringe pattern as a measure of the spatial coherence of the light source.

The visibility of the fringes in an interferometer with an optical path difference less than the coherence length of the light source is determined by the *spatial* coherence of the source. This applies when interference is obtained using light from *different parts* of the source, (but see Section 5.3 in which interference is obtained using light coming from the *same* part of the source).

Returning again to the expression $\sin(ky_0''d/2D)/(ky_0''d/2D)$ for the contrast, suppose that the slits are 2 mm apart, the slit to source distance, D, is 50 cm and the wavelength is 546 nm. Then, if we wish to ensure that the contrast of the fringe pattern is not less than 50% of the maximum, the source width y'' has to be no wider than 80 μm. This is a very important result as it demonstrates the difficulty we face when trying to obtain interference patterns using conventional light sources.

To ensure reasonable contrast in a fringe pattern, we have to restrict the source size, usually by focusing its light onto a small aperture, which is not easy if the source radiates over a large solid angle. Use of a laser enables us to overcome this difficulty as laser light is readily focused down to a very small spot using a lens because the light is very nearly collimated at the outset.

2.9 POLARIZED LIGHT

In Section 1.4, it was stated that the electric field is a vector that can change direction and amplitude with time. In the following discussion, we are not so much concerned with the polarization of light per se but with methods of controlling and changing the polarization state of light as these are of great importance in practical holography.

2.9.1 PLANE POLARIZED LIGHT

A plane, or linearly, polarized light wave has its electric field confined to a fixed plane. We can describe such a wave using Equation 1.10.

$$\mathbf{E}(\mathbf{r},t) = \mathbf{E}_0\cos(\mathbf{k}\cdot\mathbf{r} - \omega t + \phi)$$

We assume the wave propagates along the z-axis and its wavefronts are plane and normal to the z-axis. We now have

$$\mathbf{E}(z,t) = \mathbf{E}_0\cos(kz - \omega t + \phi) \tag{2.26}$$

The plane of polarization can be oriented in any direction parallel to the x,y plane so that $\mathbf{E}_0$ has components $\hat{\mathbf{x}}E_{0x}, \hat{\mathbf{y}}E_{0y}$, where $\hat{\mathbf{x}}$ and $\hat{\mathbf{y}}$ are unit vectors along the x and y axes, respectively, and

$$\mathbf{E}(z,t) = \hat{\mathbf{x}}E_{0x}\cos(kz - \omega t) + \hat{\mathbf{y}}E_{0y}\cos(kz - \omega t + \phi) \tag{2.27}$$

with $\phi = 2p\pi$, p an integer. Looking in the $-z$ direction at the approaching wave, the plane of polarization is at an angle $\tan^{-1}(E_{0y}/E_{0x})$ to the x-axis (Figure 2.18). If $\phi = (2p+1)\pi$, the plane of polarization is rotated clockwise by $2\tan^{-1}(E_{0y}/E_{0x})$. If $E_{0x} = E_{0y}$, the rotation is $\pi/2$.

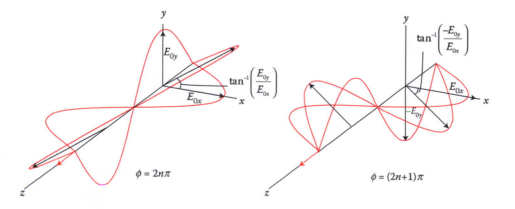

FIGURE 2.18 Plane polarization.

2.9.2 OTHER POLARIZATION STATES

In the previous section, we described linearly polarized light in terms of two orthogonally plane polarized components having a phase difference, ϕ, of $2p\pi$ or $(2p + 1)\pi$ between them.

We now consider the situation when ϕ has a value of $\pi/2$ or $-\pi/2$. In Figure 2.19a, two linearly polarized waves, $\hat{x}E_{0x}\cos(kz - \omega t)$ and $\hat{y}E_{0y}\cos(kz - \omega t - \pi/2)$, are shown *at time t = 0*. As the wave advances in the +z direction, the electric field rotates *anticlockwise* around the z-axis, its resultant electric field having values of E_{0x} at $z = 0$, E_{0y} at $z = \lambda/4$, $-E_{0x}$ at $z = \lambda/2$, $-E_{0y}$ at $z = 3\lambda/4$, and E_{0x} again at $z = \lambda$ and so on.

At any *fixed point* on the z-axis, looking in the −z direction toward the source, we see that the resultant electric field rotates *clockwise with time,* the resultant electric field having values of E_{0x} at $t = 0$, $-E_{0y}$ at $t = T/4$, $-E_{0x}$ at $t = T/2$, E_{0y} at $t = 3T/4$, and E_{0x} again at $t = T$ and so on, where $T = 2\pi/\omega$. Such a wave is called a *right elliptically polarized* wave.

If $E_{0x} = E_{0y}$ the wave is *right circularly polarized* because the amplitude of the resultant electric field vector is

$$E = \left\{ \left[\hat{x}E_0\cos(kz - \omega t) + \hat{y}E_0\cos\left(kz - \omega t - \frac{\pi}{2} \right) \right] \cdot \left[\hat{x}E_0\cos(kz - \omega t) + \hat{y}E_0\cos\left(kz - \omega t - \frac{\pi}{2} \right) \right] \right\}^{\frac{1}{2}} = E_0 \quad (2.28)$$

with the tip of the electric field vector sweeping out a circle of radius E_0.

A *left elliptically polarized* wave ($\phi = \pi/2$) is shown in Figure 2.19b. Here, the electric field rotates *clockwise* around the z-axis, its electric field progressing through values of E_{0x}

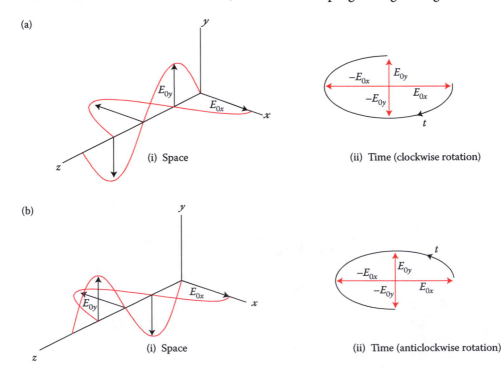

FIGURE 2.19 Elliptically polarized light. (a) Right elliptically polarized light; (b) left elliptically polarized light.

at $z = 0$, $- E_{0y}$ at $z = \lambda/4$, $-E_{0y}$ at $z = \lambda/2$, E_{0y} at $z = 3\lambda/4$, and E_{0x} again at $z = \lambda$ and so on as the wave advances in space.

At any *fixed point* on the z-axis, looking in the $-z$ direction toward the source, we see that the resultant electric field rotates *anticlockwise with time,* the resultant electric field being E_{0x} at $t = 0$, E_{0y} at $t = T/4$, $-E_{0x}$ at $t = T/2$, $-E_{0y}$ at $t = 3T/4$, and E_{0x} again at $t = T$ and so on.

2.9.3 PRODUCTION OF LINEARLY POLARIZED LIGHT BY REFLECTION AND TRANSMISSION

We saw in Section 2.8.2 that normal sources emit light in the form of wavetrains of very short duration. Each of them has state of polarization whether plane or left or right elliptical or circular. The polarization states of successive wavetrains from one atom, or from different atoms in most light sources, by which we mean nonlaser sources, differ from one another in random fashion so that the light from such sources is *randomly polarized*. From the perspective of holography, our main concern is to know what is the state of polarization of the laser light which we are using and how we may control or alter its polarization state to suit our purpose. In many holographic recording systems and in holographic applications generally, we need to change the polarization state of the laser which we are using. We also frequently need to split the beam into at least two beams and later cause these beams to interfere. For this to happen, the beams must at least have a component of electric field in the same plane and ideally they should be linearly polarized parallel to one another for most hologram recording purposes although in polarization holography (Chapter 16) such restrictions do not apply.

We begin by looking again at the reflection and transmission of light at the boundary between dielectric media having different refractive indices. Recall from Section 1.7.2.2 and Figure 1.15 that light, which is plane polarized *in the plane of incidence* and has an angle incidence of 57° (the Brewster angle) at an air–glass boundary, is not reflected at all. Only light with this incidence angle and plane polarized perpendicular to the plane of incidence is reflected. It follows that when a beam of randomly polarized light is incident at the Brewster angle, the reflected light (about 15% of the incident intensity, see Figure 1.15) is linearly polarized perpendicular to the plane of incidence. As a consequence, the transmitted light is *partially* linearly polarized in the plane of incidence. The process can be repeated at additional air–glass boundaries leading to a plane polarized reflected beam with an intensity comparable with that of the transmitted light which is also increasingly plane polarized *in the plane of incidence*. This is the principle of the so-called pile of plates polarizer, also applied in many laser systems. Without bothering for now about how a laser actually works, take an advance look at Figure 5.7 in which the end windows of the container of the lasing material are deliberately oriented so that the light traveling along the main axis is always incident on them at the Brewster angle. What we have here is really a two plate polarizer (one plate at each end), but the light reflected back and forth between the end mirrors of the laser goes through these plates many times, leading to the pile of plates effect. Light waves polarized perpendicular to the plane of incidence are always reflected out of the laser and no longer participate in the lasing process. So by this quite simple means we can ensure that the light emitted by the laser is almost perfectly plane polarized. Once we have such a linearly polarized laser light source, we can change its polarization state as we shall see.

Another application of the pile of plates principle is a cubical polarizing beam splitter consisting of two 45°, 90°, 45° glass prisms [2]. Each hypotenuse face is coated with thin films of alternating high and low refractive index and the coated prisms are then cemented together. Randomly polarized light is normally incident on a cube face. We require that light reflected out through the adjacent face be completely plane polarized as shown in Figure 2.20.

For this to happen the angle of incidence at the upper surface of the low index film must be the Brewster angle. The refractive index, n_g, of the glass must therefore satisfy the condition

$$n_g^2 = \frac{2n_H^2 n_L^2}{n_H^2 + n_L^2}$$

(2.29)

where n_H and n_L are the thin film high and low refractive indices. The light transmitted through the face opposite to the entrance face must be completely plane polarized in the plane of incidence. The two emerging beams are of equal intensity, provided we further ensure that the film thicknesses are such that the three light beams reflected at the

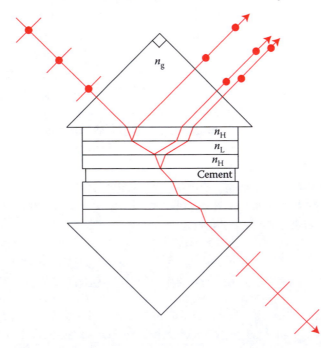

FIGURE 2.20 Pile of plates polarizing beam splitter cube. (Reproduced from M. Banning, Practical methods of making and using multilayer filters, *J. Opt. Soc. Am.*, 37, 792–97 (1947). With permission of Optical Society of America.)

boundaries of the low-index layer and at the boundary of the latter and the cement layer, are all in phase (see Section 1.7.2.3). Choosing zinc sulfide (ZnS) of refractive index 2.3 and cryolite (sodium aluminum fluoride, Na_3AlF_6) of refractive index 1.25 allows the use of glass of refractive index 1.55.

Other methods of producing linearly polarized light are by *scattering* or by *selective absorption* that is the absorption of light, which is linearly polarized in a particular plane, leaving the remaining light partially linearly polarized.

2.9.4 ANISOTROPY AND BIREFRINGENCE

The electric field associated with light propagating in a material sets the electronic charges in the atoms of the material in oscillatory motion in the same direction as the field, at optical frequencies, resulting in the emission of electromagnetic waves at those frequencies. These waves combine with the incident wave to form the refracted light beam. The atomic electrons cannot oscillate with complete freedom under the influence of the incident light but are constrained by forces exerted by the electrons in neighboring atoms. The greater the constraining force, the higher the natural frequency of oscillation of the charge and vice versa. It is these natural frequencies that determine the permittivity and hence the refractive index of the material. In many crystalline materials, the constraining forces are the same in all directions and so is the refractive index. Such crystals are *isotropic*. However, in *anisotropic* crystals, the forces are weaker in some directions than in others, and the natural frequency of the oscillating charge is accordingly lower.

Imagine a linearly polarized light wave propagating in such a medium with frequency below the natural frequencies of the charge oscillations. If the plane of polarization is parallel to a direction of weak constraining force, then the refractive index is lower. If, on the other hand, the plane of polarization is parallel to a direction of strong constraining force, the refractive index is higher. *In summary, the speed of light, which is determined by the refractive index of the material, depends on its plane of polarization.* In some of these anisotropic materials, there is one particular direction called the *optic axis* and *atomic charge experiences the same constraining force in all directions normal to this axis.* Such materials are called *uniaxial*. A light wave, which is linearly polarized *normal* to the optic axis, travels at the same speed regardless of its direction of propagation. We call this an *ordinary wave* (Figure 2.21). A light wave, which is linearly polarized in any other direction, travels at a *speed that depends on the direction of propagation* and is called an *extraordinary wave*. In so-called *negative uniaxial* crystals, the speed of the extraordinary wave is minimum along the optic axis and maximum normal to it. For this reason, the optic axis in a negative uniaxial crystal is known as the *slow axis,* while the direction normal to the optic axis is the *fast axis.* The reverse is the case for a *positive uniaxial* crystal whose optic axis is the *fast axis.*

Suppose that a linearly polarized beam of light is normally incident on a negative uniaxial crystal, which is cut so that its optic axis is parallel to the front and back faces and

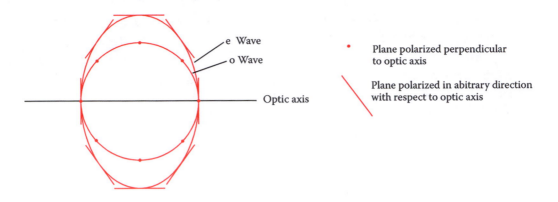

FIGURE 2.21 Birefringence: ordinary and extraordinary waves in a negative uniaxial crystal.

is aligned with the x-axis. We resolve the electric field vector into two orthogonal components, one of them normal to the optic axis, the other parallel to it (Figure 2.22). These two components are, respectively, an ordinary wave and an extraordinary wave. On emerging from the crystal, there is a phase difference between them of $\phi = (2\pi/\lambda)(n_o - n_e)t$ where t is the thickness and n_o and n_e are respectively the ordinary and extraordinary refractive indices. The value of ϕ is positive if $n_e < n_o$, that is, if light travels *faster* when linearly polarized parallel to the optic, x-axis. We refer to the direction that has the *lower index* as the *fast axis*.

In Figure 2.22, the phase difference has been set at $(2p+1)\pi$ so that the emerging wave is now plane polarized at an angle of $2\tan^{-1}(E_{0y}/E_{0x})$ with respect to its original plane. The device is called a *halfwave plate*. When the plane of polarization is originally at 45° to the optic axis, the rotation is 90°.

A phase difference of $\pi/2$ is obtained using a *quarter-wave plate* of thickness $\lambda/4(n_o - n_e)$ so that linearly polarized light can be converted to elliptically or circularly polarized light. In the latter case, the plane of polarization of the incident light must again be at 45° to the optic axis. The reader might consider how one could use the same quarter-wave plate to produce right *or* left circularly polarized light. Also, the polarization state of light, which is plane polarized either perpendicular to or parallel to the optic axis of a wave plate, is totally unaffected. Why is this?

2.9.5 BIREFRINGENT POLARIZERS AND POLARIZING BEAM SPLITTERS

It is very useful to have in the holographic laboratory a device that will enable us to identify the plane of polarization of any laser light source. Polaroid© linear polarizers produce plane polarized light by selective absorption of the electric field of the incident light. If the direction of selective absorption is known, then the plane of polarization of the laser is identified by rotating the polarizer in front of it. One should *never* look at any strong light source, much less a laser, even through a polarizer of this type. The beam should be allowed to fall on a diffusely scattering surface. When the scattered light intensity is minimum,

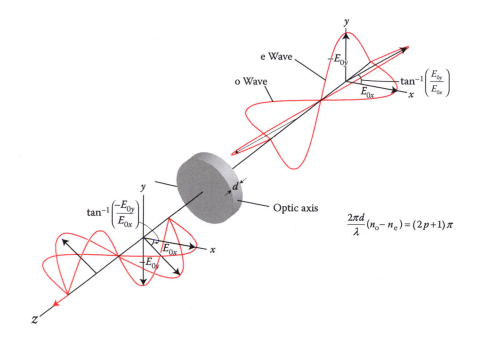

FIGURE 2.22 Halfwave plate.

then the plane of polarization of the light and the direction of selective absorption are parallel. If the direction of selective absorption is not marked on the device, it can be found by looking through it at the reflection of an ordinary light source in a flat horizontal glass plate. The reflected light is partially plane polarized in the horizontal plane so minimum intensity is observed when the direction of selective absorption is also horizontal.

A better alternative plane polarizer is a polarizing beam splitter made from a uniaxial birefringent crystal such as *calcite* whose ordinary and extraordinary refractive indices are 1.6584 and 1.4864, respectively, at a wavelength of 589.3 nm. Two identical triangular calcite prisms with apex angles of 40°, 50°, and 90° are positioned with their hypotenuse faces parallel and separated by a small air gap to form a parallel-sided block. This form is called a Glan-air polarizing beam splitter. The prisms are cut so that the optic axes are parallel to each other and perpendicular to the triangular faces. The electric field vectors associated with randomly polarized light that is incident normally on a square face of a prism can be resolved two orthogonal components, one parallel to the optic axis and therefore ordinary light (Figure 2.23).

This light is incident at the calcite–air boundary at an angle of incidence of 40°, exceeding the critical angle of 37°. The other component, normal to the optic axis and therefore the extraordinary light, is transmitted across the air gap since the critical angle for this component is 42°. The light transmitted by the device is therefore plane polarized normal to the optic axis, and the reflected light is plane polarized parallel to the optic axis.

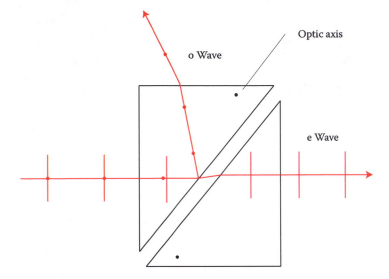

Optic axis

o Wave

e Wave

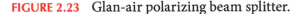

FIGURE 2.23   Glan-air polarizing beam splitter.

2.10 SUMMARY

In this chapter, we look at how light behaves because of its encounter with diffracting elements such as a single aperture, whether rectangular or circular and systems of apertures in the form of diffraction gratings. The latter are powerful tools for the study and analysis of light. We also consider the basic problem of resolution due to the limited aperture presented to light by an imaging lens. The equivalence of diffraction under Fraunhofer conditions and Fourier transformation is demonstrated.

Interference effects arising from the splitting and recombination of light beams are also examined, and the need for coherent light sources in interferometry is explained.

Finally, polarization is discussed with specific reference to holography. Methods of producing polarized light and controlling its polarization state are described.

REFERENCES

1. V. Toal and E. Mihaylova, "Double glazing interferometry," *Phys. Teacher*, 47, 295–6 (2009).
2. M. Banning, "Practical methods of making and using multilayer filters," *J. Opt. Soc. Am.*, 37, 792–7 (1947).

PROBLEMS

1. Sketch the intensity distribution due to a diffraction grating consisting of 11 slits with slit separation four times that of the slit width.
2. A transmission diffraction is 1 cm wide. In which diffraction order will the two yellow lines in the sodium spectrum, which have wavelengths 589 and 589.6 nm, be resolved if the grating spacing is 20 µm? What is the maximum possible resolving power of the grating when used in normal incidence at a wavelength of about 500 nm?

3. Calculate the blaze angle of a transmission phase grating 12.5 cm in width with 800 grooves per millimeter, made from material of refractive index 1.68, which diffracts maximum intensity at wavelength 457 nm into the third order of diffraction, assuming normal incidence. Determine whether it can resolve two modes of a 633-nm helium–neon laser separated by 500 MHz.

4. For diffraction to occur at angles that can readily be measured, the diffracting object has to have a length scale that is comparable to the wavelength. Bats emit high-frequency sound waves in order to detect their prey at night. What frequency are the sounds likely to be if the velocity of sound in air is about 340 ms⁻¹? The sounds made by smaller woodland birds are audible to the human ear and thought more likely to relate to the length scale of their environment than that of their prey. Explain why this might be the case.

5. The lens of a camera on board an orbiting satellite is required to resolve objects 30 cm apart on the Earth's surface, 200 km below. What should be the diameter of the camera lens?

6. Why do we normally use mirrors in large astronomical telescopes rather than lenses?

7. Annular apertures are sometimes used with lenses in order to improve resolution. Obviously some light is wasted so we only use them when there is plenty of light available. To see how the method works, consider an annular aperture whose inner diameter is half that of the lens. Make a rough sketch of the intensity distribution at the lens focus due to the lens aperture and then subtract the intensity distribution which is blocked by the opaque central disc of the annular aperture.

8. Another method of improving resolution is called apodization. The idea is that if an aperture has a transmissivity function which declines smoothly to zero at the edges, the side lobes of its diffraction pattern will be reduced in magnitude relative to the maximum. An ideal example (in 1-D form) is a Gaussian function whose amplitude transmission is given by

$$t(y) = \sqrt{\frac{2}{\pi a^2}} \exp\left[-2\left(\frac{y}{a}\right)^2\right]$$

and whose Fraunhofer diffraction pattern is also Gaussian in form and so has no side-lobes at all. Prove that this is so.

A more practicable transmission function is given by

$$t(y) = \cos\left(\frac{\pi y}{a}\right), \quad -a/2 \le y \le a/2$$
$$= 0, \qquad \text{everywhere else}$$

Find an expression for its Fraunhofer diffraction pattern. You will see that its central lobe is wider than that of a normal slit aperture, but compare the intensities of the side lobes with that of the central one.

9. A sheet of transparent material with surface height profile given by

$$h(y) = h_0[1 + 0.25\cos(k'y)]$$

is located in the front focal plane of a lens of focal length, f, and illuminated normally collimated monochromatic light. What is observed in the back focal plane?

10. Show that the coherence length of white light is about 1 micron. If it is passed through a spectral filter with a narrow spectral passband 5 nm wide, what is its coherence length?

11. Light of mean wavelength 500 nm filling a slit aperture 1.0 mm wide is used to illuminate an otherwise opaque screen having two parallel narrow slits 5 mm apart, at a distance of 5 cm. Interference fringes are observed on a screen 10 cm further away. If the light has a

spectral bandwidth of 0.5 nm, find the width of the fringe pattern. Is it determined by the width of the source or by its spectral bandwidth?

12. Prove Equation 2.29.

13. Often, when recording a hologram, we need to split a linearly polarized laser beam into two beams, with the same plane of polarization and with the facility to change the ratio of their intensities arbitrarily. Explain how this can be done using a polarizing beam splitter and two waveplates.

14. Consider what happens when white light is passed through a linear polarizer and a parallel-sided layer of birefringent material with the transmission axis of the polarizer set at 45° to the fast axis of the layer. The transmitted light is passed through a second polarizer with transmission axis parallel to that of the first and then spectrally analyzed. What is notable about the spectrum that is observed? How would an analysis of the spectrum enable you to measure the birefringence of the layer?

15. A Wollaston polarizing beam splitter consists of two calcite 45°, 45°, 90° prisms with hypotenuse faces in contact. The prisms can be cut so that the optic axes are as shown in the diagrams. Find the angle between the emerging beams in each case. The refractive indices of calcite are $n_o = 1.658$ and $n_e = 1.486$.

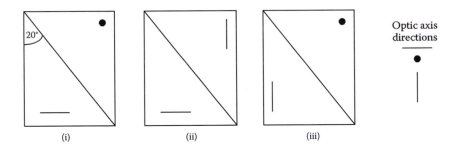

II

Principles of Holography

CHAPTER 3

Introducing Holography

3.1 INTRODUCTION: DIFFERENCE BETWEEN TWO SPATIAL FREQUENCIES

In order to get started, we consider holography in the simplest possible way and will adopt a more general approach in the chapter to follow.

Any two waves can interfere to produce a spatial frequency, which is the difference between the spatial frequencies of the waves. We saw this in Section 2.7 in which we discussed the Mach–Zehnder interferometer. Referring to Figure 3.1, suppose the wave vectors or normals to two interfering plane wavefronts are at angles ψ_1 and ψ_2 to the z-axis and at 90° to the x-axis (not shown).

The spatial periods measured along the z-axis are $\lambda/\cos\psi_2$ and $\lambda/\cos\psi_1$ and the spatial frequencies are $\cos\psi_2/\lambda$ and $\cos\psi_1/\lambda$. The spatial frequencies measured along the y-axis are $\sin\psi_2/\lambda$ and $\sin\psi_1/\lambda$ and the difference or beat spatial frequency is

$$\frac{\sin\psi_2 - \sin\psi_1}{\lambda}.$$

We define a parameter called the interference fringe vector **K**, normal to the fringes with a magnitude **K** $= 2\pi/spatial\ period$, which is

$$\mathbf{K} = 2\pi \frac{2\sin\left(\dfrac{\psi_2 - \psi_1}{2}\right)}{\lambda} \tag{3.1}$$

and which makes an angle $(\psi_2 + \psi_1)/2$ with the y-axis.

3.2 RECORDING AND RECONSTRUCTION OF A SIMPLE DIFFRACTION GRATING

Let us now consider the simplest hologram of all, in which two plane wavefronts interfere and a permanent record is made of the interference pattern in a very thin, flat layer of silver

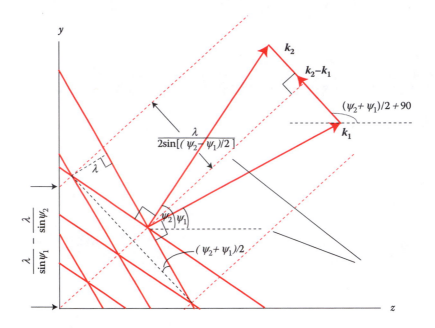

FIGURE 3.1 Interference of two plane waves. The interfering waves have wavelength λ. The spacing of fringes in the interference pattern is $\lambda/2\sin[(\psi_2-\psi_1)/2]$. The projection of the pattern along the y-axis has spatial period $\lambda/\sin\psi_1-\lambda/\sin\psi_2$.

halide photographic emulsion in the x,y plane. After chemical processing, which we will discuss in Chapter 6, we have a diffraction grating consisting of opaque silver where the maximum intensity occurred in the interference pattern, interspersed with transparent gelatin (Figure 3.2).

3.2.1 AMPLITUDE GRATINGS

The transmissivity, t, is ideally described by the equation $t(y)=t_0[1-\cos(k''y)]$, t_0 being the spatially uniform amplitude transmissivity of the plate before it is exposed to light. After exposure and processing, $t(y)$ can, at best, vary between 0 and 1 so that we must set $t(y)=1/2[1-\cos(k''y)]$ since $\cos(k''y)$ varies between -1 and $+1$.

Now, let us illuminate this sinusoidal transmission gating with one of the plane waves originally used to create it, namely

$$E_0\cos(kx\cos\alpha_1+ky\sin\psi_1+kz\cos\psi_1-\omega t)=E_0\cos(ky\sin\psi_1-\omega t)$$

in the x,y plane. We use Equation 2.3, modified to take account of the cosinusoidal transmissivity of the grating, of width L.

$$\tilde{E}_P=\frac{E_0L}{\lambda r'r_0}e^{j[k(r+r_0')-\pi/2-\omega t]}\int_{\text{grating}}e^{jky\sin\psi_1}e^{-jky\sin\theta}\frac{1}{2}[1-\cos(k''y)]dy$$

$$=\frac{E_0Le^{j[k(r+r_0')-\pi/2-\omega t]}}{\lambda r'r_0}\int_{\text{grating}}e^{-jky\sin\theta}e^{jky\sin\psi_1}\frac{1}{4}(2-e^{jk''y}-e^{-jk''y})dy$$

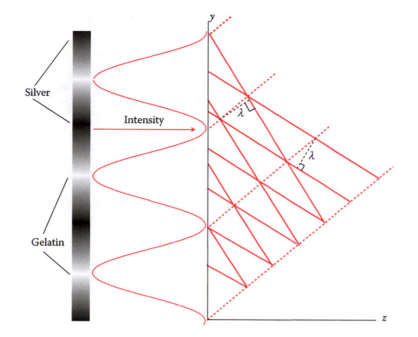

FIGURE 3.2 Interference pattern in the x,y plane.

where $k'' = 2\pi(\sin\psi_2 - \sin\psi_1)/\lambda = k(\sin\psi_2 - \sin\psi_1)$.

$$\tilde{E}_p = \frac{E_0 L e^{j[k(r+r_0')-\pi/2-\omega t]}}{\lambda r' r_0} \left\{ \frac{1}{2}\left[\frac{e^{-j(k\sin\theta - k\sin\psi_1)y}}{-j(k\sin\theta - k\sin\psi_1)}\right]_{-\frac{W}{2}}^{\frac{W}{2}} -\frac{1}{4}\left[\frac{e^{-j(k\sin\theta - k'' - k\sin\psi_1)y}}{-j(k\sin\theta - k'' - k\sin\psi_1)} + \frac{e^{-j(k\sin\theta + k'' - k\sin\psi_1)y}}{-j(k\sin\theta + k'' - k\sin\psi_1)}\right]_{-\frac{W}{2}}^{\frac{W}{2}} \right\}$$

$$= \frac{E_0 L W e^{j[k(r+r_0')-\pi/2-\omega t]}}{\lambda r' r_0} \left\{ \frac{\frac{1}{2}\sin\left[k(\sin\theta - \sin\psi_1)\frac{W}{2}\right]}{k(\sin\theta - \sin\psi_1)\frac{W}{2}} -\frac{\frac{1}{4}\sin\left[k(\sin\theta + \sin\psi_2 - 2\sin\psi_1)\frac{W}{2}\right]}{k(\sin\theta + \sin\psi_2 - 2\sin\psi_1)\frac{W}{2}} -\frac{\frac{1}{4}\sin\left[k(\sin\theta - \sin\psi_2)\frac{W}{2}\right]}{k(\sin\theta - \sin\psi_2)\frac{W}{2}} \right\}$$

(3.2)

where W is the total extent of the grating in the vertical, y, direction.

It is clear that if W is large, the peaks of intensity are confined to directions

$$\theta = \psi_1, \quad \theta = \psi_2, \quad \theta = \sin^{-1}(2\sin\psi_1 - \sin\psi_2)$$

The most important thing to note here is the *reconstruction* of the original plane wavefront at angle ψ_2 to the z-axis. Another way of looking at this is that we have *added* the spatial frequency of the ψ_1 wavefront to the *recorded beat frequency* to restore the ψ_2 spatial frequency. Note also that the process is *associative*, meaning that if we illuminate the grating using the ψ_2 wavefront, the ψ_1 wavefront is reconstructed.

Now, let us look at the intensities of the reconstructed light beams. The two diffracted beams both have amplitudes one quarter that of the incident light and so each has intensity 1/16 or 6.25% of the incident light intensity. We define an important parameter of the grating, the *diffraction efficiency, η,* which is the ratio of intensity of the reconstructed light beam to that of the incident beam. This definition is extended in later chapters to a hologram whose *diffraction efficiency is the ratio of the intensity of the reconstructed light beam to that of the incident reference beam.*

3.2.2 PHASE GRATINGS

If the photographic plate is processed so that the silver is converted into a transparent silver compound with a refractive index, which differs from that of the gelatin in which it is embedded, we have a thin, pure phase grating. Let us assume that the phase shift, $\phi(y)$, imposed by the grating on the reconstructing reference wave is proportional to the intensity during recording

$$\phi(y) \propto \phi_0 \left[1 + \cos(k''y) \right] \tag{3.3}$$

ϕ_0 being the phase shift when the plate is processed following exposure to the spatial average of the light intensity. Using Equation 2.3 again, we have

$$\tilde{E}_{\mathrm{p}} = \tilde{C} \int_y \exp\left\{ j\left[ky\sin\psi_1 - ky\sin\theta + \phi_0\cos(k''y) \right] \right\} \mathrm{d}y$$

where $\tilde{C} = (E_0 L / \lambda r' r_0) e^{j\left[k(r + r'_0) - \pi/2 - \omega t + \phi_0 \right]}$

$$\tilde{E}_{\mathrm{p}} = \tilde{C} \int_y \exp\left\{ j\left[ky(\sin\psi_1 - \sin\theta) \right] \right\} \exp\left[j\phi_0 \cos(k''y) \right] \mathrm{d}y$$

$$= \tilde{C} \int_y \exp\left\{ j\left[ky(\sin\psi_1 - \sin\theta) \right] \right\} \sum_{n=-\infty}^{n=\infty} j^n B_n(\phi_0) \exp(jnk''y) \mathrm{d}y$$

where $B_n(\phi_0)$ is a Bessel function of the first kind and of order n.

$$\tilde{E}_{\mathrm{p}} = \frac{E_0 L}{\lambda r' r_0} e^{jk\left[(r + r'_0) - \pi/2 - \omega t \right]} \sum_{n=-\infty}^{n=\infty} j^n B_n(\phi_0) \int_{\mathrm{grating}} \exp\left\{ j\left[y\left(k\sin\psi_1 - k\sin\theta + nk'' \right) \right] \right\} \mathrm{d}y$$

$$= \frac{E_0 L}{\lambda r' r_0} e^{j(k[(r+r_0')-\pi/2-\omega t]} \sum_{n=-\infty}^{n=\infty} j^n B_n(\phi_0) \left[\frac{\exp\left\{ j \left[y\left(k\sin\psi_1 - k\sin\theta + nk''\right) \right] \right\}}{j\left[\left(k\sin\psi_1 - k\sin\theta + nk''\right) \right]} \right]_{-\frac{W}{2}}^{\frac{W}{2}}$$

We consider just three light beams of nonnegligible amplitude, corresponding to $n = 0$, $n = 1$, and $n = -1$ in directions $\theta = \psi_1$, $\theta = \sin^{-1}(2\sin\psi_1 - \sin\psi_2)$, and $\theta = \psi_2$ respectively, exactly as in the case of the amplitude grating. The difference is in the diffraction efficiency, which is $B_1^2(\phi_0)$ and which has a maximum value of 0.339 (where $\phi_0 = 3$).

If the phase modulation ϕ_0 is very small, we can rewrite Equation 3.4 as

$$\tilde{E}_P = \tilde{C} \int_y \exp\left[jky\left(\sin\psi_1 - \sin\theta\right) \right] \exp\left[j\phi_0 \cos\left(k''y\right) \right] dy$$

$$\cong \tilde{C} \int_y \exp\left[jky\left(\sin\psi_1 - \sin\theta\right) \right] \left[1 + j\phi_0 \cos\left(k''y\right) \right] dy$$

$$= \tilde{C} \int_y \exp\left[jky\left(\sin\psi_1 - \sin\theta\right) \right] \left\{ 1 + \frac{j\phi_0}{2} \left[\exp\left(jk''y\right) + \exp\left(-jk''y\right) \right] \right\} dy$$

$$= \tilde{C} \left\{ \begin{aligned} &\int_y \exp\left[jky\left(\sin\psi_1 - \sin\theta\right) \right] dy + \frac{j\phi_0}{2} \int_y \exp\left[jky\left(\sin\psi_2 - \sin\theta\right) \right] dy \\ &+ \frac{j\phi_0}{2} \int_y \exp\left[jky\left(2\sin\psi_1 - \sin\psi_2 - \sin\theta\right) \right] dy \end{aligned} \right\}$$

and the diffraction efficiency is $\phi_0^2 / 4$.

3.3 GENERALIZED RECORDING AND RECONSTRUCTION

Our main objective is to record *convoluted* wavefronts and to reconstruct them *fully*, which is what holography does. Holography is completely different from any other form of optical recording, such as photography, which can only record the *intensity distribution* in the light from a scene as explained in Sections 1.2 and 1.4. To achieve our objective, we need to obtain a record of the *instantaneous* value of the electric field of the wave at every point on the wavefront. At first sight, this appears to be an impossibly daunting task, but we have already offered a clue as to how it might be done, in Section 3.2. There, we showed how a plane wavefront could be recorded and reconstructed by recording the fringe pattern that is produced when that wavefront interferes with another plane wavefront. So, we can record the simplest hologram of all and play it back again. Suppose that one of our two sets of wavefronts is convoluted in shape as in Figure 1.4. We saw in Section 1.5 that such a wavefront can be regarded as consisting of a number of plane wavefronts. Such a wavefront could be made to interfere with another plane wavefront, known as a reference wave, and the resulting interference pattern recorded in a photosensitive medium such as silver halide. Then on illumination by the reference wavefront, the convoluted wavefront could be reconstructed. At this point, we are making a conceptual argument and we need to strengthen it considerably. We will do this in the later sections of this chapter, involving some simple mathematics. But first, we take a brief look at the origins and early development of holography.

3.4 A SHORT HISTORY OF HOLOGRAPHY

3.4.1 X-RAY DIFFRACTION

In the first half of the twentieth century, a great deal of research effort was expended on finding the atomic and molecular structure of a vast number of materials, crystalline materials in particular, that is, to identify the constituent atoms of a given substance and their arrangement with respect to one another in space. This information is of crucial importance in understanding material properties required for the design of new materials, including chemicals, engineering materials, biomaterials, and pharmaceuticals. The main tool used was X-rays, which consist of electromagnetic radiation of wavelength much shorter than that of visible light, of the order of nanometers in fact, which is typical of the interatomic spacing in most crystalline materials. This means that many materials act like 3-D gratings producing diffraction patterns, which, upon careful analysis, enable us to work back to the molecular structure of the material. One of the greatest achievements that resulted was the discovery of the structure of DNA by Crick, Watson, and Wilkins using X-ray diffraction patterns of DNA obtained by Franklin.

3.4.2 DIFFRACTION AND FOURIER TRANSFORMATION

The mathematical *form* of the diffraction pattern in general, is very significant. Recall Equation 2.18

$$\tilde{A}\left(x',y'\right)=\iint_{x,y}\tilde{A}(x,y)\mathrm{e}^{-j\left(k_x x+k_y y\right)}\mathrm{d}x\mathrm{d}y$$

which has the form of a 2-D spatial Fourier transformation. If we carry out a second spatial Fourier transformation, we obtain once more the electric field distribution of the light at the diffracting plane, in fact an image of the diffracting plane. This fact led Bragg to the idea that if we could record the X-ray diffraction pattern of a crystal and then carry out a spatial Fourier transform of the pattern, we would obtain an image of the crystalline structure. Furthermore, if we used visible light to carry out the second transformation, the image would be magnified in the ratio of visible to X-ray wavelengths, in other words, we would have an X-ray microscope [1]. The major obstacle to all of this is that the X-ray diffraction pattern or the spatial Fourier transform of the light amplitude distribution at the diffracting object, the crystal, involves spatially dependent phase, which could not be recorded. Early research in this direction was therefore confined to centrosymmetrical crystal structures

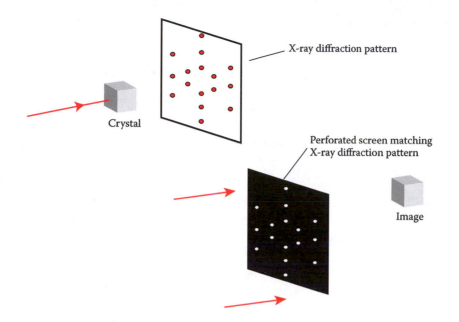

FIGURE 3.3 Bragg's double diffraction experiment.

for which the spatial Fourier transform is always real. The *sign* of the diffracted amplitude in any given direction could still be positive or negative but, by choosing structures with a massive atom at the center of symmetry, which diffracts a large amplitude, the amplitude of the Fourier transform could be always biased positive. These insights enabled Bragg to produce images [2,3] of crystalline structures. In his first experiments, he photographed a set of opaque parallel equally spaced cylindrical rods all in the same plane. The resulting transparency was projected onto photographic paper, the processed result approximating a sinusoidal fringe pattern. The spacing of the recorded pattern could be altered by changing the distance between the transparency and the paper. Forty fringe patterns were similarly recorded on the paper, each corresponding to one of the spatial frequencies in the diffraction pattern obtained from the crystal. The result was a low-contrast image of the crystal structure. Improved contrast was obtained by drilling holes in an opaque screen, the hole diameters being proportional to the intensities of the spots in the diffraction pattern. When the perforated screen was illuminated by coherent light, each pair of holes, acting as a Young's double aperture, produced one of the required sets of fringes on the photographic paper. In this way, all the fringe patterns could be recorded simultaneously (Figure 3.3). This second experiment was also a remarkable practical demonstration of Abbe's theory of image formation [4], proposed in 1873, which has as its basis the idea that all optical imaging processes are in fact double diffraction processes.

3.4.3 ELECTRON MICROSCOPY AND THE FIRST HOLOGRAMS

The next step was taken by Gabor, in the late 1940s. He was then working on the design of electron microscopes at the research laboratories of British Thompson-Houston. Electron microscopes use the wave properties of electrons to form extremely high-resolution images. The wavelength of an electron λ_e is given by

$$h/\lambda_e = mv = \sqrt{m^2v^2} = \sqrt{2m(K.E.)} = \sqrt{2meV}$$

where h is Planck's constant, 6.6×10^{-34} Js, m the electron mass, 9.1×10^{-31} kg, v its velocity, $K.E.$ its kinetic energy, e the electronic charge, 1.6×10^{-19} C, and V the potential difference through which it passes in order to reach velocity v from rest. A simple calculation shows that an electron accelerated through a potential difference of just 100 V will have a wavelength of less than 1 nm so that, from Equation 2.12, we see that the resolution limit of an electron microscope is about 500 times superior to that of an optical microscope. The problem faced by Gabor and his contemporaries was that the magnetic lenses used to focus electron waves were not very good and in fact introduced quite severe distortions. Gabor hoped to record electron waves and correct their distortions, using optical components, and he anticipated that electron holograms might be reconstructed at visible wavelengths with magnification $\lambda_{visible}/\lambda_{electron}$ (see Section 7.7), but this raised the same problem faced by Bragg, namely how to record the electron wavefront, *including its phase*. Gabor's solution [5] was ingenious and extremely simple. For his purpose, he confined his attention to the recording of optical as opposed to electron wavefronts, to the enormous benefit of optical holographers everywhere. He combined the wave to be recorded with another wave, as a reference, to form an interference pattern. The pattern contained all the information about the phase differences between the reference wave and the wave of interest. The interference pattern was recorded in the form of a spatially varying amplitude transmissivity, using a silver halide photographic emulsion. After processing, the emulsion was illuminated by the reference wave. The original wave, *including its phase*, was reconstructed by the diffraction process. The technique was given the name holography, from the Greek "holos" meaning whole or entire.

Gabor's results were convincing to the scientific community, but a number of serious obstacles stood in the way of progress with the method. Not least of them was the fact that light sources with coherence length of more than a few millimeters were simply not available. This also hindered the development of holography as a 3-D display technique since only objects of extremely restricted dimensions, such as text on a transparency or small diameter wires, could be used as objects in holography. Figure 3.4 shows the method.

A mercury lamp is filtered to isolate just one line in its spectrum so as to maximize the coherence length of the light (see Section 2.8.2). The light is then focused onto a small aperture maximizing its spatial coherence (see Section 2.8.3) and then collimated to form a parallel beam illuminating a transparency with some text on it. Some of the light is diffracted by the text letters and this diffracted light constitutes the wavefront that we wish to record. The remainder of the light is undiffracted and serves as the reference wave and the interference pattern formed by the two wavefronts is recorded at the plane of the photographic emulsion. Following processing, the finished hologram is illuminated by the light beam but with the transparency removed (Figure 3.5).

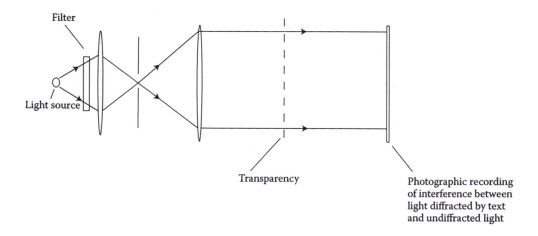

FIGURE 3.4 Gabor's experiment—recording. The object is an opaque text on a transparency.

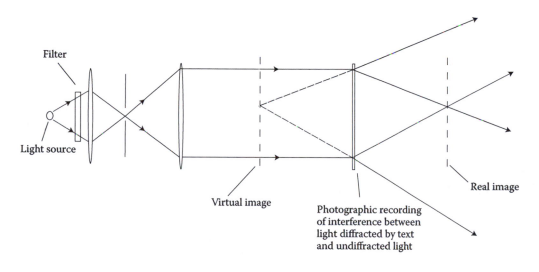

FIGURE 3.5 Gabor's experiment—reconstruction.

On looking into the transparency toward the light source, a virtual image of the original text is seen, in its original position. The drawback is that this image was partially obscured by a second real image located the same distance from the hologram but on the opposite side of the hologram from the virtual one. Efforts were made to remove one of the images or at least minimize its effect. One way relies on the fact that when viewing the real image, we see, in the same plane, the diffraction pattern from the transparency. So a *negative* of this pattern is prepared by illuminating the transparency and recording the result on a photographic emulsion located at *2f* from the transparency. This second hologram is now positioned a distance *f* to the right of the original hologram and, in principle, the real image can now be seen but without the out-of-focus light from the virtual image. However, the method did not work very well in practice.

3.4.4 PHOTOGRAPHIC EMULSIONS AND GABOR HOLOGRAPHY

Because the image in a hologram is seen against a uniform background intensity of light, it is important to process the photographic emulsion carefully to ensure that the contrast between image and background light is the same as between the object and background light.

Photographic emulsions are characterized by graphs, known as Hurter–Driffield curves, of photographic density, D, versus the logarithm of the exposure, E, which is the exposing intensity of light multiplied by the exposure time. A typical example is shown in Figure 3.6. Density is defined by the equation

$$D = \log\left(\frac{1}{T_I}\right) \tag{3.4}$$

where T_I is the intensity transmissivity of the processed emulsion. In the linear part of the curve,

$$D = \log(1/T_I) = \gamma\left[\log(It) - \log C\right] = \log\left(\frac{It}{C}\right)^{\gamma}$$

$$\frac{1}{T_I} = \left(\frac{It}{C}\right)^{\gamma} \quad \text{and} \quad T_I \propto I^{-\gamma}$$

The amplitude transmissivity t_n of a photographic negative is therefore

$$t_{\mathrm{n}} \propto I^{-\frac{\gamma_n}{2}}$$

Now, suppose we illuminate the negative to make a positive. The amplitude transmissivity of the positive is t_{p} where

$$t_{\mathrm{p}} \propto I_{\mathrm{p}}^{-\frac{\gamma_{\mathrm{p}}}{2}} \propto \left(T_{\mathrm{n}}\right)^{-\frac{\gamma_{\mathrm{p}}}{2}} \propto \left(t_{\mathrm{n}}^2\right)^{-\frac{\gamma_{\mathrm{p}}}{2}} \propto \left(I^{-\gamma_n}\right)^{-\frac{\gamma_{\mathrm{p}}}{2}} = I^{\frac{\gamma_n\gamma_{\mathrm{p}}}{2}} = I^{\frac{\Gamma}{2}}$$

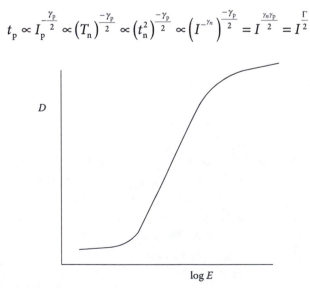

FIGURE 3.6 Hurter–Driffield curve.

We need to process negative and positive to obtain a combined Γ of 2, so that the amplitude transmissivity of the positive, that is the finished hologram, is proportional to the original light intensity.

As already explained, the hologram is illuminated by the original reference light wave and the resulting output is simply the mathematical product of the reference wave and t_p.

We use the expressions

$$E_0 \cos(kz - \omega t) \quad \text{and} \quad E(x, y) \cos[ky \sin \psi(x, y) - \omega t]$$

for the reference wave and the object wave, respectively. The reference wave is a plane wave directed parallel to the z-axis. At any point x,y in the hologram plane, the object wave travels at angle $\cos^{-1}\psi$ to the z-axis. For mathematical simplicity, the reference and object wave vectors are taken to be always at 90° to the x-axis. At the reconstruction stage, we have

$$E_0 \cos(-\omega t) \left\langle \left[\begin{array}{c} E_0^2 \cos^2(-\omega t) + E^2(x, y)\cos^2(ky \sin \psi - \omega t) \\ +2E_0 E(x, y)\cos(-\omega t)\cos(ky \sin \psi - \omega t) \end{array} \right]^{\frac{\Gamma}{2}} \right\rangle$$

$$= E_0 \cos(-\omega t) \left(\frac{E_0^2}{2} \right)^{\frac{\Gamma}{2}} \left\{ 1 + \frac{E^2(x, y)}{E_0^2} + \frac{2E(x, y)}{E_0} \cos(-ky \sin \psi) \right\}^{\frac{\Gamma}{2}}$$

$$= E_0 \cos(-\omega t) \left(\frac{E_0^2}{2} \right)^{\frac{\Gamma}{2}} \left\{ 1 + \frac{E^2(x, y)}{E_0^2} + \frac{2E(x, y)}{E_0} \cos(-ky \sin \psi) \right\}^{\frac{\Gamma}{2}}$$

$$\cong E_0 \cos(-\omega t) \left(\frac{E_0^2}{2} \right)^{\frac{\Gamma}{2}} \left\{ 1 + \frac{\Gamma}{2} \frac{E^2(x, y)}{E_0^2} + \frac{\Gamma E(x, y)}{E_0} \cos(-ky \sin \psi) \right\} \text{ if } E_0 \gg E(x, y)$$

$$= E_0 \cos(-\omega t) \left(\frac{E_0^2}{2} \right)^{\frac{\Gamma}{2}} + \left(\frac{E_0^2}{2} \right)^{\frac{\Gamma}{2}} \frac{\Gamma}{2} E(x, y)\cos(ky \sin \psi - \omega t)$$

$$+ \left(\frac{E_0^2}{2} \right)^{\frac{\Gamma}{2}} \frac{\Gamma}{2} E(x, y)\cos(ky \sin \psi + \omega t)$$

$$\propto E_0 \cos(-\omega t) + \frac{\Gamma}{2} E(x, y)\cos(ky \sin \psi - \omega t), \text{ ignoring the third term.}$$

This last equation shows that it is only when $\Gamma = 2$ that we obtain the same contrast between object and reference waves as that which pertained at the recording stage.

The need for precise control of the value of Γ, the severe limitation on the sizes and kinds of objects from which diffracted wavefronts could be recorded, and the twin-image problem were all significant obstacles that could be overcome in one go with the advent of the helium–neon laser used by Leith and Upatnieks [6] to make the first really striking holographic images. The much longer coherence length of the laser made it possible to separate object and reference light beams in space and the twin images could be disentangled from one another. A photographic negative processed to have a much less restrictive value of γ_n was quite satisfactory and one could now fully exploit holography's ability to record the object wave, *including its phase,* to produce 3-D images of startling realism.

3.5 SIMPLE THEORY OF HOLOGRAPHY

We have seen how, in principle, we can record a hologram of a plane wave and reconstruct it. We now extend the argument using some fairly simple mathematics.

3.5.1 HOLOGRAPHIC RECORDING

Let us suppose a hologram is to be recorded, using a flat glass plate coated with a layer of photosensitive material and located in the *x,y* plane (Figure 3.7).

It is illuminated simultaneously by two sets of wavefronts. The first of these is plane and is known as the reference wave. The second is the object wave and has directional cosines at the hologram plane, which depend on coordinates *x,y*. We assume that the two sets of wavefronts are mutually coherent and that there is no alteration in the interference pattern during the exposure.

The reference wave at point *x,y* is described by the expression $E_R \exp[j(\mathbf{k}_R \cdot \mathbf{r} - \omega t)]$ where E_r is its constant amplitude. The vector $\mathbf{k}_R$ is also constant as this wave is plane. The object wave at $O(x,y)$ is given by

$$\tilde{O}(x, y, t) = E_O(x, y) \exp\left[j\left(\mathbf{k}_O(x, y) \cdot \mathbf{r} - \omega t\right)\right] \tag{3.5}$$

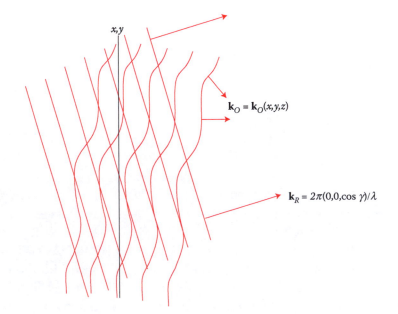

FIGURE 3.7 Interference of object and reference waves.

where the amplitude, $E_o(x,y)$, and the wave vector, $k_O(x,y)$, are locally dependent since the object wavefront shape is a convoluted one (see Section 1.5).

The total disturbance at x,y is

$$E_R \exp\left[j(\mathbf{k}_R \cdot \mathbf{r} - \omega t)\right] + E_O(x,y)\exp\left[j\left(\mathbf{k}_O(x,y)\cdot\mathbf{r} - \omega t\right)\right]$$

and the intensity, $I(x,y)$ is given by

$$
\begin{aligned}
I(x,y) &= \frac{1}{2}\left\{E_R \exp\left[j(\mathbf{k}_R\cdot\mathbf{r}-\omega t)\right]+E_O(x,y)\exp\left[j(\mathbf{k}_O(x,y)\cdot\mathbf{r}-\omega t)\right]\right\} \\
&\quad \times \left\{E_R \exp\left[j(\mathbf{k}_R\cdot\mathbf{r}-\omega t)\right]+E_O(x,y)\exp\left[j(\mathbf{k}_O(x,y)\cdot\mathbf{r}-\omega t)\right]\right\}^* \\
&= \left(\frac{E_R^2 + E_O^2}{2}\right) + \frac{1}{2}E_R E_O(x,y)\exp\left[j(\mathbf{k}_R - \mathbf{k}_O(x,y))\cdot\mathbf{r}\right] \\
&\quad + \frac{1}{2}E_R E_O(x,y)\exp\left[j(\mathbf{k}_O(x,y)-\mathbf{k}_R)\cdot\mathbf{r}\right]
\end{aligned}
$$

$$I(x,y) = \left(\frac{E_R^2 + E_O^2}{2}\right) + E_R E_O(x,y)\cos\left[(\mathbf{k}_R - \mathbf{k}_O(x,y))\cdot\mathbf{r}\right] \tag{3.6}$$

$$I(x,y) = \left(\frac{E_R^2 + E_O^2}{2}\right)\left\{1 + \frac{2E_R E_O}{E_R^2 + E_O^2}\cos\left[(\mathbf{k}_R - \mathbf{k}_O(x,y))\cdot\mathbf{r}\right]\right\} \tag{3.7}$$

The visibility or contrast of the interference pattern described by Equation 3.7 is given by

$$\frac{I_{\max} - I_{\min}}{I_{\max} + I_{\min}} = \frac{2E_R E_O}{E_R^2 + E_O^2} \tag{3.8}$$

3.5.2 AMPLITUDE HOLOGRAMS

Now, suppose that, following exposure of appropriate duration to the interference pattern and processing, the plate has an amplitude transmissivity $t(x,y)$ given by

$$t(x,y) = t_0 - \beta I(x,y)$$

where t_0 is the amplitude transmission of the processed plate in the absence of any exposure.

The plate is then an amplitude hologram. We illuminate it with the original reference wave (Figure 3.8) and obtain output

$$
\begin{aligned}
&E_R \exp\left[j(\mathbf{k}_R\cdot\mathbf{r}-\omega t)\right]t(x,y) \\
&= E_R \exp\left[j(\mathbf{k}_R\cdot\mathbf{r}-\omega t)\right]\left\{t_0 - \beta\left\{\left(\frac{E_R^2 + E_O^2}{2}\right) + E_R E_O(x,y)\cos\left([(\mathbf{k}_R - \mathbf{k}_O(x,y)]\cdot\mathbf{r}\right)\right\}\right\}
\end{aligned} \tag{3.9}
$$

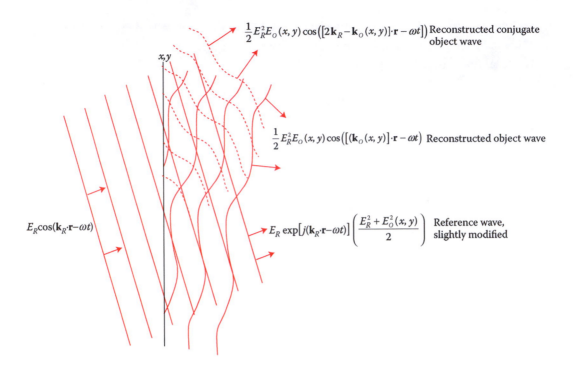

$$\frac{1}{2}E_R^2 E_O(x,y)\cos([2\mathbf{k}_R - \mathbf{k}_O(x,y)]\cdot\mathbf{r} - \omega t])$$ Reconstructed conjugate object wave

$$\frac{1}{2}E_R^2 E_O(x,y)\cos([(\mathbf{k}_O(x,y)]\cdot\mathbf{r} - \omega t)$$ Reconstructed object wave

$$E_R\cos(\mathbf{k}_R\cdot\mathbf{r}-\omega t)$$

$$E_R\exp[j(\mathbf{k}_R\cdot\mathbf{r}-\omega t)]\left(\frac{E_R^2 + E_O^2(x,y)}{2}\right)$$ Reference wave, slightly modified

FIGURE 3.8 Holographic reconstruction (amplitude hologram).

The first term of Equation 3.9,

$$E_R\exp\left[j(\mathbf{k}_R\cdot\mathbf{r}-\omega t)\right]\left(t_0 - \beta\frac{E_R^2 + E_O^2(x,y)}{2}\right)$$

is very similar to the original reference wave with some variation in the amplitude due to the presence of $E_O^2(x,y)$, which produces a halo around the reference wave with a range of spatial frequency twice that of the object wave. If $E_O(x,y)$ is much smaller than E_R, this matters very little.

From the second term, we obtain

$$-\beta E_R\cos(\mathbf{k}_R\cdot\mathbf{r}-\omega t)E_R E_O(x,y)\cos([(\mathbf{k}_R - \mathbf{k}_O(x,y)]\cdot\mathbf{r})$$

$$= \frac{-\beta}{2}E_R^2 E_O(x,y)\left\{\cos([2\mathbf{k}_R - (\mathbf{k}_O(x,y)]\cdot\mathbf{r} + \omega t]) + \cos([(\mathbf{k}_O(x,y)]\cdot\mathbf{r} - \omega t)\right\} \quad (3.10)$$

and we see that the second of the terms on the right side in Equation 3.10 is a *replica of the original object wave, including its phase*, apart from a constant.

This is a very important result because it shows that we can, in principle, record and reconstruct an object wave, even one with a convoluted shape. The drawback is that amplitude holograms are rather inefficient as we have seen in Section 3.2.1.

There are other important results to be extracted from Equation 3.10. We will deal with these in turn.

3.5.2.1 Gabor (In-Line) Holography

Suppose the plane reference wave is parallel to the z-axis, then $\mathbf{k}_R$ is zero. If, at the same time, the range of directions of the object wave vector, $\mathbf{k}_O$, is symmetrically distributed about the z-axis, then the situation is identical to Gabor's experiment as we would expect.

3.5.2.2 Off-Axis Holography

If we can ensure that the angle that the reference wave vector makes with the z-axis is *outside* the *range* of angles that the object wave vector makes with the z-axis, then all three terms in the output can be *physically* separated from one another in space. This is called off-axis holography and, as we shall see, it requires a light source of appreciable coherence length but has the considerable advantage that it overcomes the twin-image problem discussed in Section 3.4.3. It also means that the reconstructed wavefront is not viewed against the background of the reference wave and so we do not have to process the hologram to obtain a value of $\Gamma = 2$.

3.6 PHASE CONJUGACY

One of the most amazing things that we can do with holography is to make a light wave retrace its steps and return to its point of origin, and this has some very interesting and useful consequences. To see how we can make a light wave retrace its steps, we look closely at the first term in Equation 3.10, which involves $-[\mathbf{k}_O(x, y) \cdot \mathbf{r} + \omega t]$. Note the positive sign in front of the term ωt. To understand the implications of this, let us look at the simplest expression describing a plane wave traveling along the z-axis, so the directional cosines of the wave are $(0,0,1)$.

$$E = E_0 \cos(kz - \omega t) \tag{3.11}$$

with E_0 a constant.

At $t = 0$, $E = E_0 \cos(kz)$ shown in Figure 3.9.

A quarter of a period ($t = T/4$) later, the wave has moved a quarter of a wavelength further along the $+z$-axis.

Consider now the wave expression $E = E_0 \cos(kz + \omega t)$. At $t = 0$, this still has the appearance of Figure 3.9, but a quarter of a period ($t = T/4$) later has moved a quarter of a wavelength in the direction of the $-z$-axis. This is what we would expect because, if we reverse the sign of the time, t, the wave should retrace its steps, returning to its source. Such a wave is known as a *phase conjugate wave*. Similarly, a phase conjugate version of a spherical wavefront, emanating from a point source of light, is a spherical wavefront *converging to* that source point (Figure 3.10).

The first term in Equation 3.8,

$$\frac{-\beta}{2} E_R^2 E_O(x, y) \cos\left(2\mathbf{k}_R \cdot \mathbf{r} - \left[\mathbf{k}_O(x, y) \cdot \mathbf{r} + \omega t\right]\right)$$

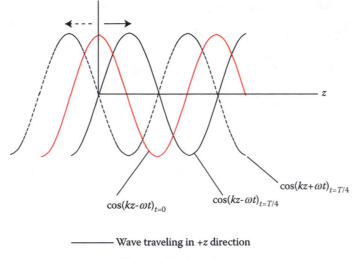

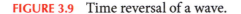

$$\cos(kz+\omega t)_{t=T/4}$$

$$\cos(kz-\omega t)_{t=0}$$

$$\cos(kz-\omega t)_{t=T/4}$$

——— Wave traveling in $+z$ direction

- - - - - - Wave traveling in $-z$ direction

FIGURE 3.9 Time reversal of a wave.

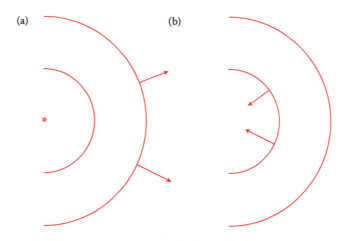

FIGURE 3.10 (a) Spherical wavefront and (b) phase conjugate (time reversed) version.

is a phase conjugate version of the original object wave superimposed on a plane wave with wave vector $2\mathbf{k}_R$. We can see this more clearly if we set $\mathbf{k}_R = 0$. If, for simplicity, we take $\mathbf{k}_R = 2\pi(0,0,\cos\gamma)$, then this wave is traveling at an angle $\cos^{-1}(2\cos\gamma)$ with the *z-axis*.

This leads to an interesting consequence with great potential for holographic applications. Suppose that, instead of using the reference wave to illuminate the hologram, we use the *phase conjugate* version of the reference wave. This is not difficult to do in the case of a plane reference wave. Let us assume we have recorded a hologram of a point source of light, which will produce a spherical object wave. Figure 3.11a shows the normal holographic reconstruction of the light source.

Note that it is a virtual image that we see since the rays or normals to the wavefronts are diverging from the image. We can reconstruct the phase conjugate of this image by directing

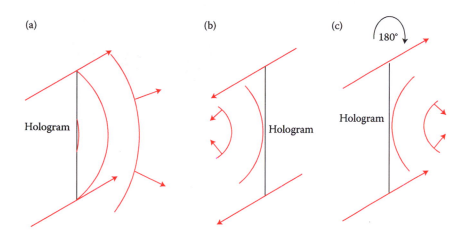

FIGURE 3.11 Reconstruction of phase conjugate spherical wavefront: (a) normal holographic reconstruction of a spherical wavefront; (b) reconstruction of phase conjugate spherical wavefront by phase conjugate reference; and (c) reconstruction of phase conjugate spherical wavefront by normal reference but with hologram rotated.

a plane wave at the hologram from the exact opposite direction as the original reference wave (Figure 3.11b). Alternatively, we can simply rotate the hologram in its plane by 180° (Figure 3.11c) and illuminate it with the *original* reference wave. The *conjugate object wave* is reconstructed, the wavefront normals *converging* to a point to produce a real image (one that we can see on a screen) located at the same distance to the right of the hologram as the original point source was to the left. Thus, provided we have a means of generating the phase conjugate version of the reference wave used in the holographic recording process, then we can reconstruct the phase conjugate version of the light wave corresponding to the original object. (Note: It is more correct to say that the conjugate image is *constructed* from the hologram rather than *reconstructed* since no conjugate object existed in the first place but we will follow convention and refer to the reconstruction of conjugate images.)

Figure 3.12 shows the consequences that this has when we reconstruct the phase conjugate version of an extended object as opposed to a single point.

Object 2 has exactly the same dimensions as object 1 but appears bigger to someone observing from the right. Let us record a hologram of these two objects using a collimated reference wave and then reconstruct the phase conjugate of the object wave by illuminating the hologram with the reference wave but with the hologram rotated by 180° in its plane. The conjugate images 1* and 2* of each of objects 1 and 2, respectively, will appear at the same distances to the right of the hologram as the objects were originally to the left. As a consequence, 1* is now closer to the observer than 2*. This means that *phase conjugation reverses the normal law of perspective.* Furthermore suppose that, in recording, illumination is from the right from the lower right so that a shadow of object 2 falls partly on object 1, then the phase conjugate image 1* is partly invisible that is it is partially obscured by 2*, even though 2* appears to be behind 1*. Phase conjugate images are often called *pseudoscopic,* to distinguish them from normal or *orthoscopic* images.

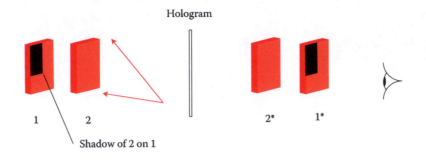

FIGURE 3.12 Objects 1 and 2 on the left and their reconstructed, virtual images obey the normal laws of perspective. The phase conjugate, real images on the right do not.

3.7 PHASE HOLOGRAMS

Amplitude holograms do not produce bright images as most of the reconstructing light is lost through absorption in the recording medium. For this reason, it is usually a good idea to process the hologram so that its thickness or its refractive index is proportional to the original intensity of illumination during recording. This means that at the reconstruction stage, the hologram only *imposes on the reference wave an additional phase, whose value depends on the recording intensity.* Then, the output becomes

$$E_R \exp j\left[(\mathbf{k}_R \cdot \mathbf{r} - \omega t + CI(x, y)\right]$$
$$= E_R \exp j\left[(\mathbf{k}_R \cdot \mathbf{r} - \omega t + C\left\{\left(\frac{E_R^2 + E_O^2(x,y)}{2}\right) + E_R E_O(x,y)\cos\left[(\mathbf{k}_R - \mathbf{k}_O(x,y)\cdot\mathbf{r}\right]\right\}\right] \quad (3.12)$$

where C is a constant. We will assume that E_O is small so that $C((E_R^2 + E_O^2)/2)$ adds a constant phase, which we can set to zero. We take the real part of

$$E_R \exp j\left[(\mathbf{k}_R \cdot \mathbf{r} - \omega t + CI(x, y)\right]$$

and rewrite it as

$$E_R \cos\left[(\mathbf{k}_R \cdot \mathbf{r} - \omega t\right]\cos[CI(x,y)] - E_R \sin[\mathbf{k}_R \cdot \mathbf{r} - \omega t]\sin[CI(x,y)]$$

If we assume the additional phase $CI(x,y)$ is small, the output is approximately

$$E_R \cos(\mathbf{k}_R \cdot \mathbf{r} - \omega t) - CE_R \sin(\mathbf{k}_R \cdot \mathbf{r} - \omega t)\left\{E_R E_O(x,y)\cos\left[(\mathbf{k}_R - \mathbf{k}_O(x,y)\cdot\mathbf{r}\right]\right\}$$
$$= E_R \cos(\mathbf{k}_R \cdot \mathbf{r} - \omega t) - CE_R^2 E_O(x,y)\sin(\mathbf{k}_R \cdot \mathbf{r} - \omega t)\cos\left[(\mathbf{k}_r - \mathbf{k}_O(x,y)\cdot\mathbf{r}\right]$$
$$= E_R \cos(\mathbf{k}_R \cdot \mathbf{r} - \omega t) - CE_R^2 E_O(x,y)\left\{\sin\left[(2\mathbf{k}_R - \mathbf{k}_O(x,y))\cdot\mathbf{r} - \omega t\right]\right\}$$
$$- CE_R^2 E_O(x,y)\sin\left[\mathbf{k}_O(x,y)\cdot\mathbf{r} - \omega t\right] \quad (3.13)$$

As in the case of an amplitude hologram, the first term is an undiffracted version of the original reference wave while the third term

$$-CE_R^2 E_O(x, y) \sin\left[\mathbf{k}_O(x, y) \cdot \mathbf{r} - \omega t\right]$$

is again a replica of the original object wave with the subtle difference that it has the sinusoidal form and is therefore out of phase with the original object wave by 90°.

So far, we have made no assumptions about the physical nature of the hologram, which we have recorded, only that either it has an amplitude transmittance that is proportional to the intensity of the light during the recording or that it imposes a phase shift proportional to the intensity of the light during the recording, on the light passing through it at reconstruction. A real hologram is rather different. It has a finite thickness and therefore we may reasonably assume that the conversion of light from the reference wave to the reconstructed object waves takes place gradually, as the reference wave propagates through the hologram. We will consider such holograms in the next chapter.

3.8 SUMMARY

In this chapter, we show how one can create a photographic record of an interference pattern between two plane waves and illuminate it with one of the waves and thus reproduce the other one. We define diffraction efficiency and obtain values for both amplitude and phase gratings.

We also look at the history of holography from its origins in X-ray diffraction placing emphasis on the concepts of double diffraction and double Fourier transformation as image forming processes. We discuss the practical difficulties in implementing holography.

We use a simple mathematical approach to holography and show that one can record and reconstruct an object wavefront of arbitrary shape. Our approach includes Gabor holography as a special case. We show also that if the object and reference waves approach the recording plane at rather different angles to the normal, the reconstructed images are physically separated so that they are viewed in different directions. The approach used here also removes the need for processing of silver halide to obtain a precise value of contrast, since the reference wave is also physically separated from the reconstructed waves. We consider the meaning of phase conjugation and show that we can reconstruct the phase conjugate of the object, from the hologram to produce a real image. We adopt this simple mathematical approach also for phase holograms.

REFERENCES

1. W. L. Bragg, "An optical method of representing the results of X-ray analyses," *Z. Kristallg., Krystallgeometrie Kristallphys. Kristallchem.*, 70, 475 (1929).
2. W. L. Bragg, "A new type of X-ray microscope," *Nature*, 143 678 (1939).
3. W. L. Bragg, "The X-ray microscope," *Nature*, 149, 470–471 (1942).
4. H. Volkmann, "Ernst Abbe and his work," *Appl. Opt.*, 5, 1720–1731 (1966).
5. D. Gabor, "A new microscope principle," *Nature*, 161, 777–778 (1948).
6. E. N. Leith and J. Upatnieks, "Wavefront reconstruction with diffused illumination and three-dimensional objects," *J. Opt. Soc. Amer.*, 54, 1295–1301 (1964).

PROBLEMS

1. Two sets of plane waves of the same wavelength are described by the expressions

$$\mathbf{E}_1(\mathbf{r},t) = \mathbf{E}_0 \cos(\mathbf{k}_1 \cdot \mathbf{r} - \omega t + \phi) \quad \text{and} \quad \mathbf{E}_2(\mathbf{r},t) = \mathbf{E}_0 \cos(\mathbf{k}_2 \cdot \mathbf{r} - \omega t + \phi)$$

   The wave vectors are in directions (2, 3, 4) and (1, 1, 2). Find the grating vector, that is the vector normal to the interference fringes formed by the two sets of waves. What is the spacing of the fringes?

2. There is no point in increasing the magnification of an optical microscope for visual observation beyond the point at which two barely resolved points are separated in the visual image by more than the resolution limit of the eye, about 0.15 mm. What should the magnification therefore be?

3. Electrons are accelerated through a potential difference of 1 kV in an electron microscope. What is their wavelength? Estimate the potentially useful magnification of the instrument if the photographic emulsion used to record the images can resolve 500 l mm^{-1}.

4. Suppose we use incoherent white light to make a hologram. What is the largest size of object we could use?

5. Assuming we do not have a problem with spatial coherence, we might be able to make a hologram using object and reference beams of temporally incoherent light, which propagate along quite different paths, in order to avoid the twin-image problem which arises in Gabor's in-line holographic method. Suggest *two* possible ways of doing this.

6. Any illuminated object can be regarded as a large set of point sources of light. The spherical wavefronts from all of these points can interfere with a collimated reference beam or indeed the light from a point source. A hologram could then be regarded as a recording of all of these interference patterns. Suppose we record a hologram of a point object using a point source of light, with both sources located on the normal to the hologram plane. Show that the recorded fringes have radii r_n given approximately by

$$\frac{1}{r_n^2} = \frac{1}{2n\lambda}\left(\frac{1}{z_O} - \frac{1}{z_R}\right)$$

   where z_O and z_R are respectively the distance of the object and the reference source from the hologram plane.

7. Show that the amplitude transmissivity of the hologram in problem 6 varies as $\cos(Cr^2)$ and find the value of the constant C.

8. Consider a thin diametrical slice of the hologram in problem 6, illuminated normally by a plane wave, and show that the hologram is a lens with foci at distances given by

$$\frac{1}{f} = \pm\left(\frac{1}{z_O} - \frac{1}{z_R}\right)$$

CHAPTER 4

Volume Holography

4.1 INTRODUCTION

Until now we have assumed that holograms are either amplitude holograms, with amplitude transmission proportional to the original intensity of illumination at the recording stage, or phase holograms, which at the reconstruction stage, impose on the reference wave an additional phase, proportional to the original intensity of illumination at the recording stage. We have also ignored the finite thickness of the hologram, which can, in fact, be substantial compared to the spacing of the recorded interference fringes in the hologram. We will discuss the unique properties of thick, or volume holograms in this chapter, using the coupled-wave theory of holographic gratings developed by Kogelnik [1] and following the treatment given by Collier, Burckhardt, and Lin [2]. The mathematics involved is not particularly difficult to handle but can obscure the physics. For this reason, the reader may wish to go straight to Section 4.5 and following to see how the important results can be obtained more simply from the optical principles, which we have already discussed in Chapters 1 and 2.

4.2 VOLUME HOLOGRAPHY AND COUPLED-WAVE THEORY

Coupled-wave theory is needed in order to understand the properties of thick holograms, which are the most common type. The theory produces results of great practical use. Not least is the fact that it can cope with very high diffraction efficiency holograms in which the reconstructing light is progressively and significantly reduced in amplitude as it traverses the hologram.

4.2.1 THICK HOLOGRAPHIC DIFFRACTION GRATINGS

We consider only a single recorded grating, thickness T, produced by the interference of two plane waves. As explained in Section 1.5, we assume that convoluted wavefront shapes can be synthesized from many plane waves having different wave vectors. An actual hologram, in which the recorded fringes are convoluted, can thus be considered a collection of diffraction gratings each of a unique spatial frequency and each recorded by the interference of two plane waves.

Figure 4.1 shows a thick, or volume, parallel-sided diffraction grating, which is parallel to the x,y plane and made by recording a set of interference fringes (such as those depicted in Figure 3.1).

The grating takes the form of a sinusoidal variation in refractive index or amplitude transmissivity with maxima represented in Figure 4.1 as planes parallel to the x-axis and making angle ϕ with the z-axis. The interference fringe vector defined in Section 3.1 is here renamed the *grating vector*, since the grating is geometrically identical to the interference pattern of which it is a record. Planes of constant refractive index or amplitude transmissivity are defined by the vector equation

$$\mathbf{K} \cdot \mathbf{r} = \text{constant} \tag{4.1}$$

where $\mathbf{K}$ is the grating vector of magnitude $K = 2\pi/d$ with d, the grating spacing.

The grating is illuminated by a plane wave, $\mathbf{E}$, polarized parallel to the x-axis, incident at the front surface of the grating and refracted to make angle ψ with the z-axis.

4.2.2 LIGHT WAVES IN A DIELECTRIC MEDIUM

We repeat the argument leading to Equation 1.8, except that now the electric and magnetic fields are in a dielectric medium ($\mu_r = 1$) whose refractive index and absorption both vary sinusoidally with y.

The scalar product $\varepsilon_0 \mathbf{E} \cdot \nabla(\varepsilon_r)$ remains zero since $\nabla(\varepsilon_r)$ only has a nonzero component parallel to the y-axis while $\mathbf{E}$ is parallel to the x-axis.

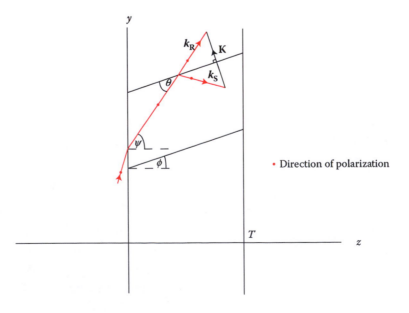

FIGURE 4.1 Geometry of plane slanted grating.

If the grating is an amplitude one and absorbs light, the conductivity, σ, is no longer zero so the wave equation now becomes

$$\nabla^2 \mathbf{E} = \mu_0 \varepsilon_0 \varepsilon_r \frac{\partial^2 E}{\partial t^2} - \mu_0 \sigma \frac{\partial E}{\partial t} \tag{4.2}$$

Since,

$$
\begin{aligned}
E &= E(y,z)\exp\left[j(ky\sin\psi + kz\cos\psi - \omega t)\right] \\
&= E(y,z)\exp\left[j(ky\sin\psi + kz\cos\psi)\right]\exp(-j\omega t) = \tilde{E}(y,z)\exp(-j\omega t)
\end{aligned}
$$

Equation 4.2 becomes

$$\nabla^2 \tilde{E} = \left(j\omega\mu_0\sigma - \omega^2\mu_0\varepsilon_0\varepsilon_r\right)\tilde{E} \tag{4.3}$$

4.2.3 LIGHT WAVES IN A DIELECTRIC MEDIUM WITH A GRATING

The sinusoidal variations of relative permittivity and of conductivity arising from the grating recording can be expressed by

$$\varepsilon_r = \varepsilon_{r,0} + \varepsilon_1 \cos(\mathbf{K} \cdot \mathbf{r}) \tag{4.4}$$

$$\sigma = \sigma_0 + \sigma_1 \cos(\mathbf{K} \cdot \mathbf{r}) \tag{4.5}$$

where $\varepsilon_{r,0}$ is the average value of the relative permittivity, $\varepsilon_{r,0}^{1/2} = n_0$ the average refractive index, and σ_0 is the average value of the conductivity. If $n_1 \ll n_0$, the refractive index, n, is given by

$$n^2 = \left[n_0 + n_1\cos(\mathbf{K}\cdot\mathbf{r})\right]^2 = n_0^2 + 2n_0 n_1\cos(\mathbf{K}\cdot\mathbf{r}) = \varepsilon_{r,0} + \varepsilon_1\cos(\mathbf{K}\cdot\mathbf{r}) \tag{4.6}$$

and $\varepsilon_1 = 2n_1 n_0$. The term n_1 denotes the refractive index modulation amplitude, that is, the maximum extent to which the refractive index varies.

Equation 4.3 becomes

$$\nabla^2\tilde{E} - j\omega\mu_0\left[\sigma_0 + \sigma_1\cos(\mathbf{K}\cdot\mathbf{r})\right]\tilde{E} + \omega^2\mu_0\varepsilon_0\left[\varepsilon_{r,0} + \varepsilon_1\cos(\mathbf{K}\cdot\mathbf{r})\right]\tilde{E} = 0 \tag{4.7}$$

$$\nabla^2\tilde{E} + \left(\omega^2\mu_0\varepsilon_0\varepsilon_{r,0} - j\omega\mu_0\sigma_0\right)\tilde{E} + \left(\omega^2\mu_0\varepsilon_0\varepsilon_1 - j\omega\mu_0\sigma_1\right)\cos(\mathbf{K}\cdot\mathbf{r})\tilde{E} = 0 \tag{4.8}$$

$$\nabla^2\tilde{E} = -\left[\left(k^2\varepsilon_{r,0} - j\omega\mu_0\sigma_0\right) + \left(k^2\varepsilon_1 - j\omega\mu_0\sigma_1\right)\cos(\mathbf{K}\cdot\mathbf{r})\right]\tilde{E} \tag{4.9}$$

with $k^2 = \omega^2\mu_0\varepsilon_0 = \omega^2/c^2 = 4\pi^2/\lambda_{vac}^2 \cong 4\pi^2/\lambda_{air}^2$.

The coefficient of $\tilde{E}$ on the right-hand side of Equation 4.9 is

$$
\begin{aligned}
& k^2\varepsilon_{r,0} - j\omega\mu_0\sigma_0 + \left(k^2\varepsilon_1 - j\omega\mu_0\sigma_1\right)\cos(\mathbf{K}\cdot\mathbf{r}) \\
&= k^2\varepsilon_{r,0} - 2j\frac{\omega\mu_0\sigma_0}{2k\varepsilon_{r,0}^{1/2}}k\varepsilon_{r,0}^{1/2} + 2k\varepsilon_{r,0}^{1/2}\left[\frac{1}{2}\left(\frac{k\varepsilon_1}{2\varepsilon_{r,0}^{1/2}} - \frac{j\omega\mu_0\sigma_1}{2k\varepsilon_{r,0}^{1/2}}\right)\right]\left\{\exp\left[j(\mathbf{K}\cdot\mathbf{r})\right] + \exp\left[-j(\mathbf{K}\cdot\mathbf{r})\right]\right\} \\
&= \beta^2 - 2j\alpha\beta + 2\beta\chi\left\{\exp\left[j(\mathbf{K}\cdot\mathbf{r})\right] + \exp\left[-j(\mathbf{K}\cdot\mathbf{r})\right]\right\}
\end{aligned}
$$

where $\alpha = (\omega\mu_0\sigma_0/2k\varepsilon_{r,0}^{1/2})$ and $\beta = k\varepsilon_{r,0}^{1/2} = kn_0$

$$\chi = \frac{1}{2}\left(\frac{k\varepsilon_1}{2\varepsilon_{r,0}^{1/2}} - \frac{j\omega\mu_0\sigma_1}{2k\varepsilon_{r,0}^{1/2}}\right) = \frac{\pi n_1}{\lambda_{air}} - \frac{j\alpha_1}{2} \quad \text{where} \quad \alpha_1 = \frac{\omega\mu_0\sigma_1}{2k\varepsilon_{r,0}^{1/2}} \qquad (4.10)$$

The constant χ describes the coupling between waves, via the grating.

We now consider a dielectric material, which is homogeneous and absorbent with refractive index everywhere equal to n_0. The expression $E\exp(-\rho z)$ with E constant, describing a plane wave propagating along the z-axis, is a solution of Equation 4.3 if

$$\rho = j\left(k^2\varepsilon_{r,0} - 2j\frac{\omega\mu_0\sigma_0}{2k\varepsilon_{r,0}^{1/2}}k\varepsilon_{r,0}^{1/2}\right)^{1/2} = j\left(\beta^2 - 2j\alpha\beta\right)^{1/2} \qquad (4.11)$$

If $\alpha \ll \beta$ then

$$\rho = j\beta\left(\frac{1-2j\alpha}{\beta}\right)^{1/2} \cong j\beta\left(1 - \frac{j\alpha}{\beta}\right) = \alpha + j\beta$$

The wave $E\exp(-\rho z)$ is now

$$E\exp(-\alpha z)\exp(-j\beta z)$$

We see now that α is an attenuation constant so the wave is absorbed as it progresses in the $+z$ direction, while β is a propagation constant.

From Equation 4.5

$$\sigma = \sigma_0 + \sigma_1\cos(\mathbf{K}\cdot\mathbf{r}) = 2k\omega\mu_0\alpha\beta + 2k\omega\mu_0\alpha_1\beta\cos(\mathbf{K}\cdot\mathbf{r}) = 2k\omega\mu_0\beta\left[\alpha + \alpha_1\cos(\mathbf{K}\cdot\mathbf{r})\right]$$

so $\alpha_1 < \alpha$, otherwise the attenuation could become *negative* because of the recording process.

Returning now to the volume holographic grating, the light illuminating it is described by the expression

$$\mathbf{E}_R = \mathbf{R}(z)\exp(-j\mathbf{k}_R\cdot\mathbf{r}) \qquad (4.12)$$

This wave is linearly polarized parallel to the x-axis and its wave vector lies in the y, z plane. The expression $\mathbf{R}(z)$ describes the variation that takes place in the complex amplitude as the wave progresses, due to variations in refractive index and absorption. The magnitude of the wave vector is $k_R = 2\pi n_0/\lambda_{air}$. The wave vector makes an angle with the z-axis that satisfies, or almost satisfies, the Bragg condition (Equation 2.9). Only in this way will there be a diffracted beam of any significant intensity. We also assume that there is only one diffracted beam.

The diffracted wave is $\mathbf{E}_S = \mathbf{S}(z)\exp(-j\mathbf{k}_S\cdot\mathbf{r})$. It is also plane polarized parallel to the x-axis with its wave vector lying in the y, z plane.

If the wave $\mathbf{E}_R$ is incident on the grating at the Bragg angle, then the wave $\mathbf{E}_S$ is diffracted at the Bragg angle and the wave vectors, $\mathbf{k}_R$ and $\mathbf{k}_S$, make equal angles, θ, with the grating planes (Figure 4.2), defined by Equation 4.1. They are of equal magnitude and

$$\mathbf{k}_S = \mathbf{k}_R - \mathbf{K} \tag{4.13}$$

Then from Figure 4.1,

$$K = \frac{2\pi}{d} = 2k_R \sin\theta = \frac{2\pi n_0}{\lambda_{air}} 2\sin\theta \tag{4.14}$$

$$2n_0 d \sin\theta = \lambda_{air} \tag{4.15}$$

which is Bragg's equation.

We now focus on the solution to Equation 4.9.

$$\nabla^2 \tilde{E} + \left[k^2 \varepsilon_{0,r} - 2j \frac{\omega\mu_0\sigma_0}{2k\varepsilon_{0,r}^{1/2}} k\varepsilon_{0,r}^{1/2} + 2k\varepsilon_{0,r}^{1/2} \chi \left\{ \exp[j(\mathbf{K}\cdot\mathbf{r})] + \exp[-j(\mathbf{K}\cdot\mathbf{r})] \right\} \right] \tilde{E}$$

$$= \nabla^2 \tilde{E} + \left\{ \beta^2 - 2j\alpha\beta + 2\beta\chi \left\{ \exp[j(\mathbf{K}\cdot\mathbf{r})] + \exp[-j(\mathbf{K}\cdot\mathbf{r})] \right\} \right\} \tilde{E} = 0$$

The total disturbance is $\tilde{E} = R(z)\exp(-j\mathbf{k}_R\cdot\mathbf{r}) + S(z)\exp(-j\mathbf{k}_R\cdot\mathbf{r})_S = E_R + E_S$ with

$$\mathbf{k}_R \cdot \mathbf{r} = \beta(y\sin\psi + z\cos\psi), \; k_R^2 = \beta^2,$$
$$\mathbf{k}_S \cdot \mathbf{r} = k_{S_y} y + k_{S_z} z \quad \text{and} \quad k_S^2 = k_{S_y}^2 + k_{S_z}^2$$

$\nabla^2 \left[R(z)\exp(-j\mathbf{k}_R\cdot\mathbf{r}) + S(z)\exp(-j\mathbf{k}_S\cdot\mathbf{r}) \right]$ is obtained as follows:

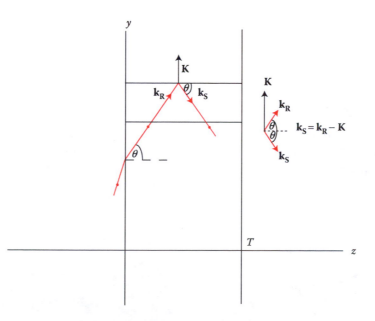

FIGURE 4.2 Bragg reflection in an unslanted transmission grating.

$$\frac{\partial E_R}{\partial z} = \frac{\partial \mathbf{R}}{\partial z}\exp(-j\mathbf{k_R}\cdot\mathbf{r}) - \mathbf{R}\,\exp(-j\mathbf{k_R}\cdot\mathbf{r})j\beta\cos\psi$$

$$\frac{\partial^2 E_R}{\partial x^2} + \frac{\partial^2 E_R}{\partial y^2} + \frac{\partial^2 E_R}{\partial z^2}$$

$$= \frac{\partial^2 \mathbf{R}}{\partial z^2}\exp(-j\mathbf{k_R}\cdot\mathbf{r}) - 2\frac{\partial \mathbf{R}}{\partial z}\exp(-j\mathbf{k_R}\cdot\mathbf{r})j\beta\cos\psi - \mathbf{R}\exp(-j\mathbf{k_R}\cdot\mathbf{r})$$

$$\beta^2\cos^2\psi - \mathbf{R}\exp(-j\mathbf{k_R}\cdot\mathbf{r})\beta^2\sin^2\psi$$

$$\frac{\partial^2 E_S}{\partial x^2} + \frac{\partial^2 E_S}{\partial y^2} + \frac{\partial^2 E_S}{\partial z^2}$$

$$= \frac{\partial^2 \mathbf{S}}{\partial z^2}\exp(-j\mathbf{k_S}\cdot\mathbf{r}) - 2\frac{\partial \mathbf{S}}{\partial z}\exp(-j\mathbf{k_S}\cdot\mathbf{r})jk_{S_z} - \mathbf{S}\exp(-j\mathbf{k_S}\cdot\mathbf{r})k_{S_z}^2 - \mathbf{S}\exp(-j\mathbf{k_S}\cdot\mathbf{r})k_{S_y}^2$$

$$= -\left\{\beta^2 - 2j\alpha\beta + 2\beta\chi\left\{\exp\left[j\left(\left[\mathbf{k_R}-\mathbf{k_S}\right]\cdot\mathbf{r}\right)\right] + \exp\left[-j\left(\left[\mathbf{k_R}-\mathbf{k_S}\right]\cdot\mathbf{r}\right)\right]\right\}\right\}\mathbf{S}\exp(-j\mathbf{k_S}\cdot\mathbf{r})$$

$$\exp(-j\mathbf{k_R}\cdot\mathbf{r})\left[\frac{\partial^2\mathbf{R}}{\partial z^2} - 2\frac{\partial\mathbf{R}}{\partial z}j\beta\cos\psi - 2j\alpha\beta\mathbf{R} + 2\beta\chi\mathbf{S}\right]$$

$$+\exp(-j\mathbf{k_S}\cdot\mathbf{r})\left[\frac{\partial^2\mathbf{S}}{\partial z^2} - 2\frac{\partial\mathbf{S}}{\partial z}jk_{S_z} - k_S^2\mathbf{S} + \beta^2\mathbf{S} - 2j\alpha\beta\mathbf{S} + 2\beta\chi\mathbf{R}\right]$$

$$+2\beta\chi\exp\left[-j\left(\left[2\mathbf{k_R}-\mathbf{k_S}\right]\cdot\mathbf{r}\right)\right]\mathbf{R} + 2\beta\chi\exp\left[-j\left(\left[2\mathbf{k_S}-\mathbf{k_R}\right]\cdot\mathbf{r}\right)\right]\mathbf{S} = 0 \qquad (4.16)$$

The last two terms of Equation 4.16 are zero as they represent waves that do not satisfy the Bragg condition. Also, the coefficients of $\exp(-j\mathbf{k_R}\cdot\mathbf{r})$ and $\exp(-j\mathbf{k_S}\cdot\mathbf{r})$ must separately be zero. Notice that each of them involves both $\mathbf{R}$ and $\mathbf{S}$ and the coupling constant χ.

$$\frac{\partial^2\mathbf{R}}{\partial z^2} - 2\frac{\partial\mathbf{R}}{\partial z}j\beta\cos\psi - 2j\alpha\beta\mathbf{R} + 2\beta\chi\mathbf{S} = 0 \qquad (4.17)$$

$$\frac{\partial^2\mathbf{S}}{\partial z^2} - 2\frac{\partial\mathbf{S}}{\partial z}jk_{S_z} - k_S^2\mathbf{S} + \beta^2\mathbf{S} - 2j\alpha\beta\mathbf{S} + 2\beta\chi\mathbf{R} = 0 \qquad (4.18)$$

Taking the third and fourth terms in Equation 4.18, namely $\mathbf{S}(\beta^2 - k_S^2)$, we consider the situation when the incident wave is not at the Bragg angle but at $\theta' = \theta + \delta$, δ being small. From Equation 4.13

$$\beta^2 - k_S^2 = \beta^2 - \left(\mathbf{k_R}-\mathbf{K}\right)^2 = \beta^2 - k_R^2 + 2k_RK\cos\left(\frac{\pi}{2}-\theta'\right) - K^2 = 2k_RK\sin\theta' - K^2$$

$$\sin\theta' = \sin\theta\cos\delta + \cos\theta\sin\delta \cong \sin\theta + \delta\cos\theta = \frac{K}{2\beta} + \delta\cos\theta$$

$$\beta^2 - k_S^2 \cong 2k_RK\left[\frac{K}{2\beta} + \delta\cos\theta\right] - K^2 = 2k_RK\delta\cos\theta = 2k_R2\beta\sin\theta\delta\cos\theta = 2\beta^2\delta\sin(2\theta) = 2\beta\Gamma$$

where

$$\Gamma = \beta\delta\sin(2\theta) \tag{4.19}$$

Returning now to Equations 4.17 and 4.18, and assuming that $\mathbf{R}(z)$ and $\mathbf{S}(z)$ vary only slowly so that $\partial^2\mathbf{R}/\partial z^2$ and $\partial^2\mathbf{S}/\partial z^2$ can be neglected, we have

$$\frac{\partial\mathbf{R}}{\partial z}\cos\psi + \alpha\mathbf{R} = -j\chi\mathbf{S} \tag{4.20}$$

$$\frac{\partial\mathbf{S}}{\partial z}\frac{k_{S_z}}{\beta} + (j\Gamma + \alpha)\mathbf{S} = -j\chi\mathbf{R} \tag{4.21}$$

These equations show that as the waves advance in the z direction, their amplitudes change through absorption as a result of terms in α, or by coupling, that is, an exchange of energy between them, as a result of terms in χ. The term in Γ introduces a change in the phase of the diffracted wave relative to the incident wave and if Γ becomes large, as it will if the incident angle is much different from the Bragg angle, the exchange of energy between the waves will stop.

Differentiating Equation 4.20

$$-j\chi\frac{\partial\mathbf{S}}{\partial z} = \cos\psi\frac{\partial^2\mathbf{R}}{\partial z^2} + \alpha\frac{\partial\mathbf{R}}{\partial z} \quad \text{and inserting into Equation 4.21}$$

$$\frac{k_{S_z}}{\beta}\frac{\partial\mathbf{S}}{\partial z} + (j\Gamma + \alpha)\mathbf{S} + j\chi\mathbf{R} = \left(\frac{k_{S_z}}{\beta}\frac{j\cos\psi}{\chi}\frac{\partial^2\mathbf{R}}{\partial z^2} + \frac{j\alpha k_{S_z}}{\chi\beta}\frac{\partial\mathbf{R}}{\partial z}\right)$$

$$+ (j\Gamma + \alpha)\frac{j}{\chi}\left(\frac{\partial\mathbf{R}}{\partial z}\cos\psi + \alpha\mathbf{R}\right) + j\chi\mathbf{R} = 0$$

$$\frac{jk_{S_z}\cos\psi}{\chi\beta}\frac{\partial^2\mathbf{R}}{\partial z^2} + \left(\frac{jk_{S_z}\alpha}{\chi\beta} + \frac{j\alpha\cos\psi}{\chi} - \frac{\Gamma\cos\psi}{\chi}\right)\frac{\partial\mathbf{R}}{\partial z} + \left(\frac{j}{\chi}(j\Gamma + \alpha)\alpha + j\chi\right)\mathbf{R} = 0$$

$$\frac{\partial^2\mathbf{R}}{\partial z^2} + \left(\frac{\alpha}{\cos\psi} + \frac{\alpha\beta}{k_{S_z}} + \frac{j\beta\Gamma}{k_{S_z}}\right)\frac{\partial\mathbf{R}}{\partial z} + \frac{\chi\beta}{jk_{S_z}\cos\psi}\left(\frac{j}{\chi}(j\Gamma + \alpha)\alpha + j\chi\right)\mathbf{R} = 0$$

$$\frac{\partial^2\mathbf{R}}{\partial z^2} + \left(\frac{\alpha}{\cos\psi} + \frac{\alpha\beta}{k_{S_z}} + \frac{j\beta\Gamma}{k_{S_z}}\right)\frac{\partial\mathbf{R}}{\partial z} + \frac{\beta}{k_{S_z}\cos\psi}(\alpha^2 + j\Gamma\alpha + \chi^2)\mathbf{R} = 0 \tag{4.22}$$

Equation 4.22 is a second-order differential equation with constant coefficients having the general solution

$$\mathbf{R}(z) = \exp(\varsigma z) \tag{4.23}$$

Substituting Equation 4.23 in Equation 4.22, we obtain the quadratic equation

$$\varsigma^2 + \left(\frac{\alpha}{\cos\psi} + \frac{\alpha\beta}{k_{S_z}} + \frac{j\beta\Gamma}{k_{S_z}}\right)\varsigma + \frac{\beta}{k_{S_z}\cos\psi}(\alpha^2 + j\Gamma\alpha + \chi^2) = 0$$

$$\varsigma = -\frac{1}{2}\left(\frac{\alpha}{\cos\psi} + \frac{\alpha\beta}{k_{S_z}} + \frac{j\beta\Gamma}{k_{S_z}}\right) \pm \frac{1}{2}\left[\left(\frac{\alpha}{\cos\psi} + \frac{\alpha\beta}{k_{S_z}} + \frac{j\beta\Gamma}{k_{S_z}}\right)^2 - \frac{4\beta}{k_{S_z}\cos\psi}\left(\alpha^2 + j\Gamma\alpha + \chi^2\right)\right] \quad (4.24)$$

$$\varsigma = -\frac{1}{2}\left(\frac{\alpha}{\cos\psi} + \frac{\alpha\beta}{k_{S_z}} + \frac{j\beta\Gamma}{k_{S_z}}\right) \pm \frac{1}{2}\left[\left(\frac{\alpha}{\cos\psi} - \frac{\alpha\beta}{k_{S_z}} - \frac{j\beta\Gamma}{k_{S_z}}\right)^2 - \frac{4\beta\chi^2}{k_{S_z}\cos\psi}\right]^{\frac{1}{2}} \quad (4.25)$$

The general solution is

$$\mathbf{R}(z) = \mathbf{R}_1 \exp(\varsigma_1 z) + \mathbf{R}_2 \exp(\varsigma_2 z) \quad (4.26)$$

where ς_1 and ς_2 are the two values of ς in Equation 4.24 and $\mathbf{R}_1, \mathbf{R}_2$ are found from the boundary conditions. By inserting $\mathbf{R}(z)$ in Equation 4.20, we can find a similar equation for

$$\mathbf{S}(z) = \mathbf{S}_1 \exp(\varsigma_1 z) + \mathbf{S}_2 \exp(\varsigma_2 z) \quad (4.27)$$

Here, we can verify that it is okay to assume $\partial^2\mathbf{R}/\partial z^2$ is negligible compared with the second term in Equation 4.22. From Equation 4.23, $\partial^2\mathbf{R}/\partial z^2 = \varsigma^2 \exp(\varsigma z)$ while $-2(\partial\mathbf{R}/\partial z)j\beta\cos\psi = -2j\beta\varsigma\cos\psi\exp(\varsigma z)$ so the requirement is that $\varsigma \ll 2\beta\cos\psi$, which requires that $\varsigma \ll \beta$ for $\psi < \pi/2$. This is indeed the case, as can be seen from Equation 4.25 since Γ, being proportional to δ, is small, while χ depends on n_1 and α_1, both of which are small, and finally, α is small.

A similar argument can be made to show that $\partial^2\mathbf{S}/\partial z^2$ in Equation 4.18 can be neglected.

4.3 CHARACTERISTICS OF THICK HOLOGRAPHIC GRATINGS

4.3.1 TRANSMISSION GRATINGS

Transmission gratings are gratings recorded by beams of light approaching the recording medium from the same side. We only consider pure phase gratings, in which refractive index varies and there are no losses, and pure amplitude gratings in which the refractive index is uniform and the absorption varies.

The beam incident on the grating, usually called the probe beam, is normalized so that $\mathbf{R}(0) = \mathbf{R}_1 + \mathbf{R}_2 = 1$ and, at the front surface of the grating, $\mathbf{S}(0) = \mathbf{S}_1 + \mathbf{S}_2 = 0$ so $\mathbf{S}_1 = -\mathbf{S}_2$.

From Equation 4.27

$$\frac{\partial\mathbf{S}(0)}{\partial z} = \varsigma_1\mathbf{S}_1 + \varsigma_2\mathbf{S}_2 \quad (4.28)$$

From Equation 4.21

$$\frac{\partial\mathbf{S}(0)}{\partial z}\frac{k_{S_z}}{\beta} + \left(j\Gamma + \alpha\right)\mathbf{S}(0) = \frac{k_{S_z}}{\beta}\left(\varsigma_1\mathbf{S}_1 + \varsigma_2\mathbf{S}_2\right) - j\chi\mathbf{R}(0) = \frac{k_{S_z}}{\beta}\mathbf{S}_1\left(\varsigma_1 - \varsigma_2\right) = -j\chi$$

$$\mathbf{S}_1 = \frac{-j\chi}{\dfrac{k_{S_z}}{\beta}\left(\varsigma_1 - \varsigma_2\right)}$$

and

$$\mathbf{S}(T) = \frac{-j\chi}{\dfrac{k_{S_z}}{\beta}(\varsigma_1 - \varsigma_2)}\left[\exp(\varsigma_1 T) - \exp(\varsigma_2 T)\right] \qquad (4.29)$$

The incident wave is at an angle θ with the grating planes, so that it is at the Bragg angle.

4.3.1.1 Phase Transmission Gratings

From Equation 4.25, with $\alpha = 0$,

$$\varsigma = -\frac{1}{2}\left(\frac{\alpha}{\cos\psi} + \frac{\alpha\beta}{k_{S_z}} + \frac{j\beta\Gamma}{k_{S_z}}\right) \pm \frac{1}{2}\left[\left(\frac{\alpha}{\cos\psi} - \frac{\alpha\beta}{k_{S_z}} - \frac{j\beta\Gamma}{k_{S_z}}\right)^2 - \frac{4\beta\chi^2}{k_{S_z}\cos\psi}\right]^{1/2}$$

$$= \frac{j\beta}{2k_{S_z}}\left\{-\Gamma \pm \left[\Gamma^2 + \frac{4k_{S_z}\chi^2}{\beta\cos\psi}\right]^{1/2}\right\}$$

From Equation 4.29

$$\mathbf{S}(T) = \frac{-j\chi}{\left(k_{S_z}/\beta\right)(\varsigma_1 - \varsigma_2)}\left[\exp(\varsigma_1 T) - \exp(\varsigma_2 T)\right]$$

$$= \frac{-j\chi\left(\beta T/2k_{S_z}\right)\exp\left(-j\beta\Gamma T/2k_{S_z}\right)}{j\left[\left(\beta\Gamma T/2k_{S_z}\right)^2 + \left(\beta(\chi T)^2/k_{S_z}\cos\psi\right)\right]^{1/2}}\,2j\sin\left\{\left[\left(\frac{\beta\Gamma T}{2k_{S_z}}\right)^2 + \frac{\beta(\chi T)^2}{k_{S_z}\cos\psi}\right]^{1/2}\right\}$$

$$\mathbf{S}(T) = -j\left(\frac{\beta\cos\psi}{k_{S_z}}\right)^{1/2}\exp(-j\xi)\frac{\sin\left[\xi^2 + \upsilon^2\right]^{1/2}}{\left[1 + (\xi^2/\upsilon^2)\right]^{1/2}} \qquad (4.30)$$

$$\text{with } \upsilon = \frac{\pi n_1 T}{\lambda_{air}}\left(\frac{\beta}{k_{S_z}\cos\psi}\right)^{1/2} \qquad (4.31)$$

$$\text{and } \xi = \frac{\beta\Gamma T}{2k_{S_z}} = \frac{\beta^2\,\delta\sin(2\theta)T}{2k_{s_z}} \quad \text{(using Equation 4.19)} \qquad (4.32)$$

Note the dependence of ξ on δ, the deviation from the Bragg angle, θ. For this reason, ξ is called the off-Bragg parameter.

The diffraction efficiency, η, is the rate of flow of diffracted energy in the z direction divided by the rate of flow of incident energy in the same direction and is, from Equation 4.30,

$$\eta = \frac{\cos(\theta - \phi)}{\cos(\theta + \phi)}\mathbf{S}\mathbf{S}^*$$

$$\eta = \frac{\sin^2 \left(\xi^2 + v^2 \right)^{1/2}}{\left(1 + \xi^2 / v^2 \right)} \tag{4.33}$$

If the Bragg condition is met, then ξ is zero since δ is zero and η is 1.0 when $v = \pi/2 = \pi n_1 T/\lambda_{air} \cos\theta$ for an unslanted grating ($\varphi = 0$) and $n_1 T/\lambda_{air} \cos \theta/2$. This means $n_1 T \cong \lambda_{air}/2$ and so a grating of thickness about 20 μm has 100% efficiency if the refractive index modulation is about 0.025. Figure 4.3 shows normalized diffraction efficiency curves for conditions ranging from on-Bragg to $\xi = 4.9$ and for three values of v namely $\pi/2$, π, and $3\pi/4$. By normalized efficiency, we mean the value of η divided by its maximum, on-Bragg value, η_0.

These curves are commonly known in the literature on holography as Bragg curves. The angular width of the curve is the difference between the angles at which η has maximum and zero values, the latter corresponding to $\xi = 2.7$ in the case of the curve with $v = \pi/2$. The angular width is often called the angular selectivity of the grating. Suppose that the recording beams make angles of 30° and −30° with the normal, the z-axis and that the grating is recorded in a material of refractive index 1.5 and is 20 μm thick. Inside the layer, the angle θ is 19°. Suppose the wavelength of the recording light is 532 nm and that the value of η is 1.0 with the index modulation n_1 such that $v = \pi/2$. From Equation 4.31, $\xi = \beta\delta \sin \theta T$ for an unslanted grating, then $\xi = \beta\delta \sin \theta T = 2.7\lambda_{air}/2\pi n_0 \sin \theta T = 0.023$ rad $= 1.34°$, measured inside the grating, and 2.01° outside.

Alternatively, we can calculate the *change in wavelength* at which the diffraction efficiency becomes zero. We call this the *wavelength selectivity* or *spectral bandwidth* of the grating. Again, $\xi = 2.7$.

We differentiate Equation 4.15 and again use the Bragg equation to obtain

$$\partial\theta = \frac{\partial\lambda_{air}}{2n_0 d \cos \theta} = \frac{\partial\lambda_{air} \tan \theta}{\lambda_{air}}$$

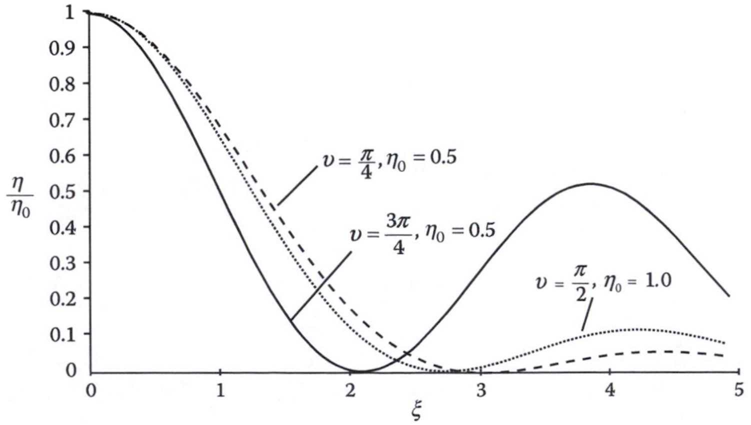

FIGURE 4.3 Normalized diffraction efficiency η/η_0 of a pure phase grating versus for different values of v.

which is the corresponding change in angle needed to continue satisfying the Bragg condition as the wavelength changes and, substituting it for δ, we obtain

$$\xi = \beta \delta \sin \theta T = \frac{2\pi n_0}{\lambda_{air}} \left(\frac{\partial \lambda_{air} \tan \theta}{\lambda_{air}} \right) \sin \theta T \qquad (4.34)$$

giving a value of 14 nm for the wavelength selectivity. The spectral bandwidth is twice this value as we take it to be the spectral width between the first zeroes on either side of the peak of the Bragg curve. We can obtain useful approximate rules for the angular and wavelength selectivity by setting

$$\partial \theta = \frac{3}{\beta T \sin \theta} = \frac{3}{(2\pi n_0 / \lambda_{air})T \sin \theta} \cong \frac{d}{T} \qquad (4.35)$$

and

$$\frac{\partial \lambda_{air}}{\lambda_{air}} \cong \frac{d \cot \theta}{T} \qquad (4.36)$$

4.3.1.2 Unslanted Amplitude Gratings

We now consider unslanted holographic gratings, in which there is no variation of refractive index ($n_1 = 0$), only of absorption and so $\chi = -j\alpha/2$. It is helpful to define $\alpha' = \alpha_1 T/(2\cos\theta)$.

From Equation 4.24

$$\varsigma = -\frac{1}{2}\left(\frac{\alpha}{\cos\theta} + \frac{\alpha\beta}{k_{S_z}} + \frac{j\beta\Gamma}{k_{S_z}} \right) \pm \frac{1}{2}\left[\left(\frac{\alpha}{\cos\theta} - \frac{\alpha\beta}{k_{S_z}} - \frac{j\beta\Gamma}{k_{S_z}} \right)^2 - \frac{4\beta\chi^2}{k_{S_z}\cos\theta} \right]^{1/2}$$

and again $k_{S_z} = \beta \cos\theta$

$$\varsigma = -\frac{1}{2}\left(\frac{2\alpha}{\cos\theta} + \frac{j\Gamma}{\cos\theta} \right) \pm \frac{1}{2}\left[\left(-\frac{j\Gamma}{\cos\theta} \right)^2 + \left(\frac{\alpha_1}{\cos\theta} \right)^2 \right]^{1/2}$$

$$\varsigma_1 - \varsigma_2 = \left[\left(\frac{\alpha_1}{\cos\theta} \right)^2 - \left(\frac{\Gamma}{\cos\theta} \right)^2 \right]^{1/2} = \frac{2}{T}\left[\left(\frac{\alpha_1 T}{2\cos\theta} \right)^2 - \left(\frac{\Gamma T}{2\cos\theta} \right)^2 \right]^{1/2}$$

$$\varsigma T = -\frac{1}{2}\left(\frac{2\alpha T}{\cos\theta} + \frac{j\Gamma T}{\cos\theta} \right) \pm \frac{1}{2}\left[\left(-\frac{j\Gamma T}{\cos\theta} \right)^2 + \left(\frac{\alpha_1 T}{\cos\theta} \right)^2 \right]^{1/2}$$

$$= -\frac{\alpha T}{\cos\theta} - j\xi \pm \left(\left[\frac{\alpha_1 T}{2\cos\theta} \right]^2 - \xi^2 \right)^{1/2}$$

From Equation 4.29

$$\mathbf{S}(T) = \frac{-j\chi}{\frac{k_{S_z}}{\beta}(\varsigma_1 - \varsigma_2)}\left[\exp(\varsigma_1 T) - \exp(\varsigma_2 T)\right]$$

$$= \frac{-\alpha_1 T}{2\cos\theta\left[\left(\alpha_1 T/2\cos\theta\right)^2 - \left(\Gamma T/2\cos\theta\right)^2\right]^{1/2}}\exp\left(-\frac{1}{2}\left(\frac{2\alpha T}{\cos\theta} + \frac{j\Gamma T}{\cos\theta}\right)\right)$$

$$\times \sinh\left[\left(-\frac{j\Gamma T}{2\cos\theta}\right)^2 + \left(\frac{\alpha_1 T}{2\cos\theta}\right)^2\right]^{1/2}$$

$$= \frac{-\exp^{(-j\xi)}\exp^{(-\alpha T/\cos\theta)}\sinh\left[\left(\alpha_1 T/2\cos\theta\right)^2 - \xi^2\right]^{1/2}}{\left[1 - \left(\frac{\xi}{\alpha_1 T/2\cos\theta}\right)^2\right]^{1/2}}$$

$$= \frac{-\exp^{(-j\xi)}\exp^{(-\alpha T/\cos\theta)}\sinh\left(\alpha'^2 - \xi^2\right)^{1/2}}{\left[1 - (\xi/\alpha')^2\right]^{1/2}}$$

where $\alpha' = \alpha_1 T/2\cos\theta$. At the Bragg angle $\xi = 0$ and

$$\mathbf{S}(T) = -\exp\left(-\frac{\alpha T}{\cos\theta}\right)\sinh\left(\frac{\alpha_1 T}{2\cos\theta}\right) \tag{4.37}$$

In Figure 4.4, the absolute value of $\mathbf{S}(T)$ is plotted against $\alpha_1 T/\cos\theta$ for three values of α/α_1.

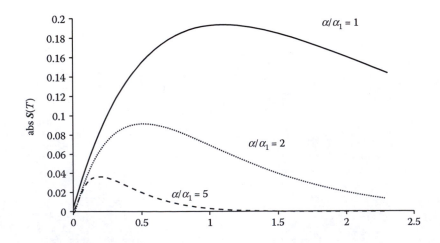

FIGURE 4.4 Amplitude of light diffracted from a pure absorption transmission grating for three values of the ratio of average absorption to absorption modulation.

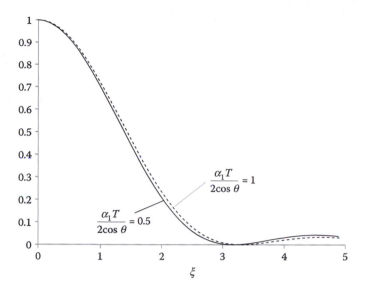

FIGURE 4.5 Normalized diffraction efficiency versus ξ, (Bragg curve) of pure amplitude holographic diffraction grating for different values of absorption modulation α_1.

While it is clear from Figure 4.4 that the diffraction efficiency increases with α_1, we obviously cannot have negative absorption. It follows that as α_1 increases, so must α, leading to reduction of the diffraction efficiency. The maximum value of α_1 is α and we can differentiate Equation 4.37 to find the value of $\alpha_1 T/\cos\theta$ (it is ln3) at which the diffraction efficiency is maximum. The value of $|\mathbf{S}(T)|_{\max}$ is $1/3\sqrt{3}$ and therefore the diffraction efficiency is 0.037, about half the value for a thin grating (see Section 3.2.1).

We can also find the transmittance of an amplitude grating whose diffraction efficiency is maximum. The amplitude transmittance, t, is simply $\exp(-\alpha T)$ and the intensity transmittance T_I is $\exp(-2\alpha T)$. From Equation 3.4,

$$D = \log(1/T_I) = \log\exp(2\alpha T) = 2\alpha T\log e = 2\ln 3\cos\theta\log e = 0.96\cos\theta \qquad (4.38)$$

and so the optimum density for an amplitude hologram should be less than 1. This is found to be the case in practice.

The normalized diffraction efficiency is

$$\frac{\eta}{\eta_0} = \frac{\sinh^2\left(\alpha'^2 - \xi^2\right)^{1/2}}{\left[1 - (\xi/\alpha')^2\right]\sinh^2(\alpha_1 T/2\cos\theta)} \qquad (4.39)$$

and is shown in Figure 4.5 for two different values of the absorption modulation α_1. (In obtaining these graphs, using a spreadsheet, care is needed in handling Equation 4.39 for the case $\xi > \alpha'$.)

4.3.2 UNSLANTED REFLECTION GRATINGS

The grating is recorded by two plane waves whose wave vectors lie in the y, z plane and make equal but opposite angles with the x, y plane as shown in Figure 4.6a.

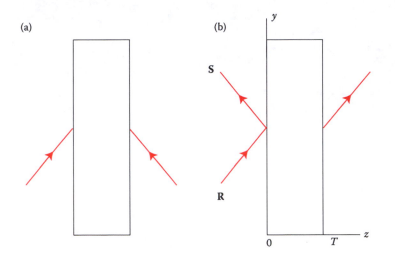

FIGURE 4.6 Reflection holographic grating: (a) recording and (b) reconstruction.

The reconstructed wavefront, **S**, begins to grow from $z = T$ (Figure 4.6b) and has zero amplitude in that plane. From Equation 4.26, we have (**R** being normalized to unity)

$$\mathbf{R}(0) = \mathbf{R}_1 + \mathbf{R}_2 = 1 \tag{4.40}$$

and from Equation 4.27

$$\mathbf{S}(T) = \mathbf{S}_1 \exp(\varsigma_1 T) + \mathbf{S}_2 \exp(\varsigma_2 T) = 0 \tag{4.41}$$

From Equation 4.21 at $z = 0$

$$\frac{\partial \mathbf{S}}{\partial z}\frac{k_{S_z}}{\beta} + (j\Gamma + \alpha)\mathbf{S} = -j\chi\mathbf{R}$$

$$-j\chi = \frac{k_{S_z}}{\beta}(\varsigma_1\mathbf{S}_1 + \varsigma_2\mathbf{S}_2) + (j\Gamma + \alpha)(\mathbf{S}_1 + \mathbf{S}_2) \tag{4.42}$$

From Equation 4.41, we can obtain

$$-\mathbf{S}_1 \exp(\varsigma_1 T) + \mathbf{S}_1 \exp(\varsigma_2 T) = (\mathbf{S}_2 + \mathbf{S}_1)\exp(\varsigma_2 T) \quad \text{and} \quad \mathbf{S}_1 = \frac{(\mathbf{S}_2 + \mathbf{S}_1)\exp(\varsigma_2 T)}{\exp(\varsigma_2 T) - \exp(\varsigma_1 T)}$$

$$\mathbf{S}_2 \exp(\varsigma_1 T) - \mathbf{S}_2 \exp(\varsigma_2 T) = (\mathbf{S}_2 + \mathbf{S}_1)\exp(\varsigma_1 T) \quad \text{and} \quad \mathbf{S}_2 = \frac{(\mathbf{S}_2 + \mathbf{S}_1)\exp(\varsigma_1 T)}{\exp(\varsigma_1 T) - \exp(\varsigma_2 T)}$$

$$\frac{k_{S_z}}{\beta}(\varsigma_1\mathbf{S}_1 + \varsigma_2\mathbf{S}_2) = \frac{k_{S_z}}{\beta}\frac{(\mathbf{S}_2 + \mathbf{S}_1)\varsigma_1 \exp(\varsigma_2 T) - \varsigma_2(\mathbf{S}_2 + \mathbf{S}_1)\exp(\varsigma_1 T)}{\exp(\varsigma_2 T) - \exp(\varsigma_1 T)} \tag{4.43}$$

and from Equation 4.42,

$$-j\chi = (\mathbf{S}_1 + \mathbf{S}_2)\left[\frac{k_{S_z}}{\beta}\frac{\varsigma_1 \exp(\varsigma_2 T) - \varsigma_2 \exp(\varsigma_1 T)}{\exp(\varsigma_2 T) - \exp(\varsigma_1 T)} + j\Gamma + \alpha\right] \tag{4.44}$$

From Equation 4.27,

$$S(0) = S_1 + S_2 = \frac{-j\chi}{\left[\dfrac{k_{S_z}}{\beta} \dfrac{\varsigma_1 \exp(\varsigma_2 T) - \varsigma_2 \exp(\varsigma_1 T)}{\exp(\varsigma_2 T) - \exp(\varsigma_1 T)} + j\Gamma + \alpha \right]} \qquad (4.45)$$

The grating vector, **K**, is normal to the surface of the hologram (Figure 4.7).

In Figure 4.7, the probe beam with wave vector $\mathbf{k}_R$ is at angle ψ to the z-axis. At Bragg incidence $\psi = \pi/2 - \theta$, where θ is the Bragg angle.

4.3.2.1 Unslanted Reflection Phase Gratings

From Figure 4.7, $k_{S_z} = \beta\cos(\pi/2 + \theta) = -\beta\sin\theta = -\beta\cos\psi$ and from Equation 4.25 with $\alpha = 0$

$$\varsigma T = -\frac{1}{2}\left(\frac{j\beta\Gamma T}{-\beta\cos\psi} \right) \pm \frac{1}{2}\left[\left(-\frac{j\beta\Gamma T}{-\beta\cos\psi} \right)^2 - \frac{4\beta T^2 \chi^2}{-\beta\cos^2\psi} \right]^{1/2}$$

$$= \frac{j\Gamma T}{2\cos\psi} \pm \left[\left(\frac{j\Gamma T}{2\cos\psi} \right)^2 + \left(\frac{T\chi}{\cos\psi} \right)^2 \right]^{1/2}$$

$$\varsigma T = j\xi' \pm \left(v'^2 - \xi'^2 \right)^{1/2} \text{ where } v' = \frac{\chi T}{\cos\psi} = \frac{\pi n_1 T}{\lambda_{air}\cos\psi} \text{ (from Equation 4.10 with } \alpha_1 = 0)$$

$$(4.46)$$

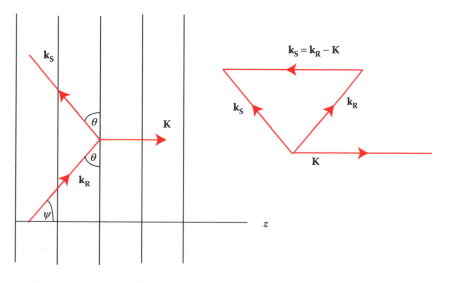

FIGURE 4.7 Reflection grating.

$$\text{and } \xi' = \frac{\Gamma T}{2\cos\psi} = \frac{\beta\delta\sin(2\theta)T}{2\cos\psi} = \delta\beta T\cos\theta, \text{ the off-Bragg parameter.} \quad (4.47)$$

$$\xi' = \frac{2\pi n_0 \partial\lambda_{\text{air}} T}{\lambda_{\text{air}}^2}\sin\theta \quad (4.48)$$

From Equation 4.44

$$S(0)$$

$$= \frac{-j\chi}{\left[\dfrac{k_{S_z}}{\beta}\dfrac{\varsigma_1\exp(\varsigma_2 T) - \varsigma_2\exp(\varsigma_1 T)}{\exp(\varsigma_2 T) - \exp(\varsigma_1 T)} + j\Gamma\right]}$$

$$= \frac{-jv'}{\left[-\dfrac{\varsigma_1 T\exp(\varsigma_2 T) - \varsigma_2 T\exp(\varsigma_1 T)}{\exp(\varsigma_2 T) - \exp(\varsigma_1 T)} + 2j\xi'\right]}$$

$$= \frac{-jv'}{\left[-\dfrac{\left[j\xi' + \left(v'^2 - \xi'^2\right)^{1/2}\right]\exp\left[j\xi' - \left(v'^2 - \xi'^2\right)^{1/2}\right] - \left[j\xi' - \left(v'^2 - \xi'^2\right)^{1/2}\right]\exp\left[j\xi' + \left(v'^2 - \xi'^2\right)^{1/2}\right]}{\exp\left[j\xi' - \left(v'^2 - \xi'^2\right)^{1/2}\right] - \exp\left[j\xi' + \left(v'^2 - \xi'^2\right)^{1/2}\right]} + 2j\xi'\right]}$$

$$= \frac{-jv'}{\dfrac{\left[j\xi' + \left(v'^2 - \xi'^2\right)^{1/2}\right]\exp\left[-\left(v'^2 - \xi'^2\right)^{1/2}\right] - \left[j\xi' - \left(v'^2 - \xi'^2\right)^{1/2}\right]\exp\left[\left(v'^2 - \xi'^2\right)^{1/2}\right]}{2\sinh\left(v'^2 - \xi'^2\right)^{1/2}} + 2j\xi'}$$

$$\mathbf{S(0)} = \frac{-j}{j\dfrac{\xi'}{v'} + \left[1 - \left(\dfrac{\xi'}{v'}\right)^2\right]^{1/2}\coth\left(v'^2 - \xi'^2\right)^{1/2}} \quad (4.49)$$

In Figure 4.8, the normalized on-Bragg diffraction efficiency, that is, with $\xi' = 0$, is plotted against the refractive index modulation parameter v'.

In Figure 4.9, the normalized diffraction efficiency is *plotted* against the off-Bragg parameter ξ' for different values ($\pi/4$, $\pi/2$, and $3\pi/4$) of the refractive index modulation parameter v'. The maximum values of diffraction efficiency are, respectively, 0.43, 0.84, and 0.96. The curve width increases with v', and it is interesting to consider the wavelength selectivity. Taking $v' = \pi/2$, we see that the normalized diffraction efficiency is zero at $\xi' = 3.5$. At $\lambda_{\text{air}} = 532$ nm, $T = 20$ μm, and $n_0 = 1.5$, and $\theta = 45°$, we find a wavelength selectivity of 17 nm. This result is of great importance as it implies that holographic images can be reconstructed using nonmonochromatic light.

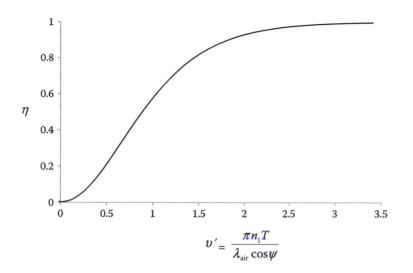

$$\upsilon' = \frac{\pi n_1 T}{\lambda_{\text{air}} \cos \psi}$$

FIGURE 4.8 Diffraction efficiency plotted against refractive index modulation parameter υ'.

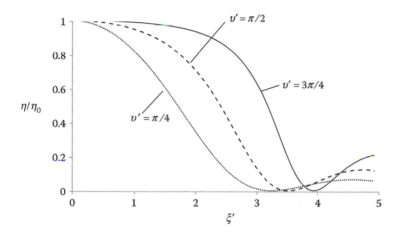

FIGURE 4.9 Volume phase reflection grating. Normalized diffraction efficiency versus off-Bragg parameter ξ' for different values of the refractive index modulation parameter υ'.

4.3.2.2 Reflection Amplitude Gratings

We assume no variation in refractive index and define

$$\upsilon'' = \frac{\alpha_1 T}{2 \sin \theta} \quad \text{and} \quad \xi'' = \frac{\alpha T}{\sin \theta} + \frac{j \Gamma T}{2 \sin \theta} \tag{4.50}$$

$$\text{while} \quad \chi = -\frac{j \alpha_1}{2} \tag{4.51}$$

Recall Equation 4.25

$$\varsigma = -\frac{1}{2} \left(\frac{\alpha}{\cos \psi} + \frac{\alpha \beta}{k_{S_z}} + \frac{j \beta \Gamma}{k_{S_z}} \right) \pm \frac{1}{2} \left[\left(\frac{\alpha}{\cos \psi} - \frac{\alpha \beta}{k_{S_z}} - \frac{j \beta \Gamma}{k_{S_z}} \right)^2 - \frac{4 \beta \chi^2}{k_{S_z} \cos \psi} \right]^{1/2}$$

From Figure 4.7 $k_{S_z} = -\beta\cos\psi = -\beta\sin\theta$

$$\varsigma T = \frac{j\Gamma T}{2\cos\psi} \pm \left(\xi''^2 - v''^2\right)^{1/2} \tag{4.52}$$

From Equations 4.45 and 4.49

$$\mathbf{S}(0) = \mathbf{S}_1 + \mathbf{S}_2 = \frac{-j\chi}{\left[\dfrac{k_{S_z}}{\beta}\dfrac{\varsigma_1\exp(\varsigma_2 T)-\varsigma_2\exp(\varsigma_1 T)}{\exp(\varsigma_2 T)-\exp(\varsigma_1 T)} + j\Gamma+\alpha\right]}$$

$$\mathbf{S}(0) = \frac{-1}{\xi''/v'' + [(\xi''/v'')^2 -1]^{1/2}\coth(\xi''^2 - v''^2)^{1/2}} \tag{4.53}$$

At Bragg incidence, $\Gamma = 0$ and $\xi''/v'' = 2\alpha/\alpha_1$

$$\mathbf{S}(0) = \frac{-1}{\dfrac{2\alpha}{\alpha_1} + \left[\left(\dfrac{2\alpha}{\alpha_1}\right)^2 -1\right]^{\frac{1}{2}}\coth\dfrac{T}{\sin\theta}\left(\alpha^2 - \dfrac{\alpha_1^2}{4}\right)^{\frac{1}{2}}}$$

If $\alpha = \alpha_1$,

$$\mathbf{S}(0) = \frac{-1}{2+3^{1/2}\coth\left[\dfrac{T}{\sin\theta}\left(\dfrac{3\alpha^2}{4}\right)^{1/2}\right]} \rightarrow \frac{-1}{2+\sqrt{3}} \quad \text{as} \quad \alpha \rightarrow \infty$$

giving a value of 0.072 for the diffraction efficiency. This result shows that we should record and (if required) process a thick amplitude reflection grating so that its absorption, α, is as high as possible, in order to obtain maximum diffraction efficiency. In Figure 4.10,

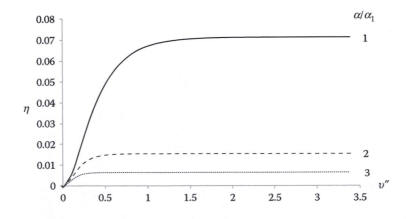

FIGURE 4.10 Diffraction efficiency of an amplitude reflection grating plotted against absorption modulation parameter $v'' = \alpha_1 T/2\sin\theta$ for various values of α/α_1.

the on-Bragg diffraction efficiency is plotted against absorption modulation parameter, or exposure if absorption is proportional to exposure, for various values of the ratio α/α_1. It is seen that a maximum value is obtained when $\nu'' = \alpha_1 T / 2\sin\theta = 1$ and from Equation 4.37 $D = \log(1/T_I) = \log\exp(2\alpha T)$ so that $D \cong 1.7$.

In Figure 4.11, the normalized diffraction efficiency is plotted against the off-Bragg parameter for different values of $\alpha_1 T / 2\sin\theta$.

Kogelnik [1] also considers an incident wave, which is plane polarized in a direction that is not parallel to the x-axis and finds that it is necessary to multiply the coupling constant χ by the cosine of the angle between the incident electric field direction and the x-axis.

4.4 RIGOROUS COUPLED-WAVE THEORY

Kogelnik's coupled-wave theory is borne out by experiment to a great extent despite the assumptions and approximations that have been made. These are

1. The second-order spatial derivatives of wave amplitudes are negligible implying that neither the probe beam nor the reconstructed beam changes rapidly with distance as they propagate through the medium.
2. Boundary diffraction effects are ignored. This means that the grating presents an abrupt or a discontinuous change in dielectric properties. The effect is that some light is diffracted back into the medium to the left of the grating.
3. Only one diffracted beam of light propagates in the grating.

These issues have been addressed by Moharam and Gaylord [3]. Their more rigorous approach involves solving the wave equation in the two media on either side of the grating, whose dielectric constants may differ and in the grating itself. Following this, one uses the fact that the electric and magnetic fields must be continuous across each of the

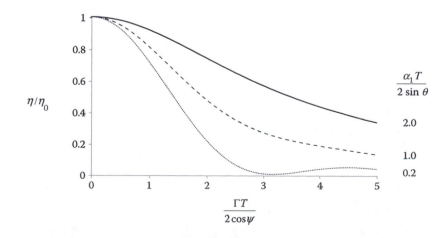

FIGURE 4.11 Absorption reflection grating: normalized diffraction efficiency plotted against the off-Bragg parameter for various absorption modulation values.

two boundaries. If for example the electric field were discontinuous, this would imply that $\nabla \cdot E$ is infinite, which in turn requires the, impossible, localization of an infinite quantity of electric charge (see also Section 1.7.1).

The most important results are

1. A reduction in the diffraction efficiency of a *slanted* pure phase reflection grating by typically 0.3
2. Reduced +1 order diffraction efficiency (by ~0.1–0.15) and the existence of –1 and +2 order diffracted beams with diffraction efficiency of up to 0.2 in the case of a pure, slanted, or unslanted phase transmission grating

The diffraction efficiency of an unslanted, pure phase, reflection grating is unaltered by use of the rigorous approach.

4.5 A SIMPLER APPROACH

The properties of thick transmission phase gratings have been analyzed by Ludman [4] using basic principles in physical optics. The grating bandwidth is obtained from considerations of path length differences between light reflected from opposite ends of fringe planes, while diffraction efficiency is obtained using the Fresnel equations (Section 1.7.2).

4.5.1 BANDWIDTH

Consider an unslanted, pure phase grating of thickness T and spacing d, as shown in Figure 4.12.

For constructive interference to take place between light waves, which are diffracted at angle θ from *adjacent* fringes, the diffraction grating equation (Equation 2.9) must be satisfied. In this case, we have

$$nd(\sin \theta_{\text{inc}} + \sin \theta) = \lambda \tag{4.54}$$

where n is the average refractive index of the holographic grating and θ_{inc} the angle at which light is incident at the fringes.

Equation 4.54 implies that rays A, B, and C with wavelength λ are all in phase while A', B', and C' at wavelength $\lambda + \Delta\lambda$ are all in phase.

Now, light rays incident at the *same angle* at different points on the *same* fringe are reflected in phase with one another so that

$$\theta_{\text{inc}} = \theta \tag{4.55}$$

This means that B and C are in phase.

Equations 4.54 and 4.55 can only be satisfied at a particular angle of incidence by light of only one wavelength λ and the *spectral bandwidth is the change* $\Delta\lambda$ *in wavelength, which causes* B' *and* C' *to be exactly out of phase*.

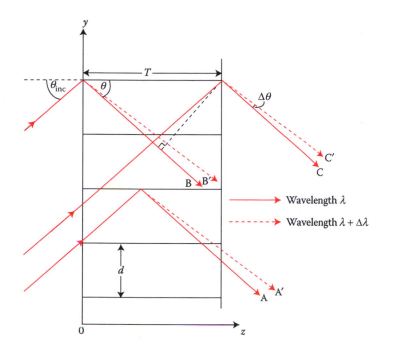

FIGURE 4.12 Spectral bandwidth of a thick grating. (Reproduced from J. E. Ludman, Approximate bandwidth and diffraction efficiency in thick holograms, *Am. J. Phys.,* 50, 244–46, 1982. © American Association of Physics Teachers. With permission.)

For A′, B′, and C′ to be in phase with one another, the condition imposed by Equation 4.54 must be strictly met because the grating thickness (~tens of microns) is normally much less than the width of the illuminating beam (~ millimeters).

Differentiating Equation 4.54 allows us to find the change in angle of the diffracted light corresponding to a change in its wavelength.

$$\Delta\lambda = nd\cos\theta\ \Delta\theta \qquad (4.56)$$

The path difference, *l*, between in-phase rays B and C, from the front and back surfaces of the grating, is

$$l = T\cos\theta \qquad (4.57)$$

and as the angle θ changes by $\Delta\theta$, the path difference, between B′ and C′ changes by

$$\Delta l = T\sin\theta\ \Delta\theta \qquad (4.58)$$

If the optical path difference between these rays changes by λ across the fringe, then there is very little diffracted light because every part of the wavefront diffracted from the fringe between $z = 0$ and $z = T/2$ interferes destructively with a corresponding part diffracted from the fringe between $z = T/2$ and $z = T$. Setting

$$n\Delta l = \lambda \qquad (4.59)$$

and combining Equations 4.56, 4.58, and 4.59

$$\frac{\Delta\lambda}{\lambda} = \frac{d\cot\theta}{T}$$

(4.60)

which is Equation 4.36 for the spectral selectivity of a phase transmission grating.

If the grating is slanted by angle ϕ from the normal, we replace T in Equation 4.60 by $T/\cos\phi$, the effective length of the fringes and

$$\frac{\Delta\lambda}{\lambda} = \frac{d\cot\theta\cos\phi}{T}$$

(4.61)

From Equations 4.56 and 4.54, we obtain

$$\frac{\Delta\theta}{\Delta\lambda} = \frac{1}{nd\cos\theta} = \frac{2\tan\theta}{\lambda}$$

(4.62)

and the angular bandwidth from Equations 4.60 and 4.62 is

$$\Delta\theta = \frac{2d}{T}$$

(4.63)

which may be compared with Equation 4.35.

In the slanted case, Equations 4.61 and 4.62 are combined, to obtain

$$\Delta\theta = \frac{2d\cos\phi}{T}$$

(4.64)

4.5.2 DIFFRACTION EFFICIENCY

The refractive index of the grating n varies according to

$$n(y) = n_0 + n_1\cos(2\pi y/d)$$

(4.65)

where n_1 is the amplitude of the refractive index modulation and n ranges from $n_0 - n_1$ to $n_0 + n_1$. In order to determine the diffraction efficiency of the grating, we can use the Fresnel equations if we assume that, at each fringe, there is an *abrupt* change in refractive index from n (= $n_0 - n_1$) to $n + 2n_1$. In Figure 4.13, $\theta_{inc} = \pi/2 - \theta_i$ and we assume n_1 is small so that $\theta_t = \theta_i - \Delta\theta$ and $\cos\theta_t = \sin(\theta_{inc} + \Delta\theta)$. Applying Equation 1.28

$$\left.\frac{E_{0r}}{E_{0i}}\right|_\perp = \frac{n_i\cos\theta_i - n_t\cos\theta_t}{n_i\cos\theta_i + n_t\cos\theta_t} = \frac{n\sin\theta_{inc} - (n+2n_1)\sin(\theta_{inc}+\Delta\theta)}{n\sin\theta_{inc} + (n+2n_1)\sin(\theta_{inc}+\Delta\theta)}$$

$$= \frac{n\sin\theta_{inc} - (n+2n_1)(\sin\theta_{inc} + \cos\theta_{inc}\Delta\theta)}{n\sin\theta_{inc} + (n+2n_1)(\sin\theta_{inc} + \cos\theta_{inc}\Delta\theta)}$$

$$\left.\frac{E_{0r}}{E_{0i}}\right|_\perp = \frac{-(n\cos\theta_{inc}\Delta\theta + 2n_1\sin\theta_{inc})}{2n\sin\theta_{inc}}$$

(4.66)

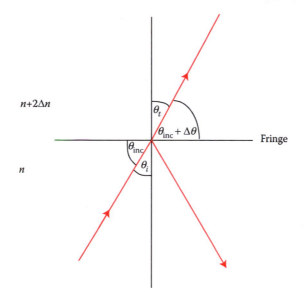

FIGURE 4.13 Angle relationships.

Applying Snell's law

$$n\cos\theta_{\text{inc}} = (n+2n_1)\cos(\theta_{\text{inc}}+\Delta\theta) \cong (n+2n_1)(\cos\theta_{\text{inc}}-\sin\theta_{\text{inc}}\Delta\theta)$$
$$2n_1\cos\theta_{\text{inc}} = n\sin\theta_{\text{inc}}\Delta\theta \tag{4.67}$$

we obtain from Equation 4.66

$$\left.\frac{\text{E}_{0r}}{\text{E}_{0i}}\right|_{\perp} = -\frac{n_1}{n\sin^2\theta_{\text{inc}}} \tag{4.68}$$

From Figure 4.14, the number of full fringes contributing to the reconstructed wave, S, is $T\tan\theta/d$ and the total electric field amplitude in the S wave is

$$\left.\text{E}_{0r}\right|_{\perp\text{total}} = -\text{E}_{0i}\big|_{\perp}\frac{T\tan\theta_{\text{inc}}}{d}\frac{n_1}{n\sin^2\theta_{\text{inc}}} \tag{4.69}$$

From Equation 4.54 with $\theta=\theta_{\text{inc}}$, we obtain

$$\left.\frac{\text{E}_{0r}}{\text{E}_{0i}}\right|_{\perp\text{total}} = -\frac{2T}{\lambda}\frac{n_1}{\cos\theta} \tag{4.70}$$

And the diffraction efficiency is given by

$$\eta=\left(\frac{2T}{\lambda}\frac{n_1}{\cos\theta}\right)^2 \tag{4.71}$$

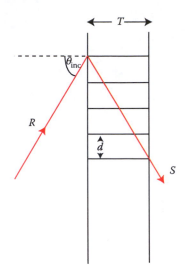

FIGURE 4.14 Number of fringes contributing to *S*.

Compare Equation 4.33 $\eta = \sin^2(\xi^2 + \upsilon^2)^{1/2}/(1 + \xi^2/\upsilon^2)$ with $\xi = 0$ and a small value of $\upsilon = \pi n_1 T / \lambda_{air} \cos\theta$ from which we obtain the very similar expression for diffraction efficiency

$$\eta = \left(\frac{\pi n_1 T}{\lambda_{air}\cos\theta}\right)^2 \tag{4.72}$$

The effect of grating slant angle ϕ on the diffraction efficiency can be included.

The number of whole fringes that contribute to the diffracted beam is $T\sin\theta/\cos(\theta + \phi)$ and Equation 4.66 is modified accordingly

$$\mathbf{E}_{0r}\big|_{\perp total} = -\mathbf{E}_{0i}\big|_{\perp}\frac{T\sin\theta}{d\cos(\theta+\phi)}\frac{n_1}{n\sin^2\theta_{inc}}$$

and from Equation 4.54 with $\theta = \theta_{inc}$

$$\mathbf{E}_{0r}\big|_{\perp total} = -\mathbf{E}_{0i}\big|_{\perp}\frac{2n_1 T}{\lambda\cos(\theta+\phi)} \tag{4.73}$$

The input beam width PQ and the diffracted beam width RS (Figure 4.15b) are related by

$$\frac{PQ}{RS} = \frac{\cos(\theta+\phi)}{\cos(\theta-\phi)} \tag{4.74}$$

and the output intensity is increased by this factor. Thus, the diffraction efficiency is given by

$$\eta = \left\{\frac{2n_1 T}{\lambda\left[\cos(\theta-\phi)\cos(\theta+\phi)\right]^{\frac{1}{2}}}\right\}^2 \tag{4.75}$$

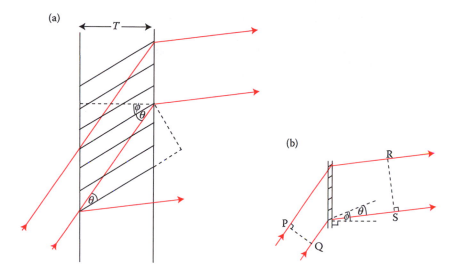

FIGURE 4.15 Effect of slant: (a) effective number of contributing fringes; (b) change in beam width. (Reproduced from J. E. Ludman, Approximate bandwidth and diffraction efficiency in thick holograms, *Am. J. Phys.*, 50, 244–46, 1982. © American Association of Physics Teachers. With permission.)

For comparison, recall Equation 4.33, obtained from the coupled-wave theory $\eta = \sin^2(\xi^2 + \upsilon^2)^{1/2}(1 + \xi^2/\upsilon^2)$ with $\upsilon = (\pi n_1 T/\lambda_{air})$ $(\beta/(k_{s_z}\cos\psi))^{1/2}$, which, with small n_1, gives the on-Bragg ($\xi = 0$) diffraction efficiency

$$\eta = \left(\frac{\pi n_1 T}{\lambda_{air}\cos(\theta+\phi)\cos(\theta-\phi)}\right)^2 \tag{4.76}$$

4.6 SUMMARY

This chapter deals with the coupled-wave theory of thick plane holographic gratings, including pure phase and pure amplitude, transmission, and reflection gratings. We derive the diffraction efficiencies of these gratings and consider the implications of some of the results. The main results emerging from rigorous coupled-wave theory are also briefly outlined.

We also present an alternative and rather simpler approach developed by Ludman [4] using basic principles of physical optics, to determine the most important results for pure phase transmission gratings.

REFERENCES

1. H. Kogelnik, "Coupled wave theory for thick holographic gratings," *Bell System Tech. J.*, 48, 2909–2947 (1969).
2. R. J. Collier, C. Burckhardt, and L. H. Lin, *Optical Holography*, Academic Press, New York (1971).

3. M. G. Moharam and T. K. Gaylord, "Rigorous coupled-wave analysis of planar-grating diffraction," *J. Opt. Soc. Am.*, 71, 811–818 (1981).
4. J. E. Ludman, "Approximate bandwidth and diffraction efficiency in thick holograms," *Am. J. Phys.*, 50, 244–246 (1982).

PROBLEMS

1. Prove Equation 4.3.
2. Following Equation 4.33, it is seen that for maximum diffraction efficiency, the grating should satisfy the condition

$$n_1 T \cong \frac{\lambda_{air}}{2}.$$

   This result was arrived at after quite a lot of mathematical argument but, on physical grounds, it is not a surprising one. Why is this?

3. What should be the thickness of an unslanted, transmission, phase, volume grating of spatial frequency 1500 mm⁻¹ in order that its spectral bandwidth be 0.1 nm? The recording laser has wavelength 633 nm and the refractive index of the recording medium is 1.5. What is the angular width of its Bragg selectivity curve?

4. An unslanted reflection grating is recorded in a photosensitive layer of thickness 10 μm and refractive index 1.5, using counter-propagating beams from a helium–neon laser of wavelength 633 nm. What is its spectral bandwidth and what is its angular selectivity?

5. For maximum reconstructed intensity over the widest possible range of the off-Bragg parameter ξ', we set value of the refractive index modulation parameter v' at π. Find the spectral bandwidth of a grating recorded as in problem 4. The result is not very different from that in problem 4, so what is the advantage of using a high value of v'?

6. An unslanted, thick, transmission, phase grating is recorded in a photosensitive medium of refractive index 1.52 using collimated laser beams of wavelength 532 nm. When illuminated by collimated light from the *side* of the grating, it is to have a high diffraction efficiency at 1650 nm. What should be the angle between the recording beams? Assume the average refractive index of the recording medium is the same at the two wavelengths.

7. An unslanted phase reflection grating is to be recorded in a layer of photosensitive material using a laser of wavelength 473 nm. If the maximum refractive index modulation that can be obtained is 0.005, what should be the thickness of the layer in order to obtain 100% diffraction efficiency?

8. Find an expression for the diffraction efficiency of a pure reflection grating using Ludman's approach and show, for small values of refractive index modulation, that it is approximately the same as predicted by the coupled-wave theory.

III

Holography in Practice

CHAPTER 5

Requirements for Holography

5.1 INTRODUCTION

In this chapter, we discuss the basic requirements for the successful practice of holography. We have seen in Chapter 3 that in recording a hologram, we are in fact recording an interference pattern so we need to ensure that the interference pattern remains stable for the duration of the recording.

The stability of the interference pattern imposes three requirements in practice. These are the *mechanical* and *thermal stability* of the interferometer used in making the holographic recording and the *coherence* of the light source.

In addition, the *spatial frequencies* of the interference patterns are normally very high and the recording medium must be capable of coping with them.

We will deal with the coherence requirement first.

5.2 COHERENCE

It would be nice if we could use normal daylight or even conventional artificial light sources for holographic recording, as we do in photography, but unfortunately, this is not possible. To see why, recall that a hologram is a recording of the interference pattern that is formed in space when two or more beams of light interfere with one another (see Section 3.5.1). The stability of the pattern can only persist for the coherence time of the light source used in recording it and if we use a white light source as we have seen in Section 2.8.2, that time is about 3×10^{-15} s. This means that the pattern will change randomly on this time scale. Putting it another way, for the pattern to be stable, the optical path difference between the (white) light beams used to record the pattern has to be less than 1 μm, the coherence length or typical length of wavetrains from a white light source.

In addition to the limitation imposed by the coherence time of the light source, a further limitation is imposed by its spatial coherence. In practice, as we saw in Section 2.8.3, this means that the source has to be of very small dimensions.

In early holography experiments, as explained in Section 3.4.3, a mercury lamp was usually employed, with a filter to isolate the green line in the mercury spectrum having a wavelength of 546 nm. Such a source has a coherence length of a few millimeters, still very short but certainly much longer than that of white light. This meant that the maximum path difference there could be in an interferometer using such a source is just a few millimeters and therefore the dimensions of any object of which a hologram was to be recorded had to be very small. Gabor used thin transparencies with text. Wires were also used in early holography. All this changed quite dramatically when the first lasers became available because they had much longer coherence lengths and could also be focused to a very small source size. Thus, the requirements of temporal and spatial coherence could be met.

Interference patterns may be observed and their stability checked using a special kind of interferometer originally developed by Michelson. We will devote some attention to the Michelson interferometer because it is a very useful practical tool in holography and is particularly useful in diagnosing problems in holographic setups.

5.3 THE MICHELSON INTERFEROMETER

The Michelson interferometer is shown schematically in Figure 5.1 The light source is an extended one, which means it is not spatially coherent but it should be fairly monochromatic, for example a mercury spectrum lamp with a green filter.

A light wave from the source is split into two parts by a beam splitter (a glass plate with a thin metallic coating on its back surface) and the two resulting waves are sent off different directions to plane mirrors. One of the mirrors M_1 can be adjusted around two orthogonal axes so that its reflecting surface is precisely orthogonal to the reflecting surface of the other, M_2, which can be translated by a micrometer screw in a direction normal to its surface. The beam splitter is fixed at 45° to M_2. The two waves are reflected so as to retrace their original paths and are recombined at the beam splitter. Obviously, light is wasted at

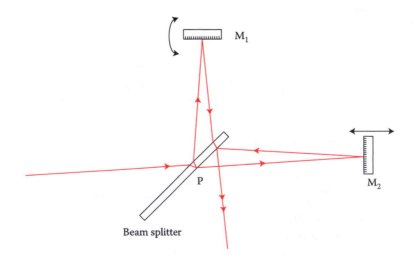

FIGURE 5.1 The Michelson interferometer (zero path difference).

the recombination stage, the wasted light going back toward the source but this is not usually a matter of concern. A glass plate, not shown, of the same thickness and type of glass as the beam splitter, but without the semireflecting coating, is inserted in the path of one of the waves and set parallel to the beam splitter to compensate for the fact that the other wave passes though the beam splitter. This ensures that any *optical path difference* is due only to paths traveled *outside* the beam splitter and compensator glass. In Figure 5.1, the compensator would be positioned in the path of the light going to M_2 because the wave going to M_1 passes through the beam splitter after the split has occurred and *again* before recombining with the wave going to M_2. In Figure 5.1, M_2 has been positioned so that the path length difference is exactly zero and M_2 is precisely orthogonal to M_1, which is why the waves that split apart at point P recombine on the same path after returning to the beam splitter.

In Figure 5.2, which we use to find the path difference, the beam splitter is represented as a thin plane.

We can set up a simple ray geometry by considering the *reflection* in the beam splitter, of the ray going to M_1. This ray is reflected from the image M_1' of M_1 in the beam splitter. Then, the path difference, $2d \cos \theta$, is simply found, knowing the angle of incidence, θ, at M_2 and the distance, d, between M_1' and M_2. When the light rays recombine at the beam splitter, they are parallel (because M_1 and M_2 are orthogonal) and therefore converge to a point in the focal plane of the lens.

It is important to note that Figure 5.2 is simply a 2-D diagram and that there are many rays of light from the source arriving at the beam splitter with the same angle of incidence θ and all lying on the surface of a cone of semiangle θ. So if there is constructive interference

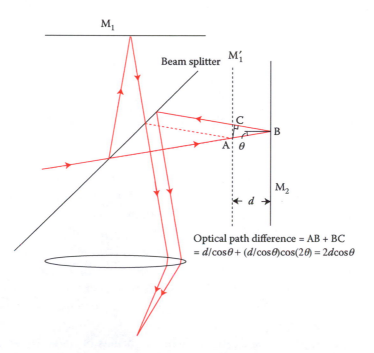

FIGURE 5.2 Path difference in the Michelson interferometer.

at the point where the rays in Figure 5.2 meet, then there must be constructive interference everywhere on a circle of radius $f\tan\theta$, where f is the focal length of the lens. Then, since θ varies, the interference pattern consists of a set of concentric, circular, alternatively bright and dark fringes. The fact that the fringes are concentric circles is an indication that the mirrors are orthogonal. Because the mirror M_1 is not at first orthogonal to M_2, we have to adjust its orientation in its mounting using fine-threaded screws, viewing the light by eye until we see some fringes usually in the form of partial circles of large radius or straight lines. The reader might consider what is the relative orientation of the mirrors in the latter case. Careful adjustment will make the fringes form complete concentric circles. Finally, we view the fringes through a small telescope, which is first focused on a distant object so it is set for viewing parallel light. The fringes should be in focus everywhere in the field of view.

It is difficult to overstate the importance of the Michelson interferometer, but it is not difficult to understand why. Imagine that we have obtained the circular fringe pattern and that the central fringe ($\theta = 0$) is a bright one. As M_2 is translated parallel to itself, this fringe becomes dark when the distance moved is one quarter of a wavelength. Thus, *mechanical displacement* can now be measured using the *wavelength of light* as our yardstick or scale of measurement. As well as in precision metrology, the interferometer is used in spectroscopy using the converse of the measurement principle just described, namely that if we count the number of peaks of intensity that appear (or disappear) in the center of the pattern as M_2 is translated through a known distance, we can calculate the wavelength of monochromatic light to an accuracy determined by the number of fringes counted. It is a tedious task but can be automated using photoelectric detection. The interferometer is also used for measuring the refractive indices of gases, liquids, and transparent solids, again by fringe counting and a variation, the Twyman–Green interferometer is used for precision testing of optical components. It was used by Michelson and Morley in their celebrated, failed attempt to confirm the existence of the so-called lumeniferous or light-bearing ether, which physicists, including Maxwell, believed was needed to support the propagation of electromagnetic waves. It was the failure of the experiment and several later refinements to detect any variation in the velocity of light relative to the interferometer that cast serious doubt on the need for the ether in the first place and led ultimately to the development of the Special Theory of Relativity by Einstein.

In practice we use the Michelson interferometer to measure the coherence length of any light source, which we intend to use for recording holograms, by measuring the change in visibility, or contrast, of the fringe pattern as the path difference in the interferometer is increased from zero. The contrast is maximum at zero path difference and becomes zero when the path difference equals the coherence length of the source.

An interference pattern can be obtained even with a white light source if one adjusts the path length difference to less than about 1 μm when a series of colored fringes is observed, each being due to constructive interference of light of that particular color. In fact the path length difference can be made practically zero, and a black fringe is seen because the path length difference for all the wavelengths in white light has been set to zero. (The reader might like to consider why a white fringe is not seen.)

This capability of obtaining zero path difference is the basis of a precise method of measuring the distance between two points. Say for example the mirror M_1 is further displaced from the beam splitter by the unknown length of a transparent or opaque gauge block. The colored fringes are immediately lost but may be restored by increasing the path to M_2 by the same length measured on the micrometer gauge. In the same way, if a transparent slab of previously measured thickness, t, and refractive index, n, is placed *between* the beam splitter and M_1 the optical path to M_2 is lengthened by $t(n-1)$ to restore the fringes and hence n can be measured.

Modern optical profilometers use this basic principle of white light interferometry to determine local optical thickness and hence complete surface profiles with nanometric precision.

5.4 LASERS

The foundations of laser theory were laid in the early years of the twentieth century by Planck and Einstein, but it was not until the late 1950s that the first visible lasers were developed. Essentially, a laser can be considered the optical equivalent of the electronic oscillator. The word laser is an acronym derived from the words *l*ight *a*mplification by *s*timulated *e*mission of *r*adiation. The device that we normally call a laser is not an amplifier but an optical oscillator and is the optical counterpart of an electronic oscillator. An optical oscillator needs to have gain provided by an optical amplifier and it needs to have positive feedback provided by a type of interferometer called a Fabry–Perot interferometer, which we will discuss first.

5.5 THE FABRY–PEROT INTERFEROMETER, ETALON, AND CAVITY

The Fabry–Perot interferometer or cavity was and still is used as a very high-resolution spectroscopic tool. It consists of two flat partially reflecting plates set parallel to one another and separated by a distance, l, which may be mechanically altered. In etalon form, the plates are held at a fixed distance apart by a cylindrical hollow cylinder of length l usually made from fused quartz or it may simply consist of a transparent, cylindrical solid of length, l, with parallel, partially reflecting end faces, called an etalon. We will not deal with the theory of Fabry–Perot devices as spectroscopic tools but only concern ourselves with their function as laser cavities.

Such a cavity is shown in Figure 5.3.

It can support electromagnetic radiation in the form of standing waves at certain wavelengths, λ, given by the condition

$$n\lambda = 2l \tag{5.1}$$

where n is any integer and l is the length of the cavity. This is because there has to be a node at each of the cavity boundaries. Standing waves obeying Equation 5.1 are called

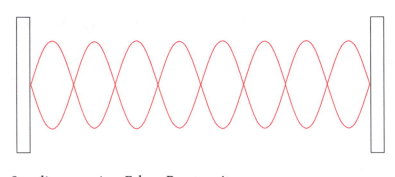

FIGURE 5.3 Standing wave in a Fabry–Perot cavity.

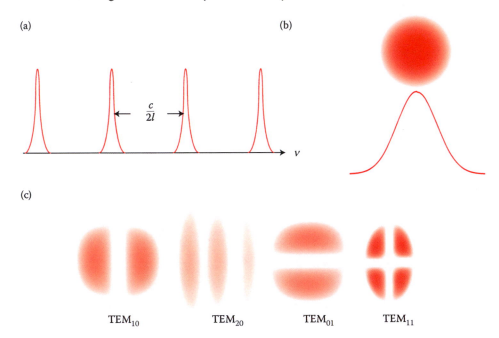

(a)

(b)

$$\frac{c}{2l}$$

v

(c)

TEM$_{10}$ TEM$_{20}$ TEM$_{01}$ TEM$_{11}$

FIGURE 5.4 Modes of a Fabry–Perot cavity: (a) longitudinal mode spacing; (b) intensity profile of longitudinal mode; and (c) transverse mode intensity profiles.

longitudinal cavity modes and each of them has frequency bandwidth Δv_{FP} (Figure 5.4) given by

$$\Delta v_{FP} = \frac{c}{\pi l \sqrt{F}} \tag{5.2}$$

where F, known as the *finesse* of the cavity, is given by $F = 4R/(1 - R)$ with R the reflectivity of the cavity mirrors. Thus, the higher the reflectivity, the smaller is the mode bandwidth. The mirrors of the cavity can be fabricated with any desired reflectivity at any wavelength, by means of multilayer coatings of different dielectric materials.

These modes have circular Gaussian intensity profiles that are symmetrical about the principal axis of the cavity (Figure 5.4b) and are known as longitudinal transverse electromagnetic (TEM) because the fields are normal to the principal axis of the cavity. A set of suffixes is added so that the longitudinal modes are designated $TEM_{0,0,n}$. This is to distinguish them from other *off-axis* modes, which may also be sustained in the cavity and are designated $TEM_{l,m,n}$, the integer numbers l and m referring to the number of nodes in the intensity distribution in the horizontal and vertical directions respectively. Examples are shown in Figure 5.4. Such modes have an adverse effect on spatial coherence because they effectively increase the physical size of the source (see Section 2.8.3). However, since they generally require larger apertures to enable them to circulate in the cavity, they can be eliminated by inserting an aperture of appropriately restricted diameter.

There are obviously many values of n that satisfy Equation 5.1 corresponding to wavelengths throughout the electromagnetic spectrum. The minimum difference in wavelength between nearest modes, $\Delta\lambda$, is given by

$$(n-1)(\lambda+\Delta\lambda) = 2l \quad \text{that is,} \quad \Delta\lambda = \frac{\lambda^2}{2l} \tag{5.3}$$

if n is a large number, which it is when we are dealing with visible wavelengths and cavity lengths of more than a few millimeters.

The difference in *frequency* between adjacent modes, known as the *free spectral range*, is

$$\Delta v_{\text{FSR}} = \frac{c}{2l} \tag{5.4}$$

where $c = v\lambda$ is the velocity of light in the cavity.

5.6 STIMULATED EMISSION AND THE OPTICAL AMPLIFIER

In Section 2.8.1, we discussed briefly how light is produced from atoms. Our model of the atom has historically gone through a number of radical transformations to the present form of a positively charged nucleus surrounded by diffuse distributions of negative charge or electrons. Generally, the atoms are in thermal equilibrium with their environment. If an atom acquires energy by electrical heating or electron bombardment, it enters an excited state, that is, one of higher energy. It then returns *spontaneously* to its original state emitting the energy that it had acquired in the form of an electromagnetic wave.

The emission from any one atom is independent of that from any other because the processes involved are all random and each light wave is emitted in a random direction.

However, Einstein showed that if an atom were *already in* an excited state, then a light wave with energy equal to the *difference* in energy between that state and one of lower energy can cause the atom to revert to the lower state in a process called *stimulated emission*. More importantly, the stimulated wave and the wave causing its emission are *in phase* with one another, have the same polarization state, and travel in precisely the *same direction*.

The process of stimulated emission is not readily established because ordinarily the atoms are in the lower energy state and a photon having energy equal to the difference in energy between the two states will normally be *absorbed* by one of the atoms.

However, if at the outset, we can arrange for a collection of atoms to be in the excited state, a single photon having the correct energy causes all the atoms to revert to the lower state. The *intensity* of the resulting light beam is *amplified* by a factor equal to the number of photons or light waves that are produced in this way. To make this work in any given atomic species, we have to look for a higher energy state that is comparatively long lived so we can populate that state with many atoms. If the higher energy state is populated by more atoms than is the lower energy state, we have *population inversion* and the process of achieving this is called *pumping*.

5.7 LASER SYSTEMS

The design of many laser systems was facilitated by the vast database of spectral data already gathered from spectroscopic studies often using high-quality diffraction gratings for precise measurements of the wavelength of spectral lines, in other words, the energy differences between different energy states of atoms and molecules.

We are concerned only with those lasers that are important for holography and holographic applications.

5.7.1 GAS LASERS

5.7.1.1 The Helium–Neon Laser

Population inversion is usually achieved by a round about route as for example in the helium–neon gas laser. We will discuss this laser in a little detail because almost 50 years after its introduction, it is still widely used in holography and many of its principles of operation apply to other lasers.

The spectra of helium and neon were well documented and as shown in Figure 5.5, some energy states, or levels, of helium coincide with some of the energy states of neon. Energy levels in atomic spectra are assigned letter and number codes such as 3p, 2s, 2^3S.

The energy scale in Figure 5.5 is the vertical axis and is measured in electron volts (eV), an electron volt being the additional energy acquired by an electron when it is accelerated through a potential difference of 1 V. Helium atoms are excited by electron bombardment into higher energy states from which they lose some energy to end up in one of the two upper energy states 2^1S and 2^3S shown. It is the helium atoms that are favored in the excitation process simply because we ensure that they considerably outnumber the neon atoms in the gaseous mixture. The ratio of helium and heon partial pressures is typically 8:1. The 2^1S and 2^3S states are long lived and when helium and heon atoms collide, energy is transferred to the energy coincident states of the latter. Thus, a state of population inversion is created in neon and is sustained so long as the helium atoms continue to be bombarded by electrons of sufficient energy. The transitions colored red in Figure 5.5 involve the emission of electromagnetic waves by stimulated emission. All other emission transitions shown are spontaneous. Although laser operation was first obtained at 1152 nm, for holography,

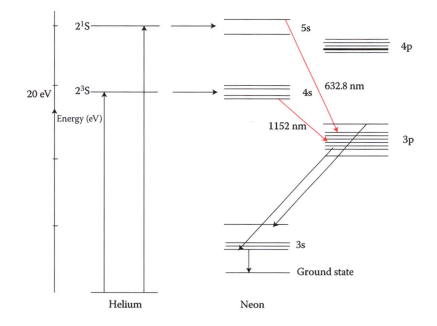

FIGURE 5.5 Energy levels of helium and neon.

the transition at 632.8 nm is the important one as this wavelength is probably the most popular, certainly for display holography. The neon atoms, which emit light at 632.8 nm, go to the relatively unpopulated 3p energy states and quickly lose energy to go further down to the 3s state. This last is finally depopulated to the ground state by collision of the neon atoms with the cylindrical glass wall of the laser tube. So, this rather roundabout process enables the creation of a state of population inversion between the 5s and the 3p energy states. Since its introduction, the helium–neon laser has been successfully operated at a number of other wavelengths in the near infrared as well as the visible, including green and yellow.

Now, we can see how optical amplification is achieved but, as already mentioned, we also need positive feedback, in order to sustain oscillation.

So, we place a laser amplifier in the form of a long cylindrical glass tube containing a mixture of helium and neon in a Fabry–Perot cavity whose mirror reflectivities are designed for the 632.8 nm transition in neon. Electrodes are located near the ends of the tube and a high voltage of about 1.5 kV applied between them to drive an electrical current through the gas and ensure that the helium atoms are sufficiently energized. The bandwidth, or spectral linewidth, of the 632.8 nm transition in neon is about 1.5 GHz. If the cavity is about 30 cm long, the free spectral range is 0.5 GHz. (The velocity of light in the cavity is still assumed to be c since the refractive index of the gas is practically 1.0.) This means that at least two longitudinal modes of the cavity can be sustained within the 632.8 nm transition (Figure 5.6).

It is worth estimating the coherence length, l, of a helium–neon laser operating in two adjacent modes of equal strength. Say the modes have frequency v and $v + \Delta v_{\text{FSR}}$. If the path

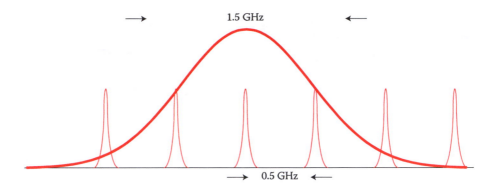

FIGURE 5.6 Longitudinal modes within the gain curve of the He–Ne amplifier.

difference in an interferometer using this laser is set equal to the coherence length, l_c, no fringes are observed and this happens because the light waves from the two modes are out of phase. This condition is simply expressed by

$$\frac{2\pi\left(\nu+\Delta\nu_{\mathrm{FSR}}\right)}{c}l_c - \frac{2\pi\nu}{c}l_c = \pi \quad \text{that is} \quad l_c = \frac{c}{2\Delta\nu_{\mathrm{FSR}}} = l \tag{5.5}$$

In the present case, l is 30 cm. If we wanted the laser to operate in just one mode, we would need a cavity with a free spectral range of 1.5 GHz, that is, one with a length of only 10 cm but the gain would not be very great because of the limited amount of gas we could have in such a short tube.

However, the two modes will *again* be fully in phase when the path difference l in the interferometer given by

$$\frac{2\pi\left(\nu+\Delta\nu_{\mathrm{FSR}}\right)}{c}l - \frac{2\pi\nu}{c}l = 2\pi$$

is twice the cavity length and fringes will be observed over a range of path difference extending from $2l - l_c$ to $2l + l_c$. In fact, as a general rule, the temporal coherence of a laser is endlessly repetitive with interference fringes occurring at path differences extending from $2nl - l_c$ to $2nl + l_c$ where n is any integer.

Spontaneous emission is randomly directed. Any spontaneously emitted light will usually escape through the walls of the tube. Waves emitted by stimulated emission behave very differently. They are in the same *mode* as the wave that causes the emission. Thus, if a wave traveling along the cavity axis stimulates the emission of another wave in the helium–neon amplifier, the new stimulated wave travels in the *same direction*, has the *same phase*, and has the same state of *plane polarization*. The geometry of the laser amplifier favors gain along the cavity axis, and it is in this direction that positive feedback occurs by reflection of light from the cavity mirrors. Thus, light passing through the gas along the axis is amplified by stimulated emission and on reflection from the end mirrors is further amplified so long as we continue to pump the helium gas by electron bombardment. The reflectivity of

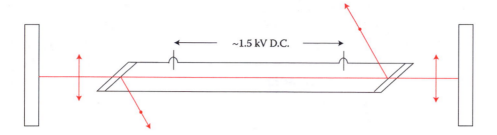

FIGURE 5.7   Vertical polarization favored by using windows set at the Brewster angle.

one of the cavity mirrors is practically 1.0. That of the other is slightly less (~0.99), allowing some light to escape from the cavity after each round trip.

The cavity is much easier to align if mirrors are concave, with the same radius of curvature and separated by their radius of curvature. We can see intuitively why, as any tendency of the beam to diverge as it passes between the mirrors, is counteracted by the focusing effect of the mirrors. The mirrors have a common focal point at the midpoint of the cavity, which is known as a *confocal cavity* for this reason. Monochromatic light from a point source located at the midpoint of the cavity will be reflected by each mirror, parallel to the axis and therefore focused back on the midpoint by the other mirror.

The 632.8-nm helium–neon laser operates over a bandwidth of 1.5 GHz, so it has a coherence length of about 20 cm.

Stimulated emission has the same polarization state as the stimulating light wave and this light may have any plane of polarization. To counteract this, the laser tube is fitted with windows set at the Brewster angle. This geometry favors transmission, through the windows, of light that is plane polarized in the plane of the diagram as discussed in Section 2.9.3. Light polarized in the plane perpendicular to the plane of the diagram is preferentially reflected out of the cavity as shown in Figure 5.7 and is therefore prevented from participating in the process of stimulated emission.

The helium–neon laser, as already stated, has a coherence length of about 20 cm and output power of between 1 and 15 mW, which is quite sufficient for many holographic applications. However, the need for higher power arises in many instances especially in optical setups involving a large number of optical components such as lenses and beam splitters. At every optical surface or boundary between media of differing refractive index encountered by a beam of light, there is always some optical loss unless that surface has an anti-reflection coating. Sometimes the losses cannot be avoided and we may need to opt for a higher power laser.

5.7.1.2 Argon Ion Lasers

Argon ion lasers are capable of producing high output powers (~Watts) albeit using water cooling of both the laser tube and its three-phase electrical power supply. A range of wavelengths is available, the most important being 488 nm (blue) and 514 nm (green). As Figure 5.8 shows, the different wavelengths are selected by means of a prism in the cavity. Rotation of the prism allows Equation 5.1 to be satisfied for the different wavelengths with *l* now the *optical* length of the cavity including the path through the prism. The bandwidth of the laser transition is typically several gigahertz because the natural laser linewidth is considerably Doppler broadened. The high powers require fairly long gas tubes (~1 m) and therefore cavity lengths. For a cavity length of 1 m, many longitudinal modes (0.15 GHz apart) will be supported within the laser transition bandwidth and so the coherence length of the laser will only be a few cm at best.

The solution is to force the laser to operate in a single longitudinal mode by inserting a solid Fabry–Perot etalon a few millimeters in thickness in the cavity. The laser can then only operate in longitudinal modes that satisfy Equation 5.1 for *both* the laser cavity and the etalon. An etalon of thickness 10 millimeters and refractive index 1.5 has a longitudinal mode separation of $c/2nl = 10$ GHz. The tuning necessary to ensure that a mode of the etalon coincides with the central frequency of the laser transition at which the intensity is maximum, is accomplished by fine control of the etalon temperature to adjust its thickness and by rotating it to alter its effective optical thickness (Figure 5.8). The coherence length of an etalon equipped Argon laser is typically several meters.

The cost of an Argon laser is considerable and to this should be added the cost of a closed-cycle chiller as, in most countries, it is no longer permissible to simply run mains water through the cooling circuits of the laser head and its power supply, into the drainage system.

Air-cooled Argon lasers are available but with powers of around 100 mW and without an etalon.

5.7.1.3 Krypton Ion Lasers

Krypton ion lasers are similar in design to Argon lasers. They can provide single longitudinal mode operation and power output of about 1 W at 647 nm and are widely used for holography. They also require a three-phase electrical power supply and water cooling.

5.7.1.4 Helium–Cadmium Lasers

Helium–cadmium lasers are used for recording at 325 and 441 nm because photoresists, which are widely used in the mass production of holograms, are more sensitive at these wavelengths than at longer wavelengths. Power outputs from 10 to 100 mW are typical with typical coherence lengths of about 10 cm.

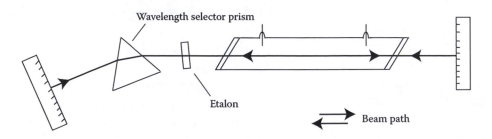

FIGURE 5.8 Argon ion laser.

5.7.1.5 Exciplex Lasers

Exciplex (the word is a contracted form of *excited complex)* lasers are more commonly, if incorrectly, known as excimer lasers. They use noble gases such as argon, xenon, and krypton, which are largely inert and do not form chemical compounds with other elements as their outer shells of orbital electrons are complete. However, when a noble gas atom is raised to an excited energy state by electron bombardment, it can form an excited complex, or quasi molecule, with a halogen gas atom such as fluorine. An exciplex emits UV light and returns to the ground state, dissociating very rapidly into the two individual ground state atoms. It is therefore possible to create a population inversion consisting of exciplexes since they cannot exist in the ground state. Typical exciplex laser wavelengths are 193 nm using argon and fluorene and 248 nm using krypton and fluorene.

An important application of these lasers is in delicate surgical procedures because the UV light provides sufficient energy to break the chemical bonds in biological tissue without burning. This allows ablation of surface tissue without damage to the remainder. Exciplex lasers are particularly well suited to reshaping the cornea of the eye to correct shortsightedness and astigmatism.

From a holography perspective, the main application for exciplex lasers is the fabrication of holographic diffraction gratings in glass fibers, also known as fiber Bragg gratings (see Section 11.3).

5.7.2 SOLID STATE LASERS

5.7.2.1 Semiconductor Diode Lasers

The drive toward the development of semiconductor diode lasers originally came from the optical communications industry because optical glass fibers became available with extremely low transmission losses at 1550 nm, enabling long haul telecommunications with very large bandwidth and therefore information capacity. The device consists of a *pn* junction between two pieces of semiconducting material. On one side (*p*) of the junction, the semiconductor is heavily doped with acceptor atoms, which are atoms with an electron missing in the valence band (a hole) while on the other side (*n*), the dopant atoms each have an excess electron in their conduction band. The conduction band is higher in energy than the valence band. Thus, when an electrical bias is applied to the junction causing excess conduction band electrons to cross the junction to the *p* side, they combine with holes and laser radiation is emitted. The oscillator cavity is formed by polishing the ends of the junction. Semiconductor diode lasers are extremely efficient, up to 100% of electrical power being converted to laser output power, compared with gas laser efficiencies of 1% or even less. Semiconductor lasers operating at around 650 nm providing optical output power of a few milliwatts, running on D-cell batteries, are now available for a few dollars. In addition, they do not require spatial filtering in order to expand the output beam, which is naturally much more divergent than that produced by other lasers. They also have naturally long coherence lengths as the solid etalon comprising the semiconductor is extremely short so that single-mode operation is normal. Generally, the cross-sectional shape of the cavity is rectangular with one of the dimensions significantly different from the other and the shape of the output beam is elliptical.

Obtaining higher power (tens or hundreds of milliwatts) from laser diodes is possible and single mode operation is obtained by a number of methods. One is the coupled cavity method in which two semiconductor diode etalons of different lengths are separated by a very narrow gap and laser operation is restricted only to those modes which are common to both etalons. Another is the distributed feedback laser in which a diffraction grating is etched into the top face of the semiconductor diode etalon, effectively serving as a mode selection device. The cavity can be maintained at a fixed length by very precise thermo-electric temperature control. This helps prevent mode hopping, a phenomenon exhibited by many semiconductor lasers, whereby the laser suddenly switches from operation in one mode to the adjacent one with an accompanying change in wavelength. Clearly, this causes instability in any interference fringe pattern produced from the laser.

High-power single longitudinal mode semiconductor lasers emitting visible light are not readily available. They are not so easy to fabricate as for the telecommunications wave-lengths and the market for them is possibly not seen to be worthwhile. One device is avail-able in which the single mode laser output of up to 35 mW at 658 nm is coupled to a volume holographic grating.

Another option is to place the diode in an eternal cavity with a Fabry–Perot etalon.

5.7.2.2 Doped Crystal Lasers

Some solid state lasers consist of a crystalline material doped with ions. The advantage is that they can be pumped by means of a flashlamp and produce an output in the form of a very short pulse, a few milliseconds in duration.

5.7.2.2.1 Ruby Lasers

The first doped crystal laser to be operated successfully was the ruby laser, which consists of a sapphire crystal (Al_2O_3) in which some of the Al ions are replaced by Cr^{3+} ions. The output wavelength is 694.3 nm, and the laser can be operated in continuous wave or pulsed mode. These lasers are used in holography particularly when very short exposure times are required for recording of fast moving objects.

5.7.2.2.2 Neodymium Yttrium Aluminum Garnet lasers

Yttrium aluminum garnet $Y_3Al_5O_{12}$ (YAG) doped with Nd^{3+} ions can provide optical amplifi-cation at 1.06 μm. Pumping to produce population inversion in the ions is by a powerful lamp such as xenon but water cooling is required. Much more efficient pumping can be provided by laser diodes with output wavelengths that precisely match that required to produce pop-ulation inversion. The wavelength is unsuitable for most holographic materials but may be halved to 532 nm by frequency doubling (see Sections 5.9 and 6.6.3).

5.7.2.2.3 Titanium Sapphire Lasers

Titanium sapphire lasers have a number of important advantages, among them the fact that they emit laser radiation, which can be tuned over an extremely wide range of wavelengths from 650 to 1100 nm. The lasing medium is triply ionized titanium (Ti^{3+}) in a sapphire crystal host and the pump is a laser operating in the green part of the spectrum. The upper state lifetime of the laser transition is very short (~3.2 μs) so that high intensity pumping light is needed to achieve population inversion. For this reason, argon ion lasers at 514 nm or, more commonly, frequency doubled Nd Yag lasers at 532 nm (Section 5.9) are used for pumping, although semiconductor lasers with increased output powers also act as very efficient pumping lasers. The output power of Ti sapphire lasers can be up several Watts with pulse power up to 1 W and pulse durations of a just a few femtoseconds have been obtained.

5.7.3 DYE LASERS

Liquid dye lasers rely for their operation on the fact that a dye molecule may be optically pumped into an excited electronic state, which in turn consists of a number closely spaced energy sublevels. The multiliplicity of these levels is a direct consequence of the close proximity of molecules in the liquid state and the lasing action may be selectively favored to take place between any one these levels and the ground state. Consequently, dye lasers may be tuned over a wide range of wavelength.

Although dye lasers are not in such widespread use as in the past, they do still have important holographic applications requiring synthesis of light sources of short coherence length, as we will see in Section 13.6 in which we discuss digital holographic microscopy and in Section 15.5.4 dealing with digital holographic imaging through scattering media, including human tissue.

5.8 Q-SWITCHED LASERS

Pulses of a few nanoseconds duration can be obtained from ruby and Nd Yag lasers by Q-switching or temporarily preventing light from propagating in the laser cavity while continuing to pump the amplifier. When the cavity is reopened, the laser radiation is emitted in a single burst. The cavity is usually temporarily closed by saturable absorber, which becomes bleached, that is colorless or transparent, when illuminated by light of appropriate wavelength and intensity for a sufficient period of time, or by rotating one of the end mirrors of the cavity, or by means of an electrooptic device called a Pockels cell, which is the most reliable method. Such a cell is positioned between one of the end mirrors of the laser cavity and a plane polarizer (Figure 5.9) oriented so as to transmit the laser light, which is plane polarized in the $+y$ direction. When a voltage is applied to the cell, it exhibits a phenomenon called the linear electrooptic effect. The cell develops birefringence (see Section 2.9.4) meaning that the refractive indices differ in directions parallel to and normal to the field direction. Consequently, the laser light electric field components parallel and perpendicular to the applied field become progressively out of phase as they travel the length of the cell along the $+z$ direction. By suitable choices of cell material (usually lithium niobate $LiNbO_3$,) applied voltage, and cell length, a phase difference of $\pi/2$ is established. Following reflection, a further phase difference of $\pi/2$ is added as the light travels in the $-z$ direction.

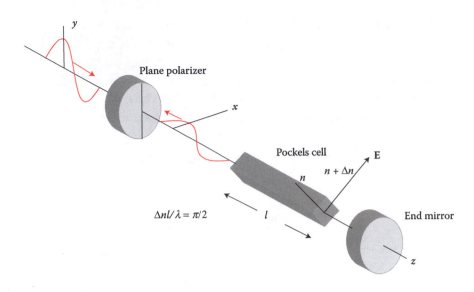

FIGURE 5.9   Pockels cell Q-switch.

On emerging from the cell, the light is now plane polarized in the x direction and cannot pass through the polarizer. Thus, the laser cavity is effectively blocked. Meanwhile, the pumping process continues so that a very large population inversion becomes established. When the cell voltage is reduced to zero, the cavity reopens and the laser radiation is emitted in the form of, usually, a single high energy pulse of very short duration, typically about 1 ns. Such pulsed lasers are used in holography when the object or indeed the optical setup itself is interferometrically unstable.

5.9 FREQUENCY DOUBLED LASERS

We can very effectively double the frequency of the Nd^{3+} YAG laser by a lithium niobate ($LiNbO_3$) or other nonlinear optical crystal to obtain an output wavelength of 532 nm with a power of several watts and very long coherence length. Efficient frequency doubling provides for a wide range of wavelengths using primary laser sources.

5.10 MODE LOCKING OF LASERS

As we have seen in Section 5.5, lasers may emit radiation in a number of cavity modes at the same time. The frequencies of longitudinal modes differ by the free spectral range of the cavity (Equation 5.5), and there is no fixed phase relationship between oscillations in different modes. It is as if there are a number of lasers all operating independently of one another in the same laser cavity. However, if we can lock the phases of the oscillations in different modes together in some way, so that the phase differences are all zero, then the oscillations in all the modes interfere constructively with one another to produce a single powerful pulse of laser light.

To see why we add the electric fields in all the modes to obtain

$$E(t) = \sum_n E_n \exp\left[j\left(\omega_0 + \frac{2\pi nc}{2l} \right)t + \phi_n \right] \tag{5.6}$$

where E_n is the electric field amplitude of the nth mode, ω_0 is the angular frequency of a mode that is sustained by the cavity, and ϕ_n is the phase, which is likely to vary in a random fashion. This is why multimode laser output power tends to fluctuate in an unpredictable way.

For stable modes,

$$E\left(t + \frac{2l}{c} \right) = \sum_n E_n \exp\left[j\left(\omega_0 + \frac{2\pi nc}{2l} \right)\left(t + \frac{2l}{c} \right) + \phi_n \right]$$

$$= \sum_n E_n \exp\left[j\left(\omega_0 + \frac{2\pi nc}{2l} \right)t + \phi_n \right] \exp j2\pi\left(n + \frac{2l\omega_0}{2\pi c} \right) = E(t) \tag{5.7}$$

since $2l\omega_0 / 2\pi c$ is an integer. We could restrict the operation of the laser to a single longitudinal mode using an intracavity etalon (Section 5.7.1.2) but instead suppose we force all the values of ϕ_n to be zero, and we set all $E_n = E_0$, then Equation 5.6 becomes

$$E(t) = E_0 e^{j\omega_0 t} \sum_{n=0}^{N-1} e^{(j2\pi nct/2l)} = E_0 \exp^{j\omega_0 t} (1 + e^{j\pi ct/l} + \cdots + e^{[j\pi(N-1)ct/l]}) = E_0 e^{j\omega_0 t} \frac{e^{j\pi Nct/l}}{e^{j\pi ct/l}}$$

where N is the number of modes within the spectral bandwidth of the lasing material.

The intensity is then given by

$$I(t) \propto \frac{\sin^2\left(N\,\pi ct/2l \right)}{\sin^2\left(\pi ct/2l \right)} \tag{5.8}$$

an expression whose form we recall from our discussion of the diffraction grating (Section 2.2.1.3) and which takes the form of a pulse repeating at intervals of $2l/c$ with a pulse duration of $4l/(Nc)$. The pulse has intensity N^2 times that of a single mode.

Now, suppose that in the cavity we insert a shutter which is opened for a time interval $4l/(Nc)$, just once in every round trip. Oscillation in individual modes cannot grow in the time for which the shutter is open. The only energy, which can pass through such a shutter is that which takes the form of pulses described by Equation 5.8. Thus, the laser energy is forced to distribute itself in this way. The shutter may take the form of a saturable absorber, which, as the name implies, absorbs low-intensity light but bleaches when illuminated by intense light, which it transmits. The shutter therefore transmits pulses of the form given by Equation 5.8.

Active mode-locking devices are also common. Amplitude modulation of any mode using an acousto-optic modulator (see Section 8.8.1) produces oscillations, which are in phase with the original mode. We can modulate at a frequency that is equal to that of the mode separation frequency $c/(2l)$. This means that the original mode and its two nearest neighbors are now in phase. If we then proceed to modulate at integer multiples of the mode separation frequency, with a total modulation bandwidth covering the spectral bandwidth of the lasing material, all of the modes are forced into phase with one another.

Frequency modulation using an electrooptic device is also possible. If a sinusoidal voltage is applied to the device, the frequency of light passing through it either increases or decreases depending on its phase on arrival at the device. We systematically apply the frequency shift to the light by using a modulation frequency equal to the mode separation frequency. Thus, the same shift is applied in the same direction (positive or negative) at each passage through the device. Ultimately, the only modes which are not frequency shifted right out of the laser spectral bandwidth are those whose phases are zero at the device at the outset.

5.11 SPATIAL COHERENCE OF LASERS

We have already discussed in Section 2.8.3 the need for spatial coherence in a light source if it is to be used to form an interference pattern having the highest contrast. Ideally, the source should be a point, which is why we try to force the laser to operate in a longitudinal rather than a transverse mode (see Section 5.5) as both normally coexist in the cavity. Let us now look at a longitudinal mode in more detail. It has a Gaussian radial intensity distribution $I(r)$ given by

$$I(r) = I_0 \exp\left(\frac{-2r^2}{w^2}\right) \tag{5.9}$$

where r is the distance from the principal axis of the cavity and I_0 the intensity at the axis. The radius of the beam, w, is defined as distance from the axis at which the intensity is reduced to $1/e^2$ of its axial value.

The diameter of a longitudinal laser mode is usually minimum, known as the beam waist, w_0, in the middle of a confocal cavity (Figure 5.10) and increases steadily as the light propagates along the cavity, z-axis according to

$$w(z) = w_0 \left[1 + \left(\frac{\lambda z}{\pi w_0^2}\right)^2\right]^{1/2} \tag{5.10}$$

At large z, this becomes simply $w(z) = \lambda z/\pi w_0$.

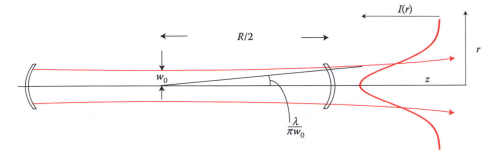

FIGURE 5.10 Growth of Gaussian beam radius.

The divergence of the laser beam is given by arctan $[w(z)/z] = \lambda/\pi w_0$ and the radius of the beam at distance z from the midpoint of the cavity is given by

$$w(z)^2 = w_0^2 + \left(\frac{\lambda z}{\pi w_0}\right)^2 \qquad (5.11)$$

If the beam radius is 1 mm near the cavity mirrors, its divergence is about 0.2 mrad. In practice, a divergence of about 0.5 mrad may be expected.

5.12 LASER SAFETY

Laser safety precautions should always be carefully observed. Unexpanded laser beams, even from very low-power laser diodes, are especially dangerous to the eyes. Particular care should be taken when changing the orientations of laser illuminated mirrors and other optical components as well as objects, even those with surfaces of comparatively low reflectivity. Steel mounting posts used for optical mounts are a particular hazard. Where possible when making adjustments to an optical setup, the laser should be operated at the lowest possible output power, using high power only if necessary for the holographic recording step. If the power cannot be adjusted directly then a neutral density filter of appropriate density should be placed in the path of the beam emerging from the laser cavity. All personnel working with lasers should be equipped with the protective eyewear specified for the wavelengths of the lasers in use and have regular eye examinations by a competent optometrist.

5.13 MECHANICAL STABILITY

The necessity for mechanical stability in holographic recording is difficult to overemphasize. Let us take another look at the Mach–Zehnder interferometer (Figure 2.15) used to record the simplest hologram, namely an interference pattern of just one spatial frequency, and imagine that one of the mirrors begins to oscillate out of plane. If the beams meet at right angles to one another, the mirror would only have to move a distance of $\lambda/2.8$ in order to change the optical path difference by $\lambda/2$ and change every bright fringe in the interference pattern to dark and vice versa. If the period of oscillation is much shorter than the time required to record the fringe pattern, the recording will not be a success as the pattern will be smeared. For this reason, it is important that we follow certain rules:

- The object itself must be mechanically stable—we cannot make holograms of objects that are not mechanically rigid.
- All optical components such as mirrors and beam splitter must be securely mounted so that they cannot move during the recording.
- The recording medium itself must be securely mounted so that it cannot bend or rotate. In this connection, even recording material on glass substrates should

be carefully supported—if mounted vertically the glass should also be firmly supported along its top and bottom edges.
- The table, on which all optical components as well as the object itself are mounted, should itself be isolated from any vibration that may be transmitted to it. For this reason, optical tables are often mounted on air-inflated rubber supports such as inner tubes of automobile or motor scooter tires. The table itself should be a rigid platform that has an extremely low natural frequency of flexural oscillation. In addition, the floor on which the table rests should be considered. A concrete basement floor is preferable to the floor of an upper storey.

The above precautions should always be revisited whenever a total failure occurs in holographic recording, that is, when no image at all is seen on reconstruction.

Many of these rules can be ignored if we can use a pulsed laser such as a Q-switched ruby or frequency doubled Nd^{3+} YAG laser. If such a laser can produce a pulse of say 1.0 ns duration and with sufficient optical energy in that time to adequately expose the photosensitive material, then any parts of the recording setup (mirrors, beam splitter, object itself) moving with a speed of up to 50 ms^{-1} would not spoil the recording since total movement during recording is only $\lambda/10$. We do not normally need to record holograms of objects moving at anywhere near this speed. Pulsed ruby lasers have been used very successfully to record holograms of people as all body motion is effectively frozen during the extremely short recording time although extreme care is needed to ensure that the optical laser energy density does not exceed the recommended limits for the human eye (5 mJ m^{-2}) for the object beam. It is also important that no stray reflected light from any part of the recording setup can exceed this limit. In particular, the Fresnel reflections (Section 1.7.2.5) of the reference beam from the surfaces of the recording medium must be directed entirely away from the subject.

5.14 THERMAL STABILITY

It must always be remembered that the light waves, which form the interference pattern to be recorded as a hologram, travel in air which has a refractive index of 1.000293 at 0°C and 1 atm pressure. Suppose that the path difference in an interferometer is 0.5 m, and a change in refractive index of 3 parts in 10^7 occurs in the air path of one of the beams due to a temperature variation (caused by flow of air such as a naturally occurring draft or air conditioning). This is sufficient to cause a change in optical path difference of about $\lambda/4$ for a helium–neon laser and the interference pattern will shift to this extent.

5.15 CHECKING FOR STABILITY

It is always good practice initially to set up the Michelson interferometer (Figure 5.11) on the optical table that is to be used for holographic recording.

The unexpanded beam from the laser is split into two parts by a cube beam splitter, although a parallel-sided glass plate with a partially reflective coating will work almost

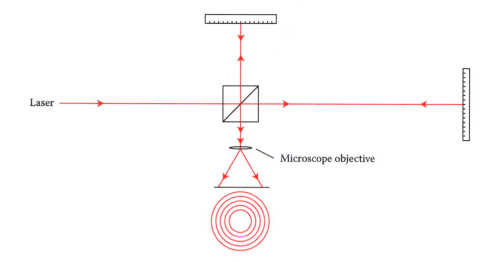

FIGURE 5.11   Fringe pattern produced by the Michelson interferometer.

as well. One of the mirrors should be capable of being translated in the direction of the normal to its surface to allow the path difference to be altered. The other should be gimbal mounted so that it can be rotated independently about horizontal and vertical axes. On reflection from the mirrors and after transmission through (beam 1) and reflection at (beam 2) the beam splitter, the recombined beams are passed through a beam expander, that is, a microscope objective of power 10× (focal length 16 mm, see Section 8.2.1) or 20× and the interference pattern observed on a white screen. The shape of the fringes whether concentric circles or straight lines is not important, so long as they are few in number and can be clearly seen. If no pattern can be seen at all, it is likely that the path difference in the interferometer exceeds the coherence length of the laser. It is best to start with near zero path difference to be sure of obtaining an interference pattern and progressively increase the path difference until the contrast of the fringe pattern noticeably decreases. The contrast may be measured by scanning a photodiode across the fringe pattern. As a rule the path difference at which the contrast has fallen to 70% of its value at zero path difference is the maximum that can be tolerated for holographic recording.

If the pattern is observed to drift slowly, this may be due to slow mechanical movement. After a component has been securely clamped, mechanical relaxation may occur, taking some time to complete. Rapid oscillatory movement of the fringes is an indication of vibration or noise.

Where the fringes change shape on time scales of about a second, this is usually an indication of temperature gradients in the laboratory and these should be excluded. One way of doing this is to erect vertical transparent sliding panels around the edges of the table along with a roof. Failing this any air conditioning or central heating units should be turned off.

If fringe instability of this type persists, it may be worth considering the use of a near common path setup (see Section 8.2) so that any air temperature variations are likely to affect reference and object beam paths in the same way and to the same extent.

A sudden deterioration in the contrast of the fringes followed by gradual restoration of the pattern after a few seconds is indicative of mode hopping. If this does not occur too frequently (say once in 15 minutes), the laser may be used but frequent mode hopping is unacceptable.

The fringe pattern should be observed over a period of at least several minutes to see what causes of instability may be present.

5.16 RESOLUTION OF THE RECORDING MATERIAL

Suppose we wish to record the interference pattern produced by two plane waves. It does not matter for now what recording material we use. The spacing of the fringes is determined by the angle between the wave vectors of the interfering waves. As shown in Section 3.1, the wave vectors or normals to the interfering plane wavefronts are at angles ψ_1 and ψ_2 to the z axis and the difference or beat spatial frequency is $\sin \psi_2 - \sin \psi_1$. The angles ψ_1 and ψ_2 are, of course, the angles *inside* the recording material, as is the wavelength, λ. For simplicity, let us assume that we record a holographic grating with angles of incidence *outside* the recording medium of 30° and −30°. If the refractive index of the medium is 1.5, then inside the medium $\psi_1 = 19.5°$ and $\psi_2 = -19.5°$ and $\lambda = 500$ nm/1.5 giving a value of 2000 lines per millimeter for the spatial frequency. *The recording material must be capable of resolving spatial frequencies of this magnitude and greater.* For comparison, we note that photographic film can normally resolve spatial frequencies of a few hundred lines per millimeter at most.

In practice, as we will see, the resolution required may be much higher.

5.17 SUMMARY

In this chapter, we explain the need for coherent light sources for holography. We explain the basic theory of lasers and discuss examples of lasers that are used in holography.

We also briefly outline the mechanical and thermal requirements for holography and explain the use of a Michelson interferometer for testing the holographic setup.

Finally, we highlight the need for high spatial resolution recording materials.

PROBLEMS

1. A Michelson interferometer is used to measure the wavelength of a monochromatic light source by counting the bright fringes, which pass a fixed point as the optical path difference is changed. If 12,500 bright fringes are counted as the mirror M_2 in Figure 5.2 is moved through a distance of 3.413 mm, what is the wavelength of the light?
2. We normally take the refractive index of air to be 1.0. In fact, its value at atmospheric pressure is 1.0029. Suppose an airtight cell with transparent windows filled with air at atmospheric pressure is placed between the beam splitter and mirror M_1 in Figure 5.2 and the cell is evacuated. What is the maximum number of fringes produced by the light source in problem 1, which pass a fixed point in the field of view?

3. A Michelson interferometer is adjusted so that one sees a circular fringe pattern in monochromatic light. The light source is exchanged for another one from which no fringes are visible. The path length from the beam splitter mirror M_2 is adjusted until fringes appear, their contrast increasing gradually and then declining until the fringes are no longer visible. The change in path length from the appearance to the disappearance of the fringes is 1.5 mm. What can you deduce from these observations?

4. The interferometer is set up to produce a pattern of colored fringes in white light. When a glass plate is inserted between the beam splitter and mirror M_1 to cover half the field of view, the fringes disappear in that half. What happens when the path length to the beam splitter mirror M_2 is increased? Eventually, colored fringes appear in the region of the plate when the path length has been increased by 0.5 mm. If the refractive index of the glass is 1.53, what is its thickness?

5. For holographic applications, we usually require a laser with a long coherence length. One way of ensuring this is to use an appropriate cavity length. What would this cavity length be in the case of a helium–neon laser and why do we not use it in practice?

6. Randomly occurring temperature fluctuations of $0.2°C$ are noted in the holographic laboratory on time scale of about 1 second. In a worst case scenario with a path difference of 50 cm, what do you anticipate would be the effect on holographic recording and what precautions should you take? The temperature coefficient of the refractive index of air is $0.87 \times 10^{-6}/°C$.

7. The figure below shows a light ray incident at a Fabry–Perot etalon. We assume that the complex amplitude of the first partially transmitted ray is $t^2\tilde{a}$ at point P. After two internal reflections, the next emerging ray has amplitude $r^2t^2\tilde{a}\exp(j\phi)$ where ϕ is the phase difference arising from the constant path difference, and r and t are respectively the reflected and transmission amplitude coefficients of each plate. Show that

$$\phi = \frac{4\pi l \cos\theta}{\lambda}$$

where t is the plate separation. Hence, prove that the total output light intensity is

$$I_{out} = \frac{I_{in}}{1 + F\sin^2(\phi/2)}$$

where $F = 4R/(1 - R)^2$ and $R = r^2$ and I_{in} is the input light intensity.

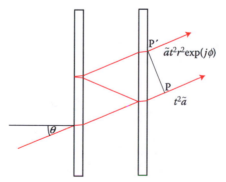

8. We can find the width between the half maximum points of the curve described by the expression for I_{out} in the previous problem. Show for large values of reflectivity R that the phase difference between these points is $4/\sqrt{F}$ and hence use the relationship

$$\frac{d\phi}{d\lambda} = \frac{-4\pi l}{\lambda^2}$$

to show that the spectral bandwidth of the etalon between the half maximum transmissivity points, that is, its mode width Δv_{mode}, is given by

$$\Delta v_{\text{mode}} = \frac{c}{\pi l \sqrt{F}}$$

9. What should be the thickness of a solid Fabry–Perot etalon of refractive index 1.5 to ensure single mode operation of an argon ion laser? The coherence length of the laser without an etalon is 1 cm. If the laser is to have a coherence length of 2.5 m, what should be the reflectivity of the etalon?

10. The etalon in problem 9 should not change dimensions, otherwise, the laser may begin to operate in a different mode during holographic recording. Given that the coefficient of thermal expansion of quartz is $0.5 \times 10^{-6} (°\text{C})^{-1}$, within what limits should the etalon temperature be controlled if the laser is operated at wavelength 514 nm?

11. A helium–neon laser has a discharge tube with internal diameter 5 mm, which we may take to be the beam waist diameter. Find the beam diameter at a range of 10 m.

12. The International Electrotechnical Commission (IEC) has specified maximum permissible exposure (MPE) levels at the cornea for collimated laser light in the visible range from 400 to 700 nm. For exposure times of indefinite duration 1mW cm^{-2} is specified, rising to 10 mW cm^{-2} for exposure time of 1 msec and to 1W cm^{-2} for exposure time of 1 μs. Under what circumstances might each MPE level be permitted in the holography laboratory?

13. Comment on the use of a pulsed Ruby laser (wavelength 694 nm) for recording holographic portraits given that the IEC specifies MPE of 10^{-6} Jcm^{-2} in the range of exposure time from 10 μs to 1 ns.

CHAPTER 6

Recording Materials

6.1 INTRODUCTION

The ideal photosensitive material for holography should have a number of features. It should have high sensitivity to a laser light source whose characteristics, particularly its coherence length, should in turn be suitable for holography. As we saw in Chapter 5, high spatial resolution is also of great importance. Linearity in response is important too and image quality implies the need for low noise, which in turn implies that the material should have no grain structure or, if there are grains, they should be extremely small to minimize light scatter. Availability in a range of formats is also desirable, as is mechanical stability, or at least the recording material should be amenable to interferometrically stable mechanical support. Self-processing is another requirement in many applications. Alternatively, chemical or other processing should involve little or no risk to health or the environment. Low cost is important for mass production of holograms. Erasability and rewrite capability are required for dynamic holographic data storage, for which we also need to assure nondestructive readout of data. For archival data storage, the recording material must be capable of very long term stability. A holographic sensor, on the other hand, is likely to require a permeable medium, whose physical and holographic characteristics change with the physical environment.

So, the choice of recording material depends very much on the application, and many research groups and individuals have painstakingly developed their own material or adapted existing ones for their particular purposes. Here, we discuss the leading materials and provide practical guidance in the use of the most popular ones.

6.2 SILVER HALIDE

Most holographers, including researchers, will want to try their hand at display holography at some time so it is appropriate to take a close look at the most consistently successful material for this purpose.

We have seen that the spatial resolution has to be of the order of several thousand cycles per millimeter, where a cycle refers to the distance between adjacent maxima of intensity in the interference pattern that is to be recorded. In some of the literature, the term line pairs per millimeter is used, where a line pair refers to a bright fringe and an adjacent dark fringe. The term lines per millimeter is also often used, with lines referring to bright *or* dark fringes. We will use the term lines per millimeter. The upper limit of resolution corresponds to the extreme situation that arises when a hologram is to be recorded with object and reference beams propagating in opposite directions, that is, with an angle of 180° between them. The spatial period is then $\lambda/2n\sin(\pi/2)$ which, for a wavelength of 632.8 nm (the helium–neon red laser) and a refractive index of 1.5, is 0.21 μm. This means that the recording material must be capable of resolving a fringe pattern with 4740 lines per millimeter. At blue wavelength (say 473 nm), we need a resolution capability of at least 6342 lines mm^{-1}.

Silver halide for holography, like its photographic counterpart, consists of transparent silver halide grains with a light-sensitive dye added to provide sensitivity to the appropriate laser wavelength, the grains being dispersed in gelatin. The big difference is that the sensitivity of holographic material is much lower than the photographic version because the silver halide grains have to be very much smaller. A great deal of research effort has been expended on producing such small grain sizes, the most notable successes being in a number of Eastern European laboratories and in Russia. Research by the Silver Cross consortium (http://silvercrossproject.org/) has led to the development of panchromatic silver halide suitable for full-color holography with a grain size of 10 nm.

Silver bromide (AgBr) is the most commonly used silver halide and, being an ionic chemical compound, is dissociated in the gelatin. Thus

$$AgBr \rightarrow Ag^+ + Br^- \tag{6.1}$$

A photon of sufficient energy will remove the excess electron from the Br^- ion, leaving it as a neutral bromine atom. The electron combines with the silver ion to produce a neutral silver atom. This is not a stable situation, and dissociation into ionic silver takes place in a time scale of a few seconds. If two silver atoms are formed in similar fashion, they form a cluster, which takes longer to dissociate. A cluster of three atoms is even more stable and, if more photons arrive at the grain and convert at least four silver ions into a cluster of silver atoms before any of them can dissociate again, the atoms in the cluster will not dissociate for a long time. Silver is opaque to light and, following exposure, the emulsion may be processed to convert all the grains, which contain stable clusters of atomic silver, entirely into silver. Thus, the spatial light intensity distribution due to the interference of object and reference waves, is recorded as a spatial variation of the transmissivity of the emulsion. We have to ensure that the recording is made permanent and stable against further illumination.

6.2.1 AVAILABLE SILVER HALIDE MATERIALS

Agfa-Gevaert formerly supplied good quality materials, as did Ilford, and the latter may again be available in the future. Integraf at http://www.Integraf.com/ supplies a range of materials, whose characteristics are given below (data reproduced from Integraf website with kind permission).

PFG-01 has resolution of more than 3000 lines/per millimeter and is available coated on glass plates and on triacetate film. The largest standard plate size available is 406 × 300 mm. The standard glass thickness is 2.5 mm. The largest standard film width is 1 m. The useful life is 12 months. The spectral sensitivity is shown in Figure 6.1a (light line) with good sensitivity in the green and in deep blue regions of spectrum, with a maximum at approximately 100 µJ/cm² (~640 nm). Figure 6.1b (light line) shows the optical density versus exposure energy using continuous wave (CW) radiation, following development. Grain size is shown in Figure 6.1c. The diffraction efficiency versus exposure for reflection holograms using a CW laser is shown in Figure 6.1d (light line).

The characteristic curve of Geola green sensitive VRP-M emulsions, showing sensitivity versus wavelength, is shown in Figure 6.1a (heavy line) with a peak of approximately 70 µJ/cm² at 530 nm.

Figure 6.1b (heavy line) shows the optical density versus exposure energy following development using CW radiation. Grain size distribution is shown in Figure 6.1c. The diffraction efficiency versus exposure for reflection holograms (Section 8.5) recorded on VRP-M (using a pulsed laser) is shown in Figure 6.1d (heavy line).

A third product PFG-03M is specially designed for reflection holography with an average grain size ranging from 8 to 12 nm and a resolution of more than 5000 lines/per millimeter and an emulsion thickness of 5–7 nm. Its spectral sensitivity, diffraction efficiency versus exposure and grain size distribution are shown in Figure 6.2.

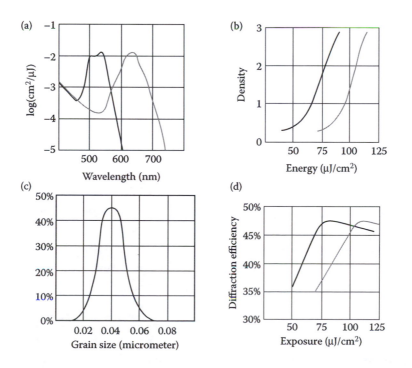

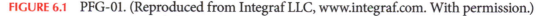

FIGURE 6.1 PFG-01. (Reproduced from Integraf LLC, www.integraf.com. With permission.)

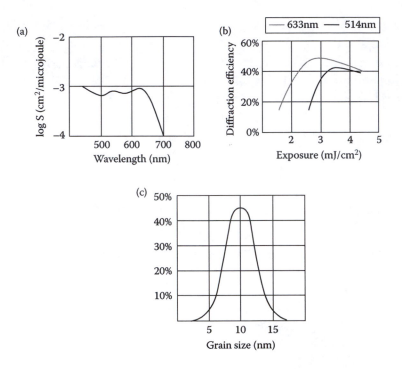

FIGURE 6.2 PFG-03. (Reproduced from Integraf LLC, www.integraf.com. With permission.)

Formats of 63×63 mm^2, 101×127 mm^2, and 300×406 mm^2 are available.

Panchromatic silver halide PFG-03C is formulated for full-color holography.

These materials are also available from Geola at http://www.holokits.com/slavich_holographic_film_plates.htm.

Another supplier of plates and film is Colour Holographics at http://www.colourholographic.com/index.htm.

6.2.2 PROCESSING OF SILVER HALIDE TO OBTAIN AN AMPLITUDE HOLOGRAM

To produce an amplitude hologram, the emulsion is first carefully developed and washed (see Figure 6.3a).

The development process reduces to silver all of the silver ions in those grains that already contain a stable silver atomic cluster. Thus, the regions of the emulsion where constructive interference between the recording beams has occurred are now occupied by opaque silver grains.

We now have an amplitude hologram but it must be stabilized against further exposure to light of the grains that were unaffected in the hologram recording stage. This is done by removing any remaining silver ions by fixing. Silver ions form soluble chemical complexes with the fixing agent (Figure 6.3b). Almost any photographic fixer will serve.

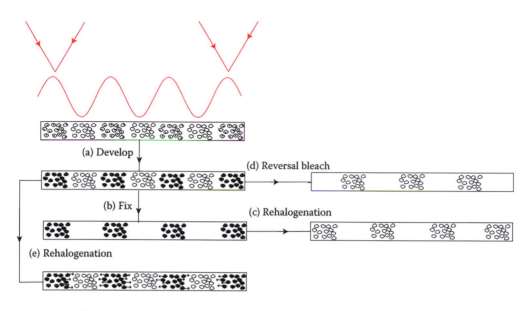

FIGURE 6.3 SH processing.

The result is an amplitude hologram whose diffraction efficiency is rather low. We saw in Section 4.3.1.2 that this can be as little as 3.7% for a thick layer. So, the next step is to convert the amplitude hologram into a phase hologram by bleaching. Bleaching is a chemical process whereby the spatial variations in transmissivity of an amplitude hologram are converted to spatial variations in thickness or refractive index (see Section 3.7) mainly the latter. There are a number of ways of doing this.

6.2.3 PROCESSING TO OBTAIN A PHASE HOLOGRAM—REHALOGENATION

A straightforward way to obtain a phase hologram after fixing is to convert all the developed silver back into silver halide, usually silver bromide, which has a higher refractive index than gelatin. Thus, we have a spatial variation or modulation of refractive index across the emulsion. Areas in which constructive interference occurred between the object and reference beams in the recording process now have a higher index than those in which destructive interference occurred. The chemicals normally used are potassium bromide and an oxidizing agent and constitute a rehalogenating bleach. The oxidizing agent, for example, potassium dichromate, removes an electron from the silver atom so that it again becomes a positive silver ion, which combines with the bromine ion in potassium bromide to form silver bromide (Figure 6.3c).

This method is not much used unless one particularly wants to study the reconstructed image from a developed and fixed amplitude hologram prior to bleaching. The reasons are that first the final phase hologram suffers from scatter noise due to growth in grain size, which occurs to some extent during all development processes, as unexposed silver ions dissolve and migrate towards exposed silver. In addition, some developers have a tanning effect, hardening the gelatin around the developing grains so that their optical scattering cross section increases. The hardened gelatin also absorbs water to a lesser extent than the surrounding gelatin, which has not been tanned by development, and shrinks more rapidly on drying. The softer gelatin in the vicinity is then readily drawn toward the harder material causing local thickening of the emulsion layer. This in turn increases the optical path difference between the bright (rehalogenized silver) and dark fringe (gelatin) regions, adding to the effect of the refractive index difference between rehalogenized silver and gelatin regions. The corresponding increase in diffraction efficiency of the hologram would ordinarily be welcome but unfortunately it is largely confined to low spatial frequencies (~100 lines per millimeter). Such low spatial frequencies are recorded when light is diffracted from closely neighboring points on the object to interfere in the hologram plane and they do not contribute to the holographically reconstructed image. The effect is called intermodulation.

6.2.4 PROCESSING TO OBTAIN A PHASE HOLOGRAM—REVERSAL BLEACHING

Alternatively, following the development process, one can use a bleach that removes the developed silver grains leaving the undeveloped silver bromide. This process is called reversal bleaching. The developed silver is chemically oxidized to form silver ions, and these combine with the bleach anions to form soluble chemical compounds, which are washed out. The phase hologram (Figure 6.3d) now consists of small silver halide grains interspersed with pure gelatin, as opposed to developed and rather larger, rehalogenated silver grains, interspersed with gelatin. It is therefore less noisy than a hologram, which has been bleached after fixing. The surface relief due to gelatin hardening in developed silver regions is still there but now it subtracts from the effect of the refractive index difference between silver halide (dark fringe) regions and gelatin (bright fringe) regions.

Some rehalogenating bleaches can be used without fixing. In this case, developed silver is converted to a silver halide as before. However, this is accompanied by diffusion of silver ions from unexposed silver halide regions into exposed regions (Figure 6.3e), a process that becomes dominant when the spatial period of the recorded fringe pattern in the hologram is comparable with the diffusion length. As one can imagine, this process is of great value for holography involving very high spatial frequencies.

6.2.5 SILVER HALIDE PROCESSING IN PRACTICE

The literature on silver halide processing for holography is vast. An excellent guide is available [1]. The Geola and Integraf websites provide detailed information on processing, including methods for ensuring faithful reconstruction of the wavelength used in recording color reflection holograms.

Details are given here of two processing schemes that work very well for white light reflection holograms. These are the pyrochrome process, originally suggested by van Renesse, and the JD-4 process. They are rapid processes and particularly useful in the teaching laboratory when time is limited. They are both easy to prepare from readily available laboratory chemicals and safe in use. Distilled, deionized water should be used to make up the solutions for storage in clean containers. Chemicals are dissolved in the order listed. *The use of laboratory rubber gloves is strongly recommended as a standard precaution primarily for skin protection but also to avoid contamination of the emulsion when handling plates and film.*

Pyrochrome process
  Developer
    Solution A

Pyrogallol	15 g
Metol	5 g
Potassium bromide	2 g
Water	1000 mL

    Solution B

Potassium hydroxide	10 g
Water	1000 mL

The solutions should be stored in bottles from which air can be excluded. Photographic stores can supply concertina plastic bottles from which most of the air can be removed.

Equal parts of A and B are mixed immediately prior to development at 20°C for about 2 minutes. The mixed developer can only be used only once as it oxidizes quite quickly.

  Bleach

Potassium dichromate	5 g
Sodium hydrogen sulphate	10 g
Water	1000 mL

Bleaching time is about 2 minutes at 20°C but depends on the developed density, which should be 2–2.5 for a reflection hologram, but the laboratory light can be turned on once bleaching has been underway for about a minute. The bleach can be used several times.

JD-4 process
  Developer
    Solution A

Metol	4 g
Ascorbic acid	25 g
Water	1000 mL

    Solution B

Sodium carbonate, anhydrous	70 g
Sodium hydroxide	15 g
Water	1000 mL

  Bleach

Copper sulfate, pentahydrate	35 g
Potassium bromide	100 g
Sodium bisulfate, monohydrate	5 g
Water	1000 mL

Part A has a limited useful life and will turn yellow although it can still be used until it turns brown. To extend useful life, it is a good idea to divide up the bulk solution into smaller lots of say 200 mL, each lot being kept in a tightly stoppered bottle and kept in a refrigerator until needed. It will keep for several months. The solution should of course be allowed to reach room temperature before use. Part B keeps well at room temperature as does the bleach.

Plastic shallow developing dishes, not much larger than the dimensions of the silver halide plates or film in use, are recommended for economy and minimal chemical waste disposal. The total volume of processing solution at each stage should be sufficient to ensure complete immersion of the emulsion in the dish. Thorough rinsing in a bath of clean, distilled, deionized water is required between development and bleaching. Two separate water baths are required after bleaching with one drop of either a photographic wetting agent or of a very dilute solution of washing up liquid added to the final wash.

In both developing and bleaching, one must ensure that the emulsion is completely covered and that no air bubbles are trapped on it. This can happen easily when the silver halide plate or film is immersed in the developer or bleach with the emulsion side down. Plates are usually packed in plastic slotted trays with emulsion sides of neighboring plates facing one another (Figure 6.4a). If the plates are always removed in order from the same end of the tray and a record kept of the number (odd or even) actually used, one should always be able to tell which is the emulsion side of the next plate on removing it from the tray. As a last resort, one can touch the plate with slightly moistened lips. The emulsion side is slightly sticky.

It is important to know which is the emulsion side for another practical reason. A reflection phase hologram is highly transparent so that, although it may have a high diffraction efficiency and produce a bright reconstructed image, objects *behind* the hologram, which are visible through it, may prove distracting. To prevent this, the emulsion side is usually

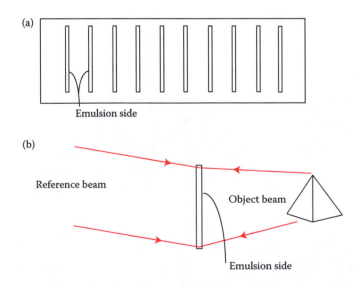

FIGURE 6.4 Emulsion side.

sprayed with matt black paint, which also helps to protect it. One needs to decide whether it is the normal or the conjugate image that is to be displayed as the paint layer will prevent illumination of the hologram from the painted side and one of the possible image reconstructions is going to be discarded. Therefore, one needs to decide at the recording stage which side of the plate is to face the object (see Figure 6.4b).

6.3 DICHROMATED GELATIN (DCG)

Dichromated gelatin has long been known as a recording material having high diffraction efficiency. Gelatin plates can be obtained from silver halide plates by soaking them in a photographic fixing solution to remove the silver halide. The gelatin is then soaked for 5 minutes in a solution of ammonium dichromate with a small amount of wetting agent and allowed to dry in total darkness. The plates should be used quite soon afterwards as reduction of hexavalent chromium ions (Cr^{6+}) to trivalent ions (Cr^{3+}) does occur, albeit slowly, even in the dark.

The hexavalent chromium ions are reduced to trivalent ions by light and form crosslinks between the gelatin molecules. This hardens the gelatin in regions of constructive interference of light during holographic recording. Typical exposure is 10 mJ/cm² at 488 nm and 50 mJ/cm² at 514 nm. Gelatin has been sensitized for use with red lasers such as helium–neon by adding methylene blue dye to the dichromate solution [2] requiring exposure of 100 mJ/cm² and increased sensitivity is obtained by the further addition of tetramethyl-guanidine [3].

The processing is simple but must be handled with care in a fume hood as it requires the use of isopropyl alcohol. The first step is a thorough soaking in water, causing the gelatin to swell to many times its original volume. The next step is a rapid drying process using isopropyl alcohol/water baths, with alcohol concentration increasing from 10% to 100% although a simpler process involving two baths of pure isopropyl alcohol at 30°C also works well. Simplified processing methods have also been published [4,5]. The procedure given in reference [4] has been found useful when time is limited.

The final alcohol bath is followed by drying, which can be accelerated using a hair dryer on the emulsion side of the plate as it is slowly removed from the alcohol. The hologram should be protected by a glass cover sheet and sealed around the edges with epoxy resin as moisture can render the reconstructed image invisible.

The mechanism of light induced refractive index change in dichromated gelatin is still not fully understood but there appears to be agreement that two processes are at work. One is the creation of voids reducing the refractive index in the soft gelatin, during the drying process. These voids are presumably small enough not to cause any scattering. The other process may be some form of chemical complexing in hardened gelatin regions involving the Cr^{3+} ions, isopropanol and gelatin leading to an increase in refractive index in those regions. In any event, a refractive index modulation of around 0.08 is readily obtained in carefully processed dichromate gelatin.

Figure 6.5 shows the characteristics of PFG-04 dichromated gelatin available from Integraf. It has a shelf life of 12 months and is designed for phase reflection hologram

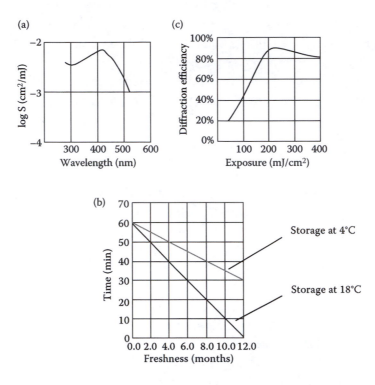

FIGURE 6.5 Integraf DCG. (Reproduced from Integraf LLC, www.integraf.com. With permission.)

recording with blue or green lasers. The resolution is greater than 5000 lines per millimeter and emulsion thickness is 16–17 μm.

The sensitivity reaches 100 mJ/cm² in the blue and 250 mJ/cm² in the green.

The processing steps are as follows:

1. Thermal hardening after exposure (100°C)—depending on the layer freshness (see Figure 6.5c).
2. Allow to cool to room temperature.
3. Soak in running filtered water for 3 minutes.
4. Soak in 50% isopropyl alcohol solution for 2–3 minutes.
5. Soak in 75% isopropyl alcohol solution for 2–3 minutes.
6. Soak in 100% isopropyl alcohol solution for 2–3 minutes.
7. Dry in a desiccator at 100⁰C for 60 minutes.

The final hologram should be sealed to protect it from moisture.

Diffraction efficiency of more than 75% can be obtained.

6.4 THERMOPLASTICS

Thermoplastic recording materials are used mainly when a holographically reconstructed image is to be compared closely and precisely with the original object. The prime example is in holographic interferometry which will be discussed in Chapter 10. The reason is that the material is processed *in situ* using a combination of heat and an electric field so that the reconstructed image is superimposed perfectly on the object itself if the object remains in place. The format is usually quite small, typically 25 cm² in area, although this is quite adequate for many applications, particularly holographic interferometry. The recorded hologram can be erased and a new one written with several hundred cycles of recording and erasure possible.

Thermoplastics are materials that when softened by heating can be distorted by a spatially varying electric field. This is the principle on which holographic recording is based.

There are four components in a holographic thermoplastic recording layer (Figure 6.6): a glass substrate coated with a thin transparent electrode made from indium tin oxide (ITO), a photoconducting layer on top of the ITO layer, and finally a thermoplastic layer on top of the photoconductor. The thermoplastic layer is first uniformly coated on top with positive electric charge, in the dark, using a corona discharge, which consists of an electrically charged line array of sharp points of metal passing over the thermoplastic layer (Figure 6.6a). This sets up a uniform electric field across the thermoplastic layer with negative charges on the ITO layer.

Illumination by the object and reference beams releases free charges in the photoconducting layer, the positive ones moving to the ITO, partly neutralizing the negative charge there (Figure 6.6b), while the negative ones move to the underside of the thermoplastic but cannot cross it as it is an insulator. A second corona charging step adds more positive charges to the upper surface of the thermoplastic in those regions below which negative charge accumulated following exposure (Figure 6.6c). The electric field across the thermoplastic layer now maps the spatial distribution of light intensity during recording. The final step is to heat the thermoplastic to its softening point, by passing current through the ITO, so that its surface buckles under the action of the spatially varying electric field becoming thinner in regions where the electric field is greater. The finished hologram takes the form of variations in thickness of the thermoplastic layer (Figure 6.6d), which are frozen when the temperature is reduced below the softening point again.

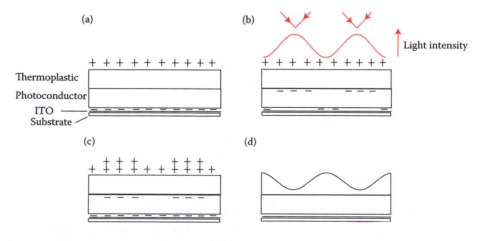

FIGURE 6.6 Thermoplastic structure and recording.

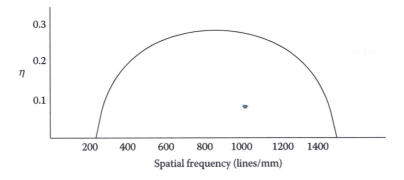

FIGURE 6.7 MTF thermoplastic.

The hologram can be erased by raising the temperature well above the softening point so that the thermoplastic can flow freely and thickness variations are smoothed out.

The range of spatial frequencies that can be recorded ranges from a few hundred to about 1500 lines per millimeter (Figure 6.7).

These limitations are acceptable for holographic interferometry, the most common holographic application for thermoplastics. Thermoplastic cameras are available from Tavex at http://www.tavexamerica.com/holocamera.htm offering 3000 write erase cycles with a hologram area of 32 × 32 mm.

6.5 PHOTORESISTS

Photoresists are organic resins of two types. One, known as negative photoresist, becomes insoluble in an appropriate developer following exposure to light while the other, positive photoresist, becomes soluble. They have long been used in the manufacture of integrated circuits containing very large numbers of electronic components (see Section 9.3.1). A pattern of photoresist laid on top of a slice of silicon allows for spatially selective doping of the silicon with another element as one step in the preparation of a very large number of different electronic devices at the same time. As a consequence, photoresists have been very well characterized.

The positive photoresists are preferred for holographic recording because they adhere to their substrates more firmly, whereas the negative ones depend on extensive exposure to light to ensure adherence. Layers of just a few micrometers in thickness are normally used.

Photoresists are only sensitive to UV and blue light so that the 458 nm line of the argon ion laser or the 442 nm line of the helium–cadmium laser are preferred even though the power available at these wavelengths is much less than from a 514 nm argon ion laser. Exposure times are usually quite long so extra care is needed to avoid mechanical or thermal instabilities during recording. Following holographic exposure, positive photoresist dissolves in an alkaline developer solution more readily in places where the exposing light intensity has been high, creating a surface relief pattern. Spatial frequencies up to 1500 lines per millimeter can be recorded. The finished hologram lends itself well to nickel plating and the resulting nickel plate surface relief hologram can be used for mass production of holograms in thin metal foil by hot stamping (see Section 9.7).

6.6 SELF-PROCESSING MATERIALS

Many materials exhibit changes when illuminated by light. For example, the polarizability may change, which in turn causes a change in the extent to which the material absorbs light in different parts of the electromagnetic spectrum and in its refractive index. Examples of the former include colored dyes bleached by light so that they becomes colorless and chalcogenide glasses, such as arsenic trisulfide (As_2O_3), which become partially opaque. An example of refractive index change is conversion of double chemical bonds in monomer molecules to single ones in a large polymer molecule. Another class of materials are the photorefractives in which changes in refractive index light are induced by exposure to light.

There are significant advantages, including both safety and convenience, in the use of many of these materials, which generally require no processing after recording.

6.6.1 PHOTOCHROMIC AND PHOTODICHROIC MATERIALS

Photochromic materials change color reversibly upon exposure to light and so are capable of recording amplitude holograms. In general, the material is bleached by light in the recording process so the optical density becomes spatially modulated although, unfortunately, the mean density decreases with consecutive recordings. The spatial modulation of the optical density is erased by spatially uniform illumination. Organic photochromics have not been much used for holography as they have a limited life and suffer from fatigue, that is, their ability to record a hologram followed by erasure and rerecording declines with age.

Certain photochromics, in which defects occur known as color centers, are also dichroic. Examples are the alkali halide crystals sodium fluoride, sodium chloride, and potassium chloride. There are various kinds of defects. One type, known as an F_A center, takes the form of an electron trapped at a vacant site where a negative ion would normally occur, in the vicinity of an impurity ion (Figure 6.8).

Such centers exhibit dichroism, meaning that they absorb light of appropriate wavelength and plane of polarization. Thus, they store the polarization state of the illuminating light. The consequence is that holographic recording is possible using object and reference waves, which are plane polarized in orthogonal planes (x,z) and (y,z) (Figure 6.9).

When the two waves are in phase, that is, the electric field is a maximum in the x and y directions at the same point z, the resultant electric field vector is at $+\pi/4$, measured clockwise from the x axis (looking in the $+z$ direction), that is, having unit vector $(\hat{x}, \hat{y})$, and the absorption of the F_A center is reduced in this electric field direction. At the same time, the absorption is increased in the orthogonal direction with unit vector $(-\hat{x}, \hat{y})$. The phase difference between the two polarizations changes from 0 through π, 2π, 3π, and so on, from left to right across the recording medium and so the absorption of light, which is

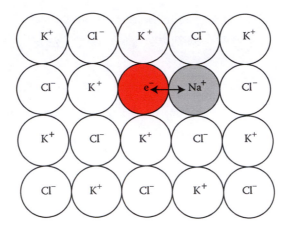

FIGURE 6.8 F$_A$ center in potassium chloride.

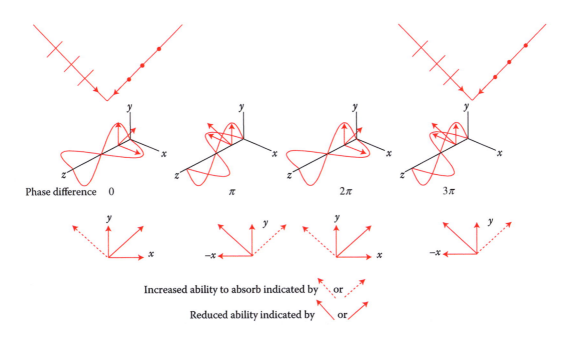

FIGURE 6.9 Photodichroic holographic recording.

plane polarized in either of these two directions, is spatially modulated and there are *two* absorption gratings, with a phase difference of π between them.

At reconstruction, a reference beam with the same plane of polarization as at recording is incident on the hologram at the Bragg angle and the object beam is reproduced in its original polarization state.

Inorganic photochromics have a resolution of more than 10,000 lines per millimeter and recyclable (over 10^9 times) [6] and they are grainless so that scatter is very low with over 100 holograms having been recorded [7]. Diffraction efficiency is however only a few percent and the sensitivity is low.

6.6.2 PHOTOREFRACTIVES

Photorefractive materials, as their name implies, undergo changes in refractive index upon exposure to light. Strictly speaking they are photo*electro*refractive as the refractive change is mediated through the electrooptic effect. They are available as high quality crystals up to several centimeters thick and have been studied extensively for holographic data storage applications. Figure 6.10 shows the recording process in a photorefractive crystal, which has a large energy bandgap between its valence and conduction bands compared with the photon energy of the illuminating light, so that it is transparent.

In the bandgap, there are energy levels occupied by electrons trapped at photoinactive sites. These electrons are excited by light so that they gain enough energy to enter the conduction band. They then diffuse through the material, either freely, or driven by an externally applied constant electric field across the crystal, until they become trapped at defects (denoted by empty circles in Figure 6.10) but mainly only in dark interference fringe regions. We thus have a spatially varying charge density with maximum negative charge density in dark fringe regions, which sets up a spatially varying electric field, E. This electric field is related to the charge distribution that produced it, see Equation 1.4 this time with ρ, the charge density included

$$\nabla \cdot D = \rho.$$

Thus, a sinusoidal spatial variation of charge density produced by the illuminating interference fringe pattern gives rise to a *cosinusoidal* spatial variation in D and therefore of E. Through the electrooptic effect, a spatially varying refractive index is set up by the spatially varying electric field and we see that this index modulation is out of phase with the original light distribution by $\pi/2$.

Photorefractive crystals are usually about 1 cm in thickness, a typical material being iron doped lithium niobate ($LiNbO_3$) although its sensitivity is quite low. Greater sensitivity is

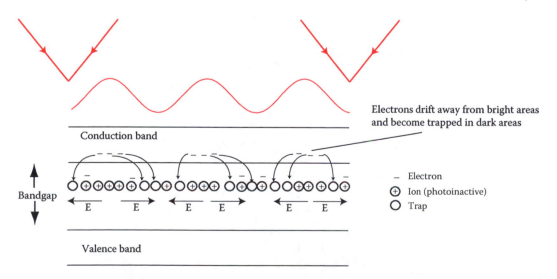

FIGURE 6.10 Photorefractive recording.

obtained from $Bi_{12} SiO_{20}$ and other photoconducting photorefractive materials by applying a potential difference of typically several kilovolts across the crystal. The crystal thickness allows for the storage of a large number of holograms because of the very high angular selectivity although reconstruction at the recording wavelength and at the appropriate Bragg angle can erase the recorded information. This is because the electrons that were trapped at the defects in dark regions are freed again and the space charge field is ultimately erased through drift. To avoid this, one can exploit the birefringence of the crystal and use a longer wavelength for reconstruction for which the Bragg angle is different from that used in the recording. Thermal stabilization is also exploited. Here, hydrogen ion charges produced at higher temperatures (120–135°C in the case of $LiNbO_3$) neutralize the original spatially modulated electronic space charge density. The crystal is then cooled to room temperature and uniformly illuminated with white light, erasing the negative space charge grating, leaving just a spatially modulated hydrogen ion charge density as H^+ ions are not affected by the illumination. The associated spatially modulated electric field and refractive index are thus also stabilized.

Frejlich [8] provides a comprehensive review of these very important materials and some of their holographic applications.

Other holographic applications of photorefractives, apart from data storage, include real-time holography, phase conjugation, and holographic interferometry, all of which will be discussed in later chapters.

Photorefractive polymers [9] are attractive for dynamic holographic image display because they can be prepared in large area format and require very low electric fields.

6.6.3 NONLINEAR OPTICAL MATERIALS

Dielectric materials such as glass and many crystalline materials do not normally conduct electrical current. However, when a dielectric material is placed in an electric field, it becomes polar. By this, we mean that the electric field displaces the bound charges (electrons and protons) of the dielectric in opposite directions thus setting up an additional electric field. Equal charges of magnitude q and of opposite sign separated by distance d in space constitute an electrical dipole, which has an *electrical dipole moment* of magnitude qd. The forces that bind the charges in dielectrics determine the magnitude of the dipole moment induced by an external electric field. The polarization, P, is the electric dipole moment per unit volume and is normally *linearly proportional* to the applied field, E through a parameter called the dielectric susceptibility, χ.

Nonlinear optical materials have the property that the polarization P induced in them by an electric field E is described by

$$P \propto \chi^{(1)} E + \chi^{(2)} E^2 + \chi^{(3)} E^3 \ldots .$$

where $\chi^{(1)}$, $\chi^{(2)}$, and so on are *linear*, second-order, and higher-order susceptibilities of the material, with $\chi^{(1)} \gg \chi^{(2)} \gg \chi^{(3)}$. Nonlinear optical effects are only obtained at the very large electric field values associated with laser light (see Section 1.4). Thus, the polarization may oscillate at the frequencies associated with E, E^2, and E^3 and therefore radiate light at those frequencies. This is how second and third harmonics of a light wave can be generated, the most ubiquitous being 532 nm laser light, which is the second harmonic of the light produced by a Neodymium Yag laser (Section 5.9). Nonlinear materials are used mainly for phase conjugation (Section 3.6) as we shall now show.

Suppose we illuminate such a material with three light beams having the same frequency and take the third order polarization.

$$P \propto \left(E_1 + E_2 + E_3 \right)^3 \tag{6.2}$$

The easiest way to deal with this expression is to write for each electric field

$$E_i = E_{0i} \cos \left(\mathbf{k} \cdot \mathbf{r} - \omega t \right) = \frac{1}{2} \left\{ E_{0i} \exp \left[j \left(\mathbf{k} \cdot \mathbf{r} - \omega t \right) \right] + E_{0i} \exp \left[-j \left(\mathbf{k} \cdot \mathbf{r} - \omega t \right) \right] \right\}$$

$$E_i = \frac{1}{2} (\tilde{E}_i + \tilde{E}_i^*) \quad i = 1, 2, 3 \ldots \tag{6.3}$$

Among other terms in the expression for P is one proportional to $\tilde{E}_1 \tilde{E}_2 \tilde{E}_3^*$, which in turn is proportional to $\tilde{E}_3^*$, if $\tilde{E}_1$ and $\tilde{E}_2$ are collimated beams traveling in opposite directions, that is, phase conjugate to each other. This process is called *degenerate four wave mixing*, because all the waves have the same temporal frequency.

6.6.4 PHOTOPOLYMERS

There is growing interest in the use of photopolymers for holographic applications because of their modest cost, high resolution, low scatter, and minimal or even no requirement for physical or chemical processing. A number of manufacturers have developed photopolymers but have not marketed them to any extent so that availability is rather restricted although some manufacturers have made their products available to research laboratories for evaluation purposes under nondisclosure agreements. The leading manufacturers are DuPont, makers of Omnidex© photopolymer, Aprilis, Bayer, and InPhase.

Generally, a photopolymer system consists of a monomer or monomers, a sensitizing dye and a cosensitizer (sometimes referred to as a catalyst or electron donor) and, optionally, a polymeric binder. If the binder is omitted, the layer is usually in liquid form and sandwiched between glass plates with spacers to ensure a specific layer thickness. Inclusion of the binder enables dry layers to be formed.

6.6.4.1 Preparation of a Photopolymer

The preparation of a dry layer self-processing photopolymer has been described [10]. Based on work by Calixto [11], it consists of a binder, polyvinylalcohol (PVA), a monomer, acrylamide, and a cross-linking monomer, bis-acrylamide, as well as a cosensitizer, triethanolamine (TEA) and a dye sensitizer chosen to match the recording laser wavelength that is to be used. All the components are water soluble.

6.6.4.2 MECHANISM OF HOLOGRAM FORMATION IN PHOTOPOLYMER

Light of appropriate wavelength causes a dye molecule (XD) to go into a singlet excited energy state, one of many in a closely spaced set (Figure 6.11).

$$XD + h\nu \rightarrow {}^1XD^*$$

After descending to the lowest state in the set, it may return quickly to the ground state emitting light of longer wavelength, a process called fluorescence, which can also be triggered by interaction with a cosensitizer molecule.

Alternatively, the excited state dye molecule may enter a comparatively long-lived, excited triplet state ($^3XD^*$) through a mechanism called intersystem crossing.

Competing processes, such as fluorescence, interaction with a dye molecule in its ground state or with oxygen, can result in the triplet state dye molecule returning to its ground state. However, it can interact with the cosensitizer molecule, TEA, which has an unused pair of electrons (denoted :) in its valence band so that the dye becomes a radical (denoted •) that is a molecule with an unpaired outer electron.

$$^3XD^* + (HOCH_2\,CH_2)_3N: \rightarrow XD^{\bullet\,-} + (HOCH_2\,CH_2)_3N^{\bullet\,+}$$

Loss of a single proton by the TEA leads to formation of a TEA radical

$$(HOCH_2\,CH_2)_3N^{\bullet\,+} \rightarrow (HOCH_2\,CH_2)_2\,N^{\bullet}\,CHCH_2OH + H^+$$

which in turn can begin the process of free radical polymerization of acrylamide monomer molecules, which link together to form larger molecules (Figure 6.12) by conversion of double bonds to single bonds. The polymer molecule continues to grow until termination occurs.

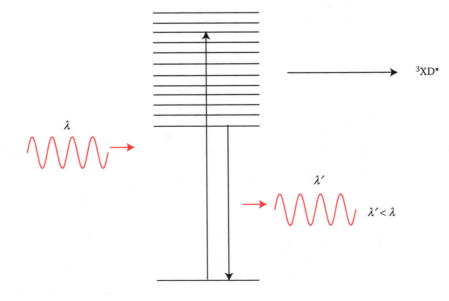

FIGURE 6.11 Dye fluorescence.

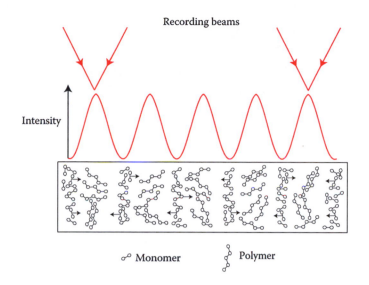

FIGURE 6.12 Photopolymer recording.

Since double bonds are converted to single ones, the polarizability of the molecules changes, with an accompanying reduction in refractive index. As monomer is used up in the formation of polymer, more monomer diffuses from dark fringe regions into bright fringe regions, driven by the gradient of monomer concentration. This naturally leads to an increase in density in the bright fringe regions with an accompanying increase in refractive index.

Short polymer molecules can escape by diffusion from the regions in which they are formed, into dark fringe regions [12–14], with adverse effect on both the refractive index modulation and the spatial resolution of which the material is capable. Photopolymer compositions and recording conditions favorable to the formation of longer, comparatively immobile polymer chains, therefore tend to facilitate the fabrication of high spatial frequency gratings as well as reflection gratings and holograms.

It is also found that a volume holographic grating is accompanied by the formation of a surface grating of the same spatial frequency with the peaks of the surface profile in the same positions as the bright interference fringes [15]. Surface gratings up to 1500 lines per millimeter have been recorded although the higher spatial frequencies (>300 lines per millimeter) only become manifest after baking [16].

6.7 HOLOGRAPHIC SENSITIVITY

It is of great importance that we have some way of comparing the holographic performance of different recording materials. In Section 3.1, we considered the interference of two plane waves with resulting intensity

$$I(x,y) = \frac{1}{2}\left\langle \left(E_R \exp\left[j(\mathbf{k}_R \cdot \mathbf{r} - \omega t)\right] + E_O(x,y) \exp\left[j(\mathbf{k}_O(x,y)\cdot\mathbf{r} - \omega t)\right]\right) \times c.c. \right\rangle \quad (6.4)$$

$$I(x, y) = \frac{1}{2}\{E_R^2 + E_O^2 + 2E_R E_O \cos(\mathbf{k}_R \cdot \mathbf{r} - \mathbf{k}_O(x, y) \cdot \mathbf{r})\} \quad (6.5)$$

The recording material is exposed to the intensity distribution $I(x,y)$ for time, τ, then processed if required, and illuminated by an exact copy of the reference wave. The term describing the reconstructed object wave, $E_{recon}(x,y,t)$, is

$$\tilde{E}_{recon}(x, y, t) = \tilde{S}\tau E_R^2 E_O \exp[j(\mathbf{k}_O(x, y) \cdot \mathbf{r}) - \omega t]/2 \quad (6.6)$$

where $\tilde{S}$ is a complex constant called the *holographic sensitivity*, which characterizes the material's holographic response whether the hologram is of the amplitude or phase type and indeed whether it is thick or thin. The diffraction efficiency η, given by the ratio of the intensities of the reconstructed and reconstructing waves, is

$$\eta = \frac{\left|\tilde{S}\right|^2 \tau^2 E_R^2 E_O^2}{4} \quad (6.7)$$

$$\sqrt{\eta} = \frac{S\tau E_R E_O}{2} \quad (6.8)$$

The contrast or visibility, V, of the interference pattern is given by

$$V = \frac{I_{max} - I_{min}}{I_{max} + I_{min}} = \frac{2E_R E_O}{E_R^2 + E_O^2} \quad (6.9)$$

$$\sqrt{\eta} = S\tau\left(E_R^2 + E_O^2\right)V/4 = SVE/2 \quad (6.10)$$

where $E = E_{av}\tau$ is the average exposure in J/m^2 during the recording.

As we shall see in Chapter 7, the visibility of the fringe pattern is always reduced *inside* the recording material. Nonetheless, for any given recording material, we can record holographic diffraction gratings using different exposures, E, and constant incident fringe pattern visibility, V, or vice versa to obtain graphs from which the holographic sensitivity, S, can be found. In this way, we can compare the sensitivities of different materials.

6.8 SUMMARY

This chapter deals with holographic recording materials. Silver halide receives special attention along with a number of chemical processing schemes of particular relevance to holography. We also describe dichromated gelatin, and photoresist recording mechanisms. Self-processing recording materials, such as photothermoplastics and photopolymers, are of particular importance in holography and many of its application, and we consider these in some detail as well as photochromic and photodichroic materials. We discuss photorefraction for holographic recording. Finally, we discuss holographic sensitivity.

REFERENCES

1. H. I. Bjelkhagen (ed.), *Silver Halide Recording Materials and Their Processing*, Springer-Verlag, Berlin (1995).
2. T. Kubota and T. Ose, "Methods of increasing the sensitivity of methylene blue sensitized dichromated gelatin," *Appl. Opt.*, 18, 2538–9 (1979).
3. J. Blythe, "Methylene blue sensitized dichromated gelatin holograms: A new electron donor for their increased photosensitivity," *Appl. Opt.*, 30, 1598–602 (1991).
4. T. G. Georgekutty and H. K. Liu, "Simplified dichromated gelatin hologram recording process," *Appl. Opt.*, 26, 372–6 (1987).
5. M. H. Jeong, J. B. Song, and I. W. Lee, "Simplified processing method of dichromated gelatin holographic recording material," *Appl. Opt.*, 30, 4172–3 (1991).
6. J. Bordogna, S. Keneman, and J. Amodei, "Optical storage and display," *RCA Rev.*, 33, 227–247 (1972).
7. H. Blume, T. Bader, and F. Bluty, "Bi-directional holographic information storage based on the optical reorientation of F_A," *Opt. Commun.*, 12, 147–51 (1974).
8. J. Frejlich, *Photorefractive Materials*, Wiley (2007).
9. R. A. Norwood, S. Tay, P. Wang, P.-A. Blanche, D. Flores, R. Voorakaranam, W. Lin, J. Thomas, T. Gu, P. St. Hilaire, C. Christenson, M. Yamamoto, and N. Peyghambarian, Photorefractive Polymers for Updatable Holographic Displays, in Organic Thin Films for Photonic Applications, ACS Symposium Series, 1039 (2010)
10. S. Martin, C. A. Feely, and V. Toal, "Holographic recording characteristics of an acrylamide-based photopolymer," *Appl. Opt.*, 36, 5757 (1997).
11. S. Calixto, "Dry polymer for holographic recording," *Appl. Opt.*, 26, 3904–10 (1987).
12. S. Martin, I. Naydenova, R. Jallapuram, R. Howard, and V. Toal, "Two way diffusion model for the recording mechanism in a self developing dry acrylamide photopolymer," Holography 2005: Intl. Conf. on Holography, Optical Recording and Processing of Information, Yu. Denisyuk, V. Sainov, and E. Stoykova (eds.), Proc. SPIE, 6252, 625025 (2006).
13. I. Naydenova, S. Martin, R. Jallapuram, R. Howard, and V. Toal, "Investigations of the diffusion processes in self-processing acrylamide-based photopolymer system," *Appl. Opt.*, 43 (14), 2900 (2004).
14. T. Babeva, I. Naydenova, D. Mackey, S. Martin, and V. Toal, "Two-way diffusion model for short-exposure holographic grating formation in acrylamide-based photopolymer," *J. Opt. Soc. Am. B*, 27, 197–203 (2009).
15. K. Pavani, I. Naydenova, S. Martin, and V. Toal, "Photoinduced surface relief studies in an acrylamide-based photopolymer," *J. Opt. A: Pure Appl. Opt.*, 9, 43–48 (2007).
16. K. Trainer, K. Wearen, D. Nazarova, I. Naydenova, and V. Toal, "Optical patterning of photopolymerisable materials," International Commission for Optics, Topical Meeting on Emerging Trends and Novel Materials in Photonics, AIP Conf. Proc., 1288, 184–187 (2010).

PROBLEMS

1. Suppose we want to record a hologram in silver halide using a helium–neon laser but the only film available has a spatial resolution of 150 mm^{-1}. What is the largest angle between object and reference beams that we can use?
2. Silver halide film, although cheaper than silver halide on glass, is flexible and must be secured in some way in order to use it in holographic recording. One way of doing this is to place it between glass plates that are then clamped together. However, very often we find that, although successful, the finished hologram is at least partly covered by bright

and dark bands. What is the reason for this? Can you suggest a better way of securing the film?

3. We wish to use a silver halide emulsion to record a single beam Denisyuk hologram, 15×10 cm in area, with a developed photographic density of 2.0, before bleaching. Our laser produces output optical power of 5 mW at 532 nm. To obtain the required density at this wavelength, the sensitivity of the emulsion is 50 μJ/cm². What should be the exposure time?

4. Silver halide has to be chemically processed and the last steps in this process are washing and drying. If a Denisyuk hologram recorded at 633 nm is allowed to dry naturally, the color of the reconstructed image is quite close to that of the recording laser light. If the hologram is dried using a hot air dryer, the reconstructed image is seen initially to be green or even blue, gradually turning red. Why is this?

5. Assuming that the refractive index modulation obtained in processed dichromated gelatin is 0.08 what thickness will result in a diffraction efficiency of 100% at a recording wavelength of 473 nm?

CHAPTER 7

Recording Materials in Practice

7. 1 INTRODUCTION

We now need to look at the characteristics of holographic recording materials and holograms in order to be able to exploit them to best advantage. This entails looking more closely at some of the simplifying assumptions that we have made up to now. First, in Chapter 3, we assumed that the amplitude transmittance of a recorded amplitude hologram is linearly proportional to the recording light intensity. Second, in the case of phase holograms, we assumed, also in Chapter 3, that the phase imposed on the reconstructing light by the hologram is proportional to the original recording intensity. We also assumed that the recording medium has no grain structure and therefore does not scatter light so that the contrast of the interference pattern is in no way affected by the recording medium. Furthermore, we assumed the reconstructing light wave to be identical to the recording light wave. We will discard each of these assumptions in turn to see what the effects are in practice. We must also look at the phenomenon of laser speckle, which has adverse effects on holographic images.

7.2 NONLINEAR EFFECTS

Except in very restricted circumstances (Figure 7.1), we cannot, in practice, assume that the amplitude transmission of an amplitude hologram is proportional to the recording light intensity, nor that in a phase hologram, an additional phase proportional to the recording light intensity is imposed on the reference wave.

The absorption or the refractive index of the material or the thickness, or any combination of them, can be affected in the recording process so that we have a *complex* amplitude transmissivity

$$\tilde{t}(x,y) = \exp[-\alpha(x,y)d(x,y)]\exp\left[j\frac{2\pi n(x,y)d(x,y)}{\lambda}\right] \tag{7.1}$$

where α is the absorption coefficient per unit length at (x,y) in the hologram plane, d is the thickness, and n the refractive index at the same point. Equation 7.1 can be written as

$$\tilde{t}(x, y) = t \exp(j\theta) \tag{7.2}$$

where $t = \exp[-\alpha(x, y)d(x, y)]$ and $\theta = 2\pi n(x, y)d(x, y)/\lambda$, both real.

In a pure amplitude hologram, only αd varies, while in a phase hologram, α is almost zero and n or d or both may vary.

7.2.1 NONLINEARITY IN AN AMPLITUDE HOLOGRAM

Although amplitude holograms are the exception rather than the rule, it is worth considering the consequences of the nonlinear dependence of pure amplitude transmissivity on exposure as shown in Figure 7.1a. In the discussion that follows, we will simply use the terms O and R for object and reference waves where $O = E_O(x, y)\exp\left[j(\mathbf{k}_O(x, y)\cdot\mathbf{r} - \omega t)\right]$ and $R = E_R \exp\left[j(\mathbf{k}_R\cdot\mathbf{r} - \omega t)\right]$, (see Section 3.5.1) and the recording intensity is

$$I = [O + R][O * + R *] = |O|^2 + |R|^2 + OR * + O * R \tag{7.3}$$

It is clear from Figure 7.1a that the amplitude transmissivity depends linearly on exposure only over a limited exposure range. In practice, we assume a nonlinear relationship between t and E,

$$t = t_0 + C_1 TE + C_2 T^2 E^2 + C_3 TE^3 \cdots \tag{7.4}$$

where $C_1 \gg C_2 \gg C_3 \ldots$ and T is the exposure time. Assuming higher powers than E^2 can be neglected, the output from the hologram at reconstruction is

$$\begin{aligned}
&R(t_0 + C_1 TE + C_2 T^2 E^2) \\
&= Rt_0 + C_1 T[R(|O|^2 + |R|^2) + O|R|^2 + O * R^2] \\
&\quad + C_2 T^2 [\underset{1}{R(|O|^4 + |R|^4 + 4|O|^2|R|^2)} + \underset{2}{R(OR*)^2} + \underset{3}{R(O*R)^2} + \underset{4}{2|O|^2 O|R|^2} \\
&\quad + \underset{5}{2|O|^2 O * R^2} + \underset{6}{2O|R|^4} + \underset{7}{2O*|R|^2 R^2}]
\end{aligned} \tag{7.5}$$

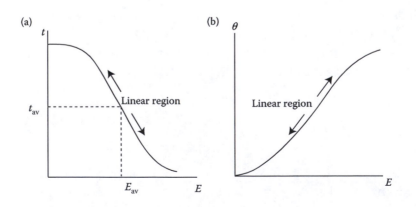

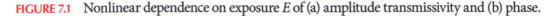

FIGURE 7.1 Nonlinear dependence on exposure E of (a) amplitude transmissivity and (b) phase.

In Section 3.5.2, we saw that a halo appears around the reference wave at the reconstruction stage because of the presence of the term $E_O^2(x, y)$. Here, we have an additional term $|O|^4$ (term **1**) accompanying the reference wave and the halo associated with it has four times the spatial frequency content of the object wave. Terms **2** and **3** are second-order terms in O and O^*, respectively. To see the effect of the former, consider an object wave of unit amplitude emitted by a single point source at $(x_1, y_1 -z_1)$. The phase difference between object light arriving at a point x, y, z on the hologram and light arriving at 0, 0, 0 on the hologram is

$$\tilde{O} = \exp\left[\frac{j2\pi}{\lambda}\left(\left\{(x-x_1)^2 +(y-y_1)^2 +(z+z_1)^2\right\}^{\frac{1}{2}} -\left\{x_1^2 + y_1^2 + z_1^2\right\}^{\frac{1}{2}}\right)\right]$$

$$= \exp\left[\frac{j2\pi}{\lambda}\left(z_1\left\{\frac{(x-x_1)^2}{z_1^2} +\frac{(y-y_1)^2}{z_1^2} +\left(1+\frac{z}{z_1}\right)^2\right\}^{\frac{1}{2}} -z_1\left\{1+\frac{x_1^2+y_1^2}{z_1^2}\right\}^{\frac{1}{2}}\right)\right]$$

$$\cong \exp\left[\frac{j2\pi}{\lambda}\left(z_1\left\{1+\frac{(x-x_1)^2}{2z_1^2} +\frac{(y-y_1)^2}{2z_1^2}\right\} -z_1\left\{1+\frac{x_1^2+y_1^2}{2z_1^2}\right\}\right)\right]$$

for small z and x_1, y_1 both $\ll z_1$

$$\tilde{O} = \exp\left[\frac{j2\pi}{\lambda}\frac{1}{2z_1}\left\{x^2 -2xx_1 +y^2 -2yy_1\right\}\right] \tag{7.6}$$

The term **2** in O^2 is therefore of the form

$$\exp\left[\frac{j2\pi}{\lambda}\frac{1}{2(z_1/2)}\left\{x^2 -2xx_1 +y^2 -2yy_1\right\}\right]$$

and represents a spherical wave emanating from a virtual point source at $-z_1/2$ and therefore having half the radius of curvature of the original object wave at the hologram plane (Figure 7.2).

A similar effect occurs in relation to term **3** in O^{*2}.

While terms **6** and **7** have no additional effect on O and O^*, respectively, terms **4** and **5** produce additional halo effects due to $|O|^2$ around the images O and O^*, respectively. Consider term **4**. We confine our attention to an object wave produced by just two point sources at x_1, y_1, z_1 and x_2, y_2, z_2 each of unit amplitude. Using Equation 7.6, the object wave is

$$O = \exp\left[\frac{j2\pi}{\lambda}\frac{1}{2z_1}\left\{x^2 -2xx_1 +y^2 -2yy_1\right\}\right] +\exp\left[\frac{j2\pi}{\lambda}\frac{1}{2z_2}\left\{x^2 -2xx_2 +y^2 -2yy_2\right\}\right] \tag{7.7}$$

Term **4** is then (assuming $2|R|^2$ is a constant)

$$|O|^2 O = \left(\exp[j\theta_1] +\exp[j\theta_2]\right)^2 \left(\exp[-j\theta_1] +\exp[-j\theta_2]\right)$$
$$= 3\exp(j\theta_1) +\exp[j(2\theta_2 -\theta_1)] +3\exp(j\theta_2) +\exp[j(2\theta_1 -\theta_2)] \tag{7.8}$$

where $\theta_1 = (2\pi/\lambda)(1/2z_1)\{x^2 -2xx_1 +y^2 -2yy_1\}$ and $\theta_2 =(2\pi/\lambda)(1/2z_2)\{x^2 -2xx_2 +y^2 -2yy_2\}$.

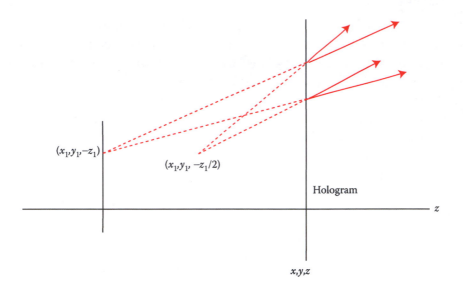

FIGURE 7.2 Effect of nonlinear recording. An object point at $(x_1, y_1, -z_1)$ is imaged in reconstruction at its original location (linear) and at $(x_1, y_{1,} -z_1/2)$ (nonlinear).

The first and third terms in Equation 7.8 are replicas of the original two point sources. The second term is

$$\exp\left[j(2\theta_2 - \theta_1)\right] = \exp\left[j\left\{2\frac{2\pi}{\lambda}\frac{1}{2z_2}\left\{x^2 - 2xx_2 + y^2 - 2yy_2\right\} - \frac{2\pi}{\lambda}\frac{1}{2z_1}\left\{x^2 - 2xx_1 + y^2 - 2yy_1\right\}\right\}\right]$$

(7.9)

Assume now the point sources are in the same plane parallel to the hologram, that is, $z_1 = z_2$. This is the case in some holographic data storage systems where the object is a 2-D array of point sources each representing binary "1."

$$\exp\left[j(2\theta_2 - \theta_1)\right] = \exp\left[j\frac{2\pi}{\lambda 2z_1}\left(x^2 - 2x(2x_2 - x_1) + y^2 - 2y(2y_2 - y_1)\right)\right]$$ (7.10)

which represents a point virtual image located at $(2x_2 - x_1, 2y_2 - y_1, z_1)$.

The fourth term in Equation 7.8 is

$$\exp\left[j(2\theta_1 - \theta_2)\right] = \exp\left[j\left\{2\frac{2\pi}{\lambda}\frac{1}{2z_1}\left\{x^2 - 2xx_1 + y^2 - 2yy_1\right\} - \frac{2\pi}{\lambda}\frac{1}{2z_1}\left\{x^2 - 2xx_2 + y^2 - 2yy_2\right\}\right\}\right]$$

$$= \exp j\frac{2\pi}{\lambda}\frac{1}{2z_1}\left\{x^2 - 2x(2x_1 - x_2) + y^2 - 2y(2y_1 - y_2)\right\}$$

(7.11)

which represents a point virtual image at $(2x_1 - x_2, 2y_1 - y_2, z_1)$.

Clearly, a direct consequence of nonlinearity is that false binary "1"s appear in the output (Figure 7.3).

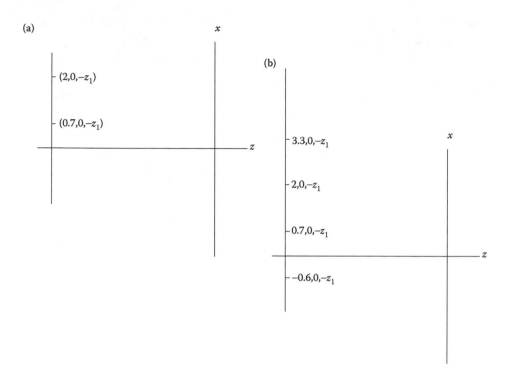

FIGURE 7.3 Effect of nonlinearity on reconstruction of point sources: (a) object points at recording and (b) reconstructed image points.

Comparing term **5**, $2|O|^2 O^* R^2$, with the term $C_1 TO^* R^2$ in Equation 7.5, we see that $|O|^2 O^*$ is superimposed on the conjugate wave O^* since it propagates in the same direction on a carrier wave, R^2 which has twice the spatial frequency of the original reference wave (see Equation 3.9).

$$
\begin{aligned}
|O|^2 O^* &= \left(\exp[j\theta_1]+\exp[j\theta_2]\right)\left(\exp[-j\theta_1]+\exp[-j\theta_2]\right)^2 \\
&= \left(\exp[j\theta_1]+\exp[j\theta_2]\right)\left(\exp[-2j\theta_1]+\exp[-2j\theta_2]+2\exp[-j(\theta_1+\theta_2)]\right) \\
&= 3\exp[-j\theta_1]+\exp[j(\theta_1-2\theta_2)]+3\exp[-j\theta_2]+\exp[j(\theta_2-2\theta_1)]
\end{aligned}
\tag{7.12}
$$

The second term in Equation 7.12 is

$$
\exp\left[j(\theta_1-2\theta_2)\right]=\exp\left\{j\frac{2\pi}{\lambda 2(-z_1)}\left[x^2-2x(2x_2-x_1)+y^2-2y(2y_2-y_1)\right]\right\} \tag{7.13}
$$

which represents an additional point real image at $2x_2-x_1, 2y_2-y_1, +z_1$ and the term

$$
\exp\left[j(\theta_2-2\theta_1)\right]=\exp\left\{j\frac{2\pi}{\lambda}\frac{1}{2(-z_1)}\left[x^2-2x(2x_1-x_2)+y^2-2y(2y_1-y_2)\right]\right\} \tag{7.14}
$$

represents an additional point real image at $(2x_1-x_2, 2y_1-y_2, +z_1)$.

We have seen that object points at $x_1, y_1, -z_1$ and $x_2, y_2, -z_1$ are reconstructed as virtual images at these points and at $(2x_1 - x_2, 2y_1 - y_2, -z_1)$ and $(2x_2 - x_1, 2y_2 - y_1, -z_1)$.

In the case of diffuse objects, we can extrapolate from the above results. Suppose the object is simply a straight line source of light extending from $(0, 0, -z_1)$ to $(x_1, 0, -z_1)$. Then, by setting $x_2, y_2, -z_1 = 0, 0, -z_1$, we find that in addition to the reconstruction of every point in the range 0 to x_1 in its original location, there are reconstructed image points in the ranges 0 to $2x_1$ and 0 to $-x_1$. The total spatial extent of the image is three times that of the original object so that a halo appears around the reconstructed image. A similar halo is produced around the reconstructed conjugate image. As these effects essentially derive from the interference of light from different object points, they are intermodulation effects.

Nonlinear effects can mitigated to a large extent by proper choice of recording geometry. Consider the simple situation in which the object consists of just two plane waves

$$\exp\left[j(ky\sin\theta_1 - \omega t)\right] \quad \text{and} \quad \exp\left[j(ky\sin\theta_2 - \omega t)\right],$$

propagating at angles θ_1 and θ_2 with the z-axis and each of unit amplitude. The complete object wave excluding time dependence is

$$O = \exp(jky\sin\theta_1) + \exp(jky\sin\theta_2).$$

Assuming a plane reference wave of unit amplitude, terms **2** and **4** in Equation 7.5 become respectively

$$O^2 = \exp\left[2jky\sin\theta_1\right] + \exp\left[2jky\sin\theta_2\right] + \exp\left[jky(\sin\theta_1 - \sin\theta_2)\right]$$
$$+ \exp\left[jky(\sin\theta_2 - \sin\theta_1)\right]$$

and $|O|^2 O = \left\{2 + \exp\left[jky(\sin\theta_1 - \sin\theta_2)\right] + \exp\left[jky(\sin\theta_2 - \sin\theta_1)\right]\right\}$
$$\times \left\{\exp(jky\sin\theta_1) + \exp(jky\sin\theta_2)\right\}$$
$$= \exp\left[jky(2\sin\theta_1 - \sin\theta_2)\right] + \exp\left[jky(2\sin\theta_2 - \sin\theta_1)\right] + 3\left\{\begin{matrix} \exp(jky\sin\theta_1) \\ + \exp(jky\sin\theta_2) \end{matrix}\right\}$$

and the associated spatial frequencies are

$$f_1, \ f_2, \ f_1 + f_2, \ f_1 - f_2, \ 2f_1, \ 2f_2, \ 2f_1 - f_2 \quad \text{and} \quad 2f_2 - f_1 \qquad (7.15)$$

where

$$f_1 = \sin\theta_1/\lambda \quad \text{and} \quad f_2 = \sin\theta_2/\lambda$$

If we include all the higher order terms in Equation 7.4, we find that additional spatial frequencies

$$mf_1 \pm nf_2 \qquad (7.16)$$

with m and n positive integers are recorded. These are well separated from the spatial frequencies due to interference between the *reference* wave and each of the two object waves if the reference and object waves are well separated in angle. Suppose that the highest spatial frequency in an object beam is f_1, then, from Equation 7.16, in the second order, the highest recorded intermodulation spatial frequency is $2f_1$. If we choose a recording geometry to ensure that the lower limit of spatial frequency due to interference between object and reference waves exceeds $2f_1$ and the angular spread of spatial frequencies in the object wave is relatively low, meaning that the object subtends a small angle at the hologram plane, then the nonlinear terms are separated from the linear.

In thick holograms, the Bragg angles corresponding to the intermodulation spatial frequencies differ significantly from those associated with the interference of the object and reference waves. We saw in Section 4.3.1.1 that, in a volume hologram, the reconstructing beam has to be at the Bragg angle in order to obtain reconstruction of the object wave with maximum diffraction efficiency. In the calculated example, the angular width of the Bragg curve outside the recording material is found to be about 2°. The two interfering beams are incident at angles of 30° and –30° to the normal and their spatial frequencies are $\pm 1/(2\lambda)$ and the holographically recorded spatial frequency is $1/\lambda$. Thus, the holographically recorded diffraction grating has an associated Bragg angle that is substantially different from the Bragg angles corresponding the intermodulation frequencies. Illumination of the grating by a beam with spatial frequency $-1/(2\lambda)$ will reconstruct a beam with spatial frequency $+1/(2\lambda)$ but not one with spatial frequency $+1/\lambda$.

7.2.2 NONLINEARITY IN PHASE HOLOGRAMS

Even if we are dealing with a pure phase hologram whose complex transmissivity is given by

$$\tilde{t}(x, y) = \exp(j\theta) = 1 + j\theta + \frac{(j\theta)^2}{2!} + \frac{(j\theta)^3}{3!} + \cdots = 1 + j\theta - \frac{\theta^2}{2} - \frac{j\theta^3}{6} + \cdots$$

we still have to modify Equation 3.13 accordingly.

$$
\begin{aligned}
E_R &\exp j\left[(\mathbf{k}_R \cdot \mathbf{r} - \omega t + CI(x, y)\right] \\
&= E_R \exp j\left[(\mathbf{k}_R \cdot \mathbf{r} - \omega t + C\left\{\left(\frac{E_R^2 + E_O^2(x, y)}{2}\right) + E_R E_O(x, y)\cos\left[(\mathbf{k}_R - \mathbf{k}_O(x, y)) \cdot \mathbf{r}\right]\right\}\right]
\end{aligned}
\tag{7.17}
$$

and, on reconstruction, higher order terms will appear as in the pure amplitude case.

7.3 GRAIN NOISE

Some holographic recording materials, especially silver halide, have a grainy structure. In the manufacture of silver halide for holographic use, considerable care is taken to reduce the grain size as much as possible. Grains scatter the light both at the recording stage and in reconstruction.

7.3.1 REDUCTION IN FRINGE CONTRAST

During recording, the pattern of interference between the object and reference beams is adversely affected since light is scattered from one beam into the other. The result is that the interference pattern becomes smeared, and its contrast decreases progressively as the interfering beams propagate through the recording medium. This reduction in contrast is in addition to that due to the finite width of the light source (see Section 2.8.3) and to its lack of temporal coherence.

The intensity of the recording light *on arrival* at the front surface of the recording medium is (using Equation 3.7)

$$I(x, y) = \left(\frac{E_R^2 + E_O^2}{2}\right) + E_R E_O(x, y) \cos\left[(\mathbf{k}_R - \mathbf{k}_O(x, y)) \cdot \mathbf{r}\right]$$

$$= \left(\frac{E_R^2 + E_O^2}{2}\right)\left[1 + \frac{2E_R E_O}{E_r^2 + E_O^2}\cos(\theta_R - \theta_O)\right] I = I_{av}\left[1 + \frac{2E_R E_O}{E_R^2 + E_O^2}\cos(\theta_R - \theta_O)\right]$$

where $\theta_R - \theta_O = \cos\left[(\mathbf{k}_R - \mathbf{k}_O(x, y)) \cdot \mathbf{r}\right]$ and the contrast of the interference pattern *on arrival* is

$$\frac{I_{max} - I_{min}}{I_{max} + I_{min}} = \frac{2E_R E_O}{E_R^2 + E_O^2} \tag{7.18}$$

The contrast is also influenced by the polarization states of the two beams. It cannot be assumed that they are plane polarized parallel to one another, because changes in the polarization states can occur by reflection from optical surfaces in the holographic recording setup and some depolarization of the object beam occurs if the object is diffusely scattering. If the angle between the planes of polarization of the interfering beams is $\phi(x, y)$, the contrast is reduced by a factor $\cos\phi(x, y)$.

The effect of source size, that is, its spatial coherence, μ, must also be included (see Section 2.8.3) so finally the incident contrast V_{in} is given by

$$V_{in} = \mu\frac{2E_R E_O}{E_R^2 + E_O^2}\cos\phi = 2\mu\frac{\sqrt{R}}{1+R}\cos\phi \tag{7.19}$$

where $R = E_O^2/E_R^2$. We assume that the path difference between object and reference beams is much less than the coherence length so that there is no reduction in contrast due to a lack of temporal coherence.

Combining Equations 7.19 and 3.7, the actual recording light intensity is

$$I = I_{av} \left\{ 1 + V_{in} \cos\left[(\mathbf{k}_R - \mathbf{k}_O(x,y)) \cdot \mathbf{r} \right] \right\} \tag{7.20}$$

The scattering effect in the recording medium reduces the fringe contrast still further so we define a parameter, M, called the modulation transfer function, as the ratio of the actual contrast in the recording material V_{mat} to the incident contrast V_{in} at spatial frequency f.

$$M(f) = \frac{V_{mat}(f)}{V_{in}(f)} \tag{7.21}$$

The exposure $E = IT$ where T is the exposure time is

$$\begin{aligned} E &= I_{av} \left\{ 1 + MV_{in} \cos\left[(\mathbf{k}_R - \mathbf{k}_O(x,y)) \cdot \mathbf{r} \right] \right\} T \\ &= E_{av} + E' \left\{ \exp\left[j(\mathbf{k}_R - \mathbf{k}_O(x,y)) \cdot \mathbf{r} \right] + \exp\left[-j(\mathbf{k}_R - \mathbf{k}_O(x,y)) \cdot \mathbf{r} \right] \right\} \end{aligned} \tag{7.22}$$

with $E' = I_{av} MV_{in} T / 2$.

If $R < 1$ then $E_O < E_R$ and assuming that the amplitude transmissivity τ can be written as a Taylor series

$$\tau = \tau\big|_{E_{av}} + E' \left.\frac{d\tau}{dE}\right|_{E_{av}} + \frac{1}{2} E'^2 \left.\frac{d^2\tau}{dE^2}\right|_{E_{av}} + \cdots$$

Neglecting higher order terms $\tau = \tau\big|_{E_{av}} + cE' \tag{7.23}$

with c a constant. Illumination of the hologram with the reference wave gives

$$E_R \exp\left[j(\mathbf{k}_R.\mathbf{r} - \omega t) \right] \tau = E_R \exp\left[j(\mathbf{k}_R.\mathbf{r} - \omega t) \right] \left\{ \tau\big|_{E_{av}} + E'c \right\}$$

and the reconstructed object wave has amplitude $cE_R E' \exp\left[j(\mathbf{k}_O(x,y) \cdot \mathbf{r} - \omega t) \right] = (1/2) cE_r I_{av} \tau MV_{in} \exp\left[j(\mathbf{k}_O(x,y).\mathbf{r} - \omega t) \right]$ and intensity $1/8(cE_R I_{av} TMV_{in})^2$.

In Section 3.2.1, we defined the diffraction efficiency, η, of a hologram as the ratio of the intensity of the reconstructed light beam to that of the incident reference beam, which here is $E_R^2 / 2$. So the diffraction efficiency

$$\eta = \frac{1}{4} (cI_{av} TMV_{in})^2 \tag{7.24}$$

The modulation transfer function is a very important parameter but more important is the diffraction efficiency dependence on spatial frequency of recording, $\eta(f)$. It is this function together with holographic sensitivity, S, discussed in Section 6.7, that enables us to compare the suitability of different recording materials for particular applications.

7.3.2 NOISE GRATINGS

Generally, when one records a simple holographic diffraction grating by interference of two collimated light beams in a photosensitive layer, each of the beams is scattered to an extent determined by the grain structure so that each of the recording beams is accompanied by (Figure 7.4) noise beams with a range of spatial frequencies.

Let us suppose for the moment that the direction of propagation of scattered light is confined to the plane of two incident beams, which have spatial frequencies f_1 and f_2 with $f_2 > f_1$, and that the spatial bandwidth of the noise beams is B. The two sets of scattered light beams have spatial frequencies extending from $f_1 - B/2$ to $f_1 + B/2$ and from $f_2 - B/2$ to $f_2 + B/2$. The spatial frequencies of the recorded gratings will extend from a minimum of $f_2 - B/2 - (f_1 + B/2)$ to a maximum of $f_2 + B/2 - (f_1 - B/2)$, that is, from $f_2 - f_1 - B$ to $f_2 - f_1 + B$ with bandwidth $2B$.

Scatter in a holographic recording medium leads to the recording of noise gratings. Figure 7.5 shows a *single* beam of light with wave vector **k** from which light is scattered.

If scatter is isotropic, then the complete set of wave vectors of the light beams scattered from the original beam is obtained by joining the center, C, of the circle in Figure 7.5 to all

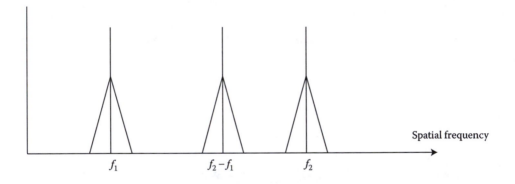

FIGURE 7.4 Noise spectra associated with spatial frequencies f_1, f_2 and the interference frequency f_2-f_1.

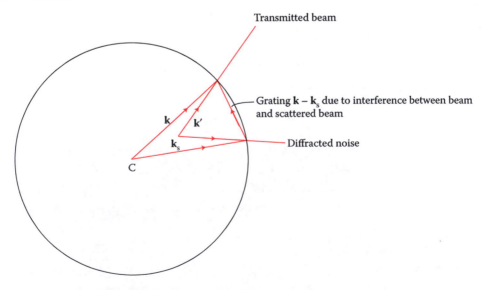

FIGURE 7.5 Recording of noise gating and its reconstruction.

points on the surface of a *sphere*, known as the Ewald sphere, of radius $|\mathbf{k}|$. In Figure 7.5, just one scattered beam with wave vector $\mathbf{k}_s$ is shown. It has the same length as $\mathbf{k}$. This wave interferes with the original one to produce a noise grating with vector $\mathbf{k} - \mathbf{k}_s$. If we now illuminate the recording with a beam of light of the same wavelength, or a different wavelength and wave vector $\mathbf{k}'$, and directed so that the Bragg condition for the noise grating is satisfied, it will be partially transmitted of course, but it will also be diffracted by the noise grating as shown.

Now, consider the situation that arises when *two* mutually coherent, collimated light beams, O and R, interfere inside a holographic recording medium, making equal angles, θ, with the normal to its plane front surface. The waves are represented by their propagation vectors $\mathbf{k}_O$ and $\mathbf{k}_R$ (Figure 7.6).

They have the same wavelength so their wave vectors are equal in length, that is, $|\mathbf{k}_R| = |\mathbf{k}_O| = |\mathbf{k}|$. The recorded grating has grating vector $\mathbf{k}_g$ and forms an isosceles triangle with $\mathbf{k}_O$ and $\mathbf{k}_R$ (see also Figure 3.1). The wave vector, $\mathbf{k}_S$, for just one scattered light beam is shown in Figure 7.6. Now, although $\mathbf{k}_O$ and $\mathbf{k}_R$ are confined to the diagram plane, as in the single beam case, $\mathbf{k}_S$ is not. Again if scatter is isotropic, then we obtain the complete set of wave vectors of the light beams, scattered from each of the incident beams, by joining the center, C, of the circle in Figure 7.7 to all points on the surface of the Ewald sphere, of radius $|\mathbf{k}|$.

The grating $\mathbf{k}_g$ is accompanied by noise gratings, each of which arises from interference between light in one recording beam and light *scattered* from the other. Any noise gratings formed by interference between light beams, *both* of which are scattered, are usually very weak and we will ignore them, along with any light scattered out of beams

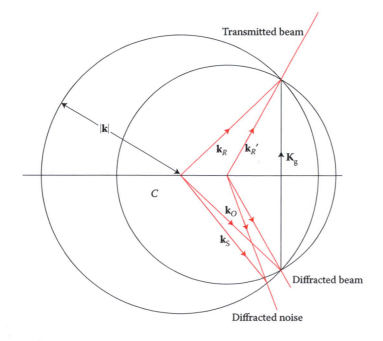

FIGURE 7.6 Ewald sphere with two recording beams.

O and R, which is diffracted by a noise grating. It follows that *the complete set of noise grating vectors is the set of vectors obtained by joining pairs of points on the surface of the Ewald sphere.*

At first, this might seem to imply that reconstruction of the object beam is likely to be impossible as it will be buried in noise. *However, in a thick holographic recording medium, reconstruction is only possible from those recorded gratings for which the reconstructing beam satisfies the Bragg condition* (see Section 4.3.1.1 for the case of a transmission phase grating) or, more correctly, is incident on the grating at an angle that lies within the limits set by the Bragg curve (Figure 4.3). In Figures 7.5 and 7.6, the beam, R', has a longer wavelength than the recording light and therefore a shorter wave vector, $\mathbf{k}'_R$. It is transmitted and diffracted by the recorded grating. For clarity, only noise accompanying the diffracted beam is shown in Figure 7.6. The beam R' is also diffracted by all of the noise gratings. Consider one such noise grating with vector at angle ϕ *with respect to the plane of incidence of the recording beams* (Figure 7.7). The wave vector of the light diffracted by this noise grating will also be at angle ϕ with respect to the plane of incidence of the recording beams, since the incident wave vector, the grating vector, and the diffracted wave vector must be coplanar. When the maximum angle of scatter, θ, is small, the maximum angle ϕ that a noise grating *satisfying the Bragg condition* can make with respect to the plane of incidence is quite small but, as scatter increases, so does ϕ. The diffracted light forms an arc, as shown by the construction in the right hand part of Figure 7.7, which is what one would observe on a translucent screen placed parallel to the hologram plane. The radius of the arc is determined by θ', the angle between $\mathbf{k}'_R$ and the normal, and the distance between the hologram plane and the screen. As scatter increases, the arc grows into a complete circle, which includes the transmitted beam (Figure 7.8).

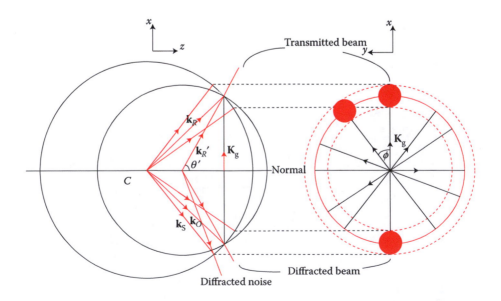

FIGURE 7.7 Ring pattern.

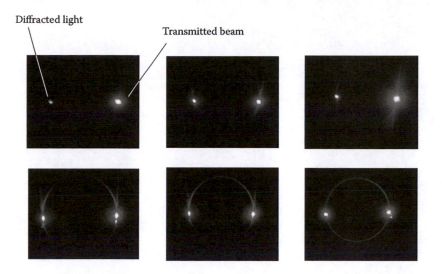

Diffracted light

Transmitted beam

FIGURE 7.8 **(See color insert.)** Diffracted and transmitted light from a grating in a photopolymer layer, as scatter increases from top left to bottom right. (Courtesy of M.S. Mahmud.)

7.3.3 MEASUREMENT OF NOISE SPECTRUM

To determine the spatial frequency spectrum of the noise produced by scatter, the photosensitive material is uniformly exposed and processed if necessary. Suppose that its complex amplitude transmissivity is $\tilde{t}(x, y)$. It is located in the (x, y) plane and illuminated normally by a collimated beam of light (Figure 7.9) of unit amplitude.

Then, the complex amplitude immediately following the lens is, using Equation. 2.19,

$$\exp\left[\frac{-j\pi}{\lambda f}\left(x_l^2 + y_l^2\right)\right]\tilde{t}(x, y)$$

and, from Equation 2.18 but replacing r_0' with f, the focal length, we can find the complex amplitude distribution in the focal plane (x', y') of the lens.

$$\tilde{A}(x',y') = \frac{-je^{jkf}}{\lambda f}\iint\limits_{x,y}\tilde{A}(x,y)e^{-j(k_x x + k_y y)}\,dxdy = \frac{-je^{jkf}}{\lambda f}\exp\left[\frac{-j\pi}{\lambda f}\left(x_l^2 + y_l^2\right)\right]\iint\limits_{x,y}\tilde{t}(x,y)e^{-j(k_x x + k_y y)}\,dxdy$$

and the intensity distribution of the noise $I(x', y')$ in the focal plane of the lens is

$$I(x', y') = \frac{1}{\lambda^2 f^2}T(x', y') \tag{7.25}$$

where $T(x', y')$ is the spatial distribution of the noise in the lens focal plane.

Thus, the intensity distribution in the focal plane, $I(x', y')$ is a direct measure of the spatial noise distribution.

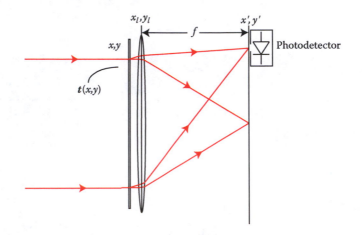

FIGURE 7.9 Measuring scatter.

7.4 THE SPECKLE EFFECT

7.4.1 THE ORIGIN OF SPECKLE

Speckle is readily observed when a surface, which is not optically smooth, is illuminated by an expanded laser. By optically smooth, we mean that the magnitude of the surface roughness is much less than the wavelength of light. The surface of the object appears covered with a random grainy pattern of very small bright and dark spots, which appear to move very rapidly as one moves one's head. The spots appear larger if one squints or looks at the surface through a small aperture. The origin of the speckles is readily explained. Because the surface is rough on the scale of optical wavelength, the path differences between light waves scattered from neighboring points vary rapidly and in a random fashion across the surface and therefore the resultant amplitude that is obtained at any point at a distance from the surface will likewise vary randomly. In Figure 7.10, a rough surface, which is either opaque or translucent, is illuminated by an expanded laser and some of the light rays scattered by the surface are shown. If we could observe the effect in the dashed plane, a, very close to the surface, that is, within about one wavelength of it, we would see that the details of the pattern are determined by the local structure of the surface. A local concavity focuses the light to a bright spot. A local convexity has the opposite effect and the space in its immediate vicinity will not be illuminated.

As one moves further away from the surface, the light intensity at any point, say in plane b, is due to contributions from a larger area of the surface and the precise details of the surface shape do not have so much influence on the pattern details. Further away again, the surface details do not matter at all and the pattern becomes completely random.

7.4.2 SPECKLE SIZE

The characteristic size, s, of a speckle can be determined by calculating the phase difference between light waves arriving at the center of a speckle, whether it is bright or dark,

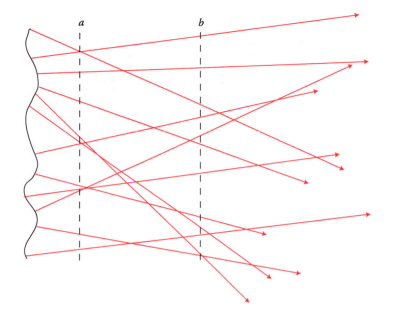

FIGURE 7.10 The speckle effect.

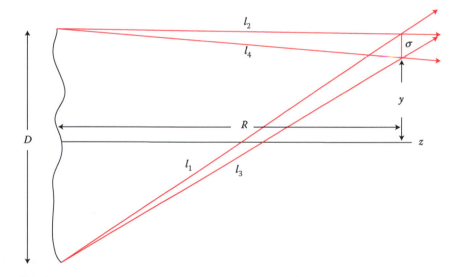

FIGURE 7.11 Speckle size.

and light waves arriving at its edge and then setting the result equal to π. In Figure 7.11, the illuminated area on a screen has width D and we consider a speckle of half width s at a distance, R, from the screen. The value of s is called the *objective speckle size*.

The phase difference between light waves arriving at one edge of the speckle from the edges of the illuminated area on the screen is $2\pi(l_1 - l_2)/\lambda$ and the phase difference between waves arriving at the middle of the speckle is $2\pi(l_3 - l_4)/\lambda$ and the change in phase, ϕ, from the center to the edge of the speckle is

$$\phi = 2\pi[(l_1 - l_2) - (l_3 - l_4)]/\lambda$$

$$= \frac{2\pi}{\lambda}\left\{ \left[R^2 + \left(\frac{D}{2} + y + \frac{s}{2}\right)^2 \right]^{\frac{1}{2}} - \left[R^2 + \left(\frac{D}{2} + y\right)^2 \right]^{\frac{1}{2}} - \left[R^2 + \left(\frac{D}{2} - y + \frac{s}{2}\right)^2 \right]^{\frac{1}{2}} \right.$$
$$\left. + \left[R^2 + \left(\frac{D}{2} - y\right)^2 \right]^{\frac{1}{2}} \right\}$$

If $R \gg D, y, \sigma$ then

$$\phi \cong \frac{2\pi}{\lambda}\left\{ \left[R + \frac{1}{2R}\left(\frac{D}{2} + y + \frac{s}{2}\right)^2 \right] - \left[R + \frac{1}{2R}\left(\frac{D}{2} - y - \frac{s}{2}\right)^2 \right] - \left[R + \frac{1}{2R}\left(\frac{D}{2} + y\right)^2 \right] \right.$$
$$\left. + \left[R + \frac{1}{2R}\left(\frac{D}{2} - y\right)^2 \right] \right\}$$

$$= \frac{\pi}{\lambda R}\left[D(2y + s) - D2y \right] = \frac{\pi D s}{\lambda R}$$

so that the speckle width

$$2s = 2\lambda R / D \tag{7.26}$$

Suppose for example a diffuser made of glass with one rough and one flat surface and of width 1 cm is fully illuminated by laser light of wavelength 633 nm. The speckles at a distance of 1 m from the diffuser are about 120 μm wide.

The situation is different when the speckle is observed through an aperture or a lens. Now, the size of the speckles is determined by the aperture itself. To see why, recall that in Section 2.2.2, that a lens cannot focus even a light beam of a single spatial frequency to a mathematical point and in practice a distant point of monochromatic light, which would give rise to a single spatial frequency, is imaged in the lens focal plane as a ring pattern with almost all the *light of spatial frequency in the range* $\pm 0.61/a$ *focused within the Airy disc* with a the radius of the lens aperture, and the speckle size is in fact the same as the Airy disc diameter, $1.22\lambda f/a$ or $2.44\lambda(F/ number)$. Speckles formed by laser light of wavelength 633 nm in the focal plane of a lens with $F/ number$ of 10 are about 22 μm in width. This kind of speckle viewed by eye or through an aperture or lens is called *subjective speckle* and it is of great importance in the technique of electronic speckle pattern interferometry as we shall see in Section 10.14. Figure 7.12 shows a speckle pattern photographed using two different lens apertures. Note the difference in speckle sizes.

7.4.3 SPECKLE CONTRAST

We now determine the contrast in a speckle pattern. Clearly, as some speckles are completely dark with zero light intensity and others are bright, the contrast defined in the usual way (Equation 7.18) as $(I_{max} - I_{min})/(I_{max} + I_{min})$ must be 1.0, but we need to take account of the frequency or probability with which a particular intensity occurs in the pattern. We

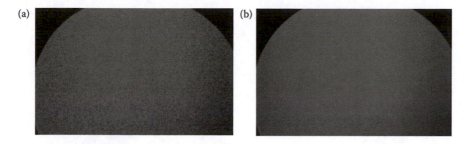

FIGURE 7.12 **(See color insert.)** Speckle patterns showing effect of aperture size, (a) f/45 (b) f/8 (author photographs).

define *contrast, C, as root mean square of fluctuations from the mean value of intensity, that is the noise, divided by the mean value of intensity.*

$$C = \frac{\sqrt{\overline{(I - \bar{I})^2}}}{\bar{I}} = \frac{\sqrt{\overline{I^2} - \bar{I}^2}}{\bar{I}} \tag{7.27}$$

The complex amplitude at any point in the speckle pattern is made up of contributions from many waves, all of the same amplitude but with different phases. This situation can be depicted as shown in Figure 7.13, which represents each wave by a *phasor*, that is, a straight line of length proportional to the amplitude of the wave and directed at an angle equal to its phase, relative to a fixed axis. The resultant phasor is found in exactly the same way as the resultant of a set of *vectors*. The term *phasor* is preferred to avoid confusion with the *wave vector*. The resultant phasor at any point is specified by its length, r, and phase angle, θ, and its probability of occurrence is given by

$$p(r, \theta) = \frac{r}{\pi\sigma^2} \exp\left(-\frac{r^2}{\sigma^2}\right) \tag{7.28}$$

Equation 7.28 is a mathematical expression of the random walk phenomenon, which can sometimes be observed late at night outside bars and night clubs. It is the probability that someone under the influence of alcohol will have positional polar coordinates, θ, relative to, say the direction of the footpath, and r relative to, say, a lamp standard, from which they appear reluctant to stray very far. Figure 7.13 describes the person's progress.

The mean value of any randomly varying quantity is given by its value multiplied by the probability of occurrence of the value, the result being integrated over all possible values. Thus, the mean intensity is given by $\bar{I} = \bar{r}^2$

$$\bar{I} = \int_0^{2\pi} \int_0^\infty r^2 \frac{r}{\pi\sigma^2} \exp\left(-\frac{r^2}{\sigma^2}\right) dr d\theta = \frac{2}{\sigma^2} \int_0^{2\pi} \int_0^\infty r^3 \exp\left(-\frac{r^2}{\sigma^2}\right) dr$$

And, using successive integration by parts the terms in Equation 7.27 are

$$\bar{I} = \sigma^2 \tag{7.29}$$

$$\overline{I^2} = 2\sigma^4 \tag{7.30}$$

and their substitution in Equation 7.27 gives a contrast of 1.0.

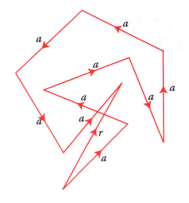

FIGURE 7.13 Effect of adding waves with the same amplitude, a, but different phases; r is the resultant.

The fact that the contrast is an absolute maximum makes speckle a serious problem in holography and especially holographic displays. Movement of the observer's viewpoint causes rapid movement of the speckles as already mentioned in Section 7.4.1. While the resulting smearing effect improves the image visibility, this is not a very convenient way to look at holographic images.

If the object is a 2-D transparency, then it is likely to be optically flat enough for speckle not to cause a problem but often there are defects such as dust particles or scratches on the transparency. These diffract light and the resulting diffraction patterns in the form or rings from dust particles and parallel fringes from scratches are recorded and reconstructed. This problem can be avoided if the transparency is illuminated through a diffuser, so that it is illuminated from many different directions and light from every point in the object reaches every point on the hologram. This in turn means that the diffraction patterns arising from the defects are spread out over the area of the hologram and the defects do not appear noticeably in the reconstructed image. The fact that light from every point in the object reaches every point on the hologram means that on reconstruction, every point on the object is visible from every point in the hologram, so that, even if we mask most of the hologram so that only a part of it is usable, the reconstructed image is still visible in its entirety, although with reduced resolution. Indeed, the hologram can be broken into fragments and the entire image can be reconstructed from each fragment.

Of course, the use of a diffuser means that speckle will once again be present and any optical system (including the eye) used to observe the reconstructed image should have a large enough aperture to ensure that the speckles are small enough not to be objectionable. Alternatively, the hologram can be reconstructed using partially coherent or non-monochromatic light. The holographic image is reconstructed at a range of wavelengths all at the same time and the speckle patterns in the images at all the wavelengths are superimposed.

Another approach is to record a number of holograms consecutively. The diffuser in the path of the beam used to illuminate the object is simply rotated in its plane between exposures. In reconstruction, the speckle patterns intensities add together and the speckle contrast is reduced by a factor $N^{1/2}$ where N is the number of exposures.

7.5 SIGNAL-TO-NOISE RATIO IN HOLOGRAPHY

The signal-to-noise ratio is a useful parameter for comparing holographic recordings of the same object under the same recording conditions, in different materials. A simple flat reflecting or transmitting surface is used as an object. If the real image is reconstructed, we can measure the light intensity at any point in it, using a photodetector. The noise is assumed to be random and it adds coherently to the signal. We assume the complex amplitude of the signal is $\tilde{E}_S$ everywhere in the image, while that due to noise, which fluctuates, is $\tilde{E}_N$. Signal and noise interfere with one another to produce a resultant intensity

$$I = (\tilde{E}_S + \tilde{E}_N)(\tilde{E}_S + \tilde{E}_N)^* / 2 = I_S + I_N + \tilde{E}_S \tilde{E}_N^* / 2 + \tilde{E}_S^* \tilde{E}_N / 2$$

$$\bar{I} = I_S + \overline{I_N} + \tilde{E}_S \overline{\tilde{E}_N^*} / 2 + \tilde{E}_S^* \overline{\tilde{E}_N^*} / 2$$

$$\overline{I^2} = I_S^2 + \overline{I_N^2} + 4I_S \overline{I_N} + I_S \tilde{E}_S \overline{\tilde{E}_N^*} + I_S \tilde{E}_S^* \overline{\tilde{E}_N} + I_N \tilde{E}_S \overline{\tilde{E}_N^*} + I_N \tilde{E}_S^* \overline{\tilde{E}_N} + \tilde{E}_S^2 \overline{\tilde{E}_N^{*\,2}} / 4 + \tilde{E}_S^{*\,2} \overline{\tilde{E}_N^2} / 4$$

We assume the amplitude of the noise has constant amplitude E_N with random phase θ so $\tilde{E}_N = E_N \exp(j\theta)$ and from Equation 7.28

$$\overline{\tilde{E}_N} = \int_0^{2\pi} \exp(j\theta)d\theta \int_0^\infty \frac{E_N}{\pi\sigma^2}\exp\left(-\frac{E_N^2}{\sigma^2}\right)dE_N = 0 \text{ and similarly } \overline{\tilde{E}_N^2}, \tilde{E}_N^*, \text{ and } \tilde{E}_N^{*\,2} \text{ are all zero.}$$

$$\bar{I} = I_S + \overline{I_N} \quad \text{and} \quad \overline{I^2} = I_S^2 + \overline{I_N^2} + 4I_S \overline{I_N} \tag{7.31}$$

We define the signal-to-noise ratio S/N as the ratio of the signal intensity to that of the root mean square intensity fluctuation.

$$\frac{S}{N} = \frac{I_S}{\sqrt{\overline{[I-\bar{I}]^2}}} = \frac{I_S}{\sqrt{\overline{I^2} - \bar{I}^2}} = \frac{I_S}{\sqrt{I_S^2 + \overline{I_N^2} + 4I_S\overline{I_N} - I_S^2 - 2I_S\overline{I_N} - \overline{I_N}^2}} \tag{7.32}$$

From Equations 7.29 and 7.30, $\overline{I_N^2} = 2\,\overline{I_N}^2$ and

$$\frac{S}{N} = \frac{I_S}{\overline{I_N}\left[1 + \dfrac{2I_S}{\overline{I_N}}\right]^{\frac{1}{2}}} \tag{7.33}$$

At large values of $I_S / \overline{I_N}$, $S/N = (1/\sqrt{2})[I_S / \overline{I_N}]^{1/2}$ meaning that the signal-to-noise ratio is proportional to the ratio of signal to noise *amplitudes*.

7.6 EXPERIMENTAL EVALUATION OF HOLOGRAPHIC CHARACTERISTICS

Probably, the two most important things one wants to know about any holographic recording material are its diffraction efficiency and its liability to dimensional change as a result of exposure to an optical interference pattern.

7.6.1 DIFFRACTION EFFICIENCY

To obtain the diffraction efficiency, a transmission holographic diffraction grating is recorded by illuminating it with two mutually coherent beams of collimated light. Usually, such gratings are unslanted, that is, the angles of incidence are equal and opposite. If the recording material is self-processing as is the case with some photopolymers, it may be probed at a laser wavelength to which it is insensitive (Figure 7.14). The probe laser is incident at the Bragg angle corresponding to its wavelength and the diffracted light is incident on a photodetector enabling the diffraction efficiency to be monitored as the grating is formed.

Otherwise, the photosensitive layer is processed and then placed in the path of the probe laser. As the fully formed holographic grating is rotated about an axis in its plane normal to the grating vector, the diffraction efficiency varies according to Equation 4.33 in

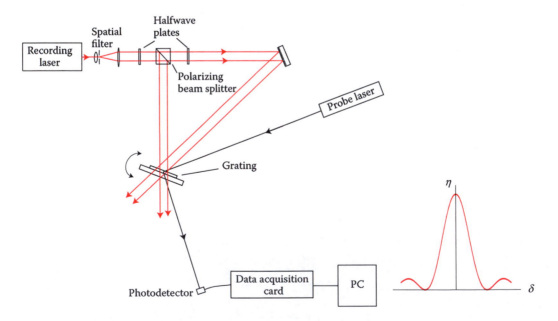

FIGURE 7.14 Monitoring growth in diffraction efficiency during recording of a grating. The probe laser has a wavelength to which the recording material is not sensitive. Rotation of the grating at the end of the recording process enables us to obtain the Bragg curve shown on the right. (The spatial filter is used to remove spatial noise in the beam while the polarizing beam splitter and halfwave plates are used to control the intensity ratio of the recording beams. See sections 8.2.1 and 8.3.1.)

the case of a pure phase transmission grating and according to Equation 4.37 for a pure amplitude grating. Diffraction efficiencies of reflection gratings may similarly be obtained and compared with the prediction of Equation 4.49 in the case of a pure phase grating and Equation 4.53 in the pure amplitude case. Detailed comparison of experimentally obtained Bragg curves with Equations 4.33 and 4.37 (transmission gratings) and Equations 4.49 and 4.53 (reflection gratings) provides much useful information about the gratings themselves and hence about the material in which they are recorded. For example, in the case of a pure phase transmission grating, the normalized diffraction efficiency data may be fitted to Equation 4.33

$$\eta = \frac{\sin^2 \left(\xi^2 + \upsilon^2 \right)^{\frac{1}{2}}}{\left(1 + \xi^2 / \upsilon^2 \right)}$$

The diffraction efficiency when the Bragg condition is optimally met ($\xi = 0$) is simply

$$\eta = \sin^2 \upsilon \tag{7.34}$$

so a measurement of its maximum value gives us the value of the modulation parameter $\upsilon = \pi n_1 T / \lambda_{air} \cos \theta$ (Equation 4.31). The values of λ_{air} and θ, the Bragg angle, are also known, so that the product of n_1, the refractive index modulation and T, the thickness of the grating, may be determined. Using υ, one can then fit the experimental data for diffraction efficiency versus the off-Bragg parameter, $\xi = \beta \delta T \sin \theta$ (Equation 4.32). Recall that $\beta = k n_0$, the propagation constant (in Equation 4.9) and δ is the off-Bragg angle so it is possible to obtain T with precision since the result is obtained from the fitting of many diffraction efficiency values. It is important to remember that the Bragg angle, θ, in Equation 4.32 and the angular deviation from the Bragg angle, δ, are *inside* the photosensitive layer and must be obtained from the corresponding measurements in air, using Snell's law.

It might at first be thought that T is simply the thickness of the photosensitive layer and could be measured directly using a micrometer gauge to measure the total thickness of the layer and its substrate and subtracting the measured value of the substrate thickness or by using an optical profilometer (see Section 5.3) in the case of soft layers. In practice, especially in the case of thick, dye sensitized photopolymer layers, the *physical thickness* and the *effective thickness* may differ since the recording light intensity is attenuated as it propagates through the layer and the contribution to diffraction efficiency due to those parts of the layer exposed to attenuated laser light may not be optimal. However, the physical thickness should be measured as a check on the upper limit on the effective thickness.

The discussion in the previous paragraph assumes that the exposure has not exceeded that required to obtain maximum diffraction efficiency. In practice, this exposure may have been exceeded. In this case, we fit the data to Equation 4.31 with υ as a freely varying parameter.

In measurements of diffraction efficiency, the transmissivity obviously changes with the angle of incidence at the air/layer and substrate/air boundaries (see Equations 1.27, 1.28, and 1.43). That these changes do not affect the intensity of light, which actually penetrates

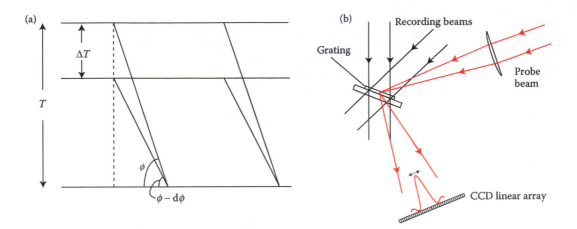

FIGURE 7.15 Change in angular position of Bragg peak due to shrinkage. (a) Geometry of shrinkage (Reproduced from J. T. Gallo and C. M. Verber, Model for the effects of material shrinkage on volume holograms, *Appl. Opt.*, 33, 6797–804, 1994. With permission of Optical Society of America.) (b) Monitoring of complete Bragg selectivity curve.

the grating, may be seen by considering the angular width of the Bragg selectivity curve. Assuming the modulation parameter $v = \pi/2$, the diffraction efficiency for a pure phase transmission grating falls to zero when the off-Bragg parameter, $\xi = \beta \delta T \sin \theta \cong 3$ (Equation 4.33 and Figure 4.3). In the grating example following Equation 4.31, we saw that the total angular width between the first minima in diffraction efficiency on either side of the maximum is 4.02° in air. Referring to Figure 1.15, we see that, at the angles of incidence of 30° in the example chosen, there is very little variation in transmissivity for probe beam light, which is plane polarized either parallel or perpendicular to the plane of incidence. For gratings involving angles of incidence of up to 70°, it is better that the plane of polarization of the probe beam be parallel to the plane of incidence. Transmissivity changes rapidly with angle of incidence above 70°. For an angle of incidence of 80°, using the same grating thickness and recording wavelength, we obtain an angular width of 2.4° in air, a range over which the transmissivity for probe light, which is plane polarized perpendicular to the plane of incidence, varies by only a few percent.

In some of the literature on holography, the diffraction efficiency is defined as the ratio of diffracted light intensity to the incident intensity minus the losses due to reflection and absorption. This definition naturally leads to higher values of diffraction efficiency, but the losses are unavoidable and cannot be simply ignored.

7.6.2 SHRINKAGE

Many holographic recording materials undergo dimensional change during or after recording or during physical or chemical processing. In a reflection hologram, the effect is a change in the spacing of the recorded fringes and therefore a change in the wavelength at which the image is reconstructed. A transmission holographic grating recorded using collimated light beams is unaffected unless the grating is *slanted*, that is, the incident angles of the recording beams are unequal. If the layer thickness is reduced due to shrinkage, the orientation of recorded fringes changes, as shown in Figure 7.15 and the Bragg peak is shifted.

Suppose the original thickness of a slanted transmission grating is T and it shrinks by ΔT (Figure 7.15a). The angle between the fringe planes and the grating plane changes from ϕ to $\phi - d\phi$ so that [2]

$$\frac{\Delta T}{T} = 1 - \frac{\tan(\phi - d\phi)}{\tan \phi} \tag{7.35}$$

from which the percentage shrinkage may be determined. Shrinkage in photopolymers begins at the outset of the recording process and may be monitored in real time. The grating is not rotated but is illuminated by a converging light beam so that it is probed at the full range of off-Bragg angles (Figure 7.15b). The spatial intensity distribution of the diffracted light may be captured using a CCD or CMOS array and rapidly analyzed at short time intervals to determine the location of the peak of the Bragg curve [3].

Dimensional change in a hologram as a consequence of the recording process may or may not be desirable depending on the application. In holographic data storage applications (Chapter 12), for example, shrinkage is to be completely avoided whereas in sensing applications (Chapter 17), the converse is true.

7.7 EFFECTS ARISING FROM DISSIMILARITIES BETWEEN REFERENCE BEAMS IN RECORDING AND RECONSTRUCTION

In our discussion of holographic recording and reconstruction, we have assumed that the reconstructing light beam is identical with the original reference beam used at the recording stage. This is often not the case and to see the consequences, we record a hologram of a point source of light at x_O, y_O, z_O, using a coherent reference wave, from another point source at x_R, y_R, z_R. At reconstruction, we may alter the position or even wavelength of the reference light source.

In Figure 7.16, the phase difference between object light arriving at x, y and that arriving at 0, 0 in the hologram plane is $(\phi_{x,y} - \phi_{0,0})_O$

$$= \frac{2\pi}{\lambda} \left[\sqrt{(x - x_O)^2 + (y - y_O)^2 + (-z_O)^2} - \sqrt{x_O^2 + y_O^2 + z_O^2} \right]$$

$$= \frac{2\pi z_O}{\lambda} \left\{ \left(1 + \frac{(x - x_O)^2 + (y - y_O)^2}{z_O^2} \right)^{\frac{1}{2}} - \left(1 + \frac{x_O^2 + y_O^2}{z_O^2} \right)^{\frac{1}{2}} \right\}$$

$$\cong \frac{2\pi}{\lambda} \left\{ \left(\frac{(x - x_O)^2 + (y - y_O)^2}{2z_O} \right) - \left(\frac{x_O^2 + y_O^2}{2z_O} \right) \right\}$$

$$\left(\phi_{x,y} - \phi_{0,0} \right)_O = \frac{\pi}{\lambda z_O} \left(x^2 + y^2 - 2xx_O - 2yy_O \right) \tag{7.36}$$

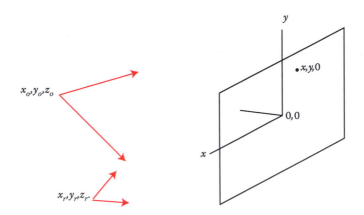

FIGURE 7.16 Recording a hologram of a point.

and we can describe the object wave, O, by

$$O = \exp\left[j\frac{\pi}{\lambda z_O}\left(x^2 + y^2 - 2xx_O - 2yy_O\right)\right] \tag{7.37}$$

and the reference wave, R, by

$$R = \exp\left[j\frac{\pi}{\lambda z_R}\left(x^2 + y^2 - 2xx_R - 2yy_R\right)\right] \tag{7.38}$$

The recorded intensity is

$$(O+R)(O+R)^* \tag{7.39}$$

We will assume as we did in Section 3.5.2 that, following exposure of appropriate duration to the interference pattern, the hologram has an amplitude transmissivity $\tilde{t}(x,y)$, which is proportional to the intensity of the pattern.

Now, assume that the reconstructing wave, R', is of wavelength λ' and that it originates from a point $x_{R'}$, $y_{R'}$, $z_{R'}$ and is again of unit amplitude.

$$R = \exp\left[j\frac{\pi}{\lambda' z_{R'}}\left(x^2 + y^2 - 2xx_{R'} - 2yy_{R'}\right)\right] \tag{7.40}$$

The two reconstructed waves are $O' = R'OR^*$

$$O' = \exp\left\{ j\left[\begin{array}{l} \dfrac{\pi}{\lambda' z_{R'}}\left(x^2 + y^2 - 2xx_{R'} - 2yy_{R'}\right) + \dfrac{\pi}{\lambda z_O}\left(x^2 + y^2 - 2xx_O - 2yy_O\right) \\[2ex] -\dfrac{\pi}{\lambda z_R}\left(x^2 + y^2 - 2xx_R - 2yy_R\right) \end{array}\right]\right\}$$

and $O'* = R'O*R$

$$O'* = \exp\left\{j\left[\begin{array}{l}\dfrac{\pi}{\lambda' z_{R'}}\left(x^2 + y^2 - 2xx_{R'} - 2yy_{R'}\right) - \dfrac{\pi}{\lambda z_O}\left(x^2 + y^2 - 2xx_O - 2yy_O\right) \\ + \dfrac{\pi}{\lambda z_R}\left(x^2 + y^2 - 2xx_R - 2yy_R\right)\end{array}\right]\right\}$$

We rearrange these last two equations in the form appropriate to a spherical wave, similar to those in Equations 7.37 and 7.38, so that the reconstructed images are points,

$$O' = \exp\left\{j\dfrac{\pi}{\lambda'\left(\dfrac{1}{z_{R'}} + \dfrac{\mu}{z_O} - \dfrac{\mu}{z_R}\right)^{-1}} \times \left[\left(x^2 + y^2\right) - 2x\dfrac{\left(\dfrac{x_{R'}}{z_{R'}} + \dfrac{\mu x_O}{z_O} - \dfrac{\mu x_O}{z_R}\right)}{\left(\dfrac{1}{z_{R'}} - \dfrac{\mu}{z_R} + \dfrac{\mu}{z_O}\right)} - 2y\dfrac{\left(\dfrac{y_{R'}}{z_{R'}} + \dfrac{\mu y_O}{z_R} - \dfrac{\mu y_R}{z_R}\right)}{\left(\dfrac{1}{z_{R'}} - \dfrac{\mu}{z_R} + \dfrac{\mu}{z_O}\right)}\right]\right\} \quad (7.41)$$

$$O'* = \exp\left\{j\dfrac{\pi}{\lambda'\left(\dfrac{1}{z_{R'}} - \dfrac{\mu}{z_O} + \dfrac{\mu}{z_R}\right)^{-1}} \times \left[\left(x^2 + y^2\right) - 2x\dfrac{\left(\dfrac{x_{R'}}{z_{R'}} - \dfrac{\mu x_O}{z_O} + \dfrac{\mu x_R}{z_R}\right)}{\left(\dfrac{1}{z_{R'}} - \dfrac{\mu}{z_O} + \dfrac{\mu}{z_R}\right)} - 2y\dfrac{\left(\dfrac{y_{R'}}{z_{R'}} - \dfrac{\mu y_O}{z_O} + \dfrac{\mu y_R}{z_R}\right)}{\left(\dfrac{1}{z_{R'}} - \dfrac{\mu}{z_O} + \dfrac{\mu}{z_R}\right)}\right]\right\} \quad (7.42)$$

with $\mu = \lambda'/\lambda$.

The coordinates of the reconstructed image points are

$$x_{O'} = \dfrac{x_{R'}z_R z_O - \mu x_R z_{R'}z_O + \mu x_O z_{R'}z_R}{z_R z_O - \mu z_{R'}z_O + \mu z_{R'}z_R}, \quad y_{O'} = \dfrac{y_{R'}z_R z_O - \mu y_R z_{R'}z_O + \mu y_O z_{R'}z_R}{z_R z_O - \mu z_{R'}z_O + \mu z_{R'}z_R},$$

$$z_{O'} = \dfrac{z_{R'}z_R z_O}{z_R z_O - \mu z_{R'}z_O + \mu z_{R'}z_R}$$

$$x_{O'}* = \dfrac{x_{R'}z_R z_O - \mu x_O z_{R'}z_R + \mu x_R z_{R'}z_O}{z_R z_O - \mu z_{R'}z_R + \mu z_{R'}z_O}, \quad y_{O'}* = \dfrac{y_{R'}z_R z_O - \mu y_O z_{R'}z_R + \mu y_R z_{R'}z_O}{z_R z_O - \mu z_{R'}z_R + \mu z_{R'}z_O},$$

$$z_{O'}* = \dfrac{z_{R'}z_R z_O}{z_R z_O - \mu z_{R'}z_R + \mu z_{R'}z_O} \quad (7.43)$$

The lateral magnifications (LaM) are

$$\text{LaM} = \dfrac{\partial x_{O'}}{\partial x_O} = \dfrac{\mu z_{R'}z_R}{z_R z_O - \mu z_{R'}z_O + \mu z_{R'}z_R} = \dfrac{\partial y_{O'}}{\partial y_O} \quad (7.44)$$

$$\text{LaM}^* = \frac{\partial x_{O'}{}^*}{\partial x_O} = \frac{-\mu z_{R'} z_R}{z_R z_O - \mu z_{R'} z_R + \mu z_{R'} z_O} = \frac{\partial y_{O'}{}^*}{\partial y_O} \tag{7.45}$$

while the angular magnifications (AM) are

$$\text{AM} = \frac{\partial (x_{O'}/z_{O'})}{\partial (x_O/z_O)} = \frac{\partial \left(\dfrac{x_{R'}}{z_{R'}} - \mu \dfrac{x_R}{z_R} + \mu \dfrac{x_O}{z_O} \right)}{\partial (x_O/z_O)} = \mu \tag{7.46}$$

$$\text{AM}^* = \frac{\partial (x_{O'}{}^*/z_{O'}{}^*)}{\partial (x_O/z_O)} = \frac{\partial \left(\dfrac{x_{R'}}{z_{R'}} + \mu \dfrac{x_R}{z_R} - \mu \dfrac{x_O}{z_O} \right)}{\partial (x_O/z_O)} = -\mu \tag{7.47}$$

Thus, a holographic image reconstructed at a longer wavelength than that used in its recording will be magnified in the ratio of reconstruction to recording wavelengths. This is the basis of Gabor's microscopic principle (Section 3.4.3) whereby holograms would be recorded at electron wavelengths and reconstructed using visible light wavelengths.

The longitudinal magnifications are

$$\text{LoM} = \frac{\partial z_{O'}}{\partial z_O} = \frac{\partial \left(\dfrac{z_{R'} z_R z_O}{z_R z_O - \mu z_{R'} z_O + \mu z_{R'} z_R} \right)}{\partial z_O} = \frac{1}{\mu} \text{LaM}^2 \tag{7.48}$$

$$\text{LoM}^* = \frac{\partial z_{O'}{}^*}{\partial z_O} = \frac{\partial \left(\dfrac{z_{R'} z_R z_O}{z_R z_O - \mu z_{R'} z_R + \mu z_{R'} z_O} \right)}{\partial z_O} = -\frac{1}{\mu} \text{LaM}^2 \tag{7.49}$$

In the derivation of Equation 7.35, we only included the first-order terms in binomial expansions. Inclusion of the higher order terms gives rise to aberrations very similar to the aberrations, which arise in all optical systems. Aberrations in holographic images have been analyzed by Meier [1]. When an *uncollimated* reference beam is used at the recording and reconstructions stages, only one of the images is reconstructed without aberration. Thus, if the reconstructing wave is a point source at x_R, y_R, z_R of wavelength λ, the image coordinates are x_O, y_O, z_O and

$$x_{O'}{}^* = \frac{x_R z_R z_O - x_O z_R^2 + x_R z_R z_O}{z_R z_O - z_R^2 + z_R z_O} = \frac{2\dfrac{x_R z_O}{z_R} - x_O}{2\dfrac{z_O}{z_R} - 1}, \; y_{O'}{}^* = \frac{2\dfrac{y_R z_O}{z_R} - y_O}{2\dfrac{z_O}{z_R} - 1}, \; z_{O'}{}^* = \frac{z_O}{2\dfrac{z_O}{z_R} - 1} \tag{7.50}$$

It is not entirely accurate to say that the conjugate image suffers from aberration since the corresponding conjugate object does not exist.

7.7.1 PHASE CONJUGATION EFFECTS

Recall that in Section 3.6, we used a conjugate version of the reference wave to obtain a reconstruction of the phase conjugate version of the object wave. There, it was stated that a reconstructed phase conjugate wave is one that retraces the steps of the original object wave. Thus, for example, the phase conjugate version of a spherical wave emanating from a point source is a spherical wave *converging* to that point. However, we do not need to use the conjugate of the reference wave, if it is collimated at the outset. Figure 7.17 shows the holographic recording and reconstruction of a point. Assuming reconstruction with precisely the original reference wave and using Equations 7.50

$$x_O{}^* = \frac{x_R z_R z_O - x_O z_R{}^2 + x_R z_R z_O}{z_R z_O - z_R{}^2 + z_R z_O} = \frac{2\dfrac{x_R z_O}{z_R} - x_O}{2\dfrac{z_O}{z_R} - 1}, \; y_O{}^* = \frac{2\dfrac{y_R z_O}{z_R} - y_O}{2\dfrac{z_O}{z_R} - 1}, \; z_O{}^* = \frac{z_O}{2\dfrac{z_O}{z_R} - 1}$$

Letting $z_R \to \infty$ for a collimated beam, we find that the conjugate image appears at $x_O, y_O, -z_O$ (Figure 7.17a).

Now, consider what happens when we make a hologram of a three-dimensional object such as a hollow cone. The reconstructed conjugate image appears as shown in Figure 7.17b with the tip of the cone nearest the viewer and the cone appears to be solid. As the viewer changes position, the conjugate image appears to rotate as if to ensure that the base of the cone is always further away and the tip always nearest. In general phase, *conjugate images tend to obscure those in front of them.*

We will discuss phase conjugation and its practical applications in later chapters.

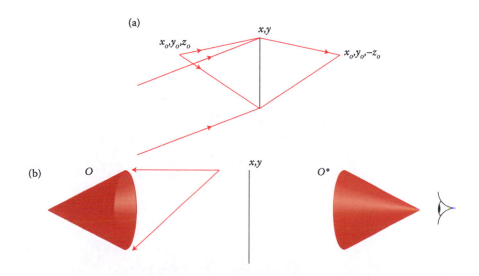

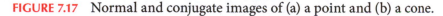

FIGURE 7.17 Normal and conjugate images of (a) a point and (b) a cone.

7.7.2 RECONSTRUCTION USING NONLASER LIGHT

It is often desirable for safety reasons to reconstruct holograms using nonlaser light. The source may not be a point source but an extended one centered at x_R, y_R, z_R. Assume that $\lambda' = \lambda$. Differentiating Equation 7.42 $(\partial x_{O'} / \partial x_{R'}) = (z_R z_O / z_R z_O - z_{R'} z_O + z_{R'} z_R)$ and, since $z_{R'} \cong z_R$, $\Delta x_{O'} = z_O \, \Delta x_R / z_R$.

The point image should not be laterally spread or blurred by more than the resolution of the eye whose angular resolution is 0.6 mr, which means that image blur of 0.6 mm at a viewing distance of 1 m is acceptable. Say the image distance from the hologram (z_O) is 20 cm and the reconstructing source is 1.5 m from the hologram (z_R). The maximum acceptable source width is therefore 4.5 mm.

The source may not be monochromatic. We can calculate the width and length of the image when it is reconstructed by a broadband light source by differentiating Equation 7.43 with respect to wavelength.

$$x_{O'} = \frac{x_R z_O / \mu z_R - x_R z_O / z_R + x_O}{z_O / (\mu z_R) - z_O / z_R + 1} = \frac{x_R z_O}{\mu z_R} - \frac{x_R z_O}{z_R} + x_O, \quad z_{O'} = \frac{z_O / \mu}{z_O / (\mu z_R) - z_O / z_R + 1}$$

We assume the recording reference and reconstructing light beams are collimated so that $z_R = z_{R'} = \infty$

$$\frac{\partial x_{O'}}{\partial \lambda_2} = -\frac{x_R z_O}{\mu^2 z_R \lambda_1}, \qquad \frac{\partial z_{O'}}{\partial \lambda_2} = -\frac{z_O}{\mu^2 \lambda_1}$$

If $\lambda_2 \cong \lambda_1$ then $\mu \cong 1$. Using a collimated light emitting diode (LED) with a typical spectral bandwidth of 50 nm and mean wavelength 500nm at an angle of 45° with the hologram normal, we obtain a lateral image blur $\Delta x_{O'} = z_O/10$. If the original object point is at 50 mm (z_O) from the hologram then the $\Delta x_{O'}$ lateral blur is 5 mm, which is not resolved by eye at an observation distance of 1 m. The longitudinal blur, $\Delta z_{O'}$, is 5 mm.

7.8 SUMMARY

In this chapter, we look at holographic materials in practice. We discard some of the simplifying assumptions that were previously made. In this way, we are able to evaluate some of the effects arising from nonlinearity in holographic recording. We also look at the effect of grain noise in holographic recording materials and its adverse effect on the contrast of the recorded interference fringe pattern.

Laser speckle arises independently of the recording material used and we consider its main properties and effects.

We discuss experimental methods for the measurement of diffraction efficiency and shrinkage.

Finally, we look at the effects that arise when the reconstructing light beam differs from the reference beam used in the holographic recording process.

REFERENCES

1. R. W. Meier, "Magnification and third order aberrations in holography," *J. Opt. Soc. Am.*, 55, 987–92 (1965).
2. J. T. Gallo and C. M. Verber, "Model for the effects of material shrinkage on volume holograms," *Appl. Opt.*, 33, 6797–804 (1994).
3. N. Pandey, I. Naydenova, S. Martin, and V. Toal, "Technique for characterization of dimensional changes in slanted holographic gratings by monitoring the angular selectivity profile," *Opt. Lett.*, 33, 1981–3 (2008).

PROBLEMS

1. We saw in Section 7.2.1.1 that intermodulation effects are due to interference between light waves scattered from different points on the object. Appropriate choice of the angle between object and reference beams helps solve this problem. Another possibility is to ensure that intermodulation effects are very weak at the outset. How can this be done? Review the main holographic recording materials discussed in Chapter 6 and suggest a further possible solution.

2. Light reflected from a diffusely reflecting object has its polarization state altered. A consequence is that some of the recording capability of the photosensitive layer is used up but does not contribute to the reconstructed image. Explain why this is the case and suggest a possible solution.

3. A single beam reflection hologram is made using silver halide on glass of refractive index 1.5. Assume a collimated beam incident at 30^0 to the normal. Find the percentage of the light actually used to record the hologram assuming an object with 100% reflectivity? What happens to the remaining light?

4. A hologram is recorded at a wavelength of 633 nm of a flat circular disc 1 cm in diameter located 140 cm from the photosensitive plate. What is the minimum size of feature on the disc, which can be observed in the reconstruction?

5. A transmission phase grating is recorded using collimated beams of wavelength 532 nm with angles of incidence of 45° and 30°. The recording layer has refractive index 1.5. What is the spacing of the interference fringes? During recording the layer shrinks by 5%. What is the slant angle of the recorded grating and at what angles of incidence should the grating be probed in order to obtain strong diffracted beams?

6. A thin hologram is recorded of a flat square plate 2×2 cm, parallel to the hologram and 40 cm from it, both being symmetrically positioned with respect to the z-axis passing through the center of the hologram. The recording wavelength is 633 nm and the reference beam is a point source located at (5, 4, –30). At reconstruction, a point source of wavelength 658 nm is used at (3, 2, –20). What is the area of the virtual image?

7. Using white light, an image is reconstructed from a transmission hologram recorded at 633 nm. The object was originally 50 cm from the hologram and the angle of incidence of the recording and reconstructing collimated light at the hologram is 30°. At what distance from the hologram should the reconstructed virtual image be viewed to ensure that lateral color blur is not resolved by eye?

8. When recording a hologram or holographic grating, which direction of plane polarization of the interfering beams is preferred, namely parallel or perpendicular to the plane of incidence, and why?

Applications

CHAPTER 8

Holographic Displays

8.1 INTRODUCTION

Holographic displays are in widespread use and often aspire, sometimes successfully, to become an art form. Small display holograms are used to provide additional authentication for a wide variety of articles including credit cards, banknotes, passports, licenses and concert and theater tickets, as well as consumer goods. Large format holographic displays are also widely used in advertising. There are two main classes of display holograms: those that require laser light in order to view the image and those from which the image can be reconstructed using white light. The former usually produce the most striking images but obviously the need for laser light imposes significant safety and other restrictions. We begin by considering simple approaches to display holography.

8.2 SINGLE-BEAM HOLOGRAPHIC DISPLAY

Figure 8.1 shows a simple holographic recording setup in which a single divergent beam of laser light provides both the object and reference beams.

This is ensured by expanding the beam using a powerful microscope objective lens. Such objectives are designated by their power when used in a microscope with a standard *optical tube length*, which is the distance between the focal point of the objective lens and that of the eyepiece in the microscope and is normally 160 mm. The focal length of the objective in mm is simply 160/power. Thus, a 10× objective has a focal length of 16 mm, whereas a 40× objective has a focal length of 4 mm. Knowing the focal length of the objective used, we can determine the angular spread of the laser. If its diameter is 1 mm and it is regarded as collimated, then using a 40× objective means that its spread is 0.125 rad. The angular spread is actually greater than this in practice because of the natural angular spread of the laser, which is typically about 0.5 mrad, (see Section 5.11). The greater the power, the greater the beam divergence and the more compact is the space necessary to ensure the objects and the recording emulsion are well illuminated.

Holograms made using this simple arrangement are fairly satisfactory if the objects are carefully chosen. Diffusely reflecting rather than specularly reflecting objects are preferable

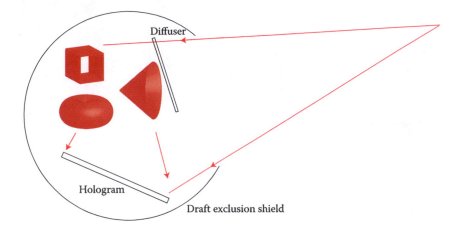

FIGURE 8.1 Single-beam holography.

as specular reflection of laser light causes glare and makes the image difficult to see well. If specularly reflecting objects are to be used, then a diffuser can be placed in the path of that part of the laser beam used to illuminate the objects. In this way, the objects are illuminated from a much wider range of directions than allowed by the beam divergence.

The setup of Figure 8.1 has the advantage of simplicity. In addition, it offers some immunity to changes in optical path differences between the object and reference beams, which arise from mechanical and thermal disturbance (see Sections 5.13 and 5.14). This is because the path differences only occur in the comparatively restricted space between the objects and the hologram plane. If the objects and the photosensitive plate are securely mounted on the same rigid platform, then mechanical disturbance should not be a problem. The area can also be surrounded by a simple shield made from any rigid plastic, timber, or metal sheeting to minimize turbulent air movement in the space between the objects and the hologram. Such shields should themselves be mechanically stable.

8.2.1 SPATIAL FILTERING

There are some disadvantages associated with the single-beam method. One problem arises with the use of an expanded laser beam. Small defects such as small air bubbles occur within the end mirrors of the laser cavity and within the components of the microscope objective. Microscopic dust particles and scratches occur on the mirror surfaces as well as on the optical surfaces of the microscope objective. All these defects diffract the laser light and produce fringe patterns usually in the form of concentric ring patterns on the object surfaces and on the hologram itself. Thus, both objects and the holographic recording plate are not uniformly illuminated. This in turn leads to significant reduction in diffraction efficiency in some parts of the hologram.

Recall from Section 2.2.2 that a lens illuminated by a distant point source of light or, equivalently, by a collimated beam of light, produces a diffraction pattern in its focal plane, consisting of a bright central disc surrounded by concentric, alternately bright and dark rings.

We can eliminate diffracted light due to the defects mentioned above by *spatially filtering* the laser, that is, by allowing only light in the central disc to propagate any further. This is accomplished using a pinhole whose diameter matches that of the central disc ($2.44\lambda f/D$) with D the beam diameter and f the focal length of the objective and that is positioned precisely at the focus of the microscope objective (Figure 8.2).

Precise positioning requires the use of a spatial filter that holds both the microscope objective and the pinhole. Three axes of translational motion are needed to position the pinhole precisely at the focus. In addition, two axes of translational motion and two axes of rotation are needed to ensure that the microscope objective is coaxial with the laser beam. For a 1-mm diameter beam of 633-nm wavelength and a 10× objective, the pinhole should be about 25 μm in diameter. In general, it is easier to spatially filter a laser using a low-power microscope objective and a larger diameter pinhole. This advantage is offset by the smaller divergence obtained. It is often better to compromise with a 20× pinhole and a 10× pinhole diameter. Higher power objectives and smaller diameter pinholes may be used, but the latter can be difficult to maintain clear of dust particles, which can seriously affect their performance. If such small pinholes are to be used, then a high-quality spatial filter is required with very precise micrometer movements, which are free of backlash.

The other problem is that we do not have any control over the relative intensities of object and reference beams reaching the photosensitive layer, and it is certain that the fringe contrast is a lot less than 1.0 since the reflectivity of the objects is likely to be rather less than 1.0. The diffraction efficiency of the hologram is usually very low if the hologram is of the amplitude type. Recall from Section 3.2.1 that a thin amplitude grating

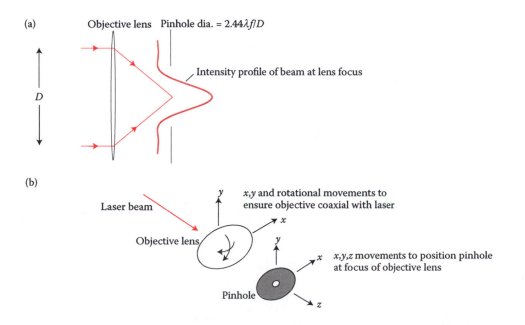

FIGURE 8.2 Spatial filtering of laser beam (a) pinhole diameter required to match Airy disc and (b) positioning/rotation of objective lens and positioning of pinhole.

cannot have diffraction efficiency more than 6.25%, while a thick amplitude grating has diffraction efficiency no greater than 3.7%.

Phase holograms produce brighter images because of their inherently greater diffraction efficiency, but their efficiency is also reduced by the less than ideal contrast of the interference pattern that is recorded. We will discuss how to deal with the problem of beam ratio in the following section.

8.3 SPLIT BEAM HOLOGRAPHIC DISPLAYS

If a beam splitter, such as a partially silvered flat glass plate or a beam splitter cube, is placed in the path of the laser beam, we obtain reflected and transmitted beams, which can be separately manipulated, one to provide the reference beam, the other to illuminate the object. We can now have much greater control over the beam ratio as well as over the illumination of the object. The arrangement (Figure 8.3) is very similar to the Mach–Zehnder interferometer of Figure 2.15 with one of the mirrors being replaced by the object and the recording layer located in the region where the two beams interfere.

Clearly, the beams travel very different routes to the recording plane, and *it is imperative, before we begin recording, to check interferometric stability by studying the interference fringes produced by a Michelson interferometer* (see Section 5.15). Some experimenters have used photodetection at two points in the interference pattern, the distance between the points being chosen so as to obtain signals which are $\pi/2$ out of phase. These signals are used in a feedback control loop to drive a piezoelectric position transducer attached to the reference beam mirror. Movement of the fringes due to instability is thus automatically compensated, and in this way, the fringe pattern can be stabilized throughout the holographic recording

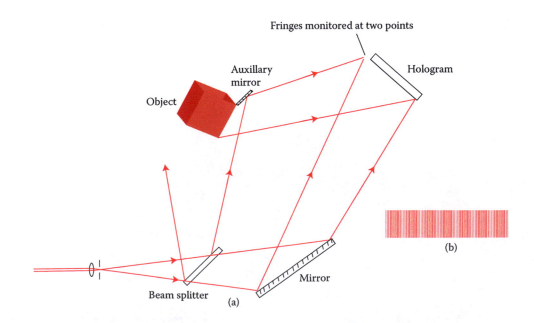

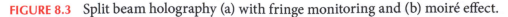

FIGURE 8.3 Split beam holography (a) with fringe monitoring and (b) moiré effect.

time. Monitoring of the fringes is facilitated by a small auxiliary mirror fixed to the object with the photodetectors placed behind the hologram or alongside it as shown in Figure 8.3. This is preferable to direct monitoring of the fringe pattern that is to be recorded in the hologram, as it is necessary to ensure good contrast in the photodetected fringe pattern. The spatial frequency of the fringes to be monitored has to be compatible with the spatial resolution of the photodetectors, so that direct monitoring only works well at very low spatial frequencies compatible with distance between the photodetectors. Alternatively, a transmission *diffraction grating*, whose spacing is about the same as that of the fringe pattern to be recorded, is placed behind the hologram plane [1]. The grating can be recorded in situ using self-processing photopolymer and the holographic recording setup with a mirror in place of the object. Photodetectors located behind the grating are used to monitor *differences* that arise between the fringe pattern spatial frequency and that of the grating, through the moiré effect. The feedback loop to the piezoelectric transducer is then used to negate such differences.

However, experience shows that it is usually better to ensure interferometric stability at the outset. All components including mirrors, beam splitters, the hologram holder, and the holographic recording medium itself should be carefully checked to see if there is any free play.

8.3.1 CONTROL OF BEAM RATIO IN SPLIT BEAM HOLOGRAPHY

The ratio of transmitted to reflected light intensity at the beam splitter is an important consideration. Usually the object reflects comparatively little light and the use of diffuse illumination can mean that the object to reference beam intensity ratio at the hologram plane is very small.

8.3.1.1 Use of Beam Splitters

To overcome this difficulty, one can use a beam splitter with a high reflectance to transmittance ratio. Beam splitters with reflectance of 95% and transmittance of 5% at a specified wavelength can be obtained. Such beam splitters usually have small apertures so must be located close to the spatial filter to ensure that the diverging beam is accommodated by the beam splitter aperture.

One can use a sheet of good quality float glass and obtain a wide range of transmissivity and reflectivity by altering the angle of incidence (see Figure 1.15). However, light beams reflected at the two surface of the glass, can interfere, producing undesirable fringes patterns on the object although careful adjustment of the angle of incidence of the laser at the glass can make these fringes so closely spaced as to be undetectable.

Whatever beamsplitting ratio is used, one can further adjust the ratio of intensities of object and reference beams at the hologram plane by attenuating the reference beam by means of a neutral density (ND) filter. Such filters are available in a wide range of optical densities. A filter with ND of 1.0 will attenuate the light passing through it by a factor of 10 while ND of 2.0 will attenuate by 100 and so on. This solution should be used with care when using materials that require some minimum *intensity* for holographic recording, unless of course the laser provides sufficient power so that the total light intensity in object and reference beams at the hologram is still high enough for recording purposes. It is for this reason that high-power lasers are often required for holography even though a great deal of the laser power is in fact wasted.

8.3.1.2 Polarizing Beam Splitters and Halfwave Plates

An elegant solution to the beam ratio problem is the use of a polarizing beam splitter with two half-wave plates. Polarized light and its control are discussed in Chapter 2. Consider a beam of light of amplitude E, plane polarized parallel to the vertical, y, axis, incident normally on a halfwave plate whose fast axis is at angle ϕ to the y-axis. The plane of polarization is rotated by 2ϕ (Figure 8.4). If the halfwave plate is followed by a polarizing beam splitter, the reflected and transmitted amplitudes are $E\cos(2\phi)$ and $E\sin(2\phi)$ respectively, the former being vertically polarized and the latter horizontally polarized. If the transmitted light is passed through a second halfwave plate whose fast axis is at 45° to the horizontal, its plane of polarization is returned to the vertical.

We thus have two beams of light plane polarized in the same plane with relative intensities $E^2\cos^2(2\phi)/2$ and $E^2\sin^2(2\phi)/2$ controlled solely by the value of the angle ϕ and any beam ratio is possible. Polarizing beam splitters and halfwave plates are expensive components especially in large sizes, so that this arrangement is normally used with an unexpanded beam. In any case, the polarizing properties of these devices vary with the angle of incidence so the incident light should be collimated. Consequently, beam expansion and spatial filtering are carried out following beamsplitting and separate spatial filters are required for object and reference beams.

A further advantage of the split beam method is that the object beam can be further split into as many beams as desired so that the objects may be illuminated from several different

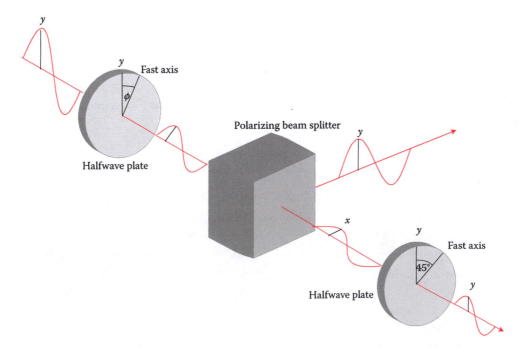

FIGURE 8.4 Use of a polarizing beam splitter and halfwave plates to control beam ratio.

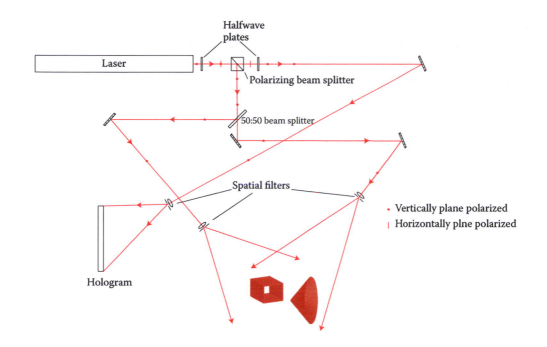

FIGURE 8.5 Split beam holography with two object illuminating beams.

directions. It is good practice to inspect the illuminated objects, by looking through the aperture defined by the area of the hologram to be recorded. The angles of incidence of the illuminating beams as well as the object positions should be adjusted to optimize the display, bearing in mind that what one observes live before the recording is the best that one can obtain from the hologram. Figure 8.5 shows a plan view of a typical split beam holographic recording setup.

8.4 BENTON HOLOGRAMS

By dispensing with perspective in one dimension, one can produce holograms that can be reconstructed using ordinary light. This idea proposed by Benton [2] led to the development of holography as an industry, primarily for security and authentication applications although impressive large format display holograms are also possible using the technique.

In its original form, Benton's technique, also known as rainbow holography for reasons that will shortly become clear, requires the reconstruction of the conjugate image from a hologram, known as the master hologram. Preferably, there should be no distortion of this image so it is best to use the conjugate reference beam for reconstruction. The easiest way to obtain the conjugate reference beam is to use a collimated reference beam at the outset. The basis of Benton's technique is shown in Figure 3.11 where a hologram is rotated 180° about a normal to its plane so that it is effectively illuminated by the conjugate reference wave and a real, conjugate image of a point is reconstructed. Figure 8.6a shows a hologram H1, usually called the master hologram, being recorded in the usual way with a diverging beam to illuminate a 3-D object and a collimated reference beam, designated *R*. Following processing if required, the hologram is inverted in its plane so that it is now effectively illuminated by the conjugate reference wave R^*, to reconstruct O^*, the conjugate of the object wave O (Figure 8.6b). We now cover the hologram with a slit, S, and record a second hologram H2, known as a copy hologram. Figure 8.6c is a perspective version of Figure 8.6b to emphasize the fact that the light diffracted by the hologram H1 is largely confined in direction to the horizontal plane. Finally, H2 is inverted in the path of R so that the conjugate of O^*, that is O, is reconstructed as a real image in the space between the viewer and the hologram. Very striking images have been produced by this method. In addition, a conjugate, real image, S^* of S is reconstructed. The viewer has to find S^* in space since this is the effective aperture. On looking through S^* the real image, O, is seen.

The purpose of the slit might seem puzzling at first, since it seriously restricts the viewer's range of perspective in the direction perpendicular to the long axis of the slit, but vertical perspective is not quite as important as horizontal since one sees the world from the perspective of one own viewing height, which does not change very much, whereas our horizontal viewpoint changes constantly as we move around. The horizontal slit eliminates vertical but retains horizontal perspective. The advantage gained is that all the reconstructed light is confined to the solid angle subtended by S^* at the plane of H2 so that the reconstructed image is much brighter than is the case when the slit is not used.

Now, consider what happens when white light is used to illuminate the copy hologram H2 in Figure 8.6d. Conjugate images of the slit are reconstructed at all the wavelengths in the visible spectrum and the image can be seen at each wavelength as the observer changes vertical perspective (see Figure 8.7). The combination of wavelength and angle of view

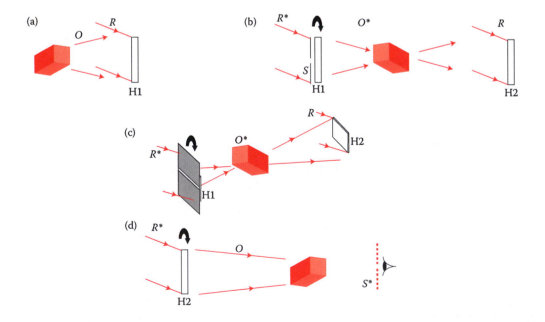

FIGURE 8.6 Benton holography: (a) recording master hologram, H1; (b) recording copy hologram, H2; (c) recording copy hologram, H2 (perspective); and (d) reconstructing the final image.

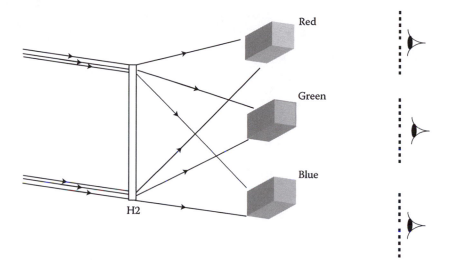

FIGURE 8.7 Rainbow holography.

satisfies the Bragg condition (Equation 2.9) for each image that we observe. Careful choice of slit width and distance between H2 and H1 ensures that the images are well separated in color and angle so that the rainbow effect is minimized if desired.

8.4.1 IMAGE PLANE HOLOGRAPHY

In Figure 8.6b, the reconstructed conjugate image is located roughly midway between the master hologram H1 and the copy H2, but in fact H2 can be positioned anywhere to the right of H1 and is often located in a plane within the conjugate image O^*. If the object itself has little depth, then $z_O \cong 0$ and from Equation 7.43, we find that $x_{O'} = x_O$, $y_{O'} = y_O$, $z_{O'} = 0$ *independently of reconstruction wavelength* so that there is no blurring even if we use a broadband light source to reconstruct the image. Obviously, blur effects occur with increasing distance from the hologram plane.

8.4.2 SINGLE STEP IMAGE PLANE RAINBOW HOLOGRAPHY

A rainbow hologram can be made in a single step [3] by using a lens to form a real image of the object in the hologram plane (see Figure 8.8). Longitudinal magnification, LoM, is the *square* of the lateral magnification, LaM (Equations 7.48, 7.49) even if the wavelength used for reconstruction is the same as for recording, so that the reconstructed image of an object with depth is distorted unless we ensure that the lateral magnification is 1.0 at the outset by locating the object at distance $2f$ from a lens of focal length f.

A slit is placed between the object and the lens at a distance greater than f from the lens so that a magnified real image of the slit appears to the right of the hologram. At reconstruction, a real image is observed through the slit.

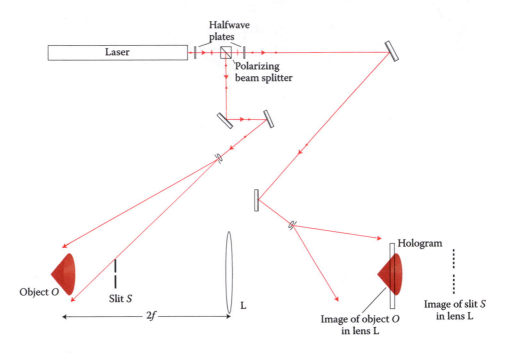

FIGURE 8.8 Single step image plane rainbow holography.

8.4.3 BLUR IN RECONSTRUCTED IMAGES FROM RAINBOW HOLOGRAMS

To obtain a simple expression for the color blur that is produced when a rainbow hologram is illuminated by white light, let us regard the final hologram as a simple diffraction grating of spacing d. It is illuminated by the conjugate of R in Figure 8.9, and we apply the grating equation Equation 2.7 with θ a small angle so that $d\theta = \lambda$.

The range of wavelengths $\Delta\lambda$ that can pass through the reconstructed image slit is given by $\Delta\lambda = d\Delta\theta$ where $\Delta\theta$ is the angle w/D subtended by the image slit of width w at the hologram at distance D. We assume that the eye is located at the image of the slit and that it is the slit width rather than the eye pupil diameter that limits the wavelength spread in the image. Again, using Equation 2.7 in the first order we have

$$\Delta\lambda = \frac{w\lambda}{D\sin\theta} \tag{8.1}$$

If w is 5 mm, $\lambda = 633$ nm and $\theta = 30°$, and $D = 500$ mm, then $\Delta\lambda = 12.5$ nm. From Equation 7.43 $\partial y_{O'}/\partial\lambda_2 = -y_R z_O/\mu^2 z_R \lambda_1$, we find the color blur is $\Delta y_{O'} = (\Delta\lambda/\mu^2\lambda)(y_R/z_R)z_O$, and from Equation 8.1, with $\mu = 1$

$$\Delta y_{O'} \cong \frac{wz_O}{D\sin\theta}\frac{y_R}{z_R} = \frac{wz_O}{D}$$

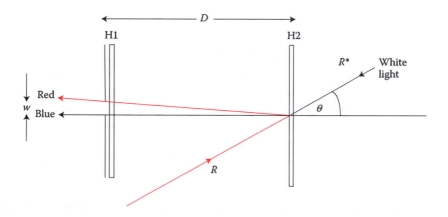

FIGURE 8.9 Image blur in rainbow holography.

Suppose the final image is 5 cm from the hologram, and it is viewed through the image slit at 500 mm from the hologram. The angular blur is 0.5 mrad, which is acceptable.

The source width must also be considered. Suppose it subtends angle $\Delta\theta$ at the hologram, then the image blur is $z_O \, \Delta\theta$, and if this is to be less than the color blur, then $\Delta\theta$ should be less than w/D, that is, about 10 mrad.

Diffraction by the slit also contributes to blur. The angular spread $\Delta\theta$ of light passing through the slit image due to diffraction is obtained from Equation 2.6 with $\beta = \pm\pi$ where we assume the light is largely confined to the central lobe of the diffraction pattern. This gives $\Delta\theta = 2\lambda/w$ and so the blur is $z_O \, \Delta\theta$, which is only about 13 μm.

8.5 WHITE LIGHT (DENISYUK) HOLOGRAMS

Denisyuk's invention of white light holography, also known as reflection holography, was perhaps the most significant in bringing holography out of the laboratory and into the realm of everyday experience. The advantages over Benton holography are that perspective is not compromised nor is there significant chromatic dispersion when the holographic image is reconstructed using white light. Denisyuk's method [4] has much in common with Lippmann's method for making color photographs [5]. The principle of Lippmann's method is that light passes through a partially transparent photosensitive layer to form a focused image at its rear surface. This light interferes with its reflection from a plane mirror in close contact with the layer, the interference fringes being essentially parallel to the mirror surface and spaced apart by half the wavelength of light in the layer. White light of course produces a multiplicity of such interference patterns. When the finished photograph is illuminated by white light, the original colors are reproduced by spectral filtering by the recorded fringe patterns. A photosensitive material of extremely high spatial resolution is required and successful Lippmann photographs have been made only relatively recently using very high-resolution panchromatic silver halide emulsions designed primarily for full color holographic recording.

The principle of Denisyuk holography is illustrated in Figure 8.10. Here, the object is illuminated through the recording medium, which acts as a beam splitter and must be highly transparent at the laser wavelength in use. Object and reference beams thus approach the recording medium from nearly opposite directions. This situation is discussed in Section 4.3.2 and the diffraction efficiency dependence on the modulation index parameter $\upsilon' = \pi n_1 T / \lambda_{air} \cos\psi$ is shown in Figure 4.8 for a phase grating. In principle, very bright reconstructions can be obtained if the refractive index modulation n_1 and the spectral bandwidth are sufficiently large. In practice, however, the transmissivity of the recording material can have a significant effect on the object illumination intensity and therefore on the reference-to-object beam ratio and the fringe contrast and finally on the recorded refractive index modulation. Transmissivity typically falls exponentially with recording layer thickness, T, so that increased layer thickness may not compensate for poor refractive index modulation in υ'. Furthermore, increased thickness reduces the spectral bandwidth (Equation 4.48). The problem can be partially alleviated by using the split beam technique discussed in Section 8.4 so that the object is independently illuminated.

A reflection hologram can be made to reconstruct a real image as shown in Figure 8.11. The image can be made to straddle the copy hologram H2 or simply positioned as close to H2 as required. This technique is widely used in the holography industry.

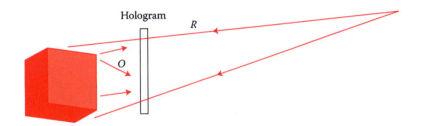

FIGURE 8.10 Denisyuk holography.

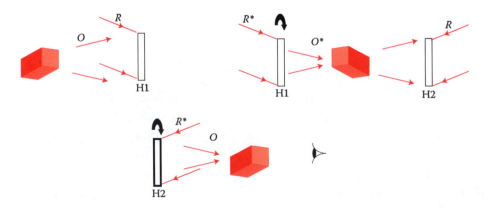

FIGURE 8.11 Denisyuk real image holography: (a) recording master hologram H1; (b) recording copy hologram H2; and (c) Reconstructing real image.

8.6 WIDE FIELD HOLOGRAPHY

The holographically reconstructed image may only be viewed from a range of angles determined by the hologram aperture. Various attempts have been made to obtain holographically reconstructed images, which may be viewed from any angle. The simplest method [6] using a single-beam holographic setup, is shown in Figure 8.12.

The object is enclosed in a rigid hollow cylinder open at one end and the holographic film is secured to the inside surface of the cylinder. It is not easy to ensure interferometric stability in such an arrangement, and it is sometimes preferable to construct a hollow cube system of holographic plates around the object using glass cement to secure them together at their edges as shown.

To make a compact flat hologram with a near 360° view [7], one first records a conventional transmission hologram of the object (Figure 8.13a) as seen from the right.

In the second stage (Figure 8.13b), a reflection hologram of the rear of the object is recorded. The object is removed in stage 3 and, using the conjugate of the reference wave used in stage 1, the conjugate image is reconstructed from the first stage hologram and recorded as a second reflection hologram. Finally, this hologram is illuminated with *both* the conjugate of the reference beam used in the recording of stage 3, so that the reconstructed orthoscopic image is seen on looking from the right, *and* with the original reference wave used in stage 2 so that a reconstructed, orthoscopic, rear view image is also seen on looking from the left.

Another method of making wide-field 3-D images is the spatially multiplexed hologram technique. If an object is rotated through 360° and photographed from the same position, after say every 0.3° of rotation, one accumulates a total of 1200 photographic transparencies collectively containing all the information required for a 360° image of the object. To produce such an image, we need to synthesize the *temporal* sequence of photographic images into a *spatial* sequence, which can be interpreted by the observer as a 3-D image. Holography provides a solution to the problem of spatial synthesis. Each photographic transparency is recorded as an image plane hologram as shown in Figure 8.14.

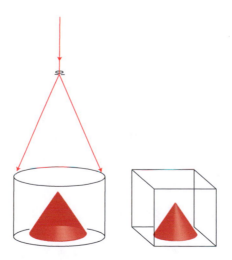

FIGURE 8.12 Single-beam 360° holography.

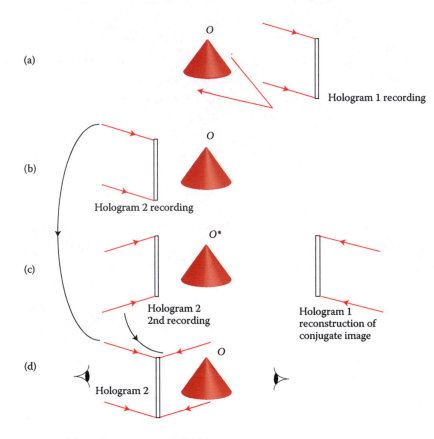

FIGURE 8.13 Double-sided near 360° hologram.

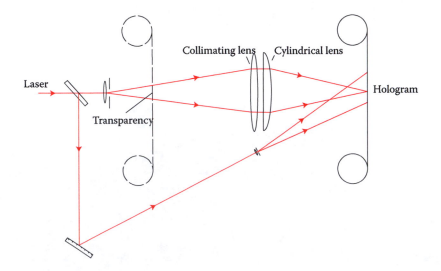

FIGURE 8.14 Multiplex holography.

A cylindrical lens effectively provides the slit aperture for each hologram so that white light may be used at the reconstruction stage. The finished roll of holograms is wound into cylindrical form. An observer looking into the cylinder, illuminated by white light from below, sees a reconstructed *stereoscopic* image. This is because several holographically reconstructed 2-D images with slightly differing horizontal perspectives are presented to the observer at the same time. The stereoscopic image rotates as the cylinder is rotated about its principal axis. There is no change in perspective as one moves one's observation point up or down.

A variation of the spatial multiplexing method is used for large format, flat stereographic imaging. A set of 2-D photographic transparencies is prepared using a camera, which is moved stepwise between photographs. These transparencies are projected in sequence by a laser onto a diffusing screen and the images are recorded as contiguous holograms, using a cylindrical lens so that the holograms are thin strips. The holograms are then masked by a horizontal slit and reconstructed using a conjugate version of the reference beam to produce a real image in the same plane as the original diffusing screen. Finally, a hologram is recorded in that plane. Upon reconstruction, a wide field flat stereoscopic image is produced.

8.7 COLOR HOLOGRAMS

With the availability of improved holographic recording materials and lasers operating at three primary color wavelengths, color holography has become a practical reality mainly using the Denisyuk method although a number of other techniques are available.

The basic principle is the same as for any color display. Three spectral colors mixed together can synthesize any color in the CIE (Commission Internationale d'Eclairage) chromaticity diagram lying within the triangle defined by the three colors (Figure 8.15). The red wavelength of 633 nm from a helium–neon laser or 647 nm from a krypton laser, green 514 nm, and blue 488 nm from an argon ion laser provide three well-spaced points in the CIE diagram. Other wavelengths are available and the advent of compact solid state diode lasers in a range of wavelengths has extended the color range available.

The first problem to be overcome is crosstalk, which is the phenomenon by which a thin hologram recorded at one wavelength can diffract light at another. Thus, if three wavelengths are used, each hologram can diffract light of the other two wavelengths producing additional undesirable images, which are subject to the effects discussed in Section 7.7 and degrade the desired color image. Full-color *transmission* holograms have been recorded in thick photosensitive layers with negligible crosstalk [8] as each hologram diffracts only light at the wavelength and in the direction of that used to record it and the other wavelengths do not satisfy the Bragg condition.

However, recently the usual approach is to prepare a set of three Denisyuk reflection holograms in the same thick layer, using each of the available laser wavelengths, all emanating from the same point source. This method is preferred for its simplicity, the availability of white light reconstruction and because again there is negligible crosstalk, since

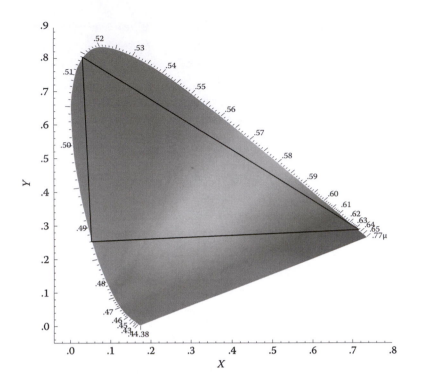

FIGURE 8.15 **(See color insert.)** CIE chromaticity diagram. Wavelengths around the periphery are in micrometers. The triangle encloses all colors that can be obtained using wavelengths of 488, 514, and 633 nm. (Redrawn from http://en.wikipedia.org/w/index. php?title=File:Chromaticity_diagram_full.pdf&page=1.)

each hologram reconstructs strongly only at the wavelength used to record it and the other wavelengths do not satisfy the Bragg condition for that hologram.

A typical arrangement for recording color reflection holograms is shown in Figure 8.16.

Dichroic filters are used to combine the laser beams so that they can be spatially filtered together. Some loss of laser power is inevitable at two of the wavelengths since a spatial filter is designed to work optimally at a single wavelength. The Argon laser has to be retuned from 514 to 488 nm between exposures unless the wavelength selector prism is removed from the cavity, in which case the laser operates at all wavelengths simultaneously and the 514 and 488 nm wavelengths can be singled out using a band-pass spectral filter. However, the intensities at these two wavelengths cannot then be independently controlled. Thus, it is preferable to use a separate 488-nm laser. If the exposure times differ, a separate shutter is required for each wavelength.

There are a number of disadvantages in the use of a single recording layer for all three holograms. The first is that the dynamic range of the recording material has to be shared and the diffraction efficiency of each is reduced by N^2 where N is the number of holograms [9]. To avoid this problem, holograms are recorded on separate photosensitive layers

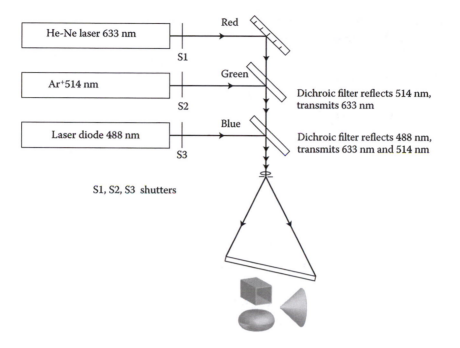

FIGURE 8.16 Color reflection holography.

since each is a phase hologram and transparent outside its own Bragg reflection wavelength range. Generally, two holograms are recorded, one for the blue and green wavelengths, and the other for red, often using dichromated gelatin.

The second difficulty is that silver halide layers tend to change dimensions during chemical processing resulting in emulsion shrinkage and change in recorded fringe spacing so that the hologram diffracts light of a different wavelength from that used in recording. Care in processing and the use of appropriate processing schemes (see Section 6.2.4) is of critical importance.

A further difficulty with silver halide is that scatter increases significantly at shorter wavelengths. As explained in Chapter 6, a great deal of effort has been expended in reducing grain size, with dramatic improvements in the color fidelity of holograms recorded in a single emulsion by the method shown in Figure 8.16 [10].

8.8 DYNAMIC HOLOGRAPHIC DISPLAYS

8.8.1 ACOUSTO-OPTIC MODULATOR-BASED SYSTEMS

Holographic display at acceptably high video rates has been the goal of considerable research, as it presents very significant technical challenges. Let us first consider the communications problem. The pattern of interference between collimated object and reference beams of wavelength 500 nm angularly separated by 60° has a spatial period of 0.5 μm. In a real hologram, this spacing is modulated by the actual object wave, but we can assume the pattern has a sinusoidal intensity whose spatial period varies only moderately around this value. In order to reproduce a pure sine wave, we obviously need to know its amplitude and spatial period and to solve for these two unknown quantities, we need two equations, that is, we need to know the precise value of light intensity at two points within a cycle. This is the well-known Nyquist criterion. We require a total of 4×10^5 samples across a single horizontal line of a hologram 100-mm wide. Approximately, the same number is required in the vertical direction. A refresh rate of about 30 holograms per second, as in current television practice, would probably be adequate. Assuming a sample is specified by a single byte of 8 bits, we therefore need a communication rate of $3.8 \times 1\,0^{13}$ bits s^{-1}, an extremely daunting task given present technological limitations. We could settle for a coarse raster system in which each hologram is represented by say 500 horizontal lines and requires 50 Gbits s^{-1}, which is rather less daunting. The huge amount of data required for a dynamic holographic display can be acquired and processed on current state-of-the-art supercomputers running large numbers of parallel processors. The remaining task is to present the data in a way that emulates holography as far as possible.

To make a dynamic holographic display system, St. Hilaire and coworkers [11] used an acousto-optic modulator (AOM). This consists of a rectangular block of solid material, which is elastically stretched or compressed when subjected to an electric field. The phenomenon is known as the piezoelectric effect, and electrical excitation is possible at frequencies up to 100 MHz producing an ultrasonic pressure wave and a corresponding refractive index modulation. A frequency-modulated signal applied to the device results in a 1-D dynamic, phase hologram, which is illuminated by a laser, and an image is reconstructed. The AOM is rather short and can only provide a correspondingly short 1-D hologram section, so the data supplied to the AOM is replaced by the data for the next section and the object wave reconstructed from the AOM is directed to a new location in space by means of a rotating polygonal mirror. Using this technique of spatial multiplexing, a 1-D hologram of acceptable width and a corresponding image space (Figure 8.17) can be synthesized.

A mirror oscillating around a horizontal axis provides the vertical scan. The velocity of sound in an AOM made from TeO$_2$ is about 1260 ms^{-1} so that a signal frequency around 100 MHz produces a hologram whose spatial period is typically around 12.6 μm and the helium–neon laser is diffracted by about 3° although a lens (not shown) is used to increase

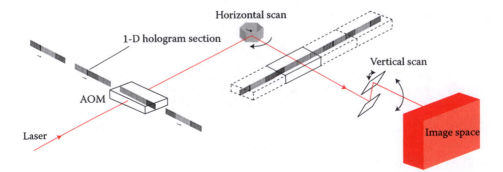

FIGURE 8.17 Use of an acousto-optic modulator to synthesize a hologram by spatial multiplexing.

this to 15°. The original design had a reconstructed image space volume of 25 × 25 × 25 mm³. This was increased in a later version [12] to 150 × 75 × 150 mm³, and a 30° field of view by using 18 AOMs in parallel so that 18 vertically stacked 1D hologram sections could be presented together and the vertical scanner shifted by 18 lines each time. In addition, small galvanometric mirrors were used instead of a polygonal mirror.

Further potential developments include the introduction of AOMs based on surface acoustic wave (SAW) devices, which can be driven at GHz frequencies. Light propagating in a slab waveguide can be diffracted by SAW device fabricated on top of the guide as shown in Figure 8.18. Crossed gratings can be implemented by placing electrodes at the four sides of the SAW device so that light can be diffracted in orthogonal planes and a 2-D hologram could be implemented although the angle of diffraction obtained from the transmission grating (Figure 8.18a) is much larger that from the reflection grating (Figure 8.18b).

In Figure 8.18, the two planes of diffraction are shown separately for clarity.

8.8.2 SPATIAL LIGHT MODULATOR-BASED SYSTEMS

At the beginning of Section 8.8.1, we outlined the fundamental problem in dynamic holographic display, the sheer amount of information in even a single hologram. The approach adopted by St-Hilaire et al. [11,12] is essentially a technique of spatial multiplexing of hologram sections and image space. An alternative spatial multiplexing method has been developed by Slinger et al. [13] using an electrically addressed spatial light modulator (EASLM) and a set of optically addressed spatial light modulators (OASLM). Both devices use liquid crystals (LC) although there are other types based on arrays of micromirrors. We need to take a short digression here to explain how SLMs work.

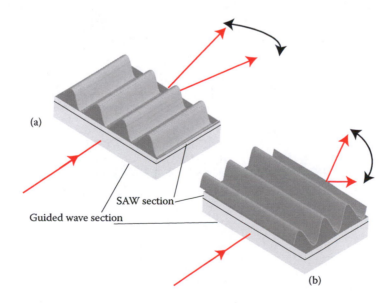

FIGURE 8.18 Surface acoustic wave (SAW)-based diffraction of a guided wave: (a) transmission grating and (b) reflection grating.

8.8.2.1 Liquid Crystal Spatial Light Modulators

Liquid crystal SLMs (LCSLM) usually consist of a large number of pixels. In one form of LCSLM, each pixel is sandwiched between crossed linear polarizers so that normally, no light can pass through. Each pixel consists of a layer of LC, which are aligned with fine parallel grooves in transparent electrodes on each face, the grooves on opposite faces being orthogonal. The result is that the liquid crystal axis rotates helically from one side of the device to the other, through 90°. Plane polarized light entering from one side of the device, with its plane of polarization parallel to the grooves and to the transmission axis of the polarizer on that side has its plane of polarization gradually twisted through 90° and can pass through the second polarizer. Application of an electric field across the device forces the LC molecules into alignment with the field, negating the polarization rotation effect, and the light can no longer pass through the second polarizer (Figure 8.19). For this reason, the device is sometimes referred to as a liquid crystal light valve (LCLV). In this type of electrically addressed SLM (EASLM), any spatial pattern of intensity may be imposed on a light beam subject to the limitations imposed by the area density of the pixels. Typically, up to 800×600 pixels occupy an area of about 2 cm².

Alternatively, one may set the alignment grooves *parallel* to one another and dispense with linear polarizers, in which case the applied electric field simply rotates the molecules from the vertical to the horizontal direction, altering the pixel's refractive index. In this case, the device is a pure phase modulator.

Another form of EASLM is a liquid crystal on silicon (LCoS) device shown in Figure 8.20. Here, the liquid crystal molecules are normally aligned perpendicular to the plane of the figure. When a pixel is electrically excited through one of the CMOS transistors in the CMOS backplane on the left, the LC molecules are rotated perpendicular to the alignment grooves, changing the local refractive index. This device can be operated as a pure phase

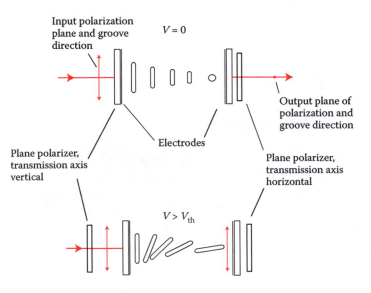

FIGURE 8.19 Twisted nematic spatial light modulator.

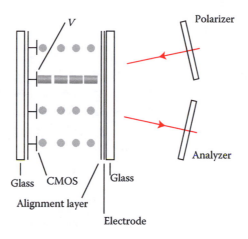

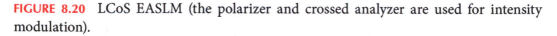

FIGURE 8.20 LCoS EASLM (the polarizer and crossed analyzer are used for intensity modulation).

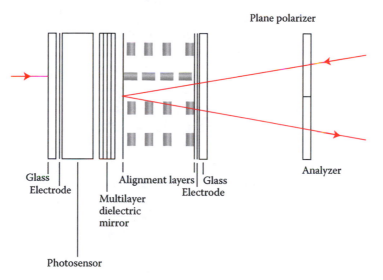

FIGURE 8.21 OASLM.

modulator. With the addition of crossed linear polarizers, it can also operate as a spatial intensity modulator since it is normally birefringent. Suitable choices of LC type and layer thickness enable the device to operate as a switchable halfwave plate. Incoming light has its plane of polarization rotated by a total of 90° on passing through the LC layer twice and can pass through the second polarizer. An appropriate voltage across a pixel orients the LC molecules normal to the electrodes, negating the birefringence, and the reflected light is blocked by the second polarizer.

An *optically addressed* SLM (OASLM) is shown in Figure 8.21. In this device, the LC molecules are aligned parallel to one another and to the electrodes. A voltage is applied to the electrodes and when light is incident from the left on one of the pixels, current flows in the photosensor, significantly reducing the voltage between the left electrode and the left side of the LC layer. Thus, most of the applied potential drop is across the LC layer, which is realigned accordingly.

As shown, the device is operated as an intensity modulator, the 90° rotation of the plane of polarization of light from the right being negated by light incident from the left. Thus, an interference fringe pattern is stored by the device even after the incident interference pattern is removed because of the bistability of the LC molecular orientations. Pure phase modulation may also be implemented by removing the polarizers. These devices do not have a pixelleted structure and are capable of very high spatial resolution depending only on the properties of the LCs and photosensor material.

8.8.2.2 Tiling Systems for Dynamic Holographic Displays

Some EASLMs used as spatial intensity modulators can operate at very high refresh rates with each pixel having only on and off states. The *visually apparent* intensity at any location on the device is controlled by the duration for which the control voltage is applied each time the pixel at that location is addressed. A computed interference fringe pattern can be represented in binary fashion (i.e., fringe = pixel on, no fringe = pixel off). A small part of a hologram can be written to the EASLM and imaged onto a small area of a larger hologram, which is to be synthesized. The EASLM can then be rapidly refreshed with the next small part of the hologram, which is imaged onto an adjacent area of the larger hologram. Thus, the EASLM is used to "tile" a large hologram. The larger hologram must store each of the small ones until it is next updated and this task is performed by an OASLM, storing each small hologram section in the appropriate location. A system consisting of a diffractive optical element, relay lenses and shutters is used to write the small hologram sections onto the larger. Figure 8.22 shows a 2 × 2 tiling arrangement.

One version of the system uses a LCoS EASLM with ferroelectric LCs to produce 1024 × 1024 binary fringe patterns at 2.5-kHz refresh rates. The patterns are tiled onto

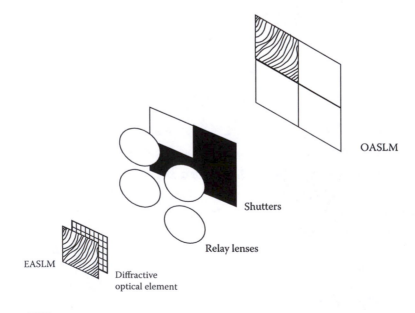

FIGURE 8.22 Tiling system.

an OASLM using switchable diffractive optics to steer each one to the correct location to form an array of 5 × 5 hologram sections and a hologram with 26 Mpixels. Systems may be stacked together to synthesize larger hologram formats.

8.8.3 PHOTOREFRACTIVE POLYMER-BASED SYSTEMS

Photorefractive polymers (Section 6.6.2) 30 × 30 cm in area have been used as recording materials in a dynamic holographic display with a refresh rate of 0.5 s^{-1} [14]. Each full size hologram is recorded as an array of small (1 × 1 mm) holographic pixels or "hogels" in an extension of the spatially multiplexed holography method (Section 8.6). The hogels are computed from images of a 3-D scene, which are synchronously captured at a rate of one per second by 16 Firewire cameras, each recording the scene from different perspective. Color is encoded in the recorded hologram using different angles between three sets of recording beams. By switching the polarizations of the interfering beams into the orthogonal plane between recordings, one can avoid crosstalk between two holograms. To avoid crosstalk with the third hologram, one uses the fact that beams whose angles of incidence are symmetrical with respect to the applied electrical field direction do not record a hologram either. LEDs are used to reconstruct the images. Recording a new hologram erases the previous one so that a dynamic display is realized.

8.9 VERY LARGE FORMAT HOLOGRAPHIC DISPLAYS

The realism of 3-D holographic displays makes them attractive for advertising and while holography has already been used quite widely in advertising, mainly in the form of rainbow holograms and small embossed holograms (see Section 9.7), the lack of full 3-D perspective and absence of true color are disadvantages. To produce a large holographic display, that is, one with full perspective and, preferably, true color, is a considerable challenge that is unlikely to be met by the conventional holographic recording process. One could envisage the use of high-energy Q-switched lasers to overcome the stability problems associated with large format and objects of significant size. However, it is difficult to satisfy the Bragg condition for reconstruction everywhere in a large hologram using a point source of white light. Besides, further applications for large format holography are the representation of landscape and cityscape imagery and data from geophysical information systems as well as computer-generated 3-D imagery.

A solution to the problem developed by Zebra Imaging© is to regard the final hologram as consisting of a large number of small holograms, also known as hogels, each 2 × 2 mm in area, recorded using the Denisyuk method, that is with object and reference beams approaching the hologram plane from opposite sides. For each hogel, scene data generated by more than 106 optical ray traces is input to a spatial light modulator (200 mm × 250 mm). A holographic optical element (see Section 11.4) in contact with the SLM forms a precisely masked image of the SLM in the front focal plane of a lens (Figure 8.23).

Thus, light from each point from the SLM becomes a collimated beam as it emerges from the lens onto the hologram plane. This has the advantage that, if a collimated reference

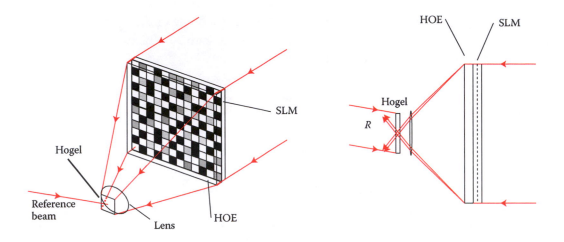

FIGURE 8.23 Recording a hogel. (Reproduced from Michael Klug, Zebra Imaging Inc. With permission.)

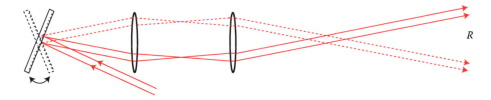

FIGURE 8.24 Adjusting reference beam angle of incidence.

beam is used, there is very little variation in recorded spatial frequency and therefore of diffraction efficiency, from one hogel to another. Dupont Omnidex© 801 photopolymer film is used to record the holograms. The film laminated onto a glass support is moved under precise spatial control in the horizontal plane so that a total of 90,000 contiguous hogels are recorded making a total hologram area of 0.6 × 0.6 m. These are then assembled as tiles into larger holograms up to 5.4 × 1.8 m in area.

It is necessary that the reference beam at the recording stage be incident on the hologram at an angle that can be replicated in reconstruction using a point source of white light. Thus, the reference beam angle is different for each hogel recording and the adjustment is made using a scanning mirror and a pair of lenses having a common focus (Figure 8.24).

8.10 QUANTUM ENTANGLEMENT HOLOGRAPHY—IMAGING THE INACCESSIBLE

When certain nonlinear optical materials, such as beta barium borate, are illuminated by a pump beam of monochromatic light, they emit two light beams [15], each of the beams having a wavelength that is twice that of the illuminating light. The process is called spontaneous optical parametric down-conversion.

The down-converted beams exhibit the phenomenon of *quantum entanglement* [16]. This means that measurement on one photon of an entangled pair of photons, such as its state of polarization, automatically and simultaneously provides us with information about the other, although the photons may be very far apart, traveling in opposite directions, so that the second photon is inaccessible to our measurement system. The phenomenon is a very strange one, not least because it would appear at first sight that it violates the basic principle of physical causality that one can not transmit information at speeds exceeding that of light. The contradictions are resolved by the fact that a quantum entangled pair of photons is a single entity.

Abouraddy and coworkers [17] have suggested that entangled beams could be used to record a hologram, even though the pump beam that produces them may not itself have sufficient coherence length for holographic recording.

The object is inside a chamber, C, which light can enter but cannot leave. Some of the light may scatter from the object to form an object beam, which is then incident on the chamber wall (Figure 8.25). The rest of the light is incident directly on the wall as a reference beam interfering with the object beam. We do not obtain a hologram because we only record the incidence of light but not the *position* of incidence.

We now suppose that the light entering the chamber is entangled with light propagating *outside* and incident on a position sensitive photosensor, D, such as a CCD array. We consider the rate at which photons arrive at each point x_D, y_D of the array *coinciding* with the arrival of photons at x_C, y_C on the chamber wall. However, since x_C, y_C is not known, we must *integrate* the rate at which photons arrive at x_D, y_D simultaneously with their entangled counterparts at the chamber wall points x_C, y_C, over the whole of the wall surface.

The result of the integration is a *hologram*, which may be transferred onto a photosensitive material or spatial light modulator to be illuminated subsequently by a coherent reference beam in order to reconstruct an image of the object.

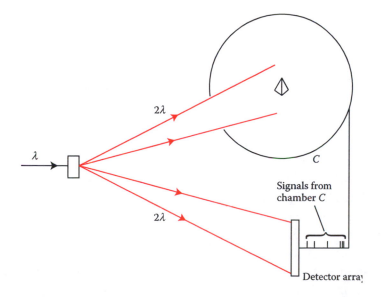

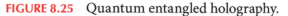

FIGURE 8.25 Quantum entangled holography.

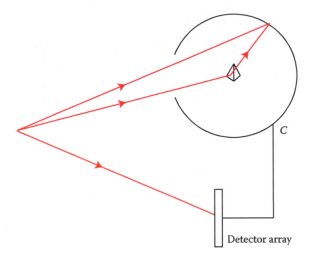

C

Detector array

FIGURE 8.26 Holography with a single entangled photon pair.

Looking at the experiment in terms of a just one entangled photon pair (Figure 8.26), suppose we detect a single photon at a known point on the photosensitive array *coinciding* with detection of a photon with which it is quantum entangled at the chamber wall. This amounts to detecting the spatial coordinates of the latter, which has *both* propagated *directly* to the chamber wall and *indirectly* to it following diffraction from the object surface. Since we have no knowledge of the path taken by the photon inside the chamber, it is *both diffracted* and *not diffracted* by the object and therefore produces an interference pattern at the chamber wall. This has been experimentally demonstrated by the production of interference fringe patterns using a light source of such low intensity that, on average, just one photon is present in the interferometer [18].

8.11 GOOD PRACTICE IN HOLOGRAM RECORDING

In all holographic works, it is important to adopt good practice to ensure best results. The recording setup should be planned in advance and care taken to ensure that there is going to be sufficient space on the optical table, especially if more than one object illuminating beam is to be employed or if a Benton hologram is to be made (Figure 8.6). It is helpful to prepare a line diagram showing the approximate positions of the laser, and of each optical component and of the hologram holder.

The interferometric stability of the optical table should be thoroughly checked using a Michelson interferometer (Section 5.15). The table should be isolated from the floor by supporting it on inflated inner tire tubes on a firm horizontal surface. Other isolation methods that have been used successfully are large sturdy wooden boxes filled with sand in which all the component holders are placed, or thick rectangular blocks of polystyrene supporting a flat metal platform. Isolation supports consisting of hollow metal cylinders, each containing a bag inflated by compressed air or oxygen-free nitrogen bottled gas, are also used. Such supports act in a similar fashion to inner tubes but are able to support massive optical tables, which reduce the transmission of floor vibration to the recording system even further. If one can locate the holography laboratory in a concrete basement well away from traffic noise, so much the better.

When the setup has been assembled, all optical path lengths are measured from the position of the first beam splitter to the hologram plane to ensure that each *optical path difference* between an object beam and the reference beam is less than the coherence length of the laser. This is particularly important when using a helium–neon laser whose coherence length is rather restrictive.

After the laser is turned on, time should be allowed to elapse for its output to stabilize, before recording a hologram.

Once the interferometric stability of the optical table has been established and all sources of noise and mechanical vibration in the laboratory eliminated as far as possible, we must also check the stability of every component by pushing it firmly with a finger. If there is any noticeable play or movement of either the component itself or its holder, this must be eliminated.

The hologram holder should also be stable. Heavy metal plateholders are best.

When materials such as photopolymers are coated onto flexible plastic substrates, one should always examine the substrate material between crossed polarizers for signs of birefringence. This is of great importance in single-beam refection holography as the state of polarization may be altered on passing through the plastic. Thus, the object and reference waves may be polarized in different planes, resulting in reduced fringe contrast (see Section 7.3.1).

Care should be exercised in mounting flexible recording material so that it cannot move during recording. Adhesive tape may be used, without stretching it, to attach the photosensitive layer around its edges to a glass support. Some time should be allowed to elapse so that mechanical stress in the adhesive material is relieved and creep movement during the exposure is avoided.

In simple single-beam holography experiments, including Denisyuk holography, a horizontal hologram plane is very useful, gravity providing the required stability.

It is difficult to overemphasize the importance of stability during holographic recording. While short exposures are certainly desirable, one can use exposures of hours duration if the proper precautions are taken. The exposure can be automated using an electronic shutter in front of the laser, with a suitable delay so that the experimenter can leave the laboratory, thus minimizing the risk of disturbance (and boredom) during long exposures.

Of course, as explained in Section 5.13, if a pulsed laser with adequate coherence length is available, then stability is not a problem, but this is a rather expensive solution.

8.12 SUMMARY

This chapter deals with holographic displays beginning with simple single-beam holographic methods. We explain spatial filtering and describe split beam techniques including setting the beam ratio using a polarizing beam and halfwave plates. We then discuss Benton holography and real image techniques using a slit aperture to obtain white light rainbow holograms.

We discuss the Denisyuk reflection holographic recording and display technique and introduce hologram copying. We also describe a double-sided holographic technique providing for near 360° field of view as well as multiplex methods for producing 360° stereograms.

We also consider full color holography using the Denisyuk method.

We discuss some techniques for dynamic holography including brief explanations of the operation of spatial light modulators.

We describe a method of producing very large format holographic displays and conclude with a brief discussion of the application of quantum entanglement to holographic imaging of inaccessible objects. Finally, we outline good practice in the recording holograms.

REFERENCES

1. D. R. MacQuigg, "Hologram fringe stabilization method," *Appl. Opt.*, 16, 291–2 (1976).
2. S. A. Benton, "Hologram reconstructions with extended incoherent sources," *J. Opt. Soc. Am.*, 59, 1545–6 (1969).
3. F. T. S. Yu, A. M. Tai, and H. Chen, "One-step rainbow holography: recent developments and application," *Opt. Eng.*, 19, 666–78 (1980).
4. Yu. N. Denisyuk, "Photographic reconstruction of the optical properties of an object in its own scattered radiation field," *Sov. Phys.-Dokl*, 7, 543 (1962). See also Yu. N. Denisyuk, "On the reproduction of the optical properties of an object by the wave field of its scattered radiation," Pt. I. *Opt. Spektrosk.*, 15, 523–32 (1963). Pt. II, *Opt. Spektrosk.*, 18, 152 (1965).
5. G. Lippmann, "La photographie des couleurs," *Hebd. Séances Acad. Sci.*, 112, 274–75 (1891).
6. T. H. Jeong, "Cylindrical holography and some proposed applications," *J. Opt. Soc. Am.*, 57, 1396–8 (1967).
7. N. George, "Full view holograms," *Opt. Commn.* 1, 457–9 (1970).
8. A. A. Friesem and R. J. Federowicz, "Recent advances in multicolor wavefront reconstruction," *Appl. Opt.*, 5, 1085–6 (1966).
9. R. J. Collier, C. B. Burckhardt, and L.H. Lin, *Optical Holography*, Academic Press, New York (1971).
10. H. Bjelkhagen and E. Mirlis, "Color holography to produce highly realistic three-dimensional images," *Appl. Opt.*, 47, A123–A133 (2008).
11. P. St.-Hilaire, S. A. Benton, M. E. Lucente, M. L. Jepsen, J. Kollin, H.Yoshikawa, and J. S. Underkoffler, "Electronic display system for computational holography," Proc. SPIE, Practical Holography IV, ed. S. A. Benton, 1212, 174–82 (1990).
12. P. St.-Hilaire, S. A. Benton, M. E. Lucente, J. D. Sutter, and W. J. Plesniak, "Advances in holographic video," Proc. SPIE, Practical Holography, VII, Imaging and Materials, ed. S. A. Benton, 1914, 188–96 (1993).
13. C. W. Slinger, C. D. Cameron, S. D. Coomber, R. J. Miller, D. A. Payne, A. P. Smith, M. G. Smith, M. Stanley, and P. J.Watson, "Recent developments in computer-generated holography: Toward a practical electroholography system for interactive 3D visualization," Proc. SPIE, Practical Holography XVIII, Materials and Applications, ed. T. H. Jeong and H. J. Bjelkhagen, 5290, 27–41 (2004).
14. P. -A. Blanche, A. Bablumian, R. Voorakaranam, C. Christenson, W. Lin, T. Gu, D. Flores, P. Wang, W.-Y. Hsieh, M. Kathaperumal, R. Rachwal, O. Siddiqui, J. Thomas, R. A. Norwood, M. Yamamoto, and N. Peyghambarian, "Holographic three-dimensional telepresence using large area phorefractive polymer," *Nature*, 468, 80–3 (2010).
15. D. Dehlinger and M. W. Mitchell, "Entangled photon apparatus for the undergraduate laboratory," *Am. J. Phys.*, 70, 898–902 (2002).
16. D. Dehlinger and M. W. Mitchell, "Entangled photons, non-locality and Bell inequalities in the undergraduate laboratory," *Am. J. Phys.*, 70, 903–10 (2002).

17. A. F. Abouraddy, B. E. A. Saleh, A.Sergienko, and M. C. Teich, "Quantum holography," *Opt. Express*, 9, 498–504 (2001).
18. R. L. Pfleegor and L. Mandel, "Interference of independent photon beams," *Phys. Rev.,* 159, 1084–8 (1967).

PROBLEMS

1. A laser beam with wavelength of 633 nm and diameter of 1.2 mm is spatially filtered using a 20× microscope objective. What diameter pinhole should be used and what is the divergence of the resulting beam? If the beam is then collimated using a lens of diameter of 15 cm and focal length 75 cm, what is its diameter?

2. A pair of photodetectors, each 1 mm in width, is placed behind the hologram holder to help stabilize the fringe pattern in a transmission holographic recording system. The recording beams of wavelength of 514 nm are incident on the hologram plane with an angle of 30° between them. What should be the spatial frequency of a diffraction grating that is located between the hologram and the photodetectors?

3. The beam ratio in a holographic recording system is controlled by a pair of halfwave plates and a polarizing beam splitter. The plane of polarization of the linearly polarized laser is set at 75° to the horizontal. What should be the orientation of the input halfwave plate to ensure that the beam ratio is 1.0? Suppose that the average reflectivity of the object is 10% and it is illuminated by the beam that is reflected from the beam splitter. Neglecting all other losses, what should be the orientation of the input halfwave plate to ensure that the intensity ratio of object to reference beam is 0.25 at the hologram plane?

4. A Benton hologram recorded using a 3-mm-wide slit is to be viewed in collimated white light. Given that the spectral resolution of the human eye is about 2 nm, what should be the distance between the master and copy holograms to ensure that the image is perceived as monochromatic? Assume the average spatial frequency in the hologram, recorded at 633 nm, is 1000 mm^{-1}.

5. A Denisyuk reflection hologram is recorded, of a spherical object 2 cm in diameter in contact with the hologram plane. A collimated beam from a diode laser, wavelength 650 nm, is used in the recording. If only about one quarter of the sphere volume can be seen in the reconstructed image, what conclusion do you draw?

6. A Q-switched ruby laser (λ = 694 nm) produces output pulses 1 ns in duration. What is the maximum velocity of an object that can be holographically recorded and what else would be observed in the reconstruction?

7. Display holograms sometimes show one of one or more of the following. What conclusions would you draw in each case and what corrective measures would you take?
   i. No image is seen at all.
   ii. No image of the object although images are seen of some of the mounts used to support various components and of part of the optical table close to the hologram.
   iii. The image of the object is covered with bright and dark bands of spatially uniform contrast.
   iv. As in (iii) but the contrast of the bands varies across the image.
   v. Only part of the object is seen in the reconstructed image.
   vi. The hologram recorded using silver halide on film (rather than on glass) is partially covered with bright and dark bands. You can view parts of the holographic image hidden by the bands by looking around them.

8. The principle of Denisyuk holography is shown in Figure 8.10. Light arriving at the photosensitive layer acts a reference beam. The layer is usually quite transparent so that some light passes through it to illuminate the object. However we need to be careful about the optical properties of the substrate on which the photosensitive layer is deposited especially if it is a plastic material. Explain why.

Other Imaging Applications

9.1 INTRODUCTION

Holography enables implementation of a range of useful imaging techniques, which cannot be implemented by other means. We have seen this in the previous chapter on displays but we now look at other imaging applications of particular importance in various branches of engineering.

9.2 HOLOGRAPHIC IMAGING OF THREE-DIMENSIONAL SPACES

Holographic reconstruction of 3-D spaces opens the way to detailed study of such spaces and the objects in them, including objects which are microscopically small. A very good example is the study of disperse collections of small particles including aerosols.

Microscopic examination of such particles is not possible using conventional microscopy since the individual particles are in free movement and the depth of focus is only about $d^2/2\lambda$ where d is the resolution limit (see Equation 2.13) of the microscope.

However, Thompson [1] showed that the whole volume occupied by the particles could be holographically recorded using a pulsed laser to freeze the particle motion. One can study the reconstructed images of the particles using a microscope, which is focused on the image of each particle in turn (Figure 9.1). If the velocity of the particles is v and we assume they must not move by more than one-tenth of their size [2], w, during the recording, then the laser pulse duration is limited to $0.1w/v$. Thus, a 1o-ns laser pulse can cope with 5 micron particles having velocities up to 50 ms^{-1}. The technique is used to measure particle sizes, not to record holographic images of the particles.

The optical arrangement is similar to that used to record a Gabor (Figure 3.4) hologram.

As we saw in Section 3.4.3, reconstruction from such a hologram produces both the virtual image and the conjugate real image, the latter obscuring the former. However, this is not a problem here because the particles are very small and, if the microscope in Figure 9.1b is focused on the real conjugate image of a particle reconstructed from the

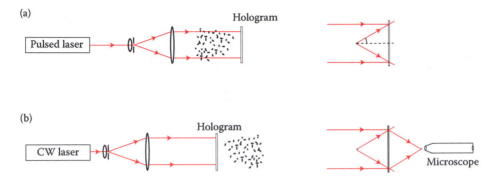

FIGURE 9.1 Holography/microscopy for particle studies: (a) recording and (b) reconstruction. Very little light from the virtual image enters the microscope.

hologram, very little of the light that contributes to the virtual image is collected by the microscope objective (Figure 9.1b).

It is the Fraunhofer diffraction pattern, in fact the spatial Fourier transform, of each particle that is recorded in the hologram plane, since for small particles the far-field condition (Equation 2.23) is met. At reconstruction, the pattern *diffracts* the light to form the Fourier transform of the diffraction pattern in the far-field, since Fraunhofer conditions are again met. Thus, a double Fourier transform is produced (see Section 3.4.2), resulting in an image.

The available depth of field is the maximum distance between a particle and the hologram plane for which a satisfactory Fraunhofer diffraction pattern can be recorded. We assume that at least the first three orders of diffraction from a particle of width, w, should be recorded. This implies that light diffracted at an angle θ to the hologram normal, given from Equation 2.7 by

$$w\sin\theta = 3\lambda$$

and from Figure 3.1, the maximum spatial frequency required is $\sin\theta/\lambda = 3/w$. For particles 5 μm wide, this means we require a recording material with resolution of 600 l mm^{-1}. The corresponding depth of field (Figure 9.1b) is $wl/3\lambda$, where l is half the width of the hologram. Thus, the depth of field of a hologram which can record the Fraunhofer diffraction pattern of particles 5 μm wide is typically 2000 times that of a conventional microscope.

A double pulse ruby or frequency doubled Nd Yag laser allows a double exposure hologram to be recorded, and the velocities of individual particles may be obtained by measurement of the distances between the reconstructed images of individual particles. The technique has been used in the study of aerosols, combustible fuels, dusts, and small marine organisms.

Another application is in studies of electrically charged fundamental nuclear particles, which create tracks of bubbles by local boiling as they pass through cryogenic liquids, such as hydrogen or helium held at temperatures above their normal boiling point in a pressurized chamber. The particles of interest normally exist for extremely short time intervals. The bubble tracks are photographed from outside the chamber and their curvatures and lengths provide momentum and lifetime data, respectively, for the particles that made

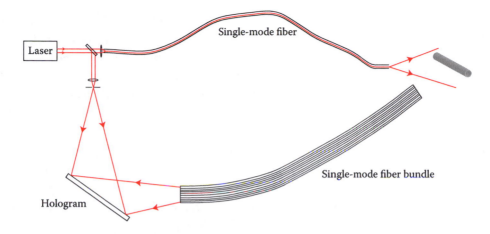

FIGURE 9.2 Optical fiber holography.

them. Clearly, depth of field and resolution are serious issues in the analysis of photographic records. Holographic recording provides a much greater depth of field and again the holographically recorded and reconstructed space can be thoroughly studied [3,4]. Bubble chambers have been largely supplanted by track detector systems, especially for the detection of particles produced in colliding particle beam experiments although they are still in use for the underground detection of dark matter [5].

Holography can also be used to record objects in comparatively inaccessible or hostile environments. Examples are nuclear reactor cores and other high-temperature and toxic environments. In principle, one only has to deliver light to the location of interest and ensure that the light reflected from the object is brought to the hologram plane. One way of doing this is by means of a single monomodal optical fiber to deliver the light and a monomodal fiber bundle to collect the light scattered by the object and ensure that it is propagated back to the hologram plane [6] (Figure 9.2). The fibers have to be monomodal to preserve the coherence of the light for hologram recording. In general, this means that the fiber diameter is small enough to ensure only axial propagation, a situation not very dissimilar to longitudinal mode propagation in a laser cavity (Section 5.5).

9.3 FURTHER APPLICATIONS OF PHASE CONJUGATION

We saw in Section 3.6 how holography allows us to reconstruct the conjugate of an object wave and we saw in Section 8.4 how phase conjugation is used in Benton and related holographic displays. We will now look at other applications of phase conjugation.

9.3.1 LENSLESS IMAGE FORMATION

All conventional optical systems and their components suffer from some degree of aberration but, if we can replace such systems by their holographic optical counterparts, the prospect of near aberration-free imaging becomes a practical possibility. A very good example is in the integrated circuit manufacturing industry. Integrated circuits are produced in stages, each stage being replicated in many integrated circuits on a single silicon slice at the same time. A typical stage in manufacture is the spatially selective diffusion of another element such as gallium or aluminium into the silicon, as a first step in the manufacture of transistors and diodes. First, the silicon is coated with a layer of photoresist (Figure 9.3a).

Then a mask, usually in the form of a spatially patterned coating of chromium metal on a quartz glass plate, is imaged onto the photoresist in UV light. Following this exposure, the unexposed, negative photoresist is removed by development (Figure 9.3b) leaving areas of silicon into which the dopant element can be diffused (Figure 9.3c). The drive toward ever higher packing density of electronic components in integrated circuits requires that the features in the mask become ever smaller and that they be faithfully imaged onto the photoresist, placing heavy demands on the lens, which must be highly corrected, that is to say free of aberrations, and consequently is a very expensive component.

Holography offers an alternative approach to the problem, through the use of conjugate image reconstruction [7] (Figure 9.4).

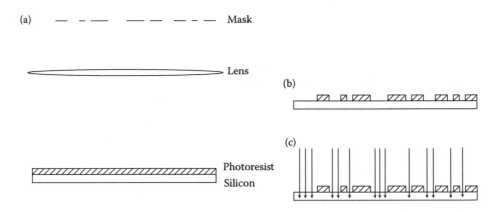

FIGURE 9.3 Stages in integrated circuit manufacture: (a) mask imaging, (b) development, and (c) diffusion.

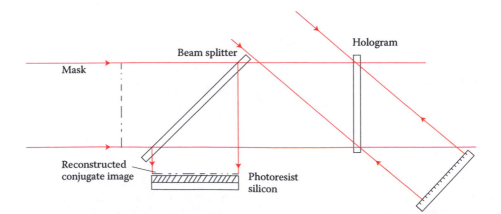
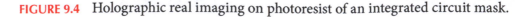

FIGURE 9.4 Holographic real imaging on photoresist of an integrated circuit mask.

A conjugate collimated reference beam is produced by reflection from a plane mirror and reconstructs a real conjugate image from the hologram. No lenses are used, minimizing the need for alignment and eliminating their associated aberrations. Laser speckle is also eliminated.

This technique has also been proposed as a way of igniting a small spherical target consisting of a mixture of deuterium and tritium in a laser-driven nuclear fusion reactor (Figure 9.5). The target, T, is isotropically illuminated by comparatively low-intensity laser light (not shown), which is scattered into a series (only one is shown) of photorefractive devices (Section 6.6.2) surrounding T, each of which produces a reconstructed conjugate wave, O^*, which inevitably converges back onto the target, both compressing its volume and raising its temperature to that required for nuclear fusion. Although, in principle, holography (Figure 9.4a) could be used with a photorefractive material to generate O^*, in practice *stimulated Brillouin scattering* is used instead. Brillouin scattering occurs when light interacts with acoustic vibrations, so-called *phonons*, in a material, and is backscattered.

When a very powerful beam of laser light propagates in the material, it both produces phonons and is simultaneously scattered by them [8]. The very strong electric field of the light wave produces the phonons by inducing charges with opposite sign on opposite walls of electric domains. These are small regions in which all the electric dipoles are oriented parallel to one another, and they become compressed in the strong electric field, an effect called *electrostriction*. The phonon distribution takes the same form as the light wavefront, which is naturally time reversed as a phase conjugate wave.

Thus, low-intensity light diffracted by the target is amplified in a laser amplifier or a series of such amplifiers, in each of which population inversion is maintained. On emerging from the final amplification stage with its wavefront distorted en route, the now very powerful light beam enters a phase conjugating mirror, which utilizes stimulated Brillouin scattering. The phase conjugate light retraces its path through the amplifier chain and the wavefront distortion is negated so that the target is now illuminated by a powerful, diffraction limited beam of light.

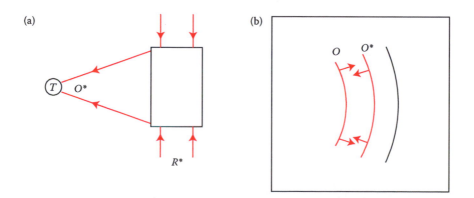

FIGURE 9.5 Laser ignition using phase conjugation: (a) holographic phase conjugation and (b) phase conjugation by stimulated Brillouin scattering.

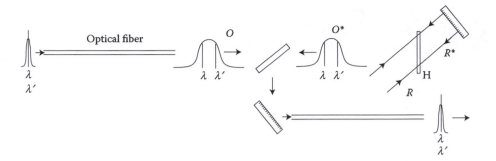

FIGURE 9.6 Optical pulse recompression. (Reproduced from A. Yariv, D. Feketa, and D. M. Pepper, Compensation for channel dispersion by nonlinear optical phase conjugation, *Opt. Lett.*, 4, 52–54, 1979. With permission of Optical Society of America.)

9.3.2 DISPERSION COMPENSATION

Yet another proposed application is in optical fiber communication links. The problem that arises here is that as optical pulses propagate along a fiber, their temporal duration increases, leading to loss of signal as pulses merge with one another. The reason is that since each pulse is of finite duration, it contains a range of temporal frequencies (see Equation 2.24) and therefore wavelengths, and each travels with a different velocity. This is the phenomenon of *optical dispersion* or the variation of refractive index with wavelength (Figure 9.6).

Suppose that the longer wavelengths in a pulse travel along the fiber faster than the shorter. Then, the longer wavelengths will emerge from the fiber leading the shorter and the pulse is stretched in time as a result, an effect known as *chirp*. If we use a photorefractive crystal to generate a phase conjugate or time-reversed version of the pulse, the situation is reversed with the shorter wavelengths leading the longer ones. If this pulse is sent along another fiber of the same length and with the same dispersion characteristic, the emerging pulse is compressed to its original duration [9].

9.3.3 DISTORTION AND ABERRATION CORRECTION

The object wave in a holographic recording may pass through a distorting medium, such as a turbulent atmosphere, or simply any optical system that has aberrations, on its way to the hologram. The image of the object can be recovered without distortion or aberration by sending the *conjugate* wave back through the distorting medium or optical system.

Suppose that an object wave passes through a diffuser so that the wavefront becomes distorted (Figure 9.7a). A hologram is recorded of the distorted wavefront and at reconstruction is illuminated by the conjugate reference wave.

The result is a phase conjugate version of the distorted object wave, which travels back through the diffuser and the distortions are removed, the final image appearing in the location of the original object (Figure 9.7b). The same diffuser must be used in the reconstruction stage in precisely the same position as in the recording stage. If the diffuser is moved or

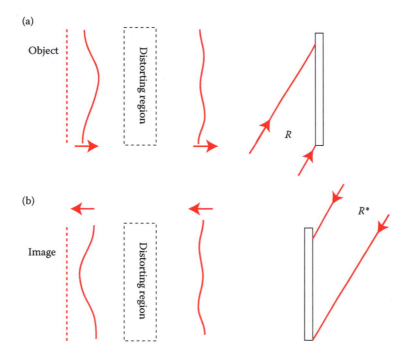

FIGURE 9.7 Removing distortion from an image: (a) recording and (b) reconstruction of image free of distortion. (Reproduced from E. N. Leith and J. Upatnieks, Holographic imagery through diffusing media, *J. Opt. Soc. Am.*, 56, 523, 1966. With permission of Optical Society of America.)

altered by mechanical or thermal stress when reconstruction takes place, the reconstructed image will be imperfect and may be completely absent. The fact that the diffuser must be in perfect registration suggests that images could be encrypted, with the diffuser acting as a decryption key [10]. Likewise, the hologram must be relocated precisely in the recording position or a self-processing recording material must be used to record the hologram.

It is possible to cope with time-varying distortion using real-time holographic recording materials such as BSO [11]. The arrangement used is as shown in Figure 9.8 with the crystal illuminated simultaneously by the distorted object wave and by the reference and conjugate reference beams.

The method is not very satisfactory however, as it produces an image on the same side of the distorting medium as the object. One would much prefer to be able to produce an image free of distortion on the opposite side of the distorting medium from the object. One solution to the problem in the case of a diffuser is to record a hologram of its image formed by a lens [12], as shown in Figure 9.9. The diffuser in the x,y plane imposes a random phase modulation $\phi(x,y)$ on a plane wavefront of unit amplitude passing through it so the complex amplitude of the disturbance in the x,y plane is $\exp[j\phi(x,y)]$, imaged by the lens as $\exp[j\phi(x_H,y_H)]$ in the hologram plane.

We assume transmittance of the recorded hologram is proportional to the intensity of light during recording $\left|R\;\exp(jky\sin\theta)+\exp[j\phi(x_H,y_H)]\right|^2$, where $R\;\exp(jky\sin\theta)$ is the complex amplitude of a plane reference wave at angle θ to the horizontal axis. Following

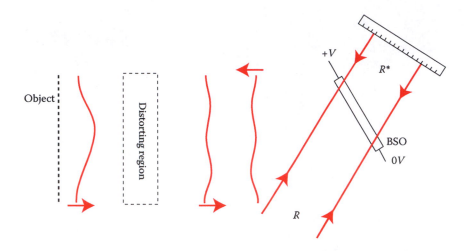

FIGURE 9.8 Real-time removal of image distortion. (Reproduced from J. P. Huignard, J. P. Herriau, P. Aubourg, and E. Spitz, Phase-conjugate wavefront generation via real-time holography in $Bi_{12}SiO_{20}$ crystals, *Opt. Lett.*, 4, 21–23, 1979. With permission of Optical Society of America.)

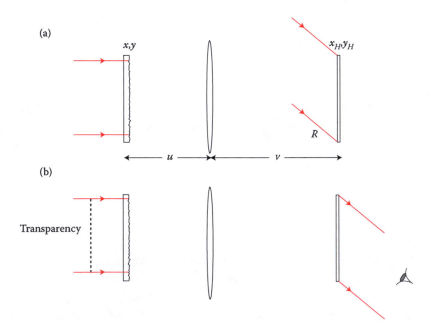

FIGURE 9.9 Distortion-free imaging with object and image on opposite sides of diffuser: (a) recording hologram of diffuser and (b) viewing undistorted image. (Reproduced from H. Kogelnik and K. S. Pennington, Holographic imaging through a random medium, *J. Opt. Soc. Am.*, 58, 273–74, 1968. With permission of Optical Society of America.)

recording, the object transparency is illuminated normally by a plane wave to produce a complex amplitude $\tilde{A}(x, y)$ at the diffuser plane and $\tilde{A}(x, y)\exp[j\phi(x,y)]$ immediately to the right of the diffuser. This disturbance is imaged on the hologram plane and the output is

$$\left| R\exp(jky\sin\theta)+ \exp[j\phi(x_H,y_H)]\right|^2 \exp[j\phi(x_H,y_H)]\tilde{A}(x_H,y_H).$$

One of the terms in this latter expression is $R\exp(jky\sin\theta)\tilde{A}(x_H,y_H)$, which is observed in the hologram plane as the image of the transparency, with the distortion removed and on the opposite side of the diffuser from the object.

A real-time version was proposed by Yariv and Koch [13] using a nonlinear optical medium. The diffusing region is again imaged in the x_H,y_H plane in a plane thin, nonlinear optical material, and we denote the complex amplitude in that plane by $\exp[j\phi(x_H,y_H)]$ (Figure 9.10).

The hologram plane is also illuminated by a plane wave, with amplitude R, introduced by a beam splitter and by a third, collimated light beam passing through the transparency. The complex amplitude of the latter in the hologram plane is $\tilde{A}(x_H,y_H)$.

The three waves overlap in the hologram producing a polarization, which includes a term

$$P \propto \tilde{A}(x_H,y_H)R\exp[-j\phi(x_H,y_H)].$$

Restoring the real parts of time dependence of $\tilde{A}(x_H,y_H)$ and R, we obtain

$$\tilde{A}(x_H,y_H)\cos(\omega t)R\cos(\omega t)\exp[-j\phi(x_H,y_H)]$$

which includes a term $(1/2)\tilde{A}(x_H,y_H)R\exp[-j\phi(x_H,y_H)]$. This polarization radiates a wave at the same frequency as the original waves. It travels in the reverse direction to that which originally passed through the distorting region and so passes back through that region with conjugate phase. Thus, the phase distortion is unravelled and an undistorted image of the transparency is observed in the $-z$ direction.

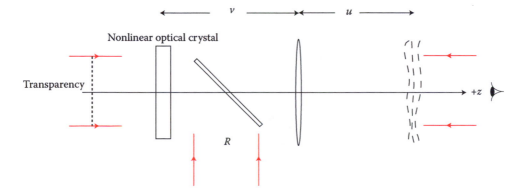

FIGURE 9.10 Real-time distortion free imaging with object and image on opposite sides of diffuser. (Reproduced from A. Yariv and T. L. Koch, One-way coherent imaging through a distorting medium using four-wave mixing, *Opt. Lett.*, 7, 113–15, 1982. With permission of Optical Society of America.)

9.4 MULTIPLE IMAGING

Here, we consider a further application of holography in the mass manufacture of integrated circuits. The problem is to make a set of exact copies of an integrated circuit mask spaced the same distance apart in the horizontal and vertical directions so that there is a rectangular $m \times n$ array. This allows for a single step in manufacture to be replicated $m \times n$ times on a silicon wafer. Usually, such a multiple mask is made by carefully photographing an original mask and then moving it to a new position and photographing it again, the process being carried out $m \times n$ times.

A holographic method allows for the production of rectangular arrays of copies of a mask in one step. We start with a planar array of point sources of monochromatic light placed in the front focal plane of a lens, the distribution described by the 2-D Kronecker delta function $\Sigma_m \Sigma_n \delta(x-ma)\delta(y-nb)$, which simply means that there are points spaced a apart in the x direction and b apart in the y direction. A recording of the spatial Fourier Transform of this array is made in the back focal plane (Figure 9.11a). We can find its form from Equation 2.22 (neglecting the term in front of the integral) as

$$\sum_m \sum_n \iint_{x,y} \delta(x-ma)\delta(y-nb)\exp\left[-\frac{jk}{f}(xx'+yy')\right]dxdy$$

$$= \sum_m \sum_n \exp\left[-\frac{jk}{f}(max'+nby')\right] \tag{9.1}$$

that is a set of plane waves with spatial frequencies $ma/\lambda f, nb/\lambda f$.

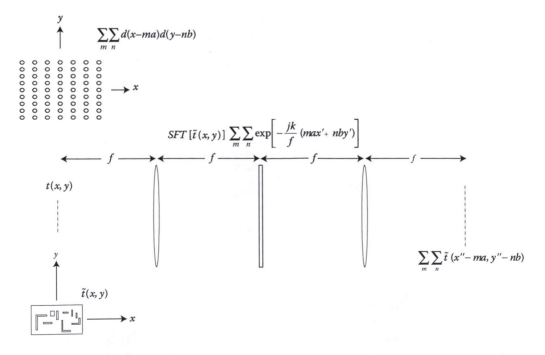

FIGURE 9.11 Multiple imaging.

We assume that the transmissivity of the hologram is proportional to the right-hand side of Equation 9.1. The array of point sources is now replaced by a mask of complex amplitude transmissivity $\tilde{t}(x, y)$ illuminated normally by a plane wave of unit amplitude. The complex amplitude incident on the hologram is given by $\iint_{x,y} \tilde{t}(x, y)\exp[-(jk/f)(xx'+yy')]dxdy$ and the complex amplitude $\tilde{t}_H$ transmitted by the hologram is this last expression multiplied by the right-hand side of Equation 9.1, that is

$$\tilde{t}_H = \iint_{x,y} \tilde{t}(x, y)\exp\left[-\frac{jk}{f}(xx'+yy')\right]dxdy \sum_m \sum_n \exp\left[-\frac{jk}{f}(max'+nby')\right]$$

$$= \sum_m \sum_n \iint_{x,y} \tilde{t}(x, y)\exp\left[\left\{-\frac{jk}{f}[(x+ma)x'+(y+nb)y']\right\}\right]dxdy$$

$$= \sum_m \sum_n \iint_{x'',y''} \tilde{t}(x''-ma, y''-nb)\exp\left[\left\{-\frac{jk}{f}[x''x'+x''y']\right\}\right]dx''dy'' \qquad (9.2)$$

where we have set $x'' = x - ma$, $y'' = y - nb$.

The spatial Fourier transform of Equation 9.2 appearing in the back focal plane of a second lens is

$$\sum_m \sum_n \tilde{t}(x''-ma, y''-nb)$$

which is the required array of mask images. In general, if x is positive, x'' is negative, and the same is true for y and y''. Also, if the second lens has the same focal length as the first, the system is called a 4f optical system since object and image are 4f apart, and $x'' = -x$ and $y'' = -y$.

9.5 TOTAL INTERNAL REFLECTION AND EVANESCENT WAVE HOLOGRAPHY

In all the techniques discussed so far, the reconstructed and the reconstructing light waves occupy space on the same side of the hologram and sometimes overlap at least for part of their paths. This is usually not a serious problem but can sometimes be inconvenient and require some thought in designing the holographic recording and reconstruction setups. Total internal reflection was employed by Stetson [14] to ensure that a reconstructed image could be produced on one side of a hologram by a reference beam, which is not transmitted through to that side. In other words the reconstructing and reconstructed light can be completely spatially separated by the hologram.

The object, a transparency, is placed on spacers to form an air gap between it and the photosensitive layer. The reference beam arrives at the upper surface of the photosensitive layer at an angle exceeding the critical angle for the layer/air boundary (Figure 9.12).

This is ensured by placing a prism in contact with the substrate of the hologram using a liquid, whose refractive index is approximately the same as those of the prism and the substrate material. Since it is reflected at the layer/air boundary, the reference beam

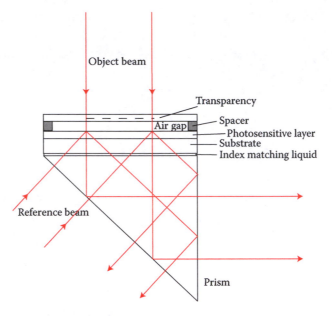

FIGURE 9.12 Total internal reflection holography. (Reprinted with permission from K. A. Stetson, Holography with totally internally reflected light, *Appl. Phys. Lett.*, 11, 225–26, 1967. Copyright 1967, American Institute of Physics.)

traverses the photosensitive layer twice so that a reflection *and* a transmission hologram are recorded. The conjugate reference wave is used for reconstruction of a real image of the transparency in its original location, and zero-order, undiffracted light cannot propagate beyond the layer/air boundary.

Even when total internal reflection occurs, some light penetrates a little way (typically about one wavelength) into the lower index medium and travels along the boundary. This light is called evanescent (see Section 1.7.2.4), and an evanescent light wave may also be holographically recorded and indeed may be used as a reference wave in holography. This is shown in Figure 9.13 in which a plane reference wave is incident on the boundary between a higher index medium and a lower index photosensitive medium at an angle exceeding the critical angle. It then becomes an evanescent wave in the lower index photosensitive medium and interferes with a normal object wave to form an evanescent wave hologram. From Equation 1.38, assuming unit amplitude, we can write the expression for the evanescent reference wave as

$$\exp\left(\frac{-z}{z_0}\right)\exp j(k_t x \sin\theta_t - \omega t) \quad \text{with } z_0 = \frac{1}{k_t}\left(\frac{n_i^2}{n_t^2}\sin^2\theta_i - 1\right)^{-(1/2)}$$

We have seen already that the penetration, or skin depth, of the evanescent wave is typically one wavelength so that the hologram is very thin. We note also that the wavelength of the evanescent wave is $\lambda/n_i\sin\theta_t$ and depends on the angle of incidence. The object wave in Figure 9.13, described by the expression $O(x, y)\exp(jk_t z - \omega t)$ with $k_t = 2\pi n_t/\lambda$, is

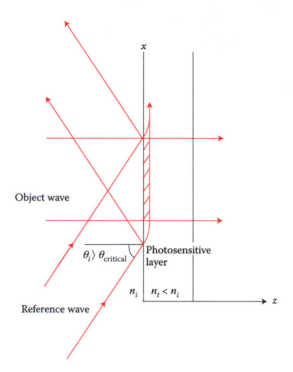

FIGURE 9.13 Evanescent wave holography.

homogeneous meaning that its amplitude is independent of z. The total recorded intensity is given by

$$\left\{ O(x,y)\exp(jk_t z - \omega t) + \exp\left(\frac{-z}{z_0}\right)\exp\left[j(k_t x\sin\theta_t - \omega t)\right]\right\} \times CC$$

$$= O^2(x,y) + \exp\left(\frac{-2z}{z_0}\right) + 2O(x,y)\exp\left(\frac{-z}{z_0}\right)\cos[k_t(z - x\sin\theta_t)] \qquad (9.3)$$

For recording, the photosensitive plate may be immersed in a liquid of higher refractive index in a hexagonal cell [15] to facilitate angles of incidence exceeding the critical angle (Figure 9.14a).

Alternatively, the photosensitive layer is coated onto a high-index substrate. These methods allow either or both waves to be evanescent. Another method (Figure 9.14b) is to use a high-refractive-index liquid between a prism and the photosensitive layer. In this method, both waves are evanescent. Note the orientation of the fringes in this case. From Figure 9.15, the fringe spacing, d, is

$$d = \frac{(\lambda/n_t \sin\theta_t)(\lambda/n_t)}{\sqrt{(\lambda/n_t \sin\theta_t)^2 + (\lambda/n_t)^2}} \qquad (9.4)$$

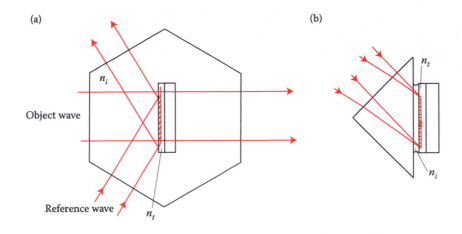

FIGURE 9.14 Recording an evanescent wave hologram: (a) evanescent reference wave and homogeneous object wave and (b) evanescent reference and object waves.

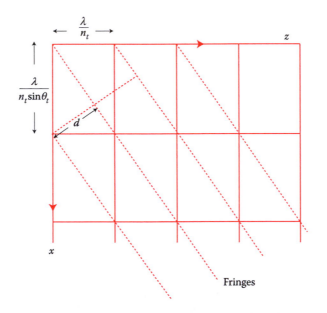

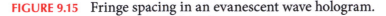

FIGURE 9.15 Fringe spacing in an evanescent wave hologram.

If we use two evanescent waves, the spacing is just $\lambda/n_i\sin\theta_t$ and $\lambda/2n_i\sin\theta_t$ if the angles of incidence are equal and opposite.

The fringe pattern shown in Figure 9.15 is the same if the object wave propagates in the opposite direction, because the hologram is typically less than 1 µm in thickness. This means that *two* object waves are reconstructed, propagating in opposite directions.

Reconstruction from the hologram shown in Figure 9.15 can only be obtained if the reconstructing evanescent light is of the correct wavelength. This means that white light can be used, the hologram acting as a narrow band spectral filter.

9.6 EVANESCENT WAVES IN DIFFRACTED LIGHT

Evanescent waves also occur when light is diffracted. To see why this is so, we recall Equation 2.18, omitting terms outside the double integral as constant

$$\tilde{A}(x',y') = \iint_{x,y} \tilde{A}(x,y)\exp\left[-\frac{jk}{z}(xx'+yy')\right]dxdy \tag{9.5}$$

Equation 9.5 expresses the complex amplitude in the x',y' plane due to a complex amplitude distribution $\tilde{A}(x,y)$ in the x,y plane. If the latter is a transparency with complex amplitude transmittance $\tilde{t}(x,y)$, at $z = 0$, illuminated by a plane wave of unit amplitude directed along the z-axis, we can write

$$\tilde{t}(x',y') = \iint_{x,y} \tilde{t}(x,y)\exp\left[-\frac{jk}{z}(xx'+yy')\right]dxdy \tag{9.6}$$

Using the inverse Fourier transform

$$\tilde{t}(x,y) = \iint_{x',y'} \tilde{t}(x',y')\exp\left[\frac{jk}{z}(xx'+yy')\right]dx'dy' \tag{9.7}$$

where $x'/\lambda z$, $y'/\lambda z$ are the spatial frequencies of the transparency, measured in the x and y directions.

We illuminate the transparency with a plane wave $\exp[j(kz - \omega t)$ to obtain a set of plane waves in the space $z > 0$, with wave vector components

$$k(\cos\alpha, \cos\beta, \cos\gamma) = k\left(\frac{x'}{z}, \frac{y'}{z}, \sqrt{1-\left(\frac{x'}{z}\right)^2 + \left(\frac{y'}{z}\right)^2}\right) \tag{9.8}$$

and directional cosines $\cos\alpha$, $\cos\beta$, $\cos\gamma$ and spatial frequencies $x'/\lambda z$, $y'/\lambda z$, $\sqrt{(1/\lambda^2)-(x'/\lambda z)^2+(y'/\lambda z)^2}$ measured, respectively, in directions x, y, and z. The total field $(z \geq 0)$ is

$$E(x,y,z,t) = \exp(-j\omega t)\iint_{\left(\frac{x'}{\lambda z}\right)^2+\left(\frac{y'}{\lambda z}\right)^2 \leq \frac{1}{\lambda^2}} \tilde{t}(x',y')\exp\left[jkz\left(1-\left[\left(\frac{x'}{z}\right)^2+\left(\frac{y'}{z}\right)^2\right]\right)^{1/2}\right]$$
$$\times \exp\left[-\frac{jk}{z}(xx'+yy')\right]dx'dy' + \exp(-j\omega t)$$
$$\times \iint_{\left(\frac{x'}{\lambda z}\right)^2+\left(\frac{y'}{\lambda z}\right)^2 > \frac{1}{\lambda^2}} \tilde{t}(x',y')\exp\left[-kz\left(\left[\left(\frac{x'}{z}\right)^2+\left(\frac{y'}{z}\right)^2\right]-1\right)^{1/2}\right]$$
$$\times \exp\left[-\frac{jk}{z}(xx'+yy')\right]dx'dy' \tag{9.9}$$

Equation 9.9 describes *two* different sets of waves. The first consists of homogeneous waves, which are diffracted by the transparency in the normal way. The maximum possible angle of diffraction, measured from the z-axis, is $\pi/2$, which means, from Equation 9.8, that $(x'/z)^2 + (y'/z)^2 \approx 1$ so that spatial frequencies in the transparency that are greater than $(x'/\lambda z), (y'/\lambda z)$, cannot produce diffracted light. Thus, features with physical scale size of less than a wavelength do not diffract.

Waves in the second set are evanescent, only traveling in the plane $z = 0$ and are attenuated with z. Such evanescent waves are often referred to as *optical near fields* as they only exist very close to the surface. The characteristic distance, z_0, from the transparency at which the evanescent wave amplitude falls to $1/e$ of its initial value is

$$z_0 = \frac{1}{k\left(\left[\left(\dfrac{x'}{z}\right)^2 + \left(\dfrac{y'}{z}\right)^2 - 1\right]\right)^{1/2}} \tag{9.10}$$

From Equation 9.9, the wavelength of an evanescent wave propagating in the x direction is $\lambda z/x'$, which is the *spatial period* in that direction so that, generally, the sizes of features on an object surface determine the wavelengths of evanescent waves. This basic principle is exploited in instruments, such as the scanning near-field optical microscope (SNOM), which uses an extremely small aperture to illuminate the object of interest or to collect the light from it. Spatial resolution is determined by the wavelength of the evanescent wave, which in turn is determined by the aperture. One method of producing a very small illuminating aperture is to taper a monomode optical fiber to a very fine tip a few nanometres in diameter and coat the tip region with metal, although the coating is not essential.

The dimensions of the extreme tip of the fiber determine the evanescent wavelength. Another approach is to use an electron beam to drill a very small aperture in the tip of an atomic force microscope (AFM) probe so that AFM and SNOM can be combined in a single instrument. Lateral resolution of the order of a few tens of nanometers is possible. Of course, in order to collect the evanescent light, the illuminating or detecting aperture, or both, must be very close to the surface of interest ($<z_0$) typically a fraction of the homogeneous wavelength (Figure 9.16). To maintain this close distance, an optical detection system is used with an auxiliary laser beam reflected from the probe tip onto a position-sensitive photodetector. The illuminating tip is scanned over the surface to obtain a detailed ultra high-resolution image.

Near-field optical techniques have expanded to incorporate spectroscopy, including Raman spectroscopy. They also have the potential for much greater surface data storage density than is possible by conventional optics, in which the minimum focused spot size is determined by the wavelength of a (homogeneous) light wave and the numerical aperture of the lens, although significant gains have been obtained using shorter wavelengths and high numerical aperture optics.

9.6.1 DIFFRACTED EVANESCENT WAVE HOLOGRAPHY

Phase conjugate wave reconstruction, which enables the effects of evanescent and homogeneous waves to be clearly distinguished [16], has been demonstrated using a photorefractive

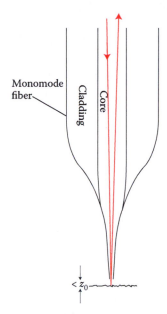

FIGURE 9.16 Scanning near-field optical microscopy (SNOM).

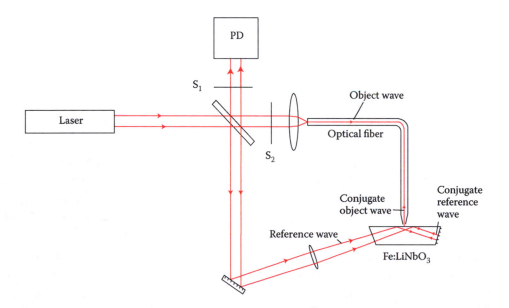

FIGURE 9.17 Phase conjugate evanescent waves. (Reproduced from S. I. Bozhevolnyi, O. Keller, and I. I. Smolyaninov, Phase conjugation of an optical near field, *Opt. Lett.*, 19, 1601–3, 1994. With permission of Optical Society of America.)

crystal of iron-doped lithium niobate in the optical arrangement shown in Figure 9.17. The crystal is illuminated by a reference wave incident at 70° to the normal and by the object light wave from the optical fiber to record a hologram. The tip of the fiber is about 100 nm in diameter and it can be maintained at a distance of 5 nm from the crystal surface during recording to ensure that evanescent waves can reach and enter the crystal. A phase conjugate reference beam, obtained by reflection of the original reference from the aluminium coated face of the crystal, reconstructs the conjugate object wave. The reference wave and its conjugate, being totally internally reflected, also give rise to evanescent waves, which may be scattered by the fiber tip to form part of the object beam.

The decay time of a hologram in the crystal is quite long (~100 s) so that, when shutter S_2 is closed and shutter S_1 opened, the intensity profile of the reconstructed conjugate wave can be obtained using an imaging detector. What is remarkable about the experiment is that the width of the conjugate object beam profile is about 500 nm when the fiber tip is at a distance of about 1 μm from the crystal surface, while at a distance of about 5 nm, it is 180 nm. This latter observation indicates a strong contribution to the conjugate beam profile from the high spatial frequencies associated with evanescent waves. With 1 μm separation during recording, evanescent waves cannot reach the crystal and are not recorded.

The optical power in the reconstructed beam was measured as a function of the distance from the fiber tip to the crystal surface. At close ($<z_0$) proximity, the power was rather low, showing that the reconstructed evanescent wave is weak, as one would expect since it only occurs close to the surface inside the crystal and only a small number of holographic fringes involving evanescent waves are recorded. At larger separations with far-field reconstructed light being dominant, the reconstructed beam power is greater since more fringes contribute.

The general concept of near-field holography involving only evanescent waves was formulated by Bozhevolnyi and Vohnsen [17]. Most holographic recording setups are only capable of recording and reconstructing the far-field object wave and information carried by evanescent waves is lost. This may be avoided by placing an object in proximity ($<z_0$) to the photosensitive surface. An evanescent wave propagating along this surface is scattered by the object and the scattered and unscattered evanescent waves interfere (Figure 9.18). A recording of the interference pattern constitutes a purely evanescent wave hologram, and we may reconstruct the scattered evanescent wave, or rather its conjugate, for detailed examination using SNOM, by using a phase conjugate version of the reference wave. The need to ensure that the object is wholly within a distance z_0 of the photosensitive surface effectively restricts application of the technique to 2-D objects. However, the coherence length of the light source need only be a few micrometers.

As shown in Figure 9.18, it is assumed that the recording takes the form of a surface profile modulation of the photosensitive material, and the analysis by Bozhevolnyi and Vohnsen [17] is restricted to the derivation of the form of the surface profile but shows nonetheless that evanescent wave reconstructions can be obtained using a photorefractive crystal, indicating as in [16] that these waves were holographically recorded within the bulk of the crystal by refractive index modulation. Figure 9.19a shows a near-field optical image of latex spheres (200 nm in diameter) obtained in this way. The image is built up by capturing *holographically reconstructed phase conjugate evanescent light* incident on the fiber tip. A topographic image is shown (Figure 9.19b) for comparison.

Although it was thought at first that the phase conjugate evanescent waves could only be reconstructed from regions very close to the crystal surface, it has been found that optical near fields can be holographically recorded in relatively remote locations [18] with the

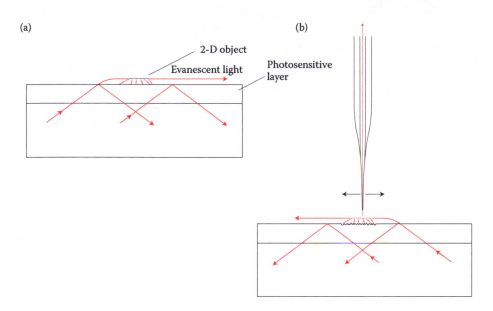

FIGURE 9.18 Pure evanescent wave holography. (Adapted with permission from S. I. Bozhevolnyi and B. Vohnsen, *Phys. Rev. Lett.*, 77, 3351, 1996. Copyright 1996 by the American Physical Society.)

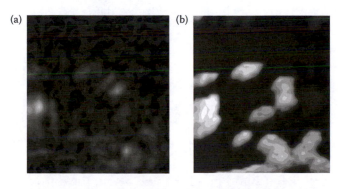

FIGURE 9.19 (a) Near-field optical image (4.9 × 4.9 μm) of latex spheres (diameter of 200 nm) on a photorefractive crystal. (b) Topographic image. (Reproduced with permission from S. I. Bozhevolnyi and B. Vohnsen, *Phys. Rev. Lett.*, 77, 3351–54, 1996. Copyright 1996 by the American Physical Society.)

potential for combining the very high spatial resolution afforded by near-field techniques with holographic optical data storage.

9.7 MASS COPYING OF HOLOGRAMS

Small holograms are now found on credit cards, banknotes, concert, and theater tickets, for authentication purposes. The most common form of mass production of holograms for authentication purposes is known as hot foil stamping (Figure 9.20). An original hologram recorded in any suitable material, such as silver halide, may be illuminated by a conjugate reference wave to produce a real image (Figure 9.20a). This image is recorded as an image plane hologram by surface relief holography in photoresist (Figure 9.20b) (see Section 6.5). The next step is to spray coat the photoresist surface with conducting metal prior to nickel metal electroplating (Figure 9.20c) so that the nickel takes up the detailed surface profile of the photoresist hologram and itself becomes a surface relief hologram. The nickel hologram is removed from the photoresist and used to stamp out a large number of copies in thermoplastic film (Figure 9.20d) covered by a thin metal layer, which reflects the illuminating light and the finished copies behave like reflection holograms. The metal layer is protected by a transparent laquer coating.

For very high volume production, hot stamping has been replaced by continuous roll embossing, involving the use of a set of identical nickel plated masters mounted on a rotating cylinder. The thermoplastic is carried on the surface of a second cylinder and the two cylinders brought into contact. In this way, a continuous roll of thermoplastic copies may be produced.

However, foil stamped holograms are not difficult to replicate, nor is it difficult to produce imitations of the original hologram. Even if the imitation holograms are rather poor, this is rarely an issue for the counterfeiter as the holograms are often not closely inspected. For this reason, significant effort is expended on producing holograms incorporating small subtle features, including machine readable features, which are harder to copy.

Volume holograms can be contact copied holographically. In the case of a transmission hologram, the copy is placed with its photosensitive side in contact with that of the original master, usually with an index matching liquid between them to minimize stray reflections. The copy is illuminated through the master by a replica of the reference wave, which was used in the original recording. The diffraction efficiency of the original should be 50% if a phase hologram copy is to be made so that the beam ratio for recording the copy is 1.0.

Reflection holograms are copied in similar fashion but the diffraction efficiency of the original should be as close to 100% as possible so that the beam ratio for recording the copy is 1.0. Mass copying is often done using a step-and-repeat process often using a pulsed laser. A continuous wave laser may be used, the beam being shaped into a thin line, which is scanned from one side of the master/copy combination to the other. For scanning purposes, a number of masters may be mounted on a rotating cylinder and the copy material brought into contact with them for the duration of recording. The advantage of this technique is that it allows for mass copying in self-processing photopolymers, whose sensitivity to short exposure is rather lower than that of silver halide.

9.8 SUMMARY

This chapter is concerned with the most important examples of uniquely holographic imaging apart from 3-D imaging. We consider first applications, which allow us to carry

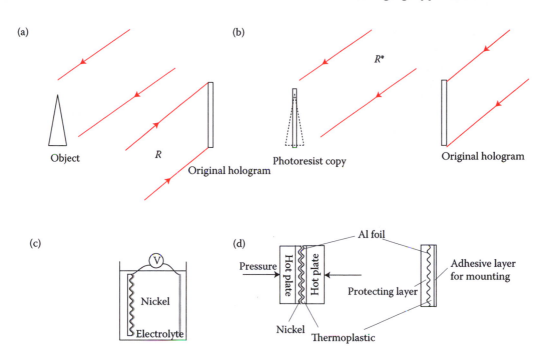

FIGURE 9.20 Mass copying of hologram: (a) recording master hologram; (b) image plane copy in photoresist; (c) nickel plating; and (d) copying into thermally softened plastic.

out microscopic examination of a 3-D space reconstructed by holography. This is followed by considering real and conjugate imaging and their applications. Of particular interest are diffraction limited real imaging and distortion or aberration-free imaging. We discuss multiple imaging and total internal reflection holography. We also consider holography using evanescent reference and object waves, produced by total internal reflection. We discuss the basic theory of diffracted evanescent waves as well as holographic methods of recording them.

The chapter concludes with a description of mass copying of holograms by the hot foil stamping process.

REFERENCES

1. B. J. Thompson, "Holographic particle sizing techniques," *J. Phys. E: Sci. Instrum.*, 7, 781–8 (1974).
2. W. R. Zinky, "Hologram techniques for particle-size analysis," *Ann. NY Acad. Sci.*, 158, 741–52 (1969).
3. W. T. Welford, "Obtaining increased focal depth in bubble chamber photography by an application of the hologram principle," *Appl. Opt.*, 5, 872–3 (1966).
4. M. Dykes, P. Lecoq, D. Güsewell, A. Hervé, H. Wenninger, H. Royer, B. Hahn, E. Hugentobler, E. Ramseyer, and M. Boratovd, "Holographic photography of bubble chamber tracks: A feasibility test," *Nuclear Instruments and Methods*, 179, 487–93 (1981).
5. E. Behnke, J. I. Collar, P. S. Cooper, K. Crum, M. Crisler, M. Hu, I. Levine, et al., "Spin-dependent WIMP limits from a bubble chamber," *Science*, 319, 933–6 (2008).

6. O. Coquoz, R. Conde, F. Taleblou, and C. Depeursinge, "Performances of endoscopic holography with a multicore optical fiber," *Appl. Opt.*, 34, 7186–93 (1995).
7. D. M. Pepper, "Applications of optical phase conjugation," *Sci. Am.*, 254, 74–83 (1986).
8. B. Ya. Zeldovich and V. V. Ragul'skii, "Wave-front inversion in induced light scattering," *Sov. Phys. Usp.*, 21, 1000–1002 (1978).
9. A. Yariv, D. Fekete, and D. M. Pepper, "Compensation for channel dispersion by nonlinear optical phase conjugation," *Opt. Lett.*, 4, 52–4 (1979).
10. E. N. Leith and J. Upatnieks, "Holographic imagery through diffusing media," *J. Opt. Soc. Am.*, 56, 523 (1966).
11. J. P. Huignard, J. P. Herriau, P. Aubourg, and E. Spitz, "Phase-conjugate wavefront generation via real-time holography in $Bi_{12}SiO_{20}$ crystals," *Opt. Lett.*, 4, 21–23 (1979).
12. H. Kogelnik and K.S. Pennington, "Holographic imaging through a random medium," *J. Opt. Soc. Am.*, 58, 273–4 (1968).
13. A. Yariv and T. L. Koch, "One-way coherent imaging through a distorting medium using four-wave mixing," *Opt. Lett.*, 7, 113–5 (1982).
14. K. A. Stetson, "Holography with totally internally reflected light," *Appl. Phys. Lett.*, 11, 225–6 (1967).
15. O. Bryngdahl, "Holography with evanescent waves," *J. Opt. Soc. Am.*, 59, 1645–50 (1969).
16. S. I. Bozhevolnyi, O. Keller, and I. I. Smolyaninov, "Phase conjugation of an optical near field," *Opt. Lett.*, 19, 1601–3 (1994).
17. S. I. Bozhevolnyi and B. Vohnsen, "Near-field optical holography," *Phys. Rev. Lett.*, 77, 3351–4 (1996).
18. K. Kyoung-Youm and B. Lee, "Recording of optical near fields in remote locations by near-field holography," *Opt. Lett.*, 26, 1800–2 (2001).

PROBLEMS

1. We wish to use in-line holography to record holograms of nuclear particle tracks in a bubble chamber using a pulsed ruby laser of wavelength 694 nm, allowing the bubbles to grow to a diameter of 100 microns before recording them. Using the criterion that at least three orders of Fraunhofer diffraction are required in order to from acceptable images, what resolution should the recording material have and what is the maximum useful size of the chamber?

2. An optical fiber is used to deliver laser light to illuminate an otherwise inaccessible object in order to record a hologram. Such fibers operate on the principle of total internal reflection within a glass core of refractive index n_1 surrounded by an outer cladding jacket of lower refractive index n_2. Find an expression for the maximum angle of incidence at the front face of a fiber, of light which is transmitted through it.

3. For holographic use, an optical fiber must be selected, which preserves the coherence of the laser light. Obtain an approximate expression for the diameter of the fiber, by considering the maximum optical path difference between light, which propagates through the fiber with, and without, reflection at the core/cladding boundary and calculate this diameter for laser light of wavelength 633 nm if the fiber has a core refractive index of 1.62 and a cladding index of 1.55.

4. A photosensitive layer of refractive index 1.48 is coated on glass of refractive index 1.5, which is index matched to an equilateral prism also of refractive index 1.5. Over what range of angles of incidence of beams at the other faces of the prism can evanescent wave gratings be recorded?

5. Find the position of the image of single point source at $x = a$ in the front focal plane of a $4f$ system, by using the double Fourier transformation.

6. Consider two coherent point sources of light and show that the interference pattern obtained in a plane parallel to one which includes the points, consists of a single spatial frequency in the region close to the plane normal the bisector of the straight line joining the points.

This result implies that a hologram of a transparency recorded using a point source in the plane of the transparency is in fact a spatial Fourier hologram and is referred to as a *lensless Fourier hologram* for that reason. However, what is the fundamental difference between a *true* Fourier hologram and a lensless one?

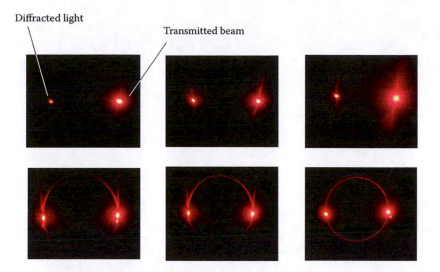

FIGURE 7.8 Diffracted and transmitted light from a grating in a photopolymer layer, as scatter increases from top left to bottom right. (Courtesy of M.S. Mahmud.)

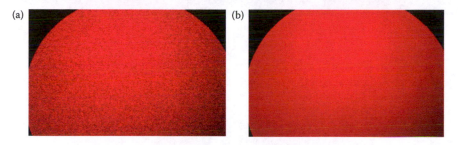

FIGURE 7.12 Speckle patterns showing effect of aperture size, (a) f/45 (b) f/8 (author photographs).

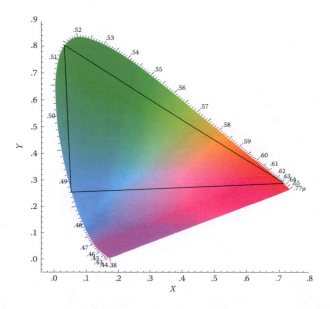

FIGURE 8.15 CIE chromaticity diagram. Wavelengths around the periphery are in micrometers. The triangle encloses all colors that can be obtained using wavelengths of 488, 514, and 633 nm. (Redrawn from http://en.wikipedia.org/w/index.php?title=File:Chromaticity_diagram_full.pdf&page=1.)

| 10% | 30% | 50% | 70% | 80% |

FIGURE 17.5 Hologram color at different levels of relative humidity (RH%). (Courtesy of I. Naydenova and R. Jallapuram.)

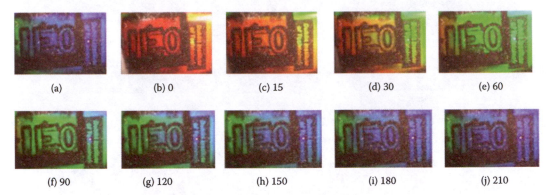

| (a) | (b) 0 | (c) 15 | (d) 30 | (e) 60 |
| (f) 90 | (g) 120 | (h) 150 | (i) 180 | (j) 210 |

FIGURE 17.6 Color change of a hologram induced by breathing (a) before, (b–j) time (sec) afterwards. (Courtesy of I. Naydenova and R. Jallapuram.)

(a)

(b)

FIGURE 17.11 Examples of dye deposition holography: (a) Left: photopolymer sample that has been sensitized but not exposed to light. The sensitized areas are not visible to the eye as the dye concentration is extremely low. Right: Following exposure to an interference pattern, transmission holographic gratings are recorded in the sensitized areas and are visible in transmitted white light. (b) In these examples, dye is deposited using a soft brush. (Courtesy of I. Naydenova.)

CHAPTER 10

Holographic Interferometry

10.1 INTRODUCTION

Holography allows us to use interferometry, with its very high precision and sensitivity, to study surfaces that are not optically smooth. For this reason, holographic interferometry (HI) has matured into a powerful, widely applied technique, driven by industrial requirements for new materials with minimum waste in their production, processing and reprocessing, and the manufacture of components made from them. Other advantages are the whole field and noncontact nature of the method, the latter facilitating the study of delicate materials. Its sensitivity using the wavelength of light as the fundamental scale of measurement allows us to carry out testing of materials and components, using loads much lower than would be applied in practical use. Qualitative HI facilitates nondestructive evaluation including the detection of defects such as delaminations in composite and multilayer materials. Increasingly quantitative methods are being applied for detailed analysis of fringe patterns by phase shifting, with unprecedented precision.

The fundamental principle of HI is that a hologram is a complete record of a wavefront and, with care, we can make direct, detailed comparisons of the wavefronts from the same object when subjected to changes in position, pressure, stress, strain, or temperature or when it is vibrating. Objects which are almost but not quite identical in shape may also be interferometrically compared. This is all possible because the wavefronts from such objects can be made to interfere, producing a fringe pattern which is due solely to the differences. The restrictions that apply are those discussed in Section 7.7 and are essential if we are to ensure that the *correct* wavefronts are compared. It is important to realize that the interference fringe patterns occupy 3-D space and correlation of fringe pattern data with the changes in the object itself is not a trivial task.

10.2 BASIC PRINCIPLE

Recall from Chapter 3 that an object wave described by Equation 3.5 may be reconstructed from a hologram in the *x,y* plane. If holograms are recorded in the same photosensitive layer, before and after the object is displaced, then simultaneous reconstruction from the two holograms produces two object waves

$$\tilde{O}_1 = E_O(x, y)\exp[j(\mathbf{k}_{O_1}(x, y)\cdot\mathbf{r}_1 - \omega t)] \text{ and } \tilde{O}_2 = E_O(x, y)\exp[j(\mathbf{k}_{O_2}(x, y)\cdot\mathbf{r}_2 - \omega t)] \quad (10.1)$$

which interfere with each other. We have assumed that the amplitude $E_O(x, y)$ does not change. The intensity of the resultant wave is $(1/2)(\tilde{O}_1 + \tilde{O}_2)(\tilde{O}_1 + \tilde{O}_2)^*$ and is given by

$$I(x, y) = 2I_0(x, y)\left\{1 + \cos\left[\mathbf{k}_{O_2}(x, y)\cdot\mathbf{r}_2 - \mathbf{k}_{O_1}(x, y)\cdot\mathbf{r}_1\right]\right\} \quad (10.2)$$

If one of the object waves is holographically recorded but the other is *live*, that is, the object is still in its original position during reconstruction of the image and illuminated as before, the resulting intensity variation has the same form as in Equation 10.2, provided that the hologram itself imposes no distortions on the *live* wavefront passing through it.

The above discussion does not take account of random variations in phase due to the surface roughness of the object. These variations give rise to the speckle effect, which was discussed in Section 7.4. The details of the speckle pattern due to illumination of a small region of the object centered at a point in its before and after positions are different, because the angle of illumination is different. So we need to change the expressions for the two object waves in Equation 10.1 to become

$$\tilde{O}_1 = E_{O_1}(x, y)\cos\left[\mathbf{k}_{O_1}(x, y)\cdot\mathbf{r}_1 - \omega t\right] \text{ and } \tilde{O}_2 = E_{O_2}(x, y)\cos\left[\mathbf{k}_{O_2}(x, y)\cdot\mathbf{r}_2 - \omega t\right] \quad (10.3)$$

Using complex notation, the resultant intensity spatially averaged over a number of speckles within a small area around an illuminated point is given by

$$I(x, y) = \langle(O + O')(O^* + O'^*)\rangle = \langle OO^*\rangle + \langle O'O'^*\rangle + \langle OO'^*\rangle + \langle O^*O'\rangle \quad (10.4)$$

$$I(x, y) = I_1 + I_2 + 2\sqrt{I_1 I_2}\cos\left[\mathbf{k}_{O_2}(x, y)\cdot\mathbf{r}_2 - \mathbf{k}_{O_1}(x, y)\cdot\mathbf{r}_1\right] \quad (10.5)$$

where I_1 and I_2 are now, respectively, the intensities of the two object waves averaged over the speckles. In many cases, they may be equal to I_O and we can write

$$I(x, y) = 2I_0(x, y)\left(1 + \cos\left[\mathbf{k}_{O_2}(x, y)\cdot\mathbf{r}_2 - \mathbf{k}_{O_1}(x, y)\cdot\mathbf{r}_1\right]\right) \quad (10.6)$$

10.3 PHASE CHANGE DUE TO OBJECT DISPLACEMENT

The expression $[\mathbf{k}_{O_2}(x, y)\cdot\mathbf{r}_2 - \mathbf{k}_{O_1}(x, y)\cdot\mathbf{r}_1]$ in Equation 10.6 is the *phase difference* between the two object waves due to the displacement of the object. The phase difference is seen to depend on position x, y in the hologram plane, which in turn depends on the *direction from which the two object waves are observed through the hologram*. It also depends on the *direction in which the object is illuminated* and of course, on the *displacement of the object*. In Figure 10.1, a point P on an object, O, illuminated by light with wave vector $\mathbf{k}_i$ is displaced by $\mathbf{d}$ so that it moves to P′.

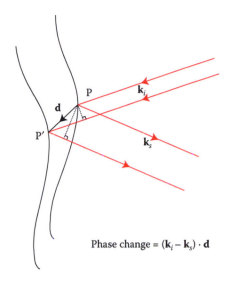

Phase change = $(\mathbf{k}_i - \mathbf{k}_s) \cdot \mathbf{d}$

FIGURE 10.1 Phase change due to general displacement (**d** variable).

The difference in phase between the light arriving at P and at P′ is $\mathbf{k}_i \cdot \mathbf{d}$. If the scattered light has wave vector $\mathbf{k}_s$, the change in its phase due to the object displacement is $-\mathbf{k}_s \cdot \mathbf{d}$ and so the total phase change $\Delta\phi$ is given by

$$\Delta\phi = (\mathbf{k}_i - \mathbf{k}_s) \cdot \mathbf{d} = \mathbf{K} \cdot \mathbf{d} \tag{10.7}$$

and Equation 10.6 may be rewritten as

$$I(x,y) = 2I_0 \left(1 + \cos[\mathbf{K} \cdot \mathbf{d}]\right) \tag{10.8}$$

where $\mathbf{K} = \mathbf{k}_i - \mathbf{k}_s$ and is known as the sensitivity vector.

10.4 FRINGE LOCALIZATION

We must consider where the interference fringes are located in space, as this is where their visibility is optimal for viewing. This in turn depends on both the precise nature of the displacement, for example, whether the object has undergone in-plane or out-of-plane displacement, and whether the displacement is accompanied by rigid body motion. All motion may be separated into pure rotational and pure translational motions, which we will consider separately.

10.4.1 PURE TRANSLATION

In Figure 10.2, the object undergoes pure translation. Assuming collimated illumination, $\mathbf{k}_i$, is constant and from Equation 10.6, the phase change, $\Delta\phi$, because the translation depends only on $\mathbf{k}_s$.

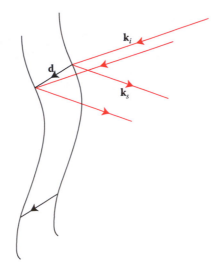

FIGURE 10.2 Phase change due to translation (**d** constant).

The phase change is constant where a fringe is localized in space so $\mathbf{k}_s$ must also be constant for that fringe. In other words, the scattered, fringe forming light is collimated and the fringe is at infinity and therefore localized in the focal plane of a lens. This applies to all the fringes so that the complete pattern is localized in the focal plane. *Variation of $\Delta\phi$ is largely confined to the plane defined by the two vectors $\mathbf{k}_i$ and $\mathbf{k}_s$, which is the plane of Figure 10.2, and declines to zero in the direction normal to this plane at a rate that depends on the microscopic structure of the surface. Thus, the fringes are parallel to one another and normal to the plane of Figure 10.2. Since the fringes are localized in the focal plane of the lens, it is not possible to assign a specific phase change due to transla-tion of any particular point *on the object surface* unless the object is effectively at infinity so that its image is also in the focal plane. A small lens aperture provides for a large depth of field (see Section 2.2.2) so that both the fringes and the object itself are in acceptably sharp focus.

In-plane translation in collimated illumination produces no phase variation.

10.4.2 IN-PLANE ROTATION

In the case of in-plane rotation, the displacement vector of any point on the surface depends on the distance of the point from the center of rotation. In Figure 10.3, the collimated illu-minating beam is assumed to be parallel to the y, z plane. A point P, with vertical coor-dinate y, whose distance from the center of rotation is r, is displaced by $ry\mathrm{d}\theta$ parallel to the y-axis, where $\mathrm{d}\theta$ is the angle of rotation and so the phase variation is purely in the y direction and the fringes are parallel to the x-axis and again localized in the focal plane of a lens. If the wave vector of the illuminating beam lies in the x, z plane, the fringes are parallel to the x-axis.

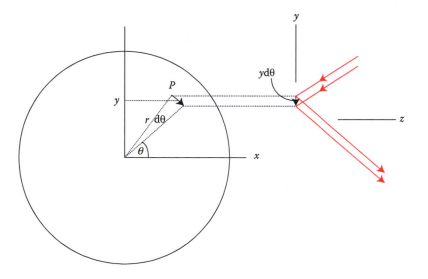

FIGURE 10.3 In-plane rotation.

10.4.3 OUT-OF-PLANE ROTATION

In Figure 10.4, a flat plate is rotated through a small angle, δ about the z-axis in its own plane. The phase change in the light scattered from the point P in its original and displaced positions is again given by Equation 10.7.

$$\Delta\phi = (\mathbf{k}_i - \mathbf{k}_s)\cdot\mathbf{d} = ky\delta(\cos\gamma + \cos\gamma') \tag{10.9}$$

and is constant for a fixed value of y so the fringes are parallel to the axis of rotation (z).

Suppose a fringe to be located at $-z_o, y_o$. We require that $\Delta\phi$ be independent of small changes in the direction of observation, γ', so that

$$\frac{d(\Delta\phi)}{d\gamma'} = k\delta(\cos\gamma + \cos\gamma')\frac{dy}{d\gamma'} - ky\delta\sin\gamma' = 0$$

$$(\cos\gamma + \cos\gamma')\frac{d(y_o - z_o\tan\gamma')}{d\gamma'} - y\sin\gamma' = (\cos\gamma + \cos\gamma')\frac{-z_o}{\cos^2\gamma'} - y\sin\gamma' = 0 \tag{10.10}$$

Thus the fringe is a distance $-z_o$ from the object given by

$$-z_o = \frac{y\sin\gamma'\cos^2\gamma'}{(\cos\gamma + \cos\gamma')} \tag{10.11}$$

If we view the object along a normal to its surface, $\sin\gamma'$ is zero and the fringes localize on the object itself. Note also that at the axis of rotation, $y = 0$ and $z_o = 0$, so the plane in which the fringes localize must include the axis of rotation. This is expected since $\Delta\phi$ must be zero, there being no displacement at the axis.

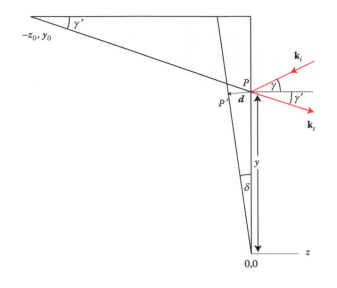

FIGURE 10.4 Out-of-plane rotation.

10.5 LIVE FRINGE HOLOGRAPHIC INTERFEROMETRY

In live fringe HI, just one hologram of the object is recorded. Following the recording, the object is maintained precisely in its original position. The hologram is either processed in situ or, following processing, is relocated exactly in its original position and the reconstructed image coincides with the object. We now have two object waves, one of them live, the other a reconstructed one. The object is now subjected to stress, strain, pressure, or thermal loading causing the live object wave to differ from the reconstructed object wave and interference fringes appear covering the object. The pattern changes depending on the viewing direction.

Thermoplastic recording materials (Section 6.4) are ideal for live fringe HI as they are processed in situ. Self-processing photopolymers are also very suitable. A silver halide hologram may be processed in situ, the hologram being mounted in a liquid tight container with transparent walls for subsequent viewing of the reconstructed image.

A simpler approach that is found to work very well is to use a kinematic holder for the hologram such as that recommended by Abramson [1], which enables the hologram to be replaced precisely in its original location. The silver halide plate is held in position by gravity, resting on three small ball bearings in a frame tilted back from the vertical. This relocates the hologram in its original plane. Position and orientation in the plane are assured by three steel cylindrical pins as shown in Figure 10.5.

One of the significant advantages of the live fringe technique is that the object can be progressively loaded while monitoring the development of the fringes. One can thus ensure that the applied load is such as to produce a fringe pattern whose spatial frequency is not too large for analysis. Fringe patterns may be photographically recorded at various loads for subsequent analysis. The *sense* of the displacement is also determined.

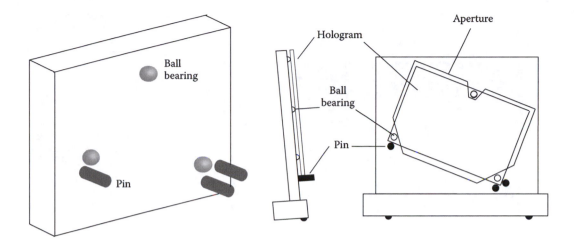

FIGURE 10.5 Kinematic plate holder.

Reconstruction from a phase hologram gives a bright image superimposed on the brightly lit object so that when the load is very small, the fringe pattern has very low contrast. This problem can be avoided by changing the phase of either the reconstructing reference wave or of the live object wave by π so that, with zero load, the field observed through the hologram is dark, since now the live and reconstructed object waves interfere destructively. The subsequent fringe patterns thus have high contrast. The phase change can be made using a halfwave plate and a linear polarizer in the path of the appropriate beam before it is expanded. (The reader may like to consider how this works in practice by reviewing Section 2.9.4.)

10.6 FROZEN FRINGE HOLOGRAPHIC INTERFEROMETRY

In the frozen fringe method, two holograms are recorded in succession, usually on the same photosensitive plate, the object being displaced between the recordings. The hologram does not have to be processed in situ but if it is not, it should nonetheless be relocated precisely in its original position for fringe analysis purposes. Very precise relocation is not necessary if the technique is simply used qualitatively to determine the presence of defects in the object. The method does not allow determination of the sense of displacement since the double exposure hologram contains no information about the order in which the recordings were made.

To facilitate determination of the sense of the displacement between exposures, a phase shift of $\pi/2$ may be introduced in the reference wave prior to the second exposure.

Since the finished double exposure hologram only provides information relating to a single displacement step, complete analysis requires that a number of double exposure holograms be recorded.

Self-processing photopolymers are particularly useful in frozen fringe holography as it is possible to choose a load on the object that produces an interference fringe pattern of moderate spatial frequency. Following the first recording, the laser power is significantly

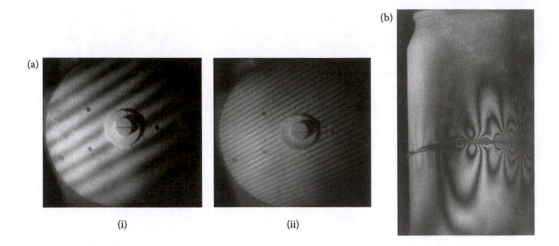

FIGURE 10.6 Double exposure holographic interferograms: (a) disk rotated in plane through different angles (i) and (ii); (b) soft drink can compressed around middle by an elastic band between exposures. (Reproduced from S. Martin, C. A. Feely, and V. Toal, Holographic recording characteristics of an acrylamide-based photopolymer, *Applied Optics*, 36, 5757, 1997. With permission of Optical Society of America.)

reduced while observing the reconstructed image, which is of course quite faint. The low power of the laser means that, the photopolymerization process is very slow and recording does not take place to any significant extent. The object is now progressively loaded so that interference fringes appear, as in live fringe HI. When a pattern considered suitable for analysis is obtained, the laser power is increased to the original recording level, and the second hologram is recorded and a frozen fringe interference pattern is seen in reconstruction from the doubly exposed hologram. This method also allows the sense of displacement to be determined.

Figure 10.6 shows some examples of holographic interferograms obtained in this way [2] using the self-processing photopolymer described in Section 6.6.4.1.

10.7 COMPENSATION FOR RIGID BODY MOTION ACCOMPANYING LOADING

When an object is loaded in a test, for example, by mechanical stress, the resulting fringes may be partly due to rigid body motion, which makes analysis difficult since one fringe looks very much like another and we normally cannot predict how many fringes we may expect to see for a given load. Abramson [1] has described a useful technique called sandwich holography, which is used to compensate for rigid body tilt. The technique requires two holograms forming a sandwich to be recorded at the same time in the same holder, for both the original and displaced states of the object. A further requirement is all the photosensitive plates have the same physical dimensions, which is normally the case with photographic plates from reputable suppliers. Figure 10.7 illustrates the method. The photosensitive sides of the plates face the object.

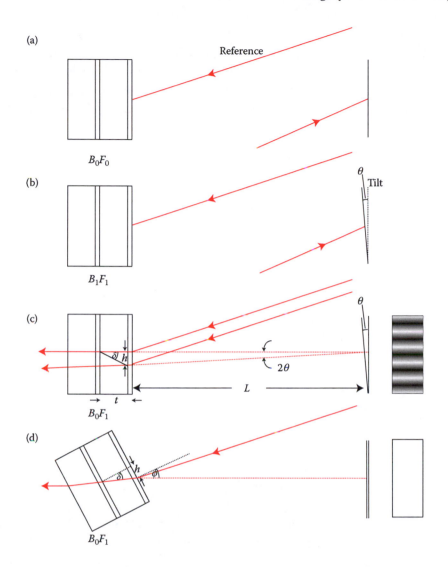

FIGURE 10.7 Sandwich holography: (a) recording with zero load; (b) recording with first load; (c) reconstruction; and (d) reconstruction following sandwich rotation.

We designate the holograms as B (before loading) and F (following loading) so that the series of hologram pairs is B_0F_0 (zero load), B_1F_1 (first load), B_2F_2 (second load), B_3F_3 (third load), and so on. After processing, the plates are replaced in the holder successively, in the pair combinations $B_0 F_1$, $B_0 F_2$, $B_0 F_2$, $B_0 F_3$, and so on and illuminated by the reference beam to display the interference patterns produced by each of the loads. Various combinations of plates may be used to determine the effect of different loads. Thus, if the loads for the recordings B_1F_1, B_2F_2, B_3F_3 are 1, 3, and 7 units, respectively, then at reconstruction, the combinations $B_0 F_1$, B_1F_2, $B_0 F_2$, B_2F_3, B_1F_3, and $B_0 F_3$ produce fringe patterns, respectively, corresponding to load changes from 0 to 1, 1 to 3, 0 to 3, 3 to 7, and 0 to 7 units. Since the *order* in which the holograms were recorded is known, the *sense* of displacement can be determined.

Consider the tilt angle of a flat plate due to bending under a mechanical load as shown in Figure 10.7. Suppose the plate is tilted toward the sandwich through an angle θ causing a reconstructed ray of light to be deflected from its original direction through an angle 2ϕ to arrive at a distance, h, below its original position at the sandwich (Figure 10.7c). This can be compensated by tilting the sandwich by an angle ϕ (Figure 10.7d). If t is the thickness of hologram F_1

$$\tan \delta = h/t = L\tan(2\theta)/t \quad \text{and, from Snell's law,} \quad \sin \phi = n \sin \delta \qquad (10.12)$$

$$\sin \phi = \frac{n}{\left[1+\left(\dfrac{t}{L\tan(2\theta)}\right)^2\right]^{\frac{1}{2}}} \cong \frac{2nL\theta}{t} \quad \text{so that} \quad \frac{\phi}{\theta} \cong \frac{2nL}{t} \qquad (10.13)$$

Let us look at the implications of Equation 10.13. Suppose in an experiment, a vertical plate of length l is illuminated almost normally by a plane wave of wavelength $\lambda = 633$ nm. In a test, its displacement is accompanied by a tilt toward the hologram plane (Figure 10.7b). If N fringes are observed in the reconstruction as shown in Figure 10.7e, then the displacement at the tip due solely to this tilt is $N\lambda/2$ and $\tan\theta = N\lambda/(2l)$. From Equation 10.13, assuming $n = 1.5$, $l = 20$ cm, $L = 100$ cm, $N = 10$, and $t = 1.5$ mm, the tilt of the plate is about 15 microradians, and the tilt of the sandwich required to compensate and remove the fringes is 30 milliradians. Thus, it is possible to measure very small object tilts by tilting the sandwich through much larger and more manageable angles.

Following holographic recording, an object may have to be removed, to be drilled for example and then replaced in its support for either live or frozen fringe holographic studies of the effects of the drilling. The effects of material fatigue in normal use may also be studied, including the initiation and growth of defects such as cracks and delaminations. Abramson [1] describes kinematic supports for precise relocation of test objects having regular shapes, using the same principles as the kinematic hologram mount shown in Figure 10.5. Alternatively, three cylindrical pins with spherical tips are attached to the base of the object. The support is a horizontal flat plate machined so that one of the pins fits into a hole whose diameter is about 0.7 of that of the pin, the second into a V-shaped groove whose width is also smaller than the pin diameter. The third pin rests on the surface of the support plate.

10.8 DOUBLE PULSE HOLOGRAPHIC INTERFEROMETRY

HI has been applied to the study of rapidly moving objects including, for example, a bullet traveling through a gas in a chamber [3]. A hologram is first recorded of the chamber in its quiescent state with a pulse of laser light, which passes through a diffuser before entering the chamber to form an object beam. The reference beam is entirely outside the chamber. Then, as the bullet passes through the chamber, it triggers a second laser pulse to record the second hologram. Reconstruction from the double pulse hologram reveals an interference pattern characteristic of the changes in gas density due to the bullet.

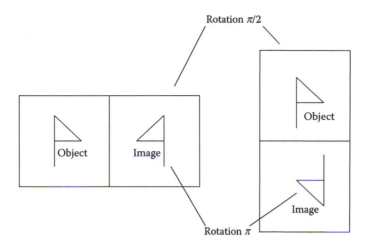

FIGURE 10.8 Image rotation.

More generally, double pulse HI may usefully be applied in many industrial contexts in which it is not possible to ensure interferometric stability of the recording setup. A very demanding application is the study of aircraft turbine blades. The effect of the turbine rotation is nullified using a simple image derotation device. The principle involved is shown in Figure 10.8 in which it is seen that as a 45°, 45°, 90° roof prism is rotated by 90°, about an axis normal to its rectangular face through the midpoint of its apex, an image retroreflected by it is rotated by 180°. The image of an object reflected by such a prism rotates at twice the angular velocity of the prism itself. Thus, if an object, rotating about its axis at angular velocity ω, is imaged by the prism rotating with angular velocity $\omega/2$ about the same axis, the image is static. Double pulse HI is used to study the motion of the blades relative to the turbine body.

10.9 HOLOGRAPHIC INTERFEROMETRY OF VIBRATING OBJECTS

It is possible to study vibrating objects by using the fact that vibratory movement involves repetitive change in the direction of motion at the extremities of the vibration cycle where the object is momentarily at rest. The phenomenon is readily observed by clamping a ruler and then pushing on and releasing its free end.

10.9.1 TIME-AVERAGED HOLOGRAPHIC INTERFEROMETRY

We already saw from Equation 10.7 that when a point on the object illuminated light of wave vector $\mathbf{k}_i$ is displaced by $\mathbf{d}$, the phase change $\Delta\phi$ observed in direction $\mathbf{k}_s$ is

$$\Delta\phi = \mathbf{K} \cdot \mathbf{d} \tag{10.14}$$

Suppose that the object vibrates with angular frequency ω_0. Then, from Equation 10.14, the phase change in light wave illuminating the point and then scattered in direction $\mathbf{k}_s$ toward the point x, y in the plane of the hologram is

$$\mathbf{K} \cdot \mathbf{d}_0 \sin(\omega_0 t + \phi_0) \tag{10.15}$$

where $\mathbf{d}_0$ is the amplitude of the displacement vector and ϕ_0 is the constant phase factor. The object wave is now described by

$$\tilde{O}(x, y, t) = E_O(x, y) \exp\left\{ j[\mathbf{k}_0 \cdot \mathbf{r} - \omega t + \mathbf{K} \cdot \mathbf{d}_0 \sin(\omega_0 t + \phi_0)] \right\} \tag{10.16}$$

Suppose the period of the vibration is much shorter than the exposure time. Effectively, a large number of object waves are holographically recorded, with the object at a different point in its vibration cycle for each one of them. To determine the result observed when all these object waves are reconstructed, we must average over the recording time. Thus, the reconstructed wave is

$$\tilde{O}(x, y, t) = E_O(x, y) \exp\left\{ \left[j\mathbf{k}_0 \cdot \mathbf{r} - \omega t \right] \right\} \lim_{T \to \infty} \frac{1}{T} \int_0^T \exp\left[j\mathbf{K} \cdot \mathbf{d}_0 \sin(\omega_0 t + \phi_0) \right] dt \tag{10.17}$$

and the intensity

$$I(x, y) = I_0(x, y) \left[\lim_{T \to \infty} \frac{1}{T} \int_0^T \exp\left[j\mathbf{K} \cdot \mathbf{d}_0 \sin(\omega_0 t + \phi_0) \right] dt \right]^2 \tag{10.18}$$

$$I(x, y) = I_0(x, y) J_0^2(\mathbf{K} \cdot \mathbf{d}_0) \tag{10.19}$$

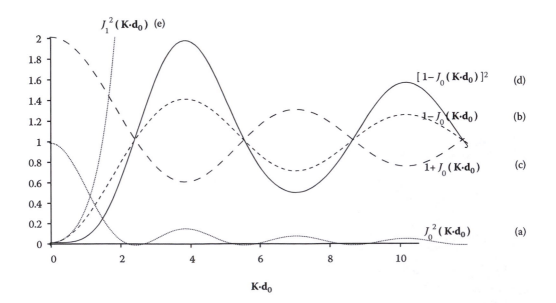

FIGURE 10.9 Intensity modulation in holographic interferometry of vibrating object: (a) time averaged; (b) real time (amplitude hologram); (c) real time (phase hologram); (d) double exposure, phase shift π; and (e) reference wave modulation at vibration frequency.

where $J_0(\mathbf{K} \cdot \mathbf{d}_0)$ is a zero-order Bessel function. The reconstructed image thus has a light intensity distribution the same as that of the original object modulated by square of the zero-order Bessel function (Figure 10.9a). The image is thus covered by contours of equal vibrational amplitude although we cannot determine the difference in phase of the vibration between any two points.

To see the effect of the modulation, suppose the object is illuminated and observed along a normal to its surface at a wavelength of 500 nm and that it is vibrating out of plane. The first zero of $J_0^2(\mathbf{K} \cdot \mathbf{d}_0)$ thus occurs for a vibration amplitude of just 0.1 nm and the intensities observed for greater amplitudes are progressively smaller as the peaks of $J_0^2(\mathbf{K} \cdot \mathbf{d}_0)$ become ever smaller in height. Thus, the method, although easy to implement, is of limited application.

10.9.2 LIVE HOLOGRAPHIC INTERFEROMETRY OF A VIBRATING OBJECT

We can record a hologram of the object at rest and then carry out live fringe HI on the vibrating object. In the case of an *amplitude* hologram (Section 3.5.2), the reconstructed object wave is

$$-E_O(x,y)\exp\left[j(\mathbf{k}_0 \cdot \mathbf{r} - \omega t)\right]$$

To this is added the live complex amplitude

$$E_O(x,y)\exp\left[j(\mathbf{k}_0 \cdot \mathbf{r} - \omega t)\right]\exp\left[j\mathbf{K} \cdot \mathbf{d}_0 \sin(\omega_0 t + \phi_0)\right]$$

giving a resultant complex amplitude

$$E_O(x,y)\exp\left[j(\mathbf{k}_0 \cdot \mathbf{r} - \omega t)\right]\left\{\exp\left[j\mathbf{K} \cdot \mathbf{d}_0 \sin(\omega_0 t + \phi_0)\right] - 1\right\}$$

and the intensity is given by

$$I(x,y) = I_0(x,y)\left\{1 - \cos\left[\mathbf{K} \cdot \mathbf{d}_0 \sin(\omega_0 t + \phi_0)\right]\right\} \tag{10.20}$$

The *observed* intensity is found by averaging $I(x, y)$ over the persistence time of the eye, which again is much longer than the period of the vibration, and we may write

$$I_{av}(x,y) = I_0(x,y)\left[1 - \lim_{T \to \infty}\frac{1}{T}\int_0^T \cos\left[\mathbf{K} \cdot \mathbf{d}_0 \sin(\omega_0 t + \phi_0)\right]\mathrm{d}t\right] \tag{10.21}$$

$$I_{av}(x,y) = I_0(x,y)\left[1 - J_0(\mathbf{K} \cdot \mathbf{d}_0)\right] \tag{10.22}$$

The form of Equation 10.22 is shown in (Figure 10.9b).

If a *phase* hologram is recorded of the static object, the sign in Equation 10.22 is positive. The result is shown in (Figure 10.9c).

10.9.3 DOUBLE EXPOSURE WITH PHASE SHIFT

In this method, a hologram is recorded of the object at rest, followed by a second recording with the object in vibratory motion but with a phase shift of π in one of the beams. As a result, the reconstructed waves are subtracted from one another giving

$$E_O(x,y)\exp\left[j(\mathbf{k}_0\cdot\mathbf{r}-\omega t)\right]\left\{1-\exp\left[j\mathbf{K}\cdot\mathbf{d}_0\sin(\omega_0 t+\phi_0)\right]\right\}$$

whose time average is

$$E_O(x,y)\exp\left[j(\mathbf{k}_0\cdot\mathbf{r}-\omega t)\right]\left\{1-\lim_{T\to\infty}\frac{1}{T}\int_0^T\exp\left[j\mathbf{K}\cdot\mathbf{d}_0\sin(\omega_0 t+\phi_0)\right]dt\right\}$$
$$=E_O(x,y)\exp\left[j(\mathbf{k}_0\cdot\mathbf{r}-\omega t)\right]\left[1-J_0(\mathbf{K}\cdot\mathbf{d}_0)\right]$$

giving an intensity

$$I(x,y)=I_0(x,y)\left[1-J_0(\mathbf{K}\cdot\mathbf{d}_0)\right]^2 \tag{10.23}$$

with rather better fringe contrast than that given by Equation 10.22 (see Figure 10.9d).

10.9.4 MODULATION OF THE REFERENCE WAVE

A number of options are available if the reference wave is modulated during recording. Again, the complex amplitude of the object wave is given by

$$\tilde{O}(x,y,t)=E_O(x,y)\exp\left\{j\left[(\mathbf{k}_0\cdot\mathbf{r}-\omega t)+\mathbf{K}\cdot\mathbf{d}_0\sin(\omega_0 t+\phi_0)\right]\right\}$$
$$=\tilde{E}_O(x,y,t)\exp\left[j\mathbf{K}\cdot\mathbf{d}_0\sin(\omega_0 t+\phi_0)\right]$$

and the reference wave is

$$E_r\exp\left[j(\mathbf{k_r}\cdot\mathbf{r}-\omega t)\right]\exp\left[j(\omega_r t+\phi_r)\right]=\tilde{E}_r(x,y,t)\exp\left[j(\omega_r t+\phi_r)\right]$$

We can set $\phi_0=0$ and ϕ_r to be the difference in phase between object and reference waves. The recorded intensity is then

$$\left\{\tilde{E}_O(x,y,t)\exp\left[j\mathbf{K}\cdot\mathbf{d}_0\sin(\omega_0 t)\right]+\tilde{E}_r(x,y,t)\exp\left[j(\omega_r t+\phi_r)\right]\right\}\times C.C.$$

of which one of the terms is

$$\tilde{E}_O(x,y,t)\exp\left[j\mathbf{K}\cdot\mathbf{d}_0\sin(\omega_0 t)\right]\tilde{E}_r{}^*(x,y,t)\exp\left[-j(\omega_r t+\phi_r)\right]$$

and on reconstruction *without* modulation of the reference wave, we obtain

$$E_r^2\tilde{E}_O(x,y,t)\exp\left[j\mathbf{K}\cdot\mathbf{d}_0\sin(\omega_0 t)\right]\exp\left[-j(\omega_r t+\phi_r)\right]$$

The time average is

$$E_r^2 \tilde{E}_O(x,y,t) \lim_{T \to \infty} \frac{1}{T} \int_0^T \exp\left[j\mathbf{K} \cdot \mathbf{d}_O \sin(\omega_O t)\right] \exp\left[-j(\omega_r t + \phi_r)\right] dt$$

$$= E_r^2 \tilde{E}_O(x,y,t) \lim_{T \to \infty} \frac{1}{T} \int_0^T \sum_{-\infty}^{\infty} J_m(\mathbf{K} \cdot \mathbf{d}_O) \exp\left[jm(\omega_O t)\right] \exp\left[-j(\omega_r t + \phi_r)\right] dt$$

with m, an integer

$$= E_r^2 \tilde{E}_O(x,y,t) \sum_{-\infty}^{\infty} J_m(\mathbf{K} \cdot \mathbf{d}_O) \lim_{T \to \infty} \frac{1}{T} \int_0^T \exp\left[j(m\omega_O - \omega_r)t\right] \exp(-j\phi_r) dt$$

$$= E_r^2 \tilde{E}_O(x,y,t) \sum_{-\infty}^{\infty} J_m(\mathbf{K} \cdot \mathbf{d}_O) \exp\left[-j\phi_r\right] \delta(m\omega_O - \omega_r)$$

Setting $\omega_r = n\,\omega_O$.

Thus, the intensity in the reconstructed image is

$$I(x,y) \propto I_O(x,y) J_n^2(\mathbf{K} \cdot \mathbf{d}_O) \cos^2 \phi_r \qquad (10.24)$$

Equation 10.24 has the consequence that when $\omega_r = \omega_o$, the observed intensity is proportional to $J_1^2(\mathbf{K} \cdot \mathbf{d}_O)$, which is plotted in Figure 10.9e. Thus, the method allows us to examine very-low-amplitude vibrations, unlike the time-averaged method described in Section 10.9.1. Compare the intensity dependence on displacement for the latter plotted in Figure 10.9a.

Note also that if ϕ_r is varied one can determine the phase difference between the reference wave and any point on the object.

10.10 STROBOSCOPIC METHODS

Frozen fringe holograms can be recorded using a Q-switched laser with a pulse duration much smaller than the period of the vibration.

If a continuous wave laser is used, the beam may be passed through a rotating disk with small apertures uniformly distributed around its circumference at equal distances from the axis, producing a regular stream of pulses of the same duration (Δt) and separated by the same time interval. If, further, the rotation of the disc is synchronized with the vibration of the object, then the object may be illuminated by laser pulses at the same two points $nT + t_1$ and $nT + t_2$ in time of its vibration cycle (Figure 10.10) where n is an integer and T the period of the vibration.

In this way, one may build up a double exposure holographic recording of the object in two positions.

Alternatively, a holographic recording is made of the static object. At the reconstruction stage, the object is illuminated by a short pulse just once in each cycle of vibration producing live fringe interference between the static object wave and the wave corresponding to the object at a single point in its displacement cycle. Furthermore, by adjusting the point in time during the vibration cycle at which the illumination pulse occurs, the full

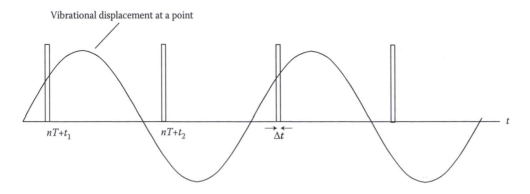

Vibrational displacement at a point

$nT+t_1$ $nT+t_2$ Δt t

FIGURE 10.10 Stroboscopic holographic interferometry.

displacement range of the vibration may be studied. The frequency of the pulses may also be altered so that a range of vibration frequencies may be studied as well.

10.11 HOLOGRAPHIC SURFACE PROFILOMETRY

Conventional profilometry involves the use of a stylus, which laterally traverses the surface usually maintaining physical contact with it. Great accuracy is obtained using a very sensitive signal transducer, which converts the vertical movements of the stylus into an electrical signal. Extreme sensitivity is obtained using the noncontact methods of atomic force microscopy (AFM), which relies on the forces between the atoms of the stylus tip and the atoms on the surface or scanning tunneling microscopy (STM), which relies on the varying tunneling current between the tip and the surface (see Figure 12.16). All these methods are slow as the complete profile is built up in raster scan fashion.

We can use a number of holographic methods of obtaining the profile of a surface. These have the advantages of being noncontact and whole field, so that a profile of even a delicate surface can be obtained in one go.

10.11.1 SURFACE PROFILING BY CHANGE IN WAVELENGTH

A hologram of the object is recorded at wavelengths λ_1 and reconstruction is carried out at wavelength λ_2, with the object remaining in its original position. The image is reconstructed if the difference in wavelength, $\lambda_1 - \lambda_2$, is less than the spectral bandwidth of the hologram. Equation 4.36 shows that a hologram recorded at 514 nm with an angle of about 90° between the reference and object wave vectors, in a photosensitive layer 10 μm thick, has a spectral bandwidth of 36 nm. Thus, we can use 514 and 488 nm wavelengths from an argon ion laser.

We assume that the reference wave is collimated and that the object is illuminated by a collimated beam propagating in a direction normal to the hologram plane (Figure 10.11). A point $P(x_O, y_O, z_O)$ on the object at z_O from the hologram is imaged at both z_O and $z_{O'}$

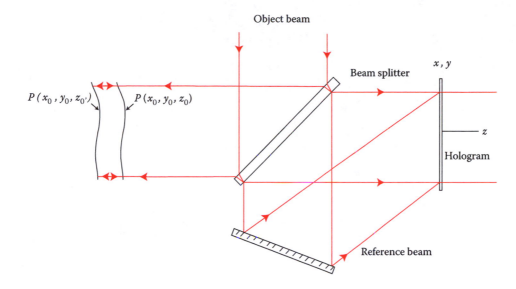

FIGURE 10.11 Two wavelength live fringe holographic contouring.

because of the change in wavelength. From Equation 7.43,

$$z_{O'} - z_O = z_O \frac{\lambda_1 - \lambda_2}{\lambda_2} \tag{10.25}$$

The y coordinate of the point P is, from Equation 7.43

$$y_{O'} = \frac{y_{R'} z_R z_O - \mu y_R z_{R'} z_O + \mu y_O z_{R'} z_R}{z_R z_O - \mu z_{R'} z_O + \mu z_{R'} z_R} \tag{10.26}$$

$$y_{O'} = \frac{y_R z_R z_O - \mu y_R z_R z_O + \mu y_O z_R^2}{z_R z_O - \mu z_R z_O + \mu z_R^2} = \frac{1 - \mu}{\dfrac{1 - \mu}{y_R} + \dfrac{\mu z_R}{y_R z_O}} + \frac{y_O}{\dfrac{z_O}{z_R}\left(\dfrac{1}{\mu} - 1\right) + 1} = y_O + z_O \frac{y_R}{z_R}\left(\frac{1}{\mu} - 1\right)$$

so that the point P is displaced laterally in the y direction by

$$y_O - y_{O'} = z_O \frac{y_R}{z_R}\left(\frac{1}{\mu} - 1\right) \tag{10.27}$$

and to the same extent in the x direction.

From Equation 7.44, the lateral magnification is

$$\mathrm{LaM} = \frac{\partial x_{O'}}{\partial x_O} = \frac{\mu z_{R'} z_R}{z_R z_O - \mu z_{R'} z_O + \mu z_{R'} z_R} = \frac{\partial y_{O'}}{\partial y_O} = \frac{\mu}{\dfrac{z_O}{z_R} - \mu \dfrac{z_O}{z_R} + \mu} = 1 \tag{10.28}$$

Returning now to Equation 10.25, the phase difference between the light wave illuminating the object and the light reconstructed by the hologram is

$$\Delta\phi(x,y) = \frac{4\pi z_o (\lambda_1 - \lambda_2)}{\lambda_2^2} \tag{10.29}$$

The surface is covered by fringes, which are contours of constant object distance from the hologram plane. Using 514 and 488 nm wavelengths, the change in height between adjacent fringes is about 5 μm.

The lateral shift, from Equation 10.27, is approximately $z_0/120$ if the reference beam makes an angle of about 10° with the z-axis, but these displacements are the *same* when the live fringe method is used. *Double exposure may be used but the lateral displacement only affects one of the reconstructions* and it must be minimized by image plane recording (see Figure 8.8) so that z_0 is zero in Equation 10.27.

10.11.2 REFRACTIVE INDEX CHANGE METHOD

In the refractive index change method, the apparent depths of the object in media of differing refractive index are interferometrically compared. Suppose that a hologram is recorded of an object that has been immersed in a medium of refractive index n_1 in a cell with transparent windows. We assume that the recording geometry is again as described in Section 10.11.1. We then replace the medium with one having a refractive index n_2 (Figure 10.12) and record a second hologram using the same vacuum wavelength λ_0 as before.

Immersion of the object in media of differing refractive index for each holographic recording has the effect of changing the path length of the object wave between recordings. If the reconstructions are viewed at an angle to the cell window normal, then points on

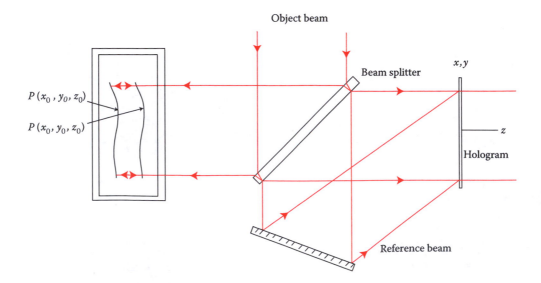

FIGURE 10.12 Two index live fringe holographic contouring.

the object surface are displaced both laterally and longitudinally. If the reconstructions are viewed *along the normal to the cell window*, only longitudinal displacement occurs and the phase change $\Delta\phi$ corresponding to the point P(x_o, y_o, z_o) is

$$\Delta\phi = \frac{4\pi z_0 (n_1 - n_2)}{\lambda_0} \tag{10.30}$$

where z_0 is the distance of P from the front window of the cell. The height difference between points on the surface overlaid by adjacent fringes is

$$\Delta z = \frac{\lambda_0}{2(n_2 - n_1)} \tag{10.31}$$

A wide range of refractive index differences, and therefore sensitivities, is available, using air and a liquid, air and another gas, or two different liquids.

10.11.3 CHANGE IN DIRECTION OF ILLUMINATION

A number of slightly different techniques come under this section. The simplest method is to record a hologram using a collimated reference beam and a collimated object beam, which is at an angle to the z-axis (see Figure 10.13a) and parallel to the y, z plane. The surface height variation is $z(x, y)$. At the reconstruction stage, with the object in place, the direction of the object beam is altered by a small angle $\Delta\theta$.

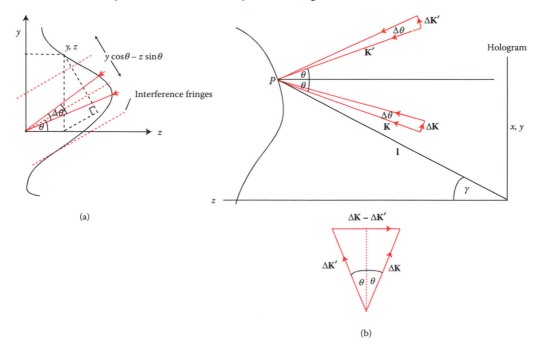

FIGURE 10.13 Change in direction of illumination: (a) change in direction of a single object illuminating beam; and (b) change in direction of two illuminating beams.

Alternatively, a double-exposure hologram is recorded with the direction of the object beam altered slightly between exposures. In both cases, the object is effectively illuminated by two beams propagating parallel to the y, z plane, which interfere with one another, producing a set of straight fringes, which are parallel to the bisector of the angle between the beams. If the reconstructed image is photographed with sufficient depth of field to ensure that its surface is everywhere in focus, the recorded fringes are effectively localized on the surface. Let us now look at these fringes. Their intensity $I(y, z)$ is

$$I(y,z) = 2I_0 \left\{ 1 + \cos \left[\frac{4\pi \sin(\Delta\theta/2)}{\lambda} (y\cos\theta - z\sin\theta) \right] \right\} \qquad (10.32)$$

The dependence on y can be obtained by placing a *plane* surface in the x, y plane. The intensity in this case is

$$I_{\text{plane}}(y,z) = 2I_0 \left\{ 1 + \cos \left[\frac{4\pi \sin(\Delta\theta/2)}{\lambda} (y\cos\theta) \right] \right\} \qquad (10.33)$$

Subtracting Equation 10.32 from Equation 10.33 leaves an intensity distribution that depends only on the surface height, z.

Then, the change in surface height between nearest neighbor fringes is

$$\Delta z = \frac{\lambda}{2\sin\theta\sin(\Delta\theta/2)} \qquad (10.34)$$

If the object beam is not parallel to the y, z plane, the resulting phase variation in the x direction can be compensated by a slight tilt of the reference beam or of the hologram.

A disadvantage of the method is that shadows always occur since the object is obliquely illuminated as shown in Figure 10.13a, and the surface profile cannot be found in shadow regions. To alleviate this problem, one can use the arrangement shown in Figure 10.13b. The object is illuminated by waves having wave vectors $\mathbf{K}$ and $\mathbf{K}'$ propagating at angles θ and $-\theta$ to the z-axis [4]. A hologram is recorded and then the object illuminating beams are rotated by the same small angle, $\Delta\theta$, in the same sense at reconstruction. Alternatively, the object illuminating beams may be rotated between *two* holographic recordings.

Suppose that the illuminating beams travel distances d and d' to reach point P on the surface and this point is viewed from x, y in the hologram plane where P is at distance l from the point x, y in the hologram. The holographically recorded phase difference between object waves arriving at x, y via P is

$$\phi = \frac{2\pi}{\lambda} \left[d - d' + l + (\mathbf{K} - \mathbf{K}') \cdot \mathbf{1} \right] \qquad (10.35)$$

The *change* in phase following rotation of the illuminating beams is

$$\Delta\phi = \frac{2\pi}{\lambda}\mathbf{l}\cdot(\Delta\mathbf{K} - \Delta\mathbf{K}') = \frac{2\pi}{\lambda}l\cos\gamma(\Delta\theta)2\sin\theta \tag{10.36}$$

where $|\Delta\mathbf{K}| = |\Delta\mathbf{K}'| = \Delta\theta|\mathbf{K}| = \Delta\theta|\mathbf{K}'| = \Delta\theta$.

$l\cos\gamma = z$ and the increment, Δz, in z between neighboring fringes is

$$\Delta z = \frac{\lambda}{2\Delta\theta\sin\theta} \tag{10.37}$$

The advantage here is that the phase has no dependence on y.

In both methods, rotation of the object itself may be substituted for rotation of the object illuminating beams.

10.12 PHASE CONJUGATE HOLOGRAPHIC INTERFEROMETRY

As explained in Chapter 6, many holographic recording materials do not require processing and this means that if the hologram is illuminated by a conjugate reference beam while recording is in progress, one may view the phase conjugate image live.

A practical application is the study of vibration as time-averaged fringes may be viewed in real time (Figure 10.14) since the recording time is typically a lot longer than the period of any vibration of interest. One method employs a BSO crystal [5] but any real-time recording material, such as the photopolymer discussed in Section 6.6.4.1, may be used.

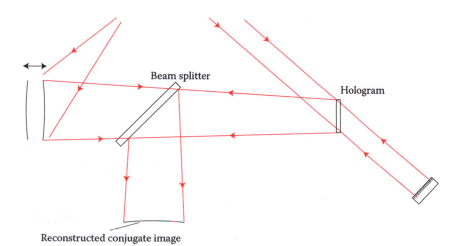

FIGURE 10.14 Real-time holographic interferometry of a vibrating object. (Reproduced from J. P. Huignard, J. P. Herreau, and T. Valentin, Time average holographic interferometry with photoconductive electrooptic $Bi_{12}SiO_{20}$ crystals, *Applied Optics*, 16, 2796–98, 1977. With permission of Optical Society of America.)

10.13 FRINGE ANALYSIS

Fringe patterns may be captured as images using a CCD or CMOS camera and the images transferred to a computer for subsequent detailed quantitative analysis. We have seen that when dealing with vibrating objects, one may obtain vibration amplitudes by evaluating the intensity of the fringe patterns. The phase of vibration may also be evaluated.

When dealing with statically loaded objects, the usual method of fringe analysis is to capture and store a number of fringe patterns, using a known change of phase of the reference beam between each, for a fixed load. The phase change is readily implemented by applying an appropriate voltage to a piezoelectric transducer attached to a mirror in the path of the reference beam thus altering the path length of the reference beam. Alternatively the reference beam can be made to propagate in an optical fiber wrapped around a hollow piezoelectric cylinder with electrodes applied to its inner and outer walls. An appropriate voltage is applied between the electrodes causing the cylinder to expand radially. This stretches the fiber and alters the optical path, and thus the phase, of the light propagating through it.

We write the intensity distribution in the interference fringes recorded by a camera as

$$I_1(x, y) = I_0 \left[1 + V \cos \phi(x, y) \right] \tag{10.38}$$

where V is the fringe visibility and ϕ the phase change at x, y (now the *camera image plane*) due to the (static) load.

Then for reference beam phase shifts of $\pi/2$, π, and $3\pi/2$, respectively, we obtain

$$
\begin{aligned}
I_2(x, y) &= I_0 \left\{ 1 + V \cos \left[\phi(x, y) + \pi/2 \right] \right\} \\
I_3(x, y) &= I_0 \left\{ 1 + V \cos \left[\phi(x, y) + \pi \right] \right\} \\
I_4(x, y) &= I_0 \left\{ 1 + V \cos \left[\phi(x, y) + 3\pi/2 \right] \right\}
\end{aligned}
\tag{10.39}
$$

$$\phi(x, y) = \frac{I_4 - I_2}{I_1 - I_3} \tag{10.40}$$

Examination of the signs of the numerator and denominator of Equation 10.40 enables us to assign $\phi(x, y)$ to its appropriate quadrant. Thus, for example, if $I_4 - I_2$ is negative and $I_1 - I_3$ positive, we know that the value of $\phi(x, y)$ must be between 270° and 360°. Using low cost, high speed data acquisition devices, we can readily obtain the *phase map*, that is, the value of $\phi(x, y)$ at every camera pixel. We need finally to deal with the fact that this phase map is *wrapped*, all values lying between 0 and 2π (Figure 10.15a) by comparing every value of $\phi(x, y)$ with those of its immediate neighbors, a procedure known as phase unwrapping (Figure 10.15b), which is based on the principle that the phase map is generally a continuous one.

If the value of ϕ at a pixel exceeds the value at the previous pixel by more than say π, then we add 2π to the phase. If the value of ϕ at a pixel is less than the value at the previous pixel by more than say $-\pi$, then we subtract 2π from the phase. The phase map is unwrapped, row by row and column by column. It must be borne in mind that phase discontinuities often arise

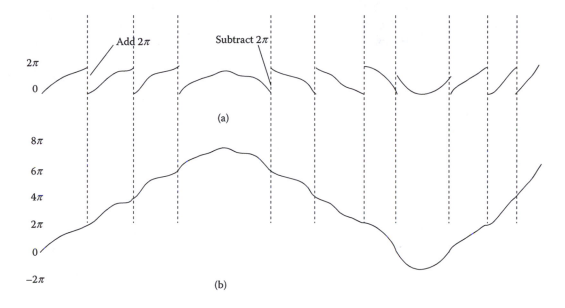

FIGURE 10.15 Phase unwrapping: (a) wrapped phase and (b) unwrapped phase.

from causes other than wrapping, for example, around cracks or discontinuous changes in surface height. For this reason, phase unwrapping requires careful intervention to exclude discontinuities, although robust phase unwrapping algorithms can be designed to cope.

A phase map is obtained in this way before the object movement and another phase map obtained afterwards. The difference between the two is a phase map due to the change, from which we can determine the movement at x, y due to the change, taking account of the illumination and observation directions.

10.14 SPECKLE PATTERN INTERFEROMETRY

A CCD or CMOS photosensitive array has a typical pixel width of 10 μm so that the resolution is only 100 lines mm^{-1} at best. This is nowhere near sufficient for holographic recording. However, it is sufficient for a powerful technique called electronic speckle pattern interferometry (ESPI), which has also been called TV holography since it shares many of holography's characteristics and requirements for successful implementation. It is a powerful technique, which is rather easy to implement and has its origins in speckle pattern correlation interferometry

10.14.1 SPECKLE PATTERN CORRELATION INTERFEROMETRY

Consider that light intensity $I_O(x,y)$ in the x,y plane scattered from an optically rough surface and a reference beam $I_R(x,y)$ interfere with a phase difference $\phi(x,y)$ to produce a resultant intensity $J(x,y)$.

$$I(x, y) = I_O + I_R + 2\sqrt{I_O I_R} \cos \phi \qquad (10.41)$$

I_O, I_R, and ϕ are all random functions of x,y due to the speckle effect. Now, the surface is displaced giving resultant intensity

$$I'(x, y) = I_O + I_R + 2\sqrt{I_O I_R}\, \cos(\phi + \psi) \tag{10.42}$$

ψ is the phase change arising from the change in the object and is not random function of x,y. We are assuming that $I_O + I_R$ does not change.

We now derive the *correlation coefficient, $R_{JJ'}$, which is a measure of the similarity of I and I'.*

$$R_{JJ'} = I \oplus I' = \frac{\overline{II'} - \overline{I}\,\overline{I'}}{\left(\overline{I^2} - \overline{I}^2\right)^{\frac{1}{2}} \left(\overline{I'^2} - \overline{I'}^2\right)^{\frac{1}{2}}} \tag{10.43}$$

where a bar over any quantity implies spatial averaging over a number of speckles. I_O, I_R, and ϕ are independent and can be treated separately.

Since $\overline{\cos \phi} = \overline{\cos(\phi + \psi)} = 0$

$$R_{JJ'} = \frac{\overline{I_O^2} + \overline{I_R^2} + 2\overline{I_O I_R} + 4\overline{I_O I_R \cos \phi \cos(\phi + \psi)} - \left(\overline{I_O}^2 + \overline{I_R}^2 + 2\overline{I_O I_R}\right)}{\left(\overline{I^2} - \overline{I}^2\right)^{\frac{1}{2}} \left(\overline{I'^2} - \overline{I'}^2\right)^{1/2}} \tag{10.44}$$

and since $\overline{2\cos \phi \cos(\phi + \psi)} = \overline{\cos \psi + \cos(2\phi + \psi)} = \cos \psi$, $\overline{4 I_O I_R \cos \phi \cos(\phi + \psi)} = 2\overline{I_O I_R} \cos \psi$.

From Equations 7.29 and 7.30 $\overline{I^2} = 2\overline{I}^2$, we have

$$R_{JJ'} = \frac{\overline{I_O^2} + \overline{I_R^2} - 2\overline{I_O I_R} + 2\overline{I_O I_R}\,(1 + \cos\psi)}{\left(\overline{I^2} - \overline{I}^2\right)^{\frac{1}{2}} \left(\overline{I'^2} - \overline{I'}^2\right)^{1/2}}$$

The numerator is $2\overline{I_O I_R}\,(1 + \cos\psi)$ if $\overline{I_O} = \overline{I_R}$ and therefore

$$R_{JJ'} = \frac{1}{2}(1 + \cos\psi) \tag{10.45}$$

which is maximum if $\psi = 2n\pi$, minimum if $\psi = (2n + 1)\pi$, n an integer. Thus, the similarity of the two interference patterns I and I' depends only on the displacement. The correlation can be done photographically [6]. The processed negative recording of $I(x, y)$ is relocated and the (altered) illuminated object is viewed through the negative. Wherever $\psi = 2n\pi$, the correlation is maximum and $I'(x, y)$ *strongly resembles $I(x, y)$ but is not transmitted by the photographic negative of $I(x,y)$.* When $\psi = (2n + 1)\pi$, the opposite is the case and $I'(x,y)$ does not resemble $I(x,y)$. Thus $I'(x,y)$ is transmitted wherever $I(x,y)$ is low and its photographic record is transparent.

10.14.2 ELECTRONIC SPECKLE PATTERN INTERFEROMETRY

The technique becomes much more practicable when I and I' are simply captured by a CCD or CMOS camera array and subtracted from one another. The first frame $I(x,y)$ is transferred from the CCD camera to a framestore in a personal computer and the incoming frame $I'(x,y)$ is subtracted, pixel by pixel, giving

$$I(x,y) - I'(x,y) = 2\sqrt{I_O I_R}\left[\cos\phi - \cos(\phi + \psi)\right] = 4\sqrt{I_O I_R}\,\sin\left(\phi + \frac{\psi}{2}\right)\sin\left(\frac{\psi}{2}\right) \quad (10.46)$$

Taking the absolute value, rectification or squaring ensures that there are no negative values of $I(x,y) - I'(x,y)$, which cannot be displayed.

The right-hand side of Equation 10.46 is a maximum when $\psi = (2n+1)\pi$ and minimum when $\psi = 2n\pi$. Thus, the result is the same as that obtained in speckle pattern correlation interferometry.

The fringe pattern described by Equation 10.46 is modulated by the speckle effect since $\phi(x,y)$ is a random function. This is not a serious problem, but it is important that the speckles be resolved by the camera lens (see Section 7.4.2).

ESPI systems are usually designed to be sensitive either to purely out-of-plane motion (Figure 10.16) or purely in-plane motion (Figure 10.17).

In the case of the former, we distinguish between a smooth reference beam system, which has the advantage that the intensity in the interference term in Equations 10.41 and 10.42 is never zero. The disadvantage is that we need to ensure that the angle between the reference wave and the object wave is zero, to avoid interference fringes that cannot be resolved by the camera. In practice, this means that the reference wave must appear to originate at the center of the aperture stop of the lens as shown in Figure 10.16a. Instead, we can use a speckled reference beam as shown in Figure 10.16b, making it rather easier to align the object and reference beams since the wide angular scatter of the two beams, one from a fixed rough surface (the reference) and the other from the object surface, ensures that the angle between the two beams will be practically zero over a field of view encompassing the object.

10.14.2.1 Fringe Analysis in Electronic Speckle Pattern Interferometry

As discussed in Section 10.13, when static loads are applied to the object, we can analyze the fringe patterns by introducing a change in phase between frames.

$$\begin{aligned}
_1(x,y) &= I_O + I_R + 2\sqrt{I_O I_R}\,\cos\phi \\
I_2(x,y) &= I_O + I_R + 2\sqrt{I_O I_R}\,\cos(\phi + \pi/2) \\
I_3(x,y) &= I_O + I_R + 2\sqrt{I_O I_R}\,\cos(\phi + \pi) \\
I_4(x,y) &= I_O + I_R + 2\sqrt{I_O I_R}\,\cos(\phi + 3\pi/2) \\
\phi(x,y) &= \frac{I_4 - I_2}{I_1 - I_3}
\end{aligned} \qquad (10.47)$$

$$\text{Similarly } \phi'(x,y) = \frac{I_4' - I_2'}{I_1' - I_3'}$$

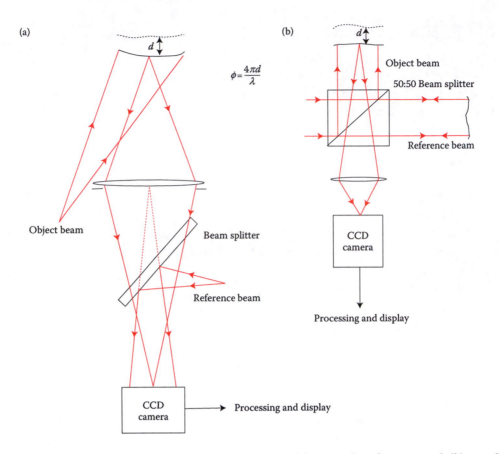

$$\phi = \frac{4\pi d}{\lambda}$$

FIGURE 10.16 Out-of-plane sensitive ESPI systems: (a) smooth reference and (b) speckle reference.

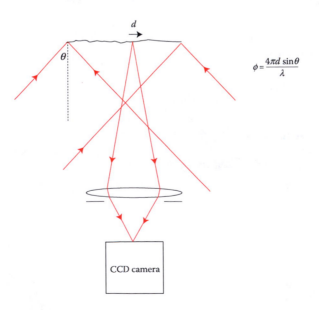

$$\phi = \frac{4\pi d \sin\theta}{\lambda}$$

FIGURE 10.17 In-plane sensitive ESPI.

and the displacement (in the out-of-plane sensitive case) is given by

$$d(x, y) = \frac{\lambda}{4\pi}(\phi' - \phi)$$

In general, the fringe patterns obtained using HI are smooth but, in speckle interferometry, since one or both of the terms $I_O(x,y)$ and $I_R(x,y)$ in Equation 10.47 are due to a random speckle pattern, some pixels in either the $I(x,y)$ or $I'(x,y)$ frames, or both, will be illuminated by very low or even zero intensity while others may be saturated. In such cases, modulation between the four phase-shifted frames is rather low or even zero and produces phase data that may be significantly in error. The value at any such a pixel is substituted, in a new array, by the median of the values of its neighbors in a window of $N \times N$ pixels. In order to decide whether a pixel should be substituted, the modulation term

$$I_m = \sqrt{(I_4 - I_2)^2 + (I_3 - I_1)^2} \tag{10.48}$$

is calculated to check whether it is greater than some preset value. The median window is then moved on to the next pixel and the process is repeated until the whole array is processed. This technique removes any features less than $(N + 1)/2$ in size. Thus is, a 3×3 median window replaces isolated single bad pixels.

Additional phase errors arise due to reduction in correlation because the speckles at a particular pixel may be replaced by others when the object is mechanically or thermally stressed or undergoes rigid body rotation. The consequence is usually that phase jumps are not handled correctly when phase unwrapping is carried out. Suppose a phase jump of $+2\pi$ should be followed by a jump of -2π but as a result of a phase error, the latter does not occur. The result is that a phase discontinuity of 2π is introduced, appearing as a streak in the final phase image. To deal with this problem, the value at each pixel is compared with those at its eight nearest neighbors in the form of a cross and, if it is very different, it is replaced by the median of those eight values [7]. Edges are retained. The assumption here is that phase should not change greatly from pixel to pixel.

Phase errors also arise if the phase step between frames is not exactly $\pi/2$, but use of a five frame algorithm greatly reduces such errors. A review of speckle fringe analysis methods is given by Huntley [8].

10.14.2.2 Vibration Studies Using Electronic Speckle Pattern Interferometry

In the case of an object vibrating out of plane, we modify Equation 10.39 and obtain

$$I(x, y, t) = I_O + I_R + 2\sqrt{I_O I_R} \cos\left[\frac{4\pi a_0}{\lambda} \cos(\omega_0 t + \theta)\right] \tag{10.49}$$

where a_0 is the amplitude of oscillation of the object and $1/\omega_0$ its period, which is assumed to be much longer than the frame time of the camera. We therefore take the time average of $I(x,y,t)$.

$$I(x,y) = I_O + I_R + 2\sqrt{I_O I_R} \left[\lim_{T \to \infty} \frac{1}{T} \int_0^T \cos\left[\frac{4\pi a_0}{\lambda} \cos(\omega_0 t + \theta) \right] dt \right] \quad (10.50)$$

$$I(x,y) = I_O + I_R + 2\sqrt{I_O I_R} J_0\left(\frac{4\pi a_0}{\lambda} \right) \cos\theta \quad (10.51)$$

Fringes are obtained by a number of methods. One is to capture a frame while the object is at rest and subtract from it another frame captured while the object is vibrating giving a result similar to that shown in Figure 10.9b.

In practice, the phase angle θ is likely to change between frames due to thermal drift and frame subtraction produces speckle fringes with the object continually vibrating. This method is rather hit and miss, and a better approach is to alter the phase of the reference beam by π at the beginning of every second frame, restoring it to zero at the beginning of every other frame. Consecutive frames are subtracted from one another, and the modulus of the result is given by

$$|I_{odd} - I_{even}| = 4\sqrt{I_O I_R} J_0\left(\frac{4\pi a_0}{\lambda} \right) \cos\theta \quad (10.52)$$

Drift at frame rates (typically 25 Hz) is much less likely to occur so that the Bessel fringe pattern is normally quite stable.

The amplitude of the zero-order Bessel function becomes small as the vibration amplitude, a_0, increases and this makes it difficult to obtain reliable data. Suppose now that the optical path of the reference wave oscillates at frequency ω_0 and amplitude a_R and phase ϕ_R with respect to the vibration of the part of the object, which is imaged at x,y. Equation 10.41 now becomes, by phasor addition

$$I(x,y,t) = I_O + I_R + 2\sqrt{I_O I_R} \cos\left\{ \frac{4\pi}{\lambda} \left[a_0^2 + a_R^2 - 2a_0 a_R \cos(\theta - \phi_R) \right]^{1/2} \right\} \quad (10.53)$$

and the time average is

$$I(x,y) = I_O + I_R + 2\sqrt{I_O I_R} J_0\left\{ \frac{4\pi}{\lambda} \left[a_0^2 + a_R^2 - 2a_0 a_R \cos(\theta - \phi_R) \right]^{1/2} \right\} \quad (10.54)$$

The zero-order Bessel function is maximum when

$$a_0^2 + a_R^2 - 2a_0 a_R \cos(\theta - \phi_R) = 0 \quad (10.55)$$

and if $\theta = \phi_R$ then $a_0 = a_R$.

Thus, if the motion of any part of the object is compensated by modulating the optical path difference in the interferometer, at the same frequency and amplitude, then the argument of the Bessel function is made zero for that part of the object and the corresponding light intensity given by Equation 10.53 is a maximum. The brightest part of the displayed interferogram indicates that part of the object that is vibrating at the same amplitude as the reference wave path length oscillation and at the same phase.

When a_0 and a_R are unequal then if $\phi_R = \theta$, the argument of the Bessel function is minimum and the brightest fringe region corresponds to that part of the object that vibrates in phase with the reference wave path length oscillation. Thus, the phase of the object vibration may be mapped out [9].

10.14.2.3 Electronic Speckle Pattern Interferometry Systems

Early ESPI systems used bulk optics (lenses, mirrors, and beam splitters) although fiber optical systems have found widespread use for two main reasons. One is the ease of alignment of object and reference beams in an out-of-plane sensitive system having a reference beam with a smooth wavefront. The fiber carrying the reference beam can be inserted in a small hole in the center of the imaging lens [7]. The other advantage is that phase changes between the reference and object waves are easily introduced by the method described in Section 10.13.

Alignment of speckle reference beam, out-of-plane sensitive systems is rather easy with the arrangement shown in Figure 10.15b but only half of the light returning to the beam splitter is detected. Phase changes require that the fixed surface be attached to a piezoelectric transducer. Alternatively, a current modulated laser diode is used to alter the phase between object and reference beams. Such lasers must be carefully chosen to be free of mode-hopping over an acceptable range of drive currents. The use of holographic beam splitters and combiners, discussed in Chapter 11, helps conserve laser light and makes for compact system design [10].

10.15 SUMMARY

This chapter is concerned with HI and its applications. We present the techniques of live fringe, frozen fringe, and time-averaged HI and show how they are underpinned by simple theory. We consider the problem of fringe localization. We discuss the technique of sandwich holography to compensate for rigid body motion, which may accompany the application of a mechanical load. We describe the double pulse method for the study of transient events.

We explain the three techniques of optical surface profilometry, namely, change in illumination wavelength, refractive index, and illumination direction.

We describe an example of real-time phase conjugate HI.

We also deal with fringe analysis for static loading conditions and vibratory motion.

Finally, we consider speckle interferometry along with some techniques for speckle interferometric fringe analysis.

REFERENCES

1. N. Abramson, *The Making and Evaluation of Holograms*, Academic Press, London (1981).
2. S. Martin, C. A. Feely, and V. Toal, "Holographic recording characteristics of an acrylamide-based photopolymer," *Appl. Opt.*, 36, 5757 (1997).

3. L. O. Heflinger, R. F. Wuerker, and R. E. Brooks, "Holographic interferometry," *J. Appl. Phys*, 37, 642–9 (1966).
4. P. K. Rastogi and L. Pflug, "A holographic technique featuring broad range sensitivity to contour diffuse objects," *J. Mod. Opt.*, 38, 1673–83 (1991).
5. J. P. Huignard, J. P. Herreau, and T. Valentin, "Time average holographic interferometry with photoconductive electrooptic $Bi_{12}SiO_{20}$ crystals," *Appl. Opt*, 16, 2796–8 (1977).
6. J. A. Leendertz, "Interferometric displacement measurement on scattering surfaces utilizing speckle effect," *J. Phys. E: Sci. Instrum.*, 3, 214–8 (1970).
7. K. Creath, "Phase-shifting speckle interferometry," *Appl. Opt*, 24, 3053–8 (1985).
8. J. M. Huntley, "Automated analysis of speckle interferograms," in *Digital Speckle Pattern Interferometry and Related Techniques*, P. K. Rastogi (ed.), John Wiley, Chichester (2001).
9. O. J. Løkberg and K. Høgmoen, "Vibration phase mapping using electronic speckle pattern interferometry," *Appl. Opt.*, 15, 2701–4 (1976).
10. V. Bavigadda, R. Jallapuram, E. Mihaylova, and V. Toal, "Electronic speckle pattern interferometer using holographic optical elements for vibration measurements," *Opt. Lett.*, 35, 3273–5 (2010).

PROBLEMS

1. A hologram of an upright steel cylinder is recorded in a self-processing recording material. With the collimated object beam still illuminating it in a direction normal to its principal axis, the cylinder is rotated about its principal axis. What form of fringes would be observed in the reconstruction? Find an expression for the angle of rotation?

2. The cylinder of problem 1 now has a mass of 20 kg carefully lowered onto its upper face. What is observed in the reconstructed image? If the radius of the cylinder is 2 cm and its height is 30 cm, how many additional fringes are seen as a result of the loading and what is their orientation? Young's modulus of elasticity, E, for steel is 2×10^{11} N · m^{-2}. The laser used in the experiment has a wavelength of 633 nm.

3. A hologram is recorded of a vertically mounted, flat, rectangular, steel plate, illuminated by a collimated beam of laser light, wavelength 633 nm, incident at an angle of 45° to the normal, in the horizontal plane, the plate being parallel to the hologram. Find the sensitivity vector if the image is viewed by looking into the hologram along its normal. The plate is now rotated about a vertical axis. What is observed in the live fringe reconstruction? Show how the rotation angle can be determined. Assuming the reconstructed image is viewed at a distance of 50 cm, find the maximum observable angle of rotation.

4. A horizontal wire attached at right angles to the back of the plate of problem 3, at the midpoint of its upper edge, passes over a fixed pulley and a mass of 0. 1 kg is carefully hung on the end. The plate is 10 cm high, 5 cm wide, and 5 mm thick. What is exactly observed in live fringe reconstruction? The maximum deflection Δ of a plate of thickness t, width w, and length L due to a force F, applied in the manner described, is given by

$$\Delta = \frac{4FL^3}{Ewt^3}$$

Young's modulus for steel is 10^{11} Nm^{-2}.

5. A flat circular disc in the vertical plane and parallel to the hologram plane is illuminated using a horizontal beam of collimated 633 nm light at an angle of incidence of 45°. Following holographic recording, the disc is rotated in its own plane and viewed through the hologram along the normal. Describe fully what is observed?

6. The disc of problem 5 is clamped around its circumference and illuminated *normally* using collimated 633-nm laser light. It is set vibrating by a piezoelectric transducer attached to it at the center, the displacement of the transducer being normal to the plate. A time-averaged hologram is recorded and viewed in the direction of the hologram normal. What is the amplitude of vibration at the second brightest fringe and roughly where would you expect to see it?

CHAPTER 11

Holographic Optical Elements

11.1 INTRODUCTION

The refraction, reflection, and diffraction processes by which one wavefront is converted to a different one are enabled by prisms, lenses, mirrors, and diffraction gratings. A hologram is essentially a recording of a wavefront and, on reconstruction, converts a reference wave into the object wave, with up to 100% efficiency. It is thus easy to see that any lightwave converter such as a lens, prism, or mirror could, in principle, be replaced by a hologram or holographic optical element (HOE). A HOE may, by necessity or design, be recorded at one wavelength for use at a different wavelength, the wavelengths being related through the angles between the recording beams and the normal to the hologram surface, the angle made by the reconstructing beam with the normal and the refractive index of the recording material at each of the two wavelengths. However, HOEs are normally designed to operate at the wavelength at which they have been recorded and their characteristics change undesirably when used at a different wavelength. This is because HOEs are wavelength sensitive since they depend for their function on *diffraction* rather than *reflection* or *refraction*. For example, the focal length and aberrations of a holographic lens vary with wavelength much more markedly than those of its refractive counterpart.

HOEs do have a number of very considerable advantages. They can be made with large aperture sizes in very lightweight materials using low cost photosensitive materials in the form of comparatively thin layers, whereas conventional optical components usually rely on comparatively expensive bulk material for the effects that they produce. For example, a convex lens is necessarily thicker at its axis than at the periphery while the converse is true for a concave lens. Similarly, cube beam splitters require significant amounts of material.

A number of HOEs can be combined in a single layer of photosensitive material, each set of recorded interference fringes occupying the same space. This feature is enabled by the Bragg angular and spectral selectivity characteristics of thick holograms and by the total phase modulation possible in the material, the latter being a combination of thickness and refractive index modulation.

Manufacturing costs are significantly lower as grinding and polishing of optical glass are not involved. HOEs are generally flat and thin and may be conveniently stacked together, whether separated or not.

The use of photorefractive materials allows for recording and erasing of HOEs and their replacement by new ones.

Some HOEs are shown in Figure 11.1 with their conventional counterparts.

However, the design of HOEs requires care and, in all but the simplest and least demanding applications, the task is best undertaken using accurate ray tracing, taking account of the wavelength dispersion of the recording material and of dimensional changes that may occur during recording.

In this chapter, we look at the range of applications of HOEs, many of which are well established.

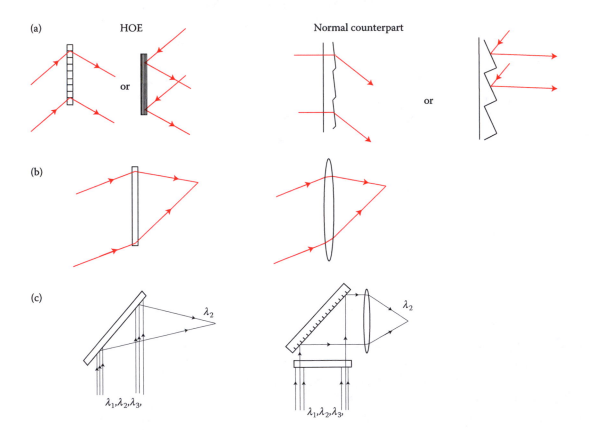

FIGURE 11.1 Holographic and conventional optical elements: (a) diffraction grating; (b) lens; and (c) combination filter, mirror, lens.

11.2 DIFFRACTION GRATINGS

We have already seen in Chapter 3 how diffraction gratings can be made holographically. If two mutually coherent plane waves are made to interfere in a photosensitive material, the recorded interference pattern, which is a hologram of a plane wave, acts as a diffraction grating so that it diffracts one of the beams to reproduce the other one. Since the gratings are prepared using fringe patterns rather than ruling by a diamond stylus, mechanical errors due to the stylus stepping mechanism do not arise and ghost diffraction orders are avoided. Once the recording system is set up, the time needed to fabricate a grating is just the recording and, where necessary, the processing time. Several weeks may be required to make a diffraction grating by scoring grooves using a diamond stylus compared with perhaps less than an hour to make a holographic grating. From Section 4.3.1.1, we can see that the angular and wavelength selectivities of a 20-μm thick phase transmission grating recorded with two mutually coherent beams incident at ±30° to the normal are 2° and 36 nm, respectively. Such a grating can be rotated in the path of the light, which is to be spectroscopically analyzed, so that the wavelength of the diffracted light is scanned across the spectral bandwidth of 36 nm. Gratings of such spectral bandwidths and even narrower bandwidths have a number of applications and we will discuss these presently.

For high-resolution spectroscopic gratings, photoresist (Section 6.5) is the recording material of choice for a number of reasons. The recorded grating may be used successfully in transmission mode since the material, being grainless, produces little scatter. Alternatively, it may be coated with metal to make a reflection grating. Photoresist can be spin coated over quite a large area of substrate to form layers 1–2 μm thick on surfaces having a variety of shapes, so that compact spectroscopic instrument designs are possible. Photoresist is rather insensitive to light especially of wavelength greater than about 470 nm. At 488 nm, typical sensitivity is 600 mJ cm^{-2} and recording time may be quite long even with a high-power laser. The necessity for interferometric stability during the long exposure times required to record large-sized gratings requires active stabilization of the interference fringe pattern using a method such as that described in Section 8.3. The finished gratings can be readily copied, usually holographically. Holographic diffraction gratings provide very high resolution. The diffraction efficiency of photoresist typically peaks at around 1500 lines per millimeter. Taking as an example the interference pattern produced by the superposition of two coherent beams of light of wavelength 488 nm derived from an argon ion laser, if the angle between the beams is 30°, the spatial frequency is over 1000 lines per millimeter and a grating of width 15 cm provides a resolution of over 150,000 in the first order (Section 2.2.1.4).

Of course, a holographic grating recorded in photoresist normally has a sinusoidal surface profile and is only usable in the first diffraction order, although high diffraction efficiencies are possible provided that the surface modulation is of sufficient amplitude. A surface modulation grating in which phase modulation arises from path differences due to the sinusoidal surface profile is described by the same mathematics as for a phase grating, which was discussed in Section 3.2.2. The maximum diffraction efficiency of a

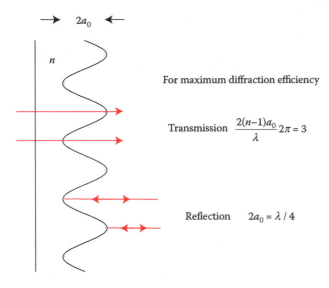

For maximum diffraction efficiency

Transmission $\dfrac{2(n-1)a_0}{\lambda}2\pi = 3$

Reflection $\quad 2a_0 = \lambda / 4$

FIGURE 11.2 Phase grating.

phase grating is 33.9% when the phase modulation $\phi_0 = 3$. Thus, the optical path difference between light passing through the peaks of the surface profile and that passing through the troughs is $(2(n-1)/a_0\lambda)2\pi = 3$ so that for $n = 1.5$ $2a_0 = 3\lambda/\pi$ (Figure 11.2), where a_0 is the surface modulation amplitude.

In practice, photoresist gratings are overcoated with a metallic layer and used in reflection mode so that the optimal surface height modulation is just $\lambda/4$, which can more readily be achieved.

The profile of a reflection holographic photoresist grating may be altered to increase the diffraction efficiency. One approach is to form higher harmonics of the basic sinusoidal surface height profile by recording fringe patterns with different angles between the recording beams. This must be accompanied by careful control of the phase relationship between each of the intensity distributions. By this means, the desired, usually triangular, profile is built up by Fourier synthesis.

Another approach [1] is to orient the photoresist layer obliquely to the interference pattern during recording (Figure 11.3). The effect is to render insoluble the positive, surface photoresist in the bright fringe regions whereas in the dark fringe areas, it remains soluble and after development, a staircase shaped profile is produced.

A disadvantage is that many of the profiles require one of the interfering beams to pass through the substrate, which may introduce defects and aberrations.

11.3 SPECTRAL FILTERS

One of the most important optical components that can be fabricated as a HOE is a spectral filter. Figure 4.9 shows the normalized diffraction efficiency of a volume phase reflection grating, as a function of the off-Bragg parameter, ξ', with refractive index modulation parameter $v' = \pi/4$. The diffraction efficiency falls to zero at $\xi' \cong 3$. From Equation 4.48,

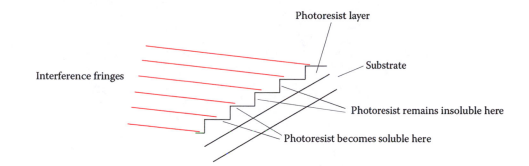

FIGURE 11.3 Blazing by oblique illumination of photoresist. (Reprinted from N. K. Sheridon, *Appl. Phys. Lett.*, 12, 316–8, (1968), copyright American Institute of Physics. With permission.)

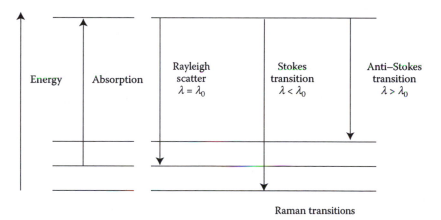

FIGURE 11.4 Raman scatter.

we have $(2\pi n_0 \partial \lambda_{air} T / \lambda_{air}^2) \sin \theta = 3$. Suppose the grating thickness T is 20 μm, n_0 is 1.5, and the fringe planes are parallel to the surface of the grating (i.e., $\theta = 0$) and $\lambda = 514$ nm. We obtain a spectral bandwidth of about 13 nm and a diffraction efficiency of 100% at 514 nm. In other words, this device acts very effectively as a notch filter. Such filters are of great importance in Raman spectroscopy, which is a very important and widely used technique for identifying molecular groups in materials. It is based on the concept that all the internal energy associated with a molecule is quantized. By this, we mean not only the energy associated with individual atoms (see Section 2.8.1) comprising the molecule but also the energy of rotation and vibration of the molecule as a whole. Thus, a molecule excited from a lower to a higher energy state by light of wavelength λ_0 (often 488 nm or 514nm light from an argon ion laser) may return to its original energy state by emission of light of the same wavelength, a process called Rayleigh scattering (Figure 11.4).

However, light of wavelength shorter or longer than that of the exciting light may be emitted instead, the molecule ultimately taking up a higher or lower energy state than the original one. The former is known as a Stokes transition, the latter an anti-Stokes transition. The *difference* between the energies of the initial and final energy states is of course the

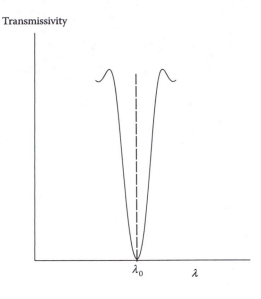

FIGURE 11.5 Spectral transmissivity of a volume phase reflection grating (notch filter).

difference between the energies of the absorbed and emitted light waves and is the difference in the energy of the molecule in two different modes of vibration. These energy differences can be assigned uniquely to specific groups of atoms within molecules. In order to detect the comparatively weak signals associated with these transitions against the background of the Rayleigh scatter, it is essential that the latter be suppressed as much as possible using a *notch filter*, which transmits all wavelengths apart from the excitation wavelength (Figure 11.5).

The need becomes even greater when using a strongly focused stimulating beam in a scanning Raman spectrometer to obtain spatially dependent Raman spectra in order to determine the spatial variations in the concentration of the molecule of interest.

Even narrower spectral rejection bandwidths are readily achieved using thicker reflection gratings. The rejection wavelength can be fine tuned by a slight rotation of the device in the path of the incident beam.

Holographic notch filters are normally manufactured in dichromated gelatin. The filter is made by placing the photosensitive layer in contact with a mirror and illuminating it with collimated monochromatic light of the wavelength to be rejected (Figure 11.6). The mirror reflects light that has been transmitted by the layer so that we have two counter propagating beams, which form planar fringes spaced a distance $\lambda/2n\sin(\theta/2)$ apart where θ is the angle between the beams in the recording layer of refractive index, n, and λ is the vacuum wavelength.

Filters that block infrared light can be made using the approach shown in Figure 11.7.

The use of a 45°, 45°, 90° prism index matched to the photosensitive layer enables total internal reflection to take place at the lower surface of the substrate so that a reflection grating is recorded using a single incident beam. When the beam is normally incident on a nonhypotenuse face of the prism, the spacing, d, of the grating is given by $d = \lambda/2n\sin 45°$. When illuminated *normally*, it reflects light of wavelength $\sqrt{2}\lambda$. By recording at a range of incident angles, a range of infrared wavelengths can be rejected. This is an example of a HOE recorded at one wavelength but intended for use at another.

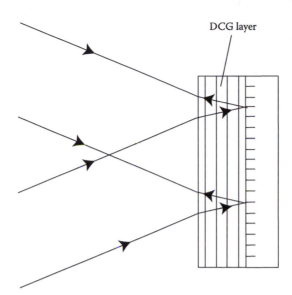

FIGURE 11.6 Making a holographic notch filter.

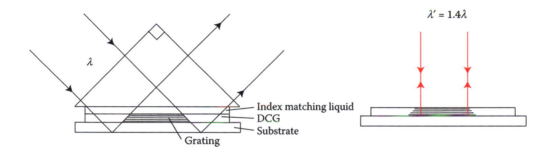

FIGURE 11.7 IR notch filter.

A similar approach enables fabrication of Bragg gratings in glass fibers as shown in Figure 11.8. using light from a UV exciplex laser, (Section 5.7.1.5) and a quartz phase grating positioned alongside the fiber. The light diffracted by the grating produces two beams, which interfere in the core of the fiber. The core is doped with germanium and has a higher refractive index than the surrounding cladding material so that light of appropriate wavelength is confined within the core by total internal reflection.

The effect of the UV light on the germanium-doped core is to induce a permanent change in refractive index in regions where the intensity is high so that a grating is formed. The spacing of the grating written into the core is given by $d = \lambda/2n\sin(\theta/2)$ where n is the refractive index of the core at UV wavelength, λ, and θ, the angle between the beams in the core. The grating *reflects* light propagating *along* the core with wavelength λ_0 given by $\lambda_0 = 2n_0 d$ with n_0 the refractive index at λ_0. Such gratings are widely used in fiber optical communication systems in order to demultiplex and multiplex signals of different wavelengths.

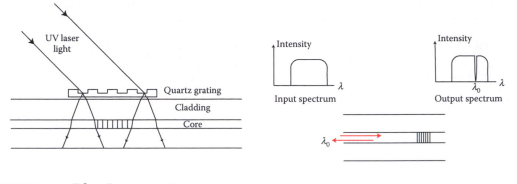

FIGURE 11.8 Fiber Bragg grating.

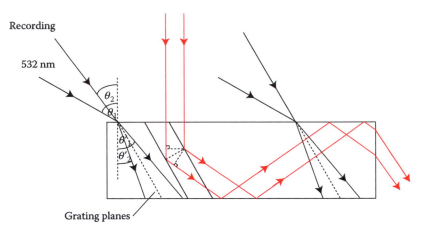

FIGURE 11.9 Slanted grating recorded at 532 nm acting as a notch filter at longer wavelength.

Fiber Bragg gratings are also used as strain gauges. Stretching the fiber causes the spacing of the grating to change and its resonant reflected wavelength is altered accordingly, the change in the reflected wavelength being directly proportional to the strain. Fibers that have Bragg gratings written into them at different locations along them can be embedded in so-called smart materials. The spatial distribution of strain in the material is continuously monitored by interrogating the fiber Bragg gratings using broadband light.

In similar fashion, gratings can be made in *photopolymers* but using *visible* light, thus avoiding the cost of a UV laser and allowing a much greater degree of flexibility in the spatial frequency of the gratings which can be recorded. Using the relationships in the previous paragraph, it is seen that a grating formed by two beams of wavelength 532 nm incident at equal and opposite angles of 30.99° to the normal at a layer of photopolymer of refractive index 1.5 has a spacing of 516.67 nm. This grating acts as a Bragg reflector at 1550 nm, which is in one of fiber optical telecommunications bands.

Slanted gratings also can be prepared in photopolymers to act as broadband blocking filters. The objective here is to ensure that the grating is at a slant angle such that light reflected from the grating planes is totally internally reflected and is wave guided through the layer as shown in Figure 11.9. Consider a grating formed by beams of 532 nm light incident at θ_1 and θ_2 with respect to the normal to a photopolymer layer. Inside the photopolymer of refractive index n, the angles are θ_1' and θ_2' and the recorded grating is slanted at $(\theta_1' + \theta_2')/2$ to the normal and has spacing $\lambda/2n\sin[(\theta_1' - \theta_2')/2]$. For light of wavelength λ' incident along the normal that satisfies the Bragg condition, we have

$$\frac{n\lambda'}{\lambda} = \frac{\sin\left[(\theta_1' + \theta_2')/2\right]}{\sin[(\theta_1' - \theta_2')/2]} = \frac{\tan(\theta_1'/2) + \tan(\theta_2'/2)}{\tan(\theta_1'/2) - \tan(\theta_2'/2)} \tag{11.1}$$

and the angle of incidence at the lower boundary of the recording medium is twice the slant angle of the grating.

11.4 LENSES

If we record the interference pattern produced by a plane wave and a spherical wave and then illuminate the hologram with the plane wave, the spherical wave will be reconstructed. Thus, a plane wave is converted to a spherical wave by the hologram, which is exactly what a lens does.

In Figure 11.10, a point source of monochromatic light is located at point x_0, y_0, z_0. We use Equation 7.36 to describe the spherical diverging wavefront at x,y in the hologram plane

$$O = \exp\left[j\frac{k}{2z_0}\left(x^2 + y^2 - 2xx_0 - 2yy_0 \right) \right]$$

Assume a plane reference wave given by $\exp(jky\sin\theta)$ propagating parallel to the y,z plane. The recording is described by the expression

$$\exp\left\{ jk\left[\frac{1}{2z_0}(x^2 + y^2 - 2xx_0 - 2yy_0) - y\sin\theta \right] \right\}$$
$$+ \exp\left\{ -jk\left[\frac{1}{2z_0}(x^2 + y^2 - 2xx_0 - 2yy_0) - y\sin\theta \right] \right\}$$

and, on reconstruction we obtain output, E given by

$$E = \exp\left[\frac{jk(x^2 + y^2 - 2xx_0 - 2yy_0)}{2z_0} \right] + \exp\left[\frac{jk(x^2 + y^2 - 2xx_0 - 2y[y_0 + 2z_0\sin\theta])}{2(-z_0)} \right] \tag{11.2}$$

The first term in Equation 11.2 is a replica of the original point source. The reconstructing plane wave is thus converted to a *diverging* wavefront so that the hologram is a negative lens. If the original reference wave propagated parallel to the z-axis, $\theta = 0$ and the second term describes a wavefront *converging* to the point $x_0, y_0, -z_0$. Thus, the hologram acts as both a negative and a positive lens.

If we record a thick hologram, then only one of the lenses is implemented, diverging when the original reference is used at reconstruction, converging when the phase conjugate reference wave is used.

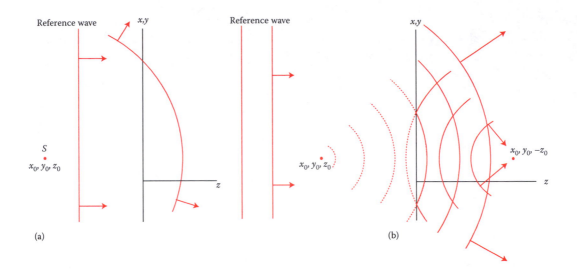

FIGURE 11.10 Holographic lens: (a) recording spherical wavefront and (b) holographic positive and negative lenses.

The resolution of a conventional lens, which is free of aberrations, is obtained from Equation 2.12 and is determined by the wavelength of the light used and the lens aperture. A holographic lens is fabricated using a point source and in reconstruction at the wavelength used in the recording, cannot form an image smaller than that source. For this reason, the pinhole used to form the source, S, in Figure 11.10 should be as small as possible. The aperture of the HOE should be as large as possible for the same reason as for a conventional lens, that is, in order to intercept as much of the wavefront as possible to minimize the effects of diffraction by its aperture.

If a holographic lens is required to form a proximate image of a proximate object, then point sources are used for object and reference beams at recording (Figure 11.11).

11.5 BEAM SPLITTERS AND BEAM COMBINERS

11.5.1 HEAD-UP DISPLAYS

High-speed, low-level flight requires an aircraft pilot to look at essential instruments and distant objects at the same time, without having to refocus or change viewpoint. Head-up displays in which instrument displays are presented to the pilot have been a feature of high-speed aircraft for some time but typically such displays are helmet mounted.

Motor vehicle drivers increasingly need the same facility because of the density of high-speed traffic as well as the growth in density of roadside information display, which must be rapidly evaluated and acted upon.

The principle of beam combining by holographic means is illustrated in Figure 11.12. The beam combiner built into or laminated to the windshield is designed to reflect and collimate the light from the monochromatically illuminated instrument panel. The observer

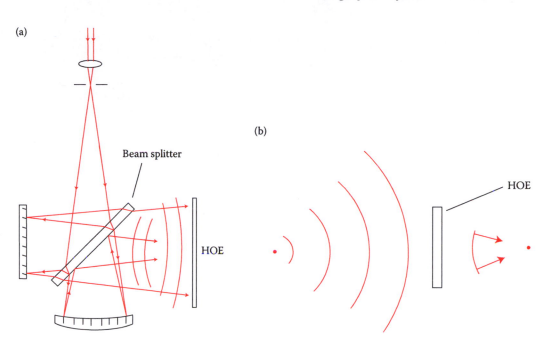

FIGURE 11.11 Holographic lens (proximate object and image): (a) recording and (b) reconstruction.

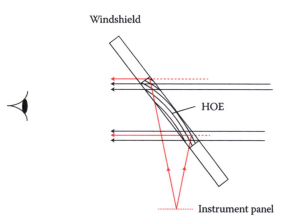

FIGURE 11.12 Beam combiner for head-up display.

then sees the instrument display and the scene outside the windshield without the need for head movement or refocusing.

11.5.2 BEAM SPLITTER AND COMBINER FOR AN ESPI SYSTEM

In Figure 10.16b, an out-of-plane sensitive speckle pattern interferometer using a speckled reference beam is shown. This arrangement is rather wasteful as 50% of the light reflected by the object surface and the fixed reference surface does not reach the camera but is reflected

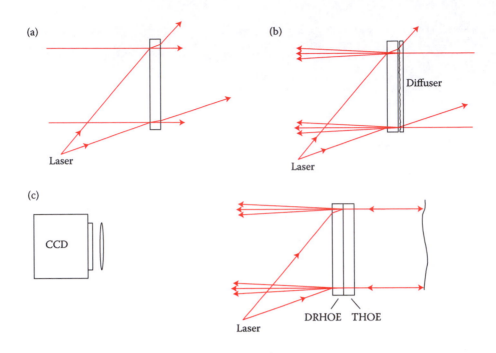

FIGURE 11.13 Beam splitter/combiner for out-of-plane sensitive ESPI system: (a) transmitting HOE; (b) diffusely reflecting HOE; and (c) ESPI system.

back toward the laser itself. Even if the loss were acceptable, laser light, which is ultimately directed back into the laser cavity, can cause instability and even damage the laser.

In Figure 11.13a, an off-axis converging lens (the transmitting HOE or THOE) provides collimated illumination for the object in an out-of-plane sensitive ESPI system. A speckled reference wave is recorded in a second, diffusely reflecting HOE (DRHOE) (Figure 11.13b).

The pair of holograms in use is shown in Figure 11.13c. An advantage of this system is that it is optically very simple with a minimum number of components. In fact the two HOEs could, in principle, be recorded in a single photosensitive layer. The system is very easy to align and is used with a laser diode, which can be current modulated to provide phase change between object and reference waves for fringe analysis purposes [10; Chapter 10].

An in-plane sensitive ESPI system is easily implemented using a holographic diffraction grating as a beam splitter (Figure 11.14).

11.5.3 HOLOGRAPHIC POLARIZING BEAM SPLITTERS

The discussion of the coupled wave theory in Chapter 4 assumes that the incident light is plane polarized perpendicular to the plane of incidence. Recall from Equations 1.5 and 1.6 when the conductivity σ is zero, that

$$\nabla(\nabla \cdot \mathbf{E}) - \nabla^2 \mathbf{E} = -\mu_0 \varepsilon_0 \varepsilon_r \frac{\partial^2 \mathbf{E}}{\partial t^2}$$

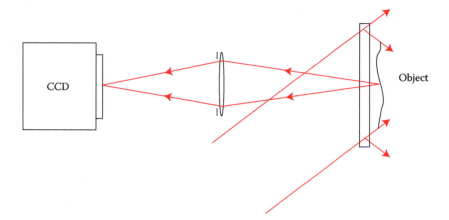

FIGURE 11.14 In-plane sensitive ESPI system using holographic diffraction grating.

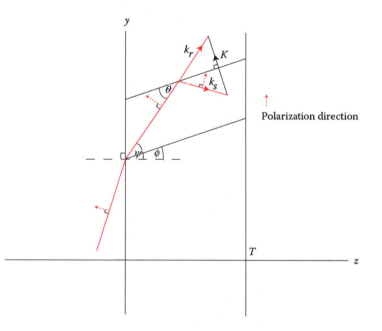

FIGURE 11.15 Polarization parallel to plane of incidence.

If the incident light is plane polarized *parallel* to the plane of incidence (Figure 11.15), then we have to modify the coupled wave equations since the term $\nabla(\nabla \cdot \mathbf{E})$ is no longer zero.

In their simplest form, neglecting second-order derivatives as before and assuming a phase only grating, Equations 4.17 and 4.18 become, respectively

$$\cos\psi \frac{\partial \mathbf{R}}{\partial z} = -j\chi \mathbf{S} \tag{11.3}$$

$$-2\frac{\partial \mathbf{S}}{\partial z} jk_{S_z} - k_S^2 \mathbf{S} + \beta^2 \mathbf{S} + 2\beta\chi \mathbf{R} = 0 \tag{11.4}$$

On-Bragg $\beta^2 - k_S^2 = 0$ and Equation 11.4 becomes

$$\frac{k_{S_z}}{\beta}\mathbf{S}' = -j\chi\mathbf{R} \qquad (11.5)$$

The mathematical steps required to modify the coupling constant, χ, in Equations 11.3 and 11.5 are rather involved and are outlined in the appendix to Kogelnik's original paper [1, Chapter 4], but one can see intuitively that, since $\mathbf{R}$ and $\mathbf{S}$ are no longer parallel, the coupling constant should be multiplied by the cosine of the angle, $2(\psi - \phi)$, between them and the diffraction efficiency modified by the same factor. A consequence is that a grating may be made polarization selective by appropriate choice of angles ψ and ϕ. A Bragg angle of 45° results in a diffraction efficiency of zero for an s-polarized beam. To achieve this requires careful design. Assuming an unslanted grating, the angles of refraction in the photosensitive layer need to be +45° and −45° with respect to the normal (Figure 11.16). To obtain such refractive angles even at grazing incidence (angle of incidence 90°), the recording material should have a refractive index of not more than 1.414.

This is not very practical because of the rather extreme angles involved and the restriction on refractive index. However, if input and output couplers are added in the form of holographic diffraction gratings, a practical polarizing beam splitter can be implemented (Figure 11.17). The couplers are slanted gratings and the orientation of the recorded fringes changes during recording if the material is subject to shrinkage (see Section 7.6.2), which must be taken into account in the design.

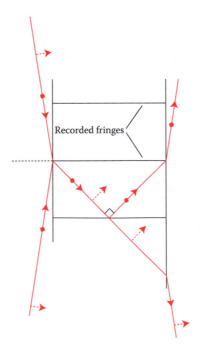

Recorded fringes

FIGURE 11.16 Plane polarizer.

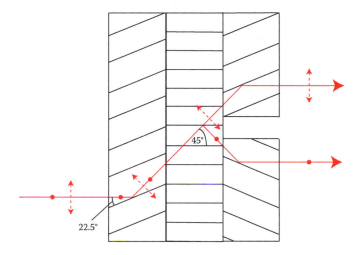

FIGURE 11.17 Polarizing beam splitter.

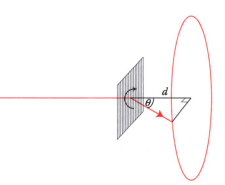

FIGURE 11.18 Basic scanner.

11.6 SCANNERS

Laser scanning systems are widely used at point of sale terminals in retail outlets. The concept is rather simple. An unexpanded laser beam is rapidly scanned over series of black and white bars, which provide a unique product code. The light reflected from the bar code is photoelectrically detected and the resulting data used to display the price to be paid by the purchaser and for stock inventory and reordering purposes. It is essential that the system reads the bar code accurately regardless of the direction in which the code is passed over the scanner. This requires that the scan be rapidly repeated in a number of different directions.

A basic scan can be implemented by spinning a diffraction grating around an unexpanded laser as shown in Figure 11.18. The diffracted beam describes a circle on a plane parallel to the grating plane, having a radius $d\tan\theta$, where d is the separation between the plane and θ the diffraction angle.

A focused beam produced by a holographic lens is preferable. Suppose that such a lens is recorded as a circular hologram on a disk, which can be rotated through an angle ϕ, in its own plane about an axis through its center normal to its plane (Figure 11.19). The focused spot moves along a circular path of length $\phi f \sin\theta$ and radius $f \sin\theta$, where f is the distance

(a)

(b)

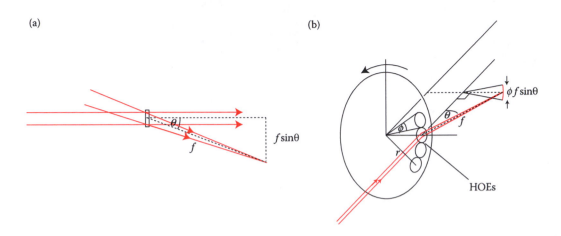

FIGURE 11.19 Scanning system: (a) recording HOE and (b) scanning. A single repeated scan line is produced if θ is the same for all HOEs. A multiple line raster is produced if θ changes by constant δ for each consecutive HOE. (Reproduced from D. H. McMahon, A. R. Franklin, and J. B. Thaxter, Light beam deflection using holographic scanning techniques, *Appl. Opt.*, 8, 399–402, 1969. With permission of Optical Society of America.)

between the HOE and the focused spot and θ the angle between the reference beam and the object beam in the recording stage [2]. If a set of identical, nonoverlapping HOES is disposed around the circumference of the disk, the scan arc is continuously repeated as the disk is rotated around its axis. To determine the bar code resolution of which the system may be capable, we need to find the number of resolvable spots in a scan line. Neglecting aberrations, the size of the focused spot is $2.44f\lambda/D$, so that the number of resolvable spots in a scan line is $D^2\sin\theta/2.44\lambda r$ where r is the distance of the HOE from the center of the disk.

The angle between the object and reference beams may be changed by a small amount, δ, for each new HOE, using the optical system of Figure 8.24 adapted to produce a focused beam (Figure 11.20). Then, as the recorded disk is rotated, a *raster* scan is produced, with a line spacing of $\delta f\cos\theta$.

The total number of resolvable spots in the raster is $\pi D\sin\theta/1.22\lambda$.

Another holographic scanning system consists of a set of holographic lenses disposed around the circumference of a cylinder, which rotates around its axis of symmetry [3]. The incident light source is located above the scanner on the axis (Figure 11.21).

11.7 LIGHTING CONTROL AND SOLAR CONCENTRATORS

HOEs offer significant advantages in lighting control and as solar concentrators. The availability of lightweight, large area elements, which can be incorporated in building windows, allows sunlight to be redirected to otherwise poorly illuminated areas of the building interior (Figure 11.22a). Long wavelength radiation can be excluded, using a reflective HOE, to prevent overheating in buildings (Figure 11.22b).

HOEs can be used as solar concentrators to improve the output from photovoltaic (PV) devices. Linear arrays of PVs can be employed to make optimal use of the radiation from a

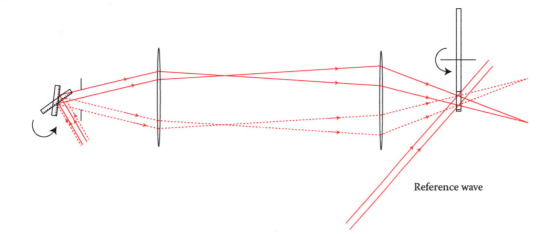

FIGURE 11.20 Recording a scanner.

(a)

(b)

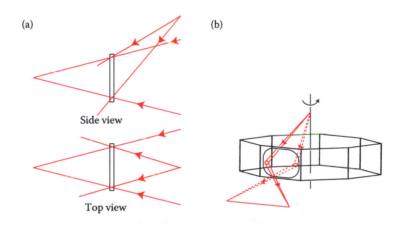

Side view

Top view

FIGURE 11.21 Straight line polygonal scanner: (a) HOE recording and (b) line scanning. (Reproduced from R. V. Pole, H. W. Werlich, and R. J. Krusche, Holographic light deflection, *Appl. Opt.*, 17, 3294–97, 1978. With permission of Optical Society of America.)

given direction, each PV having a response that is matched to the spectral passband of one of the HOEs. The HOEs are designed as converging lenses with reference and object waves *both* converging [4] (Figure 11.23a). The reference wave is designed so that each part of it approximates a monochromatic, solar wavefront at a particular time of day. This ensures that some of the sunlight, depending on the angular selectivity of the HOE and its spectral passband, is concentrated on the PV throughout the day without the need for expensive solar tracking systems. A study by Castro et al. [5] has shown that almost 50% of available solar energy can be collected by PV devices without tracking and that the optimum HOE thickness is 2 μm with refractive index modulation of 0.1 favoring dichromated gelatin as the recording material. Figure 11.23b represents the situation early in the day. Sunlight in the red part of the spectrum having the same direction of propagation as the extreme left part of the reference wavefront in Figure 11.23a is diffracted by the HOE, which was

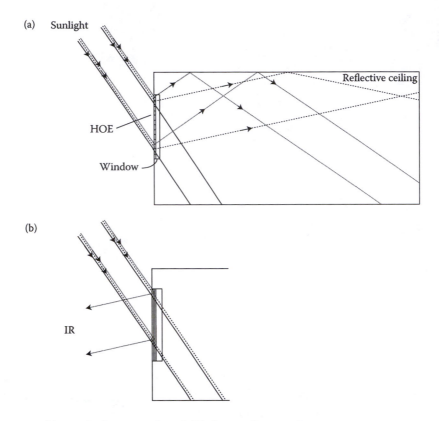

FIGURE 11.22 (a) Daylight control and (b) thermal control.

recorded in red, to the mainly red-sensitive PV. Similarly, sunlight in the remaining regions of the spectrum is diffracted toward PVs having appropriate ranges of spectral sensitivity. The system is, in effect, a holographic demultiplexer.

Photovoltaics are expensive, and an alternative is to use a series of HOEs with a single PV device (Figure 11.23c).

In designing a holographic solar concentrator, the seasonal variations in the sun's path across the sky as well as the daily path has to be taken into account. Seasonal changes in diffraction efficiency can be counteracted by using a series of HOEs but care is needed to avoid the benefit of one grating being negated by another due to crosstalk.

11.8 MULTIPLEXING AND DEMULTIPLEXING

Multiplexing and demultiplexing techniques are widely used in optical communication systems. Many closely spaced signal carrier wavelengths may be carried in a single optical fiber (multiplexing) and subsequently physically separated (demultiplexed) and individually routed by fiber Bragg gratings. Free space multiplexing and demultiplexing may also be implemented using a HOE between optical fibers as illustrated in Figure 11.24. The device also acts as a fiber coupler.

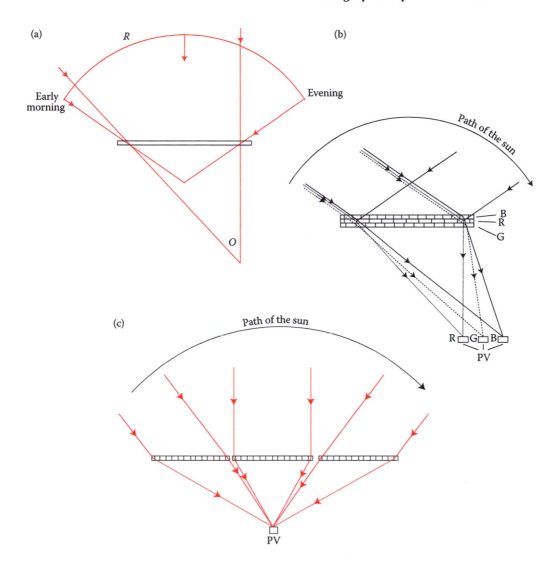

FIGURE 11.23 HOE solar concentrators: multiple HOEs and multiple photovoltaics (a) recording; (b) reconstruction; and (c) single photovoltaic scanning. (Reproduced from J. E. Ludman, Holographic solar concentrator, *Appl. Opt.*, 21, 3057–58, 1982. With permission of Optical Society of America.)

11.9 OPTICAL INTERCONNECTS

The need for optical interconnects arises because of the growing number of density of devices in integrated circuits (ICs) and the need to connect each IC with others. The surface area of an IC does not grow in proportion to the density of logic gates within the circuit, which exacerbates the connection problem.

Another reason is the increasing demand for high-speed routing of data in order to increase the rate at which calculations are made. The greater the speed at which data can be moved around a computing system, the greater the processing speed. There is also the

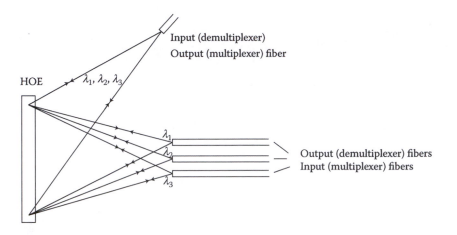

FIGURE 11.24 Holographic multiplexer/demultiplexer.

need for reliable distribution of the clock signals, which are required to ensure that logical operations are carried out synchronously and some minimum time is needed for their distribution throughout the system, inevitably setting an upper limit to processing speed. The potentially greater speed of clock distribution by optical means is therefore preferable.

11.9.1 HOLOGRAPHIC INTERCONNECTS

The use of HOEs to replace electrical connections has a number of benefits. The system can be made spatially compact since the HOEs are usually thin and flat and the use of lightweight materials helps reduce costs.

Some of the light beams may overlap while crossing the space between their origins and their destinations, but this has no adverse effect, through crosstalk, on the signals carried by the beams and in fact overlap facilitates reduction of the space required to accommodate the connections.

Photorefractive materials enable the HOEs to be reconfigured so that a complete system may be rapidly adapted to deal with different interconnection tasks.

11.9.1.1 Fanout Devices

A HOE comprising a 2-D array of small holographic lenses can convert a single light beam to a number of beams, each focused onto an individual photodetector, a process known as fan out (Figure 11.25a). The device may be used for data or clock distribution purposes. Multiple beams may also be directed by fan in to a single detector (Figure 11.25b).

The HOE requires careful design to ensure that the focused spots are smaller than the detector area. The laser diode output beam has an elliptical cross section and is usually necessary to circularize the beam using an anamorphic lens, that is, a lens whose focal length depends on the meridian, defined as the intersection of a plane containing the lens axis and a diameter of the lens. Because of the large divergence of the laser beam, it is usually easier to collimate it to obtain one with an elliptical cross section and then use a pair of prisms or a grating to circularize the beam (see problem 15, Chapter 1). The laser output wavelength changes with temperature,

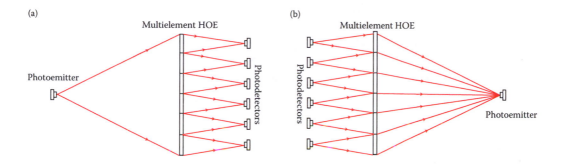

FIGURE 11.25 Holographic optical interconnects (a) fan out and (b) fan in.

which therefore must be precisely controlled using a Peltier device. Aging and accompanying wavelength change may also affect HOE performance. A further consideration is the zero-order light, which remains after both the circularizing HOE and the fanout HOE and gives rise to background noise in the photodetectors, as does the light scattered by the HOEs themselves.

11.9.1.2 Space Variant Interconnects

In addition to fanout HOES, which enable the signal from one source to be sent to a number of optical receivers, HOEs are also required enabling each one of a set of sources to connect to a particular member of a set of receivers according to some mathematical operation, for example, the transpose of a matrix so that

$$\begin{pmatrix} a_{11} & a_{12} & a_{13} & a_{14} \\ a_{21} & a_{22} & a_{23} & a_{24} \\ a_{31} & a_{32} & a_{33} & a_{34} \\ a_{41} & a_{42} & a_{43} & a_{44} \end{pmatrix} \quad \text{becomes} \quad \begin{pmatrix} a_{11} & a_{21} & a_{31} & a_{41} \\ a_{12} & a_{22} & a_{32} & a_{42} \\ a_{13} & a_{23} & a_{33} & a_{43} \\ a_{141} & a_{24} & a_{34} & a_{44} \end{pmatrix}.$$

Thus, the light output from a source with output intensity proportional to a_{23} at location 2,3 in a 4×4 array of sources is directed to location 3,2 in a corresponding 4×4 photodetector array.

Another application of space variant interconnection is a perfect shuffle and its inverse. A simple example of a 1-D perfect shuffle is the conversion of the number array

$$n_1 \quad n_2 \quad n_3 \quad n_4 \quad n_5 \quad n_6 \qquad \text{to} \qquad n_1 \quad n_4 \quad n_2 \quad n_5 \quad n_3 \quad n_6$$

A perfect shuffle operation is shown in Figure 11.26.

Each of the arrays of "1s," "2s," "3s," and "4s" is first optically fanned out to four times its original physical size, and the arrays are then optically fanned in, in the directions indicated by the arrows, to complete the shuffle.

Figure 11.27a shows a 1-D perfect shuffle implemented by a multielement HOE basically consisting of a line array of lenses. The version shown in Figure 11.27b with two HOEs, is preferred, as the output beam axes are all parallel, making alignment easier and helping avoid focusing errors.

More compact systems can be implemented using a combination of reflection and transmission HOEs (Figure 11.28a) or reflection HOEs recorded in a planar waveguide (Figure 11.28b) in which total internal reflection is also exploited.

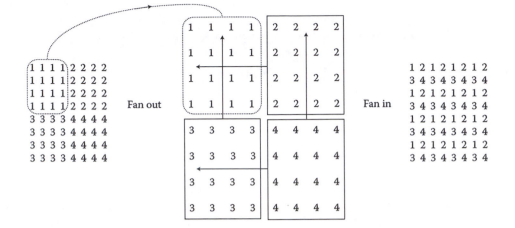

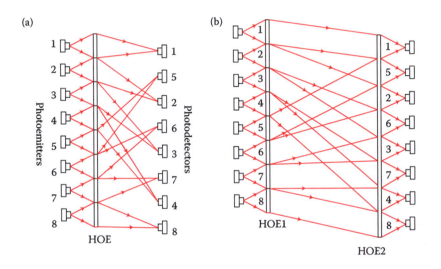

FIGURE 11.26 2-D perfect shuffle.

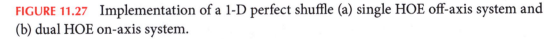

FIGURE 11.27 Implementation of a 1-D perfect shuffle (a) single HOE off-axis system and (b) dual HOE on-axis system.

11.10 HOLOGRAPHIC PROJECTION SCREENS

Rear projection display screens should direct as much image light as possible toward the viewer. This is particularly true for small portable projectors with limited light output. Holographic rear projection screens have proved very successful in meeting the requirement for efficiently directed light. Such a screen is essentially a hologram of a diffuser, recorded in dichromated gelatin, using an off-axis reference beam as shown in Figure 11.29. To avoid waste of light, the diffuser is designed to scatter light preferentially toward the middle of the audience. The projected image is focused on the holographic screen with the projector beam in the same direction as the reference beam at recording.

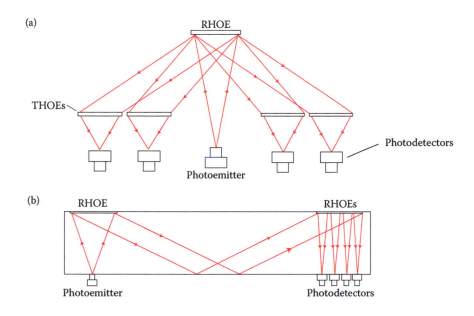

(a)

RHOE

THOEs

Photoemitter

Photodetectors

(b)

RHOE

RHOEs

Photoemitter

Photodetectors

FIGURE 11.28 Compact fan out (a) using reflection and transmission HOEs and (b) using planar waveguide and RHOEs.

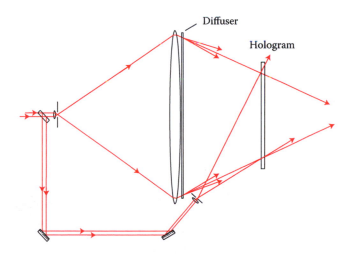

Diffuser

Hologram

FIGURE 11.29 Holographic screen (recording).

The major problem screen, of course, is that color casts appear in the projected images when the screen is used with a white light projector. A method of dealing with this problem involves the use of a reflective diffuser in the form of a long, narrow rectangle (Figure 11.30) as the object. The recorded hologram then consists of a large number of holographic lenses, each having the reference beam point source and a point on the diffuser as conjugate points. The interference fringes have the form of concentric circles centered on the intersection of the hologram plane and the major axis of the diffuser. The screen is illuminated from the side opposite to the reference source. The convergent light waves from the projector focused

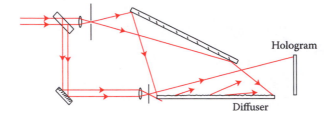

FIGURE 11.30 Holographic screen recorded with slit diffuser. (Reproduced from V. I. Bobrinev, J.-Y. Son, S.-S. Kim, and S.-K. Kim, Analysis of color distortions in a transmission-type holographic screen, *Appl. Opt.*, 44, 2943–48, 2005. With permission of Optical Society of America.)

on the screen, approximate to conjugate reference waves so that real images of the diffuser are produced, in each color, at different distances from the screen, on the opposite side. The diffuser length and position are chosen at the recording stage so that, in white light reconstruction, the distant end, with respect to the screen, of the red image, overlaps to a sufficient extent the proximate end of the blue image to ensure a satisfactory image area free of color casts. Bobrinev et al. [6] have shown that optimum performance is obtained when the screen is recorded at short wavelength using a long diffuser located close to the recording plane.

11.11 PHOTONIC BANDGAP DEVICES

We have seen that two plane coherent waves can interfere to produce a 1-D periodic variation in refractive index in, say, a photopolymer. If three plane waves with noncoplanar propagation vectors are allowed to interfere, the pattern produced is periodic in two directions. Four noncoplanar interfering plane waves produce a spatial modulation of refractive index, which is periodic in three dimensions. The interference of three or four noncoplanar beams to produce these 2-D and 3-D structures is called holographic lithography [7]. All five of the 2-D and the 14 3-D Bravais lattice structures [8] can be optically fabricated using, respectively, three and four noncoplanar beams. Such structures allow propagation of light in certain wavelength ranges, or bands, and prevent it in others, and are known as *photonic bandgap* devices. A bandgap device can be used to inhibit spontaneous emission thus enhancing the efficiency of lasers and solar cells [9]. Although antireflection coatings (Section 1.7.2.3) multilayer dielectric filters and holographic volume reflection gratings may be regarded as 1-D photonic bandgap devices, the term photonic bandgap strictly only applies to 2- and 3-D periodic structures with high refractive index modulation. The latter requirement can be met by using the lattice structure as a template. Thus, a 2-D lattice may be copied into a much higher refractive index substrate by etching. Other methods of increasing the index modulation involve either replication by atom deposition in high refractive index materials such as titanium dioxide (TiO_2) [10], or the addition of high refractive index materials such as TiO_2 particles or liquid crystals to photopolymerizable materials, as dopants. The holographic recording process results in phase separation of these materials from the photopolymerized material and enhanced refractive index modulation, the dopant being expelled from bright fringe into dark fringe regions. Randomly oriented liquid crystal dopant molecules may be aligned parallel to an externally applied electric field (see also Section 8.8.2.1) altering the refractive index and index modulation so that bandgap tuning may be implemented. Alternatively, the liquid crystal may be removed altogether using an organic solvent to produce passive optical devices with index modulation of about 0.6, the refractive index difference between air and SU-8 photoresist, the most commonly used photopolymer for holographic lithography. This negative photoresist can be coated in layers up to 200 μm thick and has a resolution of more than 10,000 lines per millimeter. It spectral sensitivity is optimal at about 325 nm.

2-D periodic structures can be created by holographic lithography using just two beams if the photosensitive layer is rotated in its own plane, which is normal to the plane of the interfering beams, between recordings (Figure 11.31a), and the recorded grating vectors are coplanar. Figure 11.31b shows the square lattice pattern obtained when the angle of rotation between two recordings is 90° with the same angle between the beams,

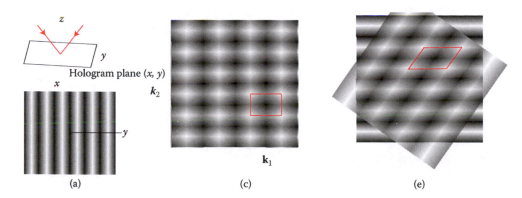

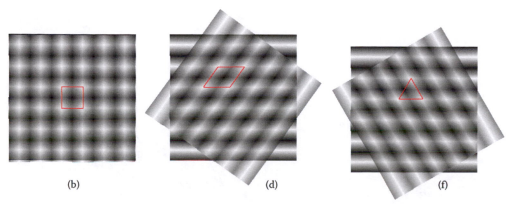

FIGURE 11.31 2-D periodic patterns: (a) single grating recording; (b) 90° rotation; $k_1 = k_2$ square; (c) 90° rotation; $k_1 \neq k_2$ rectangular; (d) arbitrary rotation; $k_1 = k_2$ rhombic; (e) arbitrary rotation $k_1 \neq k_2$ oblique; and (f) 120° rotation; $k_1 = k_2$ triangular.

that is, the same grating vector magnitude. In Figure 11.31c, the grating vector magnitude is changed between recordings to form a rectangular lattice. When the rotational angle between recordings is arbitrary and the grating vector magnitude is again the same in both recordings (Figure 11.31d), a rhombic lattice is obtained. If the grating vector magnitude is altered, an oblique lattice is the result (Figure 11.31e) and finally a rotation of 120° with fixed grating vector magnitude results in a triangular lattice (Figure 11.31f).

3-D lattice structures can be obtained by combining in-plane and out-of-plane rotations [11]. In the two beam method, plane parallel polarizations of the interfering beams can be maintained throughout the fabrication of the lattice, ensuring maximum fringe contrast and therefore refractive index modulation and lattice constants can be kept approximately the same, in all three dimensions, if required.

The range of device applications of photonic bandgap devices can be greatly expanded by introducing local defects in the structure. This can be done by two-photon photopolymerization [12] at the focus of a laser, whose photon energy is half that required to excite the dye and initiate photopolymerization. An example is shown in Figure 11.32, which shows a 2-D photonic bandgap device. A line excluded from the array by masking during recording acts as a waveguide in the plane of the array, which is sandwiched between lower refractive index substrates, so that propagation out of the device plane is prevented by total internal reflection. Local defects with the correct dimensions and distance from the main propagation channel provide coupling out of the device plane at chosen wavelengths. The defects act as optical traps for these wavelengths in a manner analogous to the trapping of charge carriers at defects in crystals.

The combination of holographic lithography and two-photon photopolymerization holds considerable promise for the future development of microphotonic engineering. Higher refractive index photopolymerizable materials would enable templating to be eliminated while problems associated with lattice boundary matching between different devices, for large-scale integration, also need to be addressed.

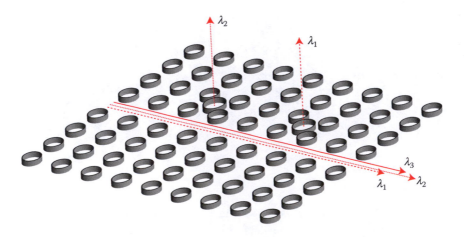

FIGURE 11.32 Local defects in a 2-D lattice providing demultiplexing.

11.12 HOLOGRAPHIC POLYMER DISPERSED LIQUID CRYSTAL DEVICES

A range of switchable optical devices can be fabricated from mixtures of liquid crystalline and photopolymerizable materials, known as polymer dispersed liquid crystals (PDLC).

An example is an electrooptical switch, which is shown in Figure 11.33. A thin layer of PDLC is contained between glass plates coated on their inner surfaces with transparent indium tin oxide electrodes. The device is made by illuminating it with an interference pattern as shown in Figure 11.33a causing phase separation of the liquid crystals from the polymerized regions, leaving the liquid crystal molecules randomly oriented in the dark fringe regions. If the average refractive index of the randomly oriented liquid crystals matches that of the photopolymerized regions, then the refractive index modulation is zero and the diffraction efficiency of the holographically recorded diffraction grating is zero. When an electric field is applied, the LC molecules become aligned, presenting an altered refractive index to incident light, which is then diffracted with high efficiency.

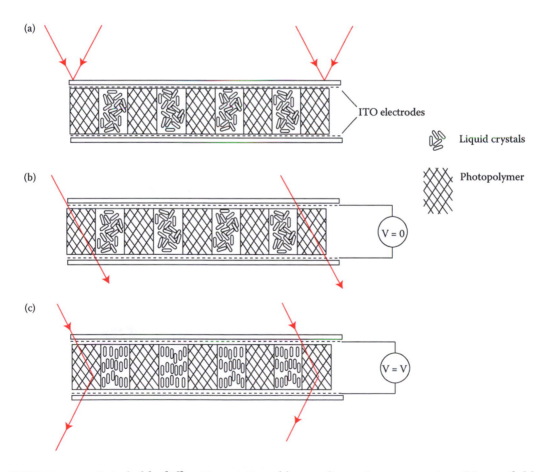

FIGURE 11.33 Switchable diffraction grating: (a) recording, phase separation; (b) zero field, zero index modulation; and (c) field applied (ON state).

Alternatively, the refractive indices of the polymer and LC regions may differ but become matched when the field is applied.

11.13 SUMMARY

This chapter surveys a range of optical devices that can be fabricated using holographic methods, beginning with diffraction gratings for spectroscopic applications and continuing with various spectral filters including notch filters. We discuss lenses, holographic beam combiners, and beam splitters for applications in display and speckle interferometry. We outline the basic principles of holographic polarizing devices. Other devices considered are scanners, holographic projection screens, solar concentrators, and daylight and thermal control devices for building interiors. Holographic optical interconnecting devices are included here and finally we briefly discuss holographic lithography for the fabrication of photonic bandgap devices.

REFERENCES

1. N. K. Sheridon, "Production of blazed holograms," *Appl. Phys. Lett.*, 12, 316–8 (1968).
2. D. H. McMahon, A. R. Franklin, and J. B. Thaxter, "Light beam deflection using holographic scanning techniques," *Appl. Opt.*, 8, 399–402 (1969).
3. R. V. Pole, H. W. Werlich, and R. J. Krusche, "Holographic light deflection," *Appl. Opt.*, 17, 3294–7 (1978).
4. J. E. Ludman, "Holographic solar concentrator," *Appl. Opt.*, 21, 3057–8 (1982).
5. J. M. Castro, Z. Deming, B. Myer, and R. K. Kostuk, "Energy collection efficiency of holographic planar solar concentrators," *Appl. Opt.*, 49, 858–70 (2010).
6. V. I. Bobrinev, J-Y. Son, S-S. Kim, and S-K. Kim, "Analysis of color distortions in a transmission-type holographic screen," *Appl. Opt.*, 44, 2943–8 (2005).
7. M. Campbell, D. N. Sharp, M. T. Harrison, R. G. Denning, and A. J. Turberfield, "Fabrication of photonic crystals for the visible spectrum by holographic lithography," *Nature*, 404, 53–6 (2000).
8. M. J. Escuti and G. P. Crawford, "Holographic photonic crystals," *Opt. Eng.*, 43, 1973–87 (2004).
9. E. Yablonovitch, "Inhibited spontaneous emission in solid-state physics and electronics," *Phys. Rev. Lett.*, 58, 2059–62 (1983).
10. J. S. King, E. Graugnard, O. M. Roche, D. N. Sharp, J. Scrimgeour, R. G. Denning, A. J. Turberfield, and C. J. Summers, "Infiltration and inversion of holographically defined polymer photonic crystal templates by atomic layer deposition," *Adv. Mater.*, 18, 1561–5 (2006).
11. N. D. Lai, W. P. Liang, J. H. Lin, C. C. Hsu, and C. H. Lin, "Fabrication of two- and three-dimensional periodic structures by multi-exposure of two beam interference technique," *Opt. Express*, 13, 9605–11 (2005).
12. J. H. Strickler and W. W. Webb, "Two-photon excitation in laser scanning fluorescence microscopy," *Proc. SPIE.*, CAN-AM Eastern '90, Antos R. L., Krisiloff A. J. (eds.), 1398, 107–118 (1990).

PROBLEMS

1. Two light beams of wavelength λ are used to record a slanted diffraction grating in a photopolymer whose average refractive index is n. One of the beams is incident along the normal to the recording layer. Find an expression for the angle of incidence of the second

beam that ensures that all light of wavelength greater than λ' normally incident on the grating is blocked. How exactly is it blocked?

2. Find an expression for the spatial frequency of an unslanted transmission diffraction grating of thickness T recorded in a photopolymer of refractive index n, at wavelength λ. The grating is required to resolve spectral lines, which are $\Delta\lambda$ apart.

   It was seen in Chapter 2 that diffraction gratings can have resolution powers as high as 10^5, which means that, in the visible spectrum around 500 nm, the wavelength selectivity of such a grating is about 5 pm. By making reasonable assumptions, show that a volume transmission holographic grating inevitably has a much lower resolving power.

3. We saw how a notch filter can be fabricated in the form of a reflection holographic grating. Consider the opposite, that is, a narrowband transmission filter, and suggest ways in which this could be implemented holographically. Could it be done by the method of problem 1?

4. Find an expression for the minimum angle of rotation of a disk in its own plane, which could be detected using a parallel thick diffraction grating of spacing d, whose refractive index varies sinusoidally and which is illuminated at angle θ to its normal.

5. Figure 11.17 shows a polarizing beam splitter consisting of four gratings. One of them is not strictly necessary and in fact one could dispense with a second grating. Explain why.

6. A holographic laser scanner consists of a rotating disk with a set of holographic diffraction lenses disposed around the circumference of a circle concentric with the disk. The system has to cope with a 1-D product code 1 cm in width. If the average angle between recording beams is 30° and the focal length of the holographic lenses is 100 cm, how many raster lines can the scanner produce and what is the maximum bar code length that it can handle? What should be the minimum separation between elements of the product code?

CHAPTER 12

Holographic Data Storage and Information Processing

12.1 INTRODUCTION

Requirements for information storage and processing capacity are growing inexorably, fueled by the demands of advanced mathematical modeling techniques in the sciences and engineering, climatology, bioinformatics, economics, the security industries, as well as entertainment and consumer demand.

Developments in data storage using optical discs have resulted in massive capacity with 25 Gbytes on a single disc now commonplace and larger capacities possible using multiple layers and smaller wavelengths. Ultimately, however, the surface area required to store a single bit of data on an optical disc cannot be reduced below a limit set by the wavelength of the light and the numerical aperture used in recording. Blue lasers and very high numerical apertures have been introduced in recent years, leading to further reduction in focused spot size. Near-field optical methods can reduce the limit further but require subwavelength separation between the writing laser and the surface of the recording medium, and this presents significant technical challenges.

Holography promises to deliver much greater storage capacity by utilizing the full volume of the recording material to store a large number of holograms in the same space. Combining volume storage with associative recall will facilitate versatile retrieval capability.

As we will see also in this chapter, a number of information processing operations can be implemented using optical methods including holography. In some instances, the data can be processed directly as it is retrieved from holographic memory. The combination of such operations with page format holographic data storage (HDS) and retrieval has great potential for parallel processing on a very large scale, with spatial light modulators (SLMs) with up to a million pixels to write data and megapixel CCD or CMOS cameras for data readout.

12.2 HOLOGRAPHIC DATA STORAGE CAPACITY

We begin by comparing the data storage capacities of a layer of photosensitive material when the data is recorded at the surface only and when it is holographically recorded. The data could be recorded in the form of holographic diffraction gratings, each grating representing a single data bit. In Figure 12.1, two collimated beams of width W and wavelength λ directed at 90° to one another form an interference fringe pattern, which is recorded in a photosensitive layer as a holographic grating representing a single data bit. The thickness of the grating is $\sqrt{2}W$, its spacing, d, is $\lambda/\sqrt{2}$, and it occupies a volume W^3. Suppose we record additional gratings by rotating each of the beams in the plane of the diagram, though an angle equal to the angular selectivity, $\lambda/(2W)$, of the grating (see Equation 4.35), a procedure is known as *angular multiplexing*, the total rotational angle being, say, $\pi/2$. In this way, we could record $\pi W/\lambda$ gratings (Figure 12.2).

We could also rotate the photosensitive layer in its own plane, through the same angular selectivity angle (Figure 12.2c), at each of the angularly multiplexed settings, to record $\pi W/\lambda$ gratings, making a total of $(\pi W/\lambda)^2$. (It is assumed that the photosensitive material is capable of recording all these gratings.) The volume density of bits is $(\pi/\lambda)^2/W$. If we

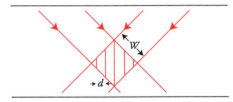

FIGURE 12.1 Data bit represented by a single holographic grating.

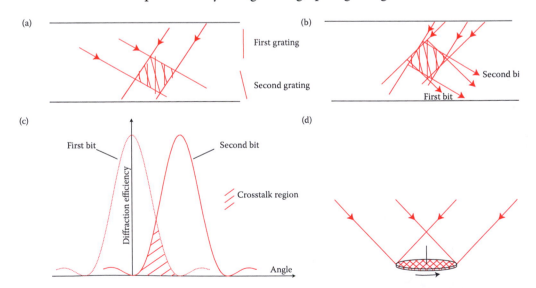

FIGURE 12.2 Angular multiplexing. (a) The recording beams are rotated with respect to the hologram plane (or vice versa) between recordings; (b) data readout; (c) crosstalk ; and (d) angular multiplexing by in-plane rotation of photosensitive layer.

assume a diffraction limited beam width of λ, the density is about 10 bits per cubic wavelength or about 80 Gbits per cubic mm. For comparison, the areal density of bits recorded by conventional optical means, assuming a single data bit to be represented by a diffraction limited spot of radius $1.22\lambda f/D$, produced by a lens of focal length, f, and diameter, D, is about 1 Mbit per mm^2.

12.3 BIT FORMAT AND PAGE FORMAT

In the preceding discussion, it was assumed that data is recorded in the form of bits, each represented by a single holographic diffraction grating record of the interference of two plane waves, one of which is a reference beam and the other a data-bearing object beam. Holographic bit format data storage systems have been the subject of much research effort. One such system [1] uses counterpropagating mutually coherent beams, having a common focus within the holographic storage medium, to record a reflection grating occupying an extremely small volume determined essentially by the resolution limit of the lenses used (Equation 2.12). Such gratings can be recorded throughout the volume of the recording medium by moving the common focus of the interfering beams, providing for massive capacity. The data retrieval process also has potential for compatibility with present-day optical disk technology.

However, it is perfectly possible for the object beam to carry a great deal more data if it is expanded and transmitted through a SLM of the type shown in Figure 8.19. Each pixel in the SLM is normally opaque to light, due to the crossed linear polarizers on either side of the device but can be turned on so that it transmits light, by applying an electrical voltage to it. Each pixel that transmits light represents a binary "1" and each nontransmitting pixel represents binary "0" so that a 2-D page of data can be written into the SLM. In Figure 12.3a, the SLM is imaged in the holographic medium, which is also illuminated by the

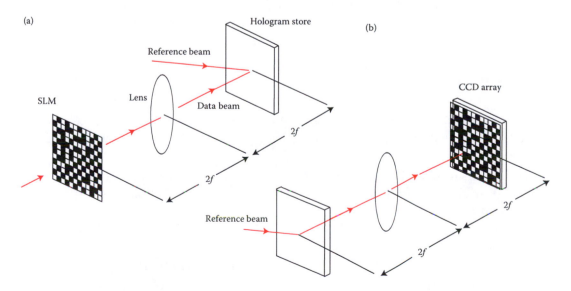

FIGURE 12.3 Page format data storage: (a) recording and (b) data readout.

reference beam, and a hologram is recorded. A new data page can be written into the SLM and recorded by changing the angle of incidence of the reference beam. At data readout (Figure 12.3b), each page is reconstructed using the appropriate reference beam angle, to be imaged onto a CCD array.

12.4 STORAGE MEDIA

Ferroelectric crystals such as iron-doped lithium niobate can be used as rewritable storage media. Fixing is required to retain the data that is otherwise erased upon readout (see Section 6.6.2). The advantage is that the ferroelectric crystals can be prepared with very high optical quality.

Photopolymers (Section 6.6.4) are cheap to make, but thickness is restricted to about 1 mm because of scatter and the difficulty in ensuring high optical surface quality. They exhibit greater scatter than photorefractives. Another significant problem is the inevitable shrinkage that occurs on polymerization, which results in a change in the Bragg angle. The use of high-molecular weight monomers helps to alleviate the problem, and shrinkage has been reduced to 0.02% using ring-shaped monomers, which open on polymerization [2].

12.5 MULTIPLEXING

There are a number of multiplexing methods available for use in HDS systems.

12.5.1 ANGULAR MULTIPLEXING

As already indicated in Section 12.2, angular multiplexing allows one to store a series of data bits or pages in the same volume, by altering the angle between each recording beam and the normal to the surface of the holographic memory, between recordings.

This is done by rotating the recording layer out of its own plane or, as shown in Figure 12.2a, by rotating both of the recording beams through the same angle. Either method ensures the same angular selectivity for the recorded gratings since the average fringe spacing is the same for all of them. To reconstruct, or retrieve, a specific data bit or page, the layer is rotated to the correct orientation in the path of the reference, or readout, beam (Figure 12.2b) or the readout beam rotated so that it is incident at the correct Bragg angle. The recorded gratings are slanted in this case, and it is essential that shrinkage be minimized.

As stated in Section 12.2, the rotational angle between recordings is equal to the angular selectivity so that the maximum diffraction efficiency of each grating occurs at the first zero of that of its nearest neighbor. In fact, a greater angular rotation is required because there is some overlap between the efficiency curves (Figure 12.2c). Also, the Bragg curve has less than the ideal $\sin^2 \theta$ form because of absorption and scatter so that the efficiency does not quite reach zero anywhere. In the case of page format recording, the rotational angle has to be greater still because of the finite range of spatial frequencies involved. Hence, there is always some crosstalk, that is, reconstructed data is contaminated by data from neighboring holograms albeit with lower intensity. For these reasons, angular multiplexing is usually implemented with rotations that position the Bragg peak of a reconstructed hologram at the second or even the third zero of the Bragg curve of its nearest neighbor.

Angular multiplexing may also include rotation of the recording layer about a normal as shown in Figure 12.2d. Note that the axis of rotation passes through the center of the overlap region of the beams, providing an element of compatibility with existing optical disk technology.

12.5.2 PERISTROPHIC MULTIPLEXING

In peristrophic multiplexing [3], the photosensitive layer is simply rotated in its own plane between recordings. Here, however, the overlap region of the recording beams may be displaced from the axis of rotation. Peristrophic multiplexing also offers the prospect of some backward compatibility with existing optical storage disk technology. The principle is shown in Figure 12.4. A SLM in which data is written is located in the front focal plane of a lens of focal length, f, and illuminated normally by a collimated beam. A hologram of the spatial Fourier transform of the SLM is recorded at the hologram plane, x', y', using a plane reference wave of unit amplitude incident at angle $-\theta_R$ and with complex amplitude $\tilde{R} = \exp(-jky'\sin\theta_R)$. We consider light from a single point of the SLM at $(0, -y_O)$, which

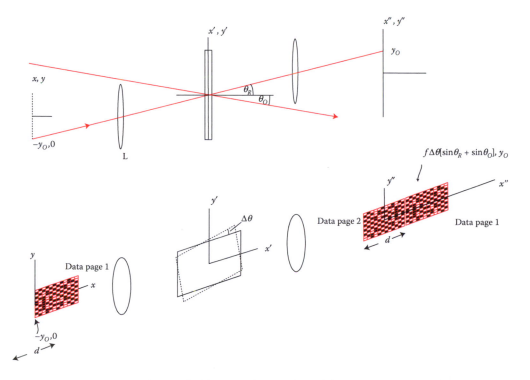

FIGURE 12.4 Peristrophic multiplexing. (Reproduced from K. Curtis, A. Pu, and D. Psaltis, Method for holographic storage using peristrophic multiplexing, *Opt. Lett.*, 19, 993–94, 1994. With permission of Optical Society of America.)

produces a plane wave of unit amplitude incident at angle θ_O to the z-axis in the hologram plane and complex amplitude $\tilde{O} = \exp(jky'\sin\theta_O)$. We only consider the recorded term

$$\tilde{R}*O = \exp(+jky'\sin\theta_R)\exp(jky'\sin\theta_O) \tag{12.1}$$

If the recording layer is rotated by a small angle $\Delta\theta$ around the z-axis, to allow another hologram to be recorded, Equation 12.1 becomes

$$\tilde{R}*O = \exp\left[\left(+jk([y'+x'\Delta\theta])\sin\theta_R\right)\right]\exp\left[jk(y'+x'\Delta\theta)\sin\theta_O\right]$$
$$= \exp(+jky'\sin\theta_R)\exp(jky'\sin\theta_O)\exp\left[+jkx'\Delta\theta(\sin\theta_R+\sin\theta_O)\right] \tag{12.2}$$

which, on reconstruction by illumination with $\tilde{R}$, becomes

$$\tilde{R}\tilde{R}*O = \exp(jky'\sin\theta_O)\exp[-jkx'\Delta\theta(\sin\theta_R+\sin\theta_O)] \tag{12.3}$$

When this is Fourier transformed by a second lens, also of focal length f, using Equation 2.22, we obtain an expression for the complex amplitude in the image plane

$$\tilde{A}(x'',y'') = \int_{x'}\int_{y'}\exp(jky'\sin\theta_O)\exp\left[jkx'\Delta\theta(\sin\theta_R+\sin\theta_O)\right]\exp\left[-\frac{jk}{f}(x''x'+y''y')\right]dx''dy''$$
$$= \int_{x'}\int_{y'}\exp(jky'[\sin\theta_O-y''/f])\exp\left(-jkx'\{\Delta\theta[\sin\theta_R+\sin\theta_O]+x''/f\}\right)dx''dy'' \tag{12.4}$$

and the image of the original object point appears at

$$x'' = f\Delta\theta(\sin\theta_R+\sin\theta_O), \quad y'' = f\sin\theta_O = y_O \tag{12.5}$$

If the reconstructed data pages are not to overlap at the detector, then x'' has to be at least the width of the detector, d, implying a rotation of

$$\Delta\theta \geq \frac{d}{f(\sin\theta_R+\sin\theta_O)} \tag{12.6}$$

between consecutive recordings. Inserting values of d = 10 mm, $\theta_R = \theta_o$ = 30°, and f = 200 mm, we obtain $\Delta\theta$ = 0.05 radian, and 60 holograms could be recorded in a rotational range of 180°. The angular rotation required between recordings is much larger than the angular selectivity which, from Equation 4.35, at a wavelength of 532 nm and thickness of 100 μm, is about 0.005 radian.

12.5.3 POLYTOPIC MULTIPLEXING

A difficulty with angular multiplexing is simply geometrical. Although storage capacity increases with the layer thickness, the surface area required so that the recording beams can access the whole thickness of the layer in angular multiplexing also increases. Polytopic multiplexing [4] has been developed to overcome this problem. As shown in Figure 12.5a,

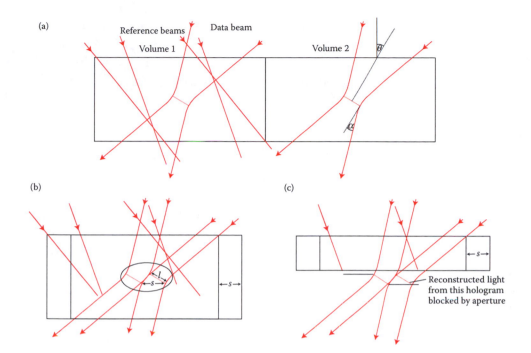

FIGURE 12.5 Principle of polytopic multiplexing: (a) angular multiplexing of Fourier holograms; (b) close packing; (c) aperturing to separate pages. (Reproduced from K. Anderson and K. Curtis, Polytopic multiplexing, *Opt. Lett.*, 29, 1402–4, 2004. With permission of Optical Society of America.)

data in the form of a 2-D array is imposed on an object beam by a SLM (not shown), and the beam is focused in a photosensitive layer. The complex amplitude distribution of light in the hologram plane is the spatial Fourier transform of the data page. By interfering this light with a reference beam, a Fourier hologram is recorded. The angle of incidence of the reference wave is altered for the recording of each Fourier hologram, until a complete *volume* of data is recorded by angular multiplexing. Each volume is assigned a separate space or block in the recording layer, and when a new volume is to be recorded, the whole layer is shifted by one block in its own plane. As already stated, the range of angles of the reference beams used to record all the pages in a block determines the physical size of the block, even though the physical space occupied by the recorded volume may be comparatively small. However, as shown in Figure 12.5b, the blocks can be packed together so that the Fourier transforms are almost contiguous. This is achieved by shifting the layer after recording each volume by a distance s given by

$$s = l \sin\theta + l \tan (\alpha + \theta)$$

where l is the width of the Fourier plane, θ the angle between the object beam axis and the normal to the hologram surface, and α is half the angle subtended by the Fourier transform lens at its focus. At readout of the data, a reference beam illuminates holograms from neighboring volumes at the same time. To solve this problem, the position of the Fourier transforming lens is adjusted so that the Fourier plane is below the photosensitive layer, where an aperture is located so as to isolate the Fourier transform of a single reconstructed data page. Alternatively, if the Fourier plane is inside the layer, a lens is used to image an aperture at the correct location. Tenfold increase in storage capacity is possible by making more efficient use of thick photosensitive layers through polytopic multiplexing.

12.5.4 SHIFT MULTIPLEXING

Shift multiplexing involves a spatial shift of the holographic recording layer to a new position between recordings. The method requires the use of a reference beam consisting of a line array of point sources equally spaced by distance, a, along the y-axis in the front focal plane of a lens [5]. This means that the holographic memory, located in the back focal plane, is illuminated by a set of collimated beams (Figure 12.6). The SLM in the front focal plane of a second lens whose axis at angle θ to that of the first is illuminated normally by a collimated beam and the Fourier transform of the data page in the SLM is recorded as a hologram.

The reference beam is given by Equation 2.22

$$\tilde{R} \propto \sum_{n=1}^{N} \int_{y} \partial(y - na) \exp\left(-\frac{jkyy'}{f}\right) dy = \sum_{n=1}^{N} \exp\left(-\frac{jkna}{f}\right) \tag{12.7}$$

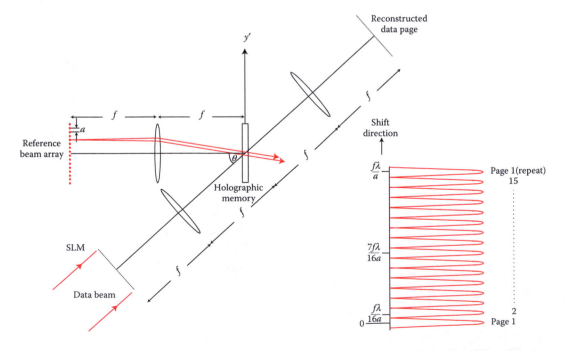

FIGURE 12.6 Shift multiplexing (a) holographic recording system (b) with 16 sources, 15 data pages can be recorded within a shift range of $f\lambda/a$. (Reproduced from G. Barbastathis, G. Levene, and D. Psaltis, Shift multiplexing with spherical reference waves, *Appl. Opt.*, 35, 2403–17, 1996. With permission of Optical Society of America.)

where N is the number of sources in the array and $\partial(y - na)$ is a delta function denoting a point at $y = na$. Equation 12.7 represents a set of plane waves, making angles of

$$\alpha = \frac{a}{f}, \frac{2a}{f}, \frac{3a}{f}, \ldots, \frac{Na}{f} \tag{12.8}$$

with the z-axis.

Each of these waves interferes with the data page so that N *angularly multiplexed holograms* of the *same* data page are recorded.

On reconstruction, each of the N plane waves reconstructs the hologram, which it recorded, along with each of the others. The diffraction efficiency of each of the latter reconstructions depends on the mismatch, δ, in angle between the reconstructing beam and the recording beam and, for pure phase transmission holograms, is obtained from Equation 4.33, modified for the small refractive index modulation normally used in HDS systems, to give

$$\eta = \frac{\upsilon^2 \sin^2 \left(\xi^2 + \upsilon^2 \right)^{\frac{1}{2}}}{\left(\upsilon^2 + \xi^2 \right)} \propto \mathrm{sinc}^2 \xi \tag{12.9}$$

with $\xi = \beta\delta \sin \theta T$ (Equation 4.32) assuming the recorded fringes are unslanted.

The diffraction efficiency is zero for values of δ given by

$$\delta = \frac{n\lambda}{2T \sin\theta} \tag{12.10}$$

and the Bragg angular selectivity is

$$\delta_{\mathrm{Bragg}} = \frac{\lambda}{T \sin \theta} \tag{12.11}$$

From Equations 12.8 and 12.10, setting

$$a = \frac{f\lambda}{2T \sin\theta} \tag{12.12}$$

ensures that all of the non-Bragg-matched reconstructions have zero diffraction efficiency.

Taking the object wave as just one point on the SLM producing a single plane wave of unit amplitude and wave vector making angle θ with the hologram normal so that

$$\tilde{O} = \exp(jky' \sin\theta) \tag{12.13}$$

we now consider what happens when the recorded Fourier hologram is shifted by a distance y_0' in the y' direction. We then have as the reconstruction

$$\tilde{R}\tilde{R}^* \tilde{O} = \sum_{n=1}^{N} \exp\left[-\frac{jknay'}{f} \right] \sum_{m=1}^{N} \exp\left[\frac{jkna(y' - y_0')}{f} \right] \exp\left[\frac{jk(y' - y_0')\sin\theta}{f} \right] \tag{12.14}$$

$$\tilde{R}\tilde{R}^* \tilde{O} = \exp\left[\frac{jk(y' - y_0')\sin\theta}{f} \right] \sum_{n=1}^{N} \exp\left[\frac{-jknay_0'}{f} \right] \tag{12.15}$$

The reconstructed intensity is therefore modulated by the term

$$\left[\sum_{n=1}^{N} \exp\left(\frac{-jknay_0'}{f} \right) \right]\left[\sum_{n=1}^{N} \exp\left(\frac{-jknay_0'}{f} \right) \right]^* \tag{12.16}$$

So the reconstructed intensity is given by

$$I(y_0') \propto \frac{\sin^2\left(Nkay_0'/2f \right)}{\sin^2\left(kay_0'/2f \right)} \tag{12.17}$$

which is zero at $Nkay_0'/2f = n\pi$ with $n = 1, 2, 3, \ldots$, that is, at

$$y_0' = \frac{nf\lambda}{Na} \tag{12.18}$$

This is in effect an expression of the shift selectivity and therefore a separate hologram can be recorded at each of the shift values given by Equation 12.18.

Maxima occur at

$$y_0' = \frac{pf\lambda}{a} \tag{12.19}$$

where $p(=n/N)$ and the reconstructed series of holograms begin repeating at shift intervals of $f\lambda/a$. Thus, the total number of holograms that can be multiplexed within the shift range of $f\lambda/a$ is $N - 1$. If $f = 100$ mm, $a = 1$ mm, and $\lambda = 532$ nm, the shift is about 50 μm, which is considerably smaller than the width of the Fourier transformed data page. Using 16 sources, the selectivity is about 3 μm.

The line array of reference beams can be replaced by a single point source, so the reference wavefront is spherical, and it has been shown that shift selectivity of a few micrometers is still possible. In this situation, the diffraction efficiency is [6]

$$\eta = \sin c^2\left(\frac{y_0'T\sin\theta}{\lambda z_0} \right) \tag{12.20}$$

where z_0 is the distance from the point source to the midpoint of the recording layer of thickness T. The shift required to ensure zero diffraction efficiency is then

$$y_0' = \frac{\pi\lambda z_0}{T\sin\theta} \tag{12.21}$$

Taking values of $\lambda = 532$ nm, $T = 5$ mm, $\theta = 30°$, and $z_0 = 10$ mm, we obtain a shift selectivity of about 6 μm, to which must be added the range of positions of the point source along the y'-axis due to its diffraction limited width of $2.44\lambda f/D$.

Shift multiplexing can readily be implemented by rotating a holographic disk in-plane about its center again offering compatibility with existing optical disk technology. Shift in the x direction could also be implemented by moving the recording optics over the disk in the radial direction.

As in angular multiplexing, in order to reduce crosstalk between data pages, shifts are used which position each Bragg peak at the second or third null of that of its nearest neighbor.

Angular, shift, and polytopic multiplexing methods have been implemented in practical HDS systems. However, no HDS system has yet emerged with very clear superiority over existing optical disk technology in terms of likely cost and storage capacity.

12.5.5 WAVELENGTH MULTIPLEXING

We saw in Section 4.3.1.1 that a wavelength selectivity of 28 nm could be obtained in a 20 μm thick photosensitive layer. We could record about 100 holograms by wavelength multiplexing assuming a laser source, which could be tuned through the visible spectrum, which is not a realistic prospect. A practical wavelength multiplexed system would require a much smaller spectral selectivity and increased recording layer thickness, as well as significant advances in laser device technology. The advantage of wavelength multiplexing is that no moving parts are required in the holographic recording system.

Wavelength multiplexing has been proposed for a data storage system using Lippmann photography [7] (Section 8.5), which, although not holographic, has strong similarities and offers the potential of similar data capacity to HDS systems. The Lippmann-based system also has some similarity to the microholographic storage system of Eichler [2]. A data page written into a SLM in the form of 1 and 0 is imaged onto a plane mirror surface. A thick photosensitive layer coated on the mirror is able to record small reflection gratings, which result from interference between normally incident and reflected light. The beam used to read the data is obtained by opening all the SLM pixels and is thus different from either of the two counter-propagating recording beams and consequently the retrieved data is distorted. The system however has the advantages of great simplicity and a very high degree of interferometric stability.

12.5.6 PHASE-CODED REFERENCE BEAM MULTIPLEXING

There are two ways of implementing phase-code multiplexing. One involves the use of a diffuser [8] in the path of the reference beam to impose a random, spatial phase distribution, or phase code, on the reference wavefront, which can be altered for the recording of each new data page by simply translating the diffuser in its plane by a precisely known distance.

The other, rather elegant method [9] requires no moving parts in the system. The reference beam is transmitted through a phase-only SLM, which imposes a deterministic spatial phase pattern, or code, on the wavefront. The phase pattern is altered for the recording of each new data page.

Phase coding by either method has the added advantage that a particular data page can only be recovered if the correct phase code is imposed on the reference beam, so the data is effectively encrypted at the recording stage. The phase code itself represents the address of location of the corresponding data page in the holographic memory.

12.5.6.1 Diffuser-Based Random Phase Coding

In the discussion of laser speckle size in Section 7.4.2, we saw from Equation 7.26 that objective speckle size is $2\lambda R/D$ where R is the distance at which a surface of width D illuminated by light of wavelength λ is observed. A collimated laser beam of wavelength 532 nm and width 10 mm transmitted through a diffuser produces speckles of about 1 μm in width. If the diffuser is translated or rotated in its plane, the speckle pattern is translated to the same extent. Thus, a diffuser translation of >1 μm causes the speckle pattern to move by more than the width of a speckle. Speckle patterns, which are shifted with respect to each other, by more than a speckle width, are uncorrelated and the second cannot be used for successful reconstruction of the hologram recorded using the first as reference. The normalized diffraction efficiency [8] is given by

$$\eta(\delta) = \left\{ \frac{2B_1\left[k(\mathrm{NA})\delta\right]}{k(\mathrm{NA})\delta} \right\}^2 \tag{12.22}$$

which has its first zero at $k(\mathrm{NA})\,\delta = 3.83$ with the numerical aperture (NA) appropriate to the reference beam and therefore

$$\delta = \frac{3.83}{k(\mathrm{NA})} = \frac{0.61\lambda}{(\mathrm{NA})} \tag{12.23}$$

which means that the shift selectivity, δ, is directly related to the size of the objective speckle.

Each data page is effectively encrypted by the unique speckle pattern, which was used as a reference beam to record it.

12.5.6.2 Deterministic Phase Coding

To see how deterministic phase coding works, we start with a set of M reference beams, with different phases and the same amplitude, which are all made to interfere with a data beam O_n, in the recording medium. For each fresh data page, the reference beam phases are altered so that the complete holographic recording of N data pages is given by

$$\left(\sum_{n=1}^{N} O_n \right)\left(\sum_{n=1}^{N} \sum_{m=1}^{M} \exp(j\phi_{n,m}) \right)^* + \text{complex conjugate } n,m \text{ both integers} \tag{12.24}$$

On reconstruction, the holographic memory is addressed by a set of reference beams $\sum_{m=1}^{M} \exp(j\phi_{k,m})$ to produce a reconstructed wave

$$\left(\sum_{m=1}^{M} \exp\left(j\phi_{k,m}\right) \right)\left(\sum_{n=1}^{N} O_n \right)\left(\sum_{n=1}^{N} \sum_{m=1}^{M} \exp\left(j\phi_{n,m}\right) \right)^* \tag{12.25}$$

The single data page O_n is retrieved, without crosstalk, provided that

$$\left(\sum_{m=1}^{M} \exp\left(j\phi_{k,m}\right) \right)\left(\sum_{n=1}^{N} \sum_{m=1}^{M} \exp\left(j\phi_{n,m}\right) \right)^* = 0, \quad k \neq n$$

$$\text{and} \quad \left(\sum_{m=1}^{M} \exp\left(j\phi_{k,m}\right)\right)\left(\sum_{m=1}^{M} \exp\left(j\phi_{k,m}\right)\right)^{*} = 1 \quad \text{for all } k \tag{12.26}$$

Equation 12.26 means that in order to retrieve a data page, one must use the phase-coded reference beam with which it was recorded. The use of any other code produces zero output.

Equation 12.26 means also that $\exp(j\phi_{k,m})$ is a square matrix and

$$\begin{pmatrix} \exp\left(j\phi_{1,1}\right) & \cdots & \exp\left(j\phi_{1,M}\right) \\ \vdots & \ddots & \vdots \\ \exp\left(j\phi_{M,1}\right) & \cdots & \exp\left(j\phi_{M,M}\right) \end{pmatrix} \begin{pmatrix} \exp\left(-j\phi_{1,1}\right) & \cdots & \exp\left(-j\phi_{1,M}\right) \\ \vdots & \ddots & \vdots \\ \exp\left(-j\phi_{M,1}\right) & \cdots & \exp\left(-j\phi_{M,M}\right) \end{pmatrix} = M(I) \tag{12.27}$$

(I) being an identity matrix. The number of data pages that can be stored is equal to M. The phase code for the kth data page is $[\phi_{k,1}, \phi_{k,2}, \phi_{k,3}, \ldots\ldots, \phi_{k,M}]$.

Binary phase code is usually chosen with phase values 0 and π, and the phase code is imposed on the reference beam by transmitting it through a phase-only SLM (see Section 8.8.2.1). (The use of simple binary phase codes means that the matrix

$$\begin{pmatrix} \exp\left(j\phi_{1,1}\right) & \cdots & \exp\left(j\phi_{1,M}\right) \\ \vdots & \ddots & \vdots \\ \exp\left(j\phi_{M,1}\right) & \cdots & \exp\left(j\phi_{M,M}\right) \end{pmatrix}$$

is also orthogonal.)

The appropriate phase codes are obtained using the Walsh–Hadamard algorithm [10]

$$H^{(0)} = 1, \; H^{(1)} = \begin{pmatrix} 1 & 1 \\ 1 & -1 \end{pmatrix}, \; H^{(p+1)} = \begin{pmatrix} H^p & H^p \\ H^p & -H^p \end{pmatrix} \tag{12.28}$$

where p is a positive integer. From Equation 12.28, we can see that

$$H^{(2)} = \begin{pmatrix} 1 & 1 & 1 & 1 \\ 1 & -1 & 1 & -1 \\ 1 & 1 & -1 & -1 \\ 1 & -1 & -1 & 1 \end{pmatrix} \tag{12.29}$$

Zero phase shift is represented by 1, and a phase shift of π is represented by -1, and for example, the storage of four data pages requires the phase-coded reference beams shown in Figure 12.7.

When a data page is retrieved by the appropriate phase-coded reference beam, the reconstructions from all the beams in the set interfere constructively. If an incorrect phase-coded reference beam is used, half of the reconstructions are out of phase by π with the others and

0	0	0	0
0	π	0	π
0	0	π	π
0	π	π	0

FIGURE 12.7 Phase codes for storing four data pages.

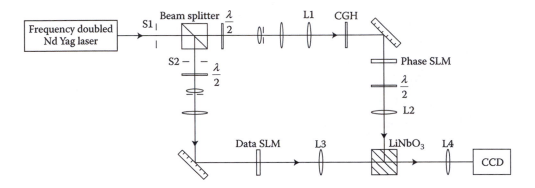

FIGURE 12.8 Phase-coded reference beam HDS. (Reproduced with permission from C. Denz, K-O. Muller., T. Heimann, and T. Tschudi, "Volume holographic storage demonstrator based on phase-coded multiplexing," *IEEE J. Sel. Top. Quant. Electron.*, 4, 832–9, © 1998 IEEE.)

interfere destructively with them. It should be stressed that the technique requires that *each data page must interfere with all* of the beams in a particular phase-coded reference beam, that is, complete overlap is required. To ensure this in practice, data pages are recorded as Fourier transform holograms and focused reference beams are employed as shown in Figure 12.8 [11]. At the recording stage, both shutters S1 and S2 are open. Uniformity of amplitude across the reference beam is assured by lens L1 and a computer-generated hologram (see Chapter 14), which converts the incident beam into 480 equal intensity beams at the phase-only modulator. Phases are maintained within 1% of the required values of 0 and π to minimize the crosstalk in the reconstructed data pages. To read out the data, S2 is closed while S1 remains open.

The crosstalk between reconstructed data pages is significantly less than with angular or wavelength multiplexing. In addition, various *arithmetic* operations can be carried out on retrieved analog data pages by introducing an *amplitude* SLM in the reference beam. This SLM is imaged onto the phase-only SLM and enables phase-coded reference beams of *reduced* amplitude to reconstruct different data pages at the same time, also with reduced amplitude. The pages are thus coherently added together. Subtraction of one data page from another is implemented by changing the phase code of the first so that π is replaced everywhere by zero phase and vice versa. Inversion is simply subtraction of a data page from a recorded plane wave. Furthermore, if the data pages are *binary*, then logical operations are available including OR (addition), XOR (subtraction), and NOT (inversion).

We can dispense with the need for an amplitude SLM by using *partial* reference beam codes, which are common *only* to the data pages on which the operation is to be performed so that the pages are reconstructed simultaneously. The data pages are simultaneously reconstructed and added, or subtracted if one of the partial reference beam codes is inverted.

Data can be encrypted at the recording stage by various logical operations between data pages. Even if the data pages are recovered, they are meaningless, unless the sequence of logical operations is known so that each pixel value can be restored to its original setting. Encryption may be enhanced by a spatially random phase mask in the form of a ground glass plate in the path of the reference beams, the plate being rotated to a new position for each data page to be recorded. The data can only be recovered using an identical plate in the correct position.

A disadvantage of phase-coded reference beam multiplexing, as described so far, is that the use of simple binary phase codes based on Hadamard matrices means that the number of phase shifts is either just 2 or an even integer power of 2, that is, 4, 16, and 64 (see Equation 12.28). This means that the phase-only SLM cannot always be fully utilized. To solve this problem, we look again at Equation 12.27 and note that the matrix is *unitary* and the binary phase codes, obtained by using Hadamard matrices, form a subset of all the possible phase codes [12]. The possible elements of the unitary matrix are given by

$$\exp\left(j\phi_{k,m}\right) = \exp\left(j\frac{2\pi km}{M}\right) \tag{12.30}$$

To verify that this is correct, we substitute Equation 12.30 into Equation 12.27

$$\begin{pmatrix} \exp(j2\pi/M) & \cdots & \exp(j2\pi) \\ \vdots & \ddots & \vdots \\ \exp(j2\pi) & \cdots & \exp(j2\pi M) \end{pmatrix} \begin{pmatrix} \exp(-j2\pi/M) & \cdots & \exp(-j2\pi) \\ \vdots & \ddots & \vdots \\ \exp(-j2\pi) & \cdots & \exp(-j2\pi M) \end{pmatrix} = \begin{pmatrix} a_{11} & \cdots & a_{1M} \\ \vdots & \ddots & \vdots \\ a_{M1} & \cdots & a_{MM} \end{pmatrix}$$

$a_{k,m} = [\exp(j2\pi[k-m]/M) + \exp(j4\pi[k-m]/M) + \cdots + \exp(j2\pi[k-m][M-1])/M + \exp(j2\pi[k-l])]$
whose value is 0 when $k \neq m$,   and M when $k = m$

The fact that a unitary matrix may have odd or even numbers of rows and columns allows us to make full use of the phase-only SLM.

The effect of crosstalk on signal-to-noise ratio has been analyzed [13] for the binary phase $(0,\pi)$ case and for the more general case [12]. It is found that certain codes especially those with high spatial frequency, such as the repeating sequence $0, \pi, 0, \pi, \ldots$, give rise to significantly increased crosstalk and they are not used.

12.6 PHASE-CODED DATA

Data pages may be coded as 2-D arrays of SLM pixels, each of which either transmits light (ON) or does not (OFF). This does not work well when the data pages are recorded as Fourier holograms in, for example, phase-coded reference beam multiplexed HDS because the large zero spatial frequency component (or zero order of diffraction, see, for example, Figure 2.3 where it is seen that the zero order of diffraction by a grating has the greatest intensity) uses a large amount of the available refractive index modulation range in a photopolymer layer or photorefractive crystal and higher spatial frequency components may not be faithfully recorded.

If the data is presented as a *binary phase* array in which zero phase represents binary 0 and phase π represents binary 1, the zero spatial frequency component has very low amplitude (zero if the number of 1s and 0s are equal) due to destructive interference.

We may alternatively use *binary phase modulation* so that zero is represented by a two-pixel phase code, $(0,\pi)$ and binary 1 by the code $(\pi,0)$ so that in any data page, the number of zero-phase pixels exactly balances the number of π-phase pixels, and cancellation of the zero spatial frequency component of the Fourier transform is always complete [14].

On reconstruction, the data page must be converted into amplitude-modulated form in order to be readable, so the reconstructed wavefront representing the phase-coded data page is superimposed on another reconstructed wavefront whose phase everywhere is 0.

As in phase-coded reference beam multiplexing, high spatial frequency codes are avoided to reduce the danger of crosstalk.

An alternative data coding system called *hybrid ternary modulation* [15] uses three states. In the first state, two adjacent pixels are *on* with phase modulation code $0,\pi$. In the second, they are *on* with phase modulation code $\pi,0$. In the third state, they are both *off*. This coding system allows for both binary phase-only modulation and binary intensity-only modulation (Figure 12.9).

12.7 ERROR AVOIDANCE

There are a number of sources of error in data recording and retrieval using holography. Aberrations in lenses cause pixels to increase in size or to become asymmetrically distorted, the latter effects becoming worse with increasing distance from the lens axis. Errors can arise due to the resulting smearing of the pixels. For example, a SLM pixel which is off but surrounded by pixels that are on, may ultimately be interpreted as being on.

FIGURE 12.9 Hybrid ternary modulation.

Scratches or pits on the surfaces of mirrors, beam splitters, and polarizing components, including SLMs, introduce coherent noise due to diffraction. Noise also arises from scatter in the recording material particularly in photopolymers. Electronic noise occurs in the detection system, as well as in the control electronics of the SLM.

Errors can be avoided by *redundancy*, that is, by assigning more than one pixel to each data bit.

We could assign an array of, say, 3×3 pixels to each bit and set a threshold intensity. An average intensity in the array, which is below this threshold is interpreted as binary 0 whereas an average above the threshold is interpreted as binary 1. Thus, even if as many as four pixels are in error, the output is still correctly interpreted.

This procedure cannot be applied globally, that is, to the whole data page because a binary 1 at the edge of the page could well be of lower intensity than a binary 0 near the center when the Gaussian intensity profile of the laser used to illuminate the SLM is taken into consideration. *Local* thresholding is applied to counter this difficulty. The page is divided into blocks of pixels, and the threshold value in a particular block, is normalized with respect to the spatial average of the laser intensity illuminating that block.

A simple error avoidance scheme is to divide each data page into a number of equally sized blocks. Included with each block is a data word coded in highly redundant form and containing parity information, namely the number of pixels in the block, which are on. This number at data reconstruction is used to determine a threshold intensity for the block, enabling us to decide which pixels, in the block are on and which are off.

Alternatively, *binary amplitude modulation coding*, is used to code binary 0 as the binary *sequence* 0,1, that is, the first of two consecutive pixels is off and the second is on. Binary 1 is coded as the binary sequence 1,0. When the data is retrieved, if the first pixel has lower intensity than the second, the data bit is a 0, otherwise it is 0. This method also helps to overcome errors due to less than perfect overlap between the reference and data beams, as well as the effects of spurious interference fringes arising from multiple reflections in the optical components of the system.

Transient or burst electronic noise causes very local errors, and to counter these, we can reassign each of the bits in a data word to different designated blocks, which are spatially separated in the page or even in different pages, a technique known as *interleaving*.

12.8 EXPOSURE SCHEDULING

Photosensitive recording materials for HDS systems are usually those that undergo a change in refractive index upon exposure to light. Materials are not normally characterized by the refractive index modulation that is possible, but by the *number of holograms of 100% diffraction efficiency*, which can be recorded in the same volume of the material, usually written *M/#*. However, in HDS applications, the diffraction efficiency of each recorded hologram must necessarily be small to allow for large data capacity. This is very different from the situation in which a single hologram is to be recorded with high diffraction efficiency, say for display purposes. So, the question is how to ensure that the total range of refractive index modulation available is uniformly distributed among all of the holograms to be recorded, so that they can all be reconstructed with the same diffraction efficiency. Following the recording of one hologram, the available range of refractive index modulation is reduced. In general, we must increase the exposure time for each subsequent recording to compensate for the reduced refractive index modulation, or dynamic range, which is now available. Exposure scheduling is the task of determining the exposure time, t_n, for the nth hologram in the sequence so that all holograms will have the same diffraction efficiency.

The decrease in diffraction efficiency for each consecutively recorded hologram is known as *holographic reciprocity law failure* [16] and, *in the case of silver halide, is due to the time delay between recordings.* The reason is the requirement that a cluster of four silver atoms be produced by exposure to light in a silver halide grain in order for the grain to be capable of chemical development (see Section 6.2). If a silver cluster of four atoms is formed in the first exposure, this is more likely to happen in the bright fringes, and the subsequently developed silver grain contributes to the reconstruction of the first image and contributes scatter noise to the second. Similarly, a four atom cluster formed in the second recording contributes to the reconstruction of the second image and noise to the first. Clusters of two atoms formed in the first exposure and gaining another two atoms in the second contribute equally to the signal and noise in the two reconstructions. However, three atom clusters formed in the first exposure are long lived and have a high probability of gaining a fourth atom in a subsequent exposure. The developed grain contributes to the reconstruction of the first image but not to the second. Thus the second and subsequent exposures contribute to the first but the converse does not apply. Single atoms formed in a grain in the first exposure *dissociate* in a time ~2 s. If the time between exposures exceeds 2 s, the grain cannot be developed even though a three-atom cluster may subsequently be formed in it. This problem was addressed in research aimed at sequential holographic recording of up to four 2-D medical tomographic images on the same photographic plate, using successive exposure times chosen to compensate for holographic reciprocity law failure. Each image is located in its correct spatial location with respect to the others so that a 3-D image is seen in the reconstruction.

For HDS, when a much larger number of holograms is to be recorded, an exposure scheduling method has been developed [17]. The diffraction efficiency scales as $1/N^2$, where N is the number of multiplexed holograms to be stored in memory. Angular multiplexing was used to record a sequence of 30 multiplexed plane diffraction gratings all under the same conditions of exposure time and beam intensities in a photopolymer layer 160 μm in thickness [18] and formulated as described in Section 6.6.1 The resulting diffraction efficiency dependence on grating number in the sequence is shown in Figure 12.10a. The *cumulative* grating strength A is plotted against cumulative exposure energy E (Figure 12.10b), where the nth data point (E_n, A_n) is given by $A_n = \sum_{i=1}^{n} \sqrt{\eta_i}$, $E_n = \sum_{i=1}^{n} E_i$.

The data in Figure 12.10b can be represented by a sixth-order polynomial expression

$$A = a_0 + a_1 E + a_2 E^2 + a_3 E^3 + a_4 E + a_5 E^5 + a_6 E^6 \qquad (12.31)$$

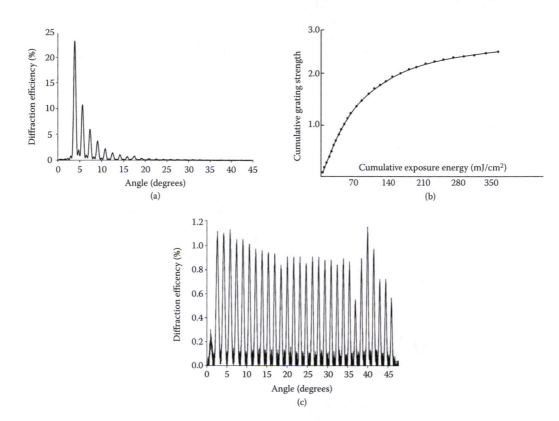

FIGURE 12.10 Exposure scheduling: (a) 30 gratings recorded with equal energies; (b) cumulative grating strength versus cumulative exposure energy; and (c) 30 gratings recorded using one iteration of exposure scheduling. (Reproduced from H. Sherif, Characterisation of an Acrylamide-Based Photopolymer for Holographic Data Storage, MPhil thesis, Dublin Institute of Technology, 2005, by kind permission of the author.)

obtained using a least squares fit. Differentiation gives the grating strength growth rate as a function of cumulative exposure energy, and the condition that all gratings are of equal strength is

$$\frac{A_{sat}}{N} = \left.\frac{\partial A}{\partial E}\right|_{E=\sum_{i=1}^{n-1} E_n} E_n \qquad (12.32)$$

and so the exposure time, t_n, for the nth recording should be

$$t_n = \frac{A_{sat}}{N\left[a_1 + 2a_2 \sum_{i=1}^{n-1} E_i + 3a_3 \left(\sum_{i=1}^{n-1} E_i\right)^2 + 4a_4 \left(\sum_{i=1}^{n-1} E_i\right)^3 + 5a_5 \left(\sum_{i=1}^{n-1} E_i\right)^4 + 6a_6 4a_4 \left(\sum_{i=1}^{n-1} E_i\right)^6\right] I} \qquad (12.33)$$

Using the exposure times generated in this way, 30 new gratings were recorded and their efficiencies, shown in Figure 12.10c, are more uniform. The procedure may be repeated to further equalize the efficiencies and $\eta = \alpha / N^2$. The parameter, $M/\#$, given by $\sqrt{\alpha}$ has been introduced [17,19] to facilitate comparison of different holographic memory systems and enable us to determine the diffraction efficiency of each of a multiplexed number of holograms recorded in the same volume. Alternatively, as already stated, it is the number of holograms that we could record, each with diffraction efficiency of 100%, in the same volume of recording material. Clearly, $M/\#$ depends on the thickness of the recording material and its optical characteristics, as well as the recording geometry and beam ratio.

12.9 DATA AND IMAGE PROCESSING

We have already seen in Section 12.5.6.2 that various logical and arithmetic operations are possible on data recorded using phase-coded reference beams. There are additional possibilities for data processing using holographic techniques.

12.9.1 ASSOCIATIVE RECALL

If the holographic memory is illuminated by one of the data pages, the corresponding reference beam, or beams in the case of phase-coded reference beam multiplexing, used to record the data page is reconstructed. This amounts to recall of the *address* at which the particular data page is stored, and the presence of specific data in the memory is verified if a valid address is recalled. Let us assume that we are using phase-coded reference beam multiplexed HDS. The complex conjugate term in Equation 12.24 is $(\sum_{n=1}^{N} O_n)^* (\sum_{n=1}^{N} \sum_{m=1}^{M} \exp(j\phi_{n,m}))$. Normally, the data pages are recorded as Fourier holograms, and the reference beams are focused in the Fourier plane. The complex conjugate term becomes $FT[(\sum_{n=1}^{N} O_n)^*] FT(\sum_{n=1}^{N} \sum_{M=1}^{M} \exp(j\phi_{n,m}))$. If this is illuminated by a data page $FT(O_j)$, the output is

$$
FT(O_j) \left\{ FT\left[\left(\sum_{n=1}^{N} O_n \right)^* \right] \left(\sum_{n=1}^{N} \sum_{m=1}^{M} \delta_{n,m} \right) \right\} = FT\left[(O_j) \otimes \left(\sum_{n=1}^{N} O_n \right)^* \right] \left(\sum_{n=1}^{N} \sum_{m=1}^{M} \delta_{n,m} \right)
$$

with $\oplus$ denoting convolution (Appendix B), which is retransformed to become

$$
(O_j) \oplus \left(\sum_{n=1}^{N} O_n \right)^* \left(\sum_{n=1}^{N} \sum_{m=1}^{M} \exp(j\phi_{n,m}) \right)
$$

where $\oplus$ denotes the *correlation* (again see Appendix B) of (O_j) with $(\sum_{n=1}^{N} O_n)$. Perfect correlation has a maximum value when $n = j$ and the reference beam with phase code $\phi_{j,1}, \phi_{j,2}, \phi_{j,3}, \ldots, \phi_{j,M}$ is produced. The *similarity* between O_j and any other data page O_n is determined by the strength of the correlation between them, that is, by the intensity of the reconstructed set of reference beams having phase code $\phi_{n,1}, \phi_{n,2}, \phi_{n,3}, \ldots, \phi_{n,M}$.

The preceding discussion equally well applies to searching a page for a particular data word, but care is needed in coding the data. Suppose that we look for particular data word, say $\overline{1011}$, in a page of data, having used binary phase modulation to code the data. Each matching bit adds 1 to the correlation amplitude (constructive interference), and a perfect match results in a correlation intensity of 16. Mismatch (destructive interference) of one or of three of the bits reduces the correlation intensity to 4 and mismatch of two of the bits reduces it to 0. However, if the data is searched for 1011, that is, 0100, the (anti)correlation intensity is again 16. For this reason, when choosing data code words, we discard any which are logical NOT versions of those already selected.

A further difficulty arises when the memory is searched using short code words because very low correlation intensities may be swamped by the pixels in the SLM, which are not off, leading to false matches, but hybrid ternary modulation coding (Section 12.6) allows for searches using short data words with reduced risk of false matches. The data is coded only by binary phase modulation using the two on states while the third off state is used to block SLM pixels, which do not form part of the search word.

12.9.2 DATA PROCESSING WITH OPTICAL FOURIER TRANSFORMS

Since many HDS systems record the data pages in the form of holograms of their spatial Fourier transforms, it is worthwhile exploring data and image processing applications, which exploit the properties of the Fourier transform and which can be readily implemented in conjunction with the holographic memory. It should be borne in mind throughout this section that an image formed by double Fourier transformation using positive lenses is always inverted (see Section 9.4).

As in Sections 9.4, we make use of the function Kronecker delta function $\delta(y)$ (see also Section 12.5.4) to describe a point source of light. It has a value of unity, or unit amplitude, only at $y = 0$ and is zero everywhere else. Similarly, $\delta(y - b)$ has a value only at $y = b$. Using Equation 2.22, the Fourier transform of the delta function is given by

$$F(y') \propto \int_y \delta(y - b) \exp(-jkyy' / f) \, dy = \exp(-jkby' / f) \tag{12.34}$$

which is a plane wave of unit amplitude propagating at angle, $\sin^{-1}(b/f)$ to the z-axis and whose spatial frequency is $b/\lambda f$.

12.9.2.1 Defect Detection

Masks containing a number of identical patterns with complex amplitude transmissivity $\tilde{t}(x, y)$ are extensively used in the manufacture of integrated electronic circuits. Scratches and dust on one of the patterns in the array may cause a circuit made using that pattern to fail. Although small numbers of defects randomly scattered over the pattern mask may not present a serious obstacle to economic production, it is important that the mask be microscopically inspected for defects. Alternatively, we can use the properties of the spatial Fourier transform to locate the defects. We start in a manner similar to that discussed in Section 9.4 with a pattern mask in the front focal plane of a lens illuminated normally by a plane monochromatic wave of unit amplitude. The amplitude transmission of the mask is described by the function $\sum_{m=1}^{M}\sum_{n=1}^{N}\tilde{t}(x-ma,y-nb)$, meaning that there are $M \times N$ identical patterns $\tilde{t}(x,y)$ spaced a apart in the x direction and b apart in the y direction. The spatial Fourier transform of this array is holographically recorded in the back focal plane of a lens and has the form (from Equation 2.22):

$$\sum_{m=1}^{M}\sum_{n=1}^{N}\iint_{x,y}\tilde{t}(x-ma,y-nb)\exp\left[-\frac{jk}{f}(xx'+yy')\right]dxdy = \sum_{m=1}^{M}\sum_{n=1}^{N}\tilde{T}(x',y')\exp\left[-\frac{jk}{f}(max'+nby')\right]$$

(12.35)

Suppose that the mask was originally prepared, as described in Section 9.4, using an array of point sources $\sum_{n=1}^{N}\sum_{m=1}^{M}\delta(x-ma)\delta(y-nb)$. The function $\delta(y-nb)$ is again the Kronecker delta function used to describe a point source of light. It has a value only at $y = nb$ and is zero everywhere else so that the Fourier transform has a value obtained by setting $y = b$ in the expression $\exp(-jkyy'/f)$. A *negative* copy of this array, although not 1 with *zero* transmissivity, properly located in the Fourier plane of the lens, filters or attenuates, the spatial frequencies of the mask array, transmitting those associated with random defects, and the defects are enhanced on retransformation by a second lens. This is because defects, which are typically smaller in size than the pattern width, have bandwidths extending beyond that of the pattern $\tilde{t}(x,y)$. Although the image of the array mask is greatly reduced in brightness by the negative filter, it is possible to identify the patterns that have defects and thus to determine the suitability of the mask.

12.9.2.2 Optical Character Recognition

Optical character recognition can readily be implemented by using a hologram of the Fourier transform of the character to be recognized, that is, an optical matched filter. The task is to determine whether a scene includes a particular object matching one in an image data base. Some examples are text searches, radar signatures, fingerprint matching, and personal identification.

We start by recording a Fourier hologram of the search pattern, that is, the object that is to be recognized, using the optical arrangement shown in Figure 12.11a. The search pattern is represented by a transparency having complex amplitude transmittance $\tilde{t}(x,y)$, and the recording is made with the transparency normally illuminated by a plane wave of unit amplitude, which originates from a point source in the front focal plane at $x = 0$, $y = b$.

Using Equation 12.34, the complex amplitude appearing in the Fourier plane includes the terms

$$\exp\left(\frac{-jkby'}{f}\right)T(x',y')+T^{*}(x',y')\exp\left(\frac{jkby'}{f}\right)$$

so that we have a *hologram of* $T^{*}(x',y')$, the conjugate spatial Fourier transform of $\tilde{t}(x,y)$, which is known as an *optical matched filter*. A transparency with transmittance $\tilde{g}(x,y)$ representing the scene to be searched is now placed in the front focal plane of the lens (Figure 12.11b) so that the hologram is illuminated by $G(x',y')$ to give the product, $P(x',y')$,

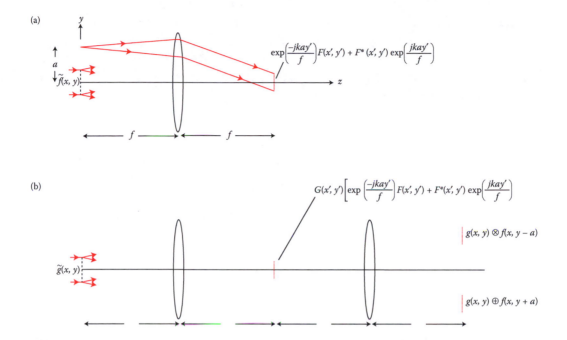

FIGURE 12.11 Optical pattern recognition: (a) recording matched filter and (b) pattern recognition.

with

$$P\left(x',y'\right)=G\left(x',y'\right)\left[\exp\left(\frac{-jkby'}{f}\right)T\left(x',y'\right)+T^*\left(x',y'\right)\exp\left(\frac{jkby'}{f}\right)\right]\quad(12.36)$$

which is then Fourier transformed by a second lens to give as output

$$p\left(x,y\right)=\tilde{g}\left(x,y\right)\otimes\tilde{t}\left(x,y-b\right)+\tilde{g}\left(x,y\right)\oplus\tilde{t}\left(x,y+b\right)\quad(12.37)$$

The first term in Equation 12.37 is the convolution of $\tilde{g}(x,y)$ and $\tilde{t}(x,y)$ centered on $y=b$. The second term, centered on $y=-b$, is the correlation of $\tilde{g}(x,y)$ and $\tilde{t}(x,y)$, which is a measure of their *similarity*. If the search pattern $\tilde{t}(x,y)$ is included anywhere within $\tilde{g}(x,y)$, then the second term above includes the *autocorrelation* of $\tilde{t}(x,y)$ with maximum intensity at the location of the search pattern. Another way to look at it is that if the search pattern $\tilde{t}(x,y)$ appears anywhere within $\tilde{g}(x,y)$, then $T(x',y')$ is included in $G(x',y')$, and Equation 12.36 includes $|T(x',y')|^2$, which is *real*. A point of light therefore appears in the output image of $\tilde{g}(x,y)$ at the location of the search pattern. In other words, the search pattern is *highlighted* wherever it appears.

It is important that the reference beam point source be at a sufficient distance, b, from the input transparency $\tilde{t}(x,y)$ so that the convolution and correlation terms do not overlap in the final output.

In this implementation, it is also important to ensure that $\tilde{g}(x,y)$ has the same orientation as $\tilde{t}(x,y)$ and that it has the same scale. However, the search can be made independent of orientation by converting from rectangular to polar coordinates, in which case rotation by ϕ simply results in a multiplication of the Fourier transform by a phase factor $\exp(j\phi)$. The search can also be carried out independent of changes in scale if we use the Mellin transform, which is the Fourier hologram of the natural logarithm $\ln[\tilde{t}(x, y)]$, so that scale changes also transform as constant phase shifts.

12.9.2.3 Joint Transform Correlation

Joint transform correlation is a simple variation on the theme of optical matched filtering, enabling a feature of interest to be identified in a scene. The search term $\tilde{t}(x, y - a)$ and the scene, or data page to be searched, $\tilde{g}(x, y + a)$, are located in the front focal plane of a lens and the intensity distribution in the Fourier plane includes the terms

$$T(x', y')G^*(x', y')\exp\left(\frac{-2jkay'}{f}\right) + T^*(x', y')G(x', y')\exp\left(\frac{2jkay'}{f}\right).$$

If the search term $\tilde{t}(x, y)$ appears in the scene $\tilde{g}(x, y)$, then a strong correlation peak appears at the appropriate location in the Fourier plane of a second lens, that is, in the image of $\tilde{g}(x, y)$ centered at $y = -2a$.

12.9.2.4 Addition and Subtraction

We have seen in Section 12.5.6.2 that addition and subtraction of data can be implemented simultaneously with retrieval from holographic memory. Another approach involves the use of a spatial filter, which has the complex amplitude transmittance $\exp[j(kby'/f + \phi)] + \exp[-j(kby'/f + \phi)] + 1$, which is a pure phase grating. The filter is located in the Fourier plane. The terms to be added are $\tilde{t}(x, y)$ and $\tilde{g}(x, y)$, which are placed in the front focal plane at $y = b$ and $y = -b$, respectively. The filter is thus illuminated by $T(x', y')\exp(jkby'/f) + G(x', y')\exp(jkby'/f)$, and the output from it is

$$\left\{T(x', y')\exp(-jkby'/f) + G(x', y')\exp(jkby'/f)\right\}\left\{\exp\left[j(kby'/f + \phi)\right] + \exp\left[-j(kby'/f + \phi)\right] + 1\right\}$$

$$= T(x', y')\exp(-jkby'/f) + G(x', y')\exp(jkby'/f) + \exp(j\phi)\left[T(x', y') + G(x', y')\exp(-2j\phi)\right]$$

$$+ T(x', y')\exp(-j[2kby'/f + \phi]) + G(x', y')\exp(j[2kby'/f + \phi])$$

Retransformation by a second lens produces an output *on the axis*, resulting from the third and fourth terms in the above expression, which is the *sum* $\tilde{t}(x, y) + \tilde{g}(x, y)$ when ϕ is 0, and the *difference* $\tilde{t}(x, y) - \tilde{g}(x, y)$ when ϕ is $\pi/2$.

12.9.2.5 Edge Enhancement

Edge enhancement can be implemented using a spatial filter, which consists of two, amplitude-only gratings with slightly differing spatial frequencies and superimposed in antiphase. The filter transmissivity is

$$T(y') = \sin\left[\left(k' + \delta k'\right)y'\right] - \sin\left(k'y'\right) = 2\cos\left(\frac{2k' + \delta k'}{2}y'\right)\sin\frac{\delta k'y'}{2} \qquad (12.38)$$

For small values of $\delta k'$ and close to the optical axis, we have

$$T(y') \propto y' \qquad (12.39)$$

If the filter is illuminated by the Fourier transform $G(y')$ and the result retransformed, we obtain

$$\int_{-\infty}^{\infty} G(y')y'\exp\left(\frac{-jkyy'}{f}\right)dy' = \frac{jf}{ky}\int_{-\infty}^{\infty} G(y')d\left[\exp\left(\frac{-jkyy'}{f}\right)\right]$$

$$= \frac{jf}{ky}G(y')\exp\left(\frac{-jkyy'}{f}\right)\Bigg|_{-\infty}^{\infty} - \frac{jf}{ky}\int_{-\infty}^{\infty}\frac{\partial G(y')}{\partial y'}\left[\exp\left(\frac{-jkyy'}{f}\right)\right]dy'$$

and when this is Fourier transformed, we obtain an output proportional to the spatial derivative $\partial \tilde{g}(y)/\partial y$. This means that the edges of an image lying parallel to the x-axis are enhanced, and the procedure can be extended to include enhancement of edges parallel to the y-axis.

12.9.2.6 Image Recovery

No lens system can form a perfect image because it presents a spatially limited aperture to incoming light and because of its aberrations, although the latter can usually be corrected to any desired extent by carefully optimized optical design. A lens system is characterized by its so-called point spread function $\tilde{h}(x, y)$ which describes the complex amplitude in the image of a point in the image plane x,y of the system. It can be shown that any image formed by the lens system is in fact the convolution of the input complex amplitude distribution with the point spread function, that is,

$$g(x'', y'') = \iint f(x, y)h(x'' - x, y'' - y)dxdy \qquad (12.40)$$

Using the convolution theorem (Appendix B),

$$G(x', y') = F(x', y')H(x', y') \qquad (12.41)$$

$$F(x', y') = \frac{G(x', y')}{H(x', y')} = G(x', y')H^*(x', y')\frac{1}{|H(x', y')|^2} \qquad (12.42)$$

In order to correct the image so that it is diffraction limited, we record a spatial filter, which is the conjugate spatial Fourier transform $H^*(x', y')$ of the point spread function $\tilde{h}(x, y)$. We also record a spatial filter, which has complex amplitude transmittance $1/|H(x', y')|^2$.

When these filters are cascaded together and illuminated by the spatial Fourier transform of an image obtained using the lens system, we obtain the Fourier transform $F(x', y')$ of the diffraction limited image, which is Fourier transformed to obtain the image $f(x, y)$.

12.10 OPTICAL LOGIC

Conventional logical networks use simple binary logic gates such as AND, NAND, OR, EOR, NOR, and NOT, implemented on a massive scale. Such logic operations can be optically implemented in a number of ways. A simple example is shown in Figure 12.12a of an optical gate implemented by means of a beam combiner. The combined beams are incident on a photodetector, which produces an output if the input intensity exceeds a threshold value. The device can function either as a logical OR gate or as an AND gate. The photodetection process is followed by conversion back to an optical signal, which is passed to the next logic gate of the processor. Such optical/electronic hybrid devices do not confer any significant advantage over their all-electronic counterparts because the bandwidth and the speed are still limited by the electronic circuitry. This bottleneck can be somewhat alleviated using the self-electrooptic effect whereby photodetection causes a change in the electric field across a semiconductor laser device, altering its optical absorption through quantum Stark confinement, which in turn alters the laser output [20]. Self-electrooptic devices (SEED) devices can act as optical switches and can be used to implement almost all-optical logic.

Another device that can contribute to all-optical logic is one consisting of an electrically pumped gain medium through which a light beam passes and increases in power through stimulated emission.

A holographic diffraction grating recorded in a photorefractive optical material can diffract a third light beam onto a detector. This is the method by which we monitor the formation of a holographic diffraction grating in real time (see Section 7.6.1). The detector produces an output if the grating is present, which is the case only if *both* of the beams are present and overlap each other in the recording medium. These are the conditions necessary for a logical AND gate, an example of which is shown in Figure 12.12b. Here, a photorefractive crystal is illuminated by mutually coherent collimated light beams, A and

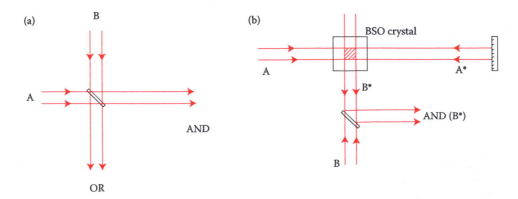

FIGURE 12.12 Optical logic (a) OR /ANDgate and (b) holographic AND gate.

B, and by the phase conjugate of A obtained by reflection at a plane mirror. A beam splitter in the path of B allows the phase conjugate of B, that is, B*, to be physically separated from B, and B* is only produced if A and B are both present. The arrangement therefore behaves as an all-optical AND gate.

12.11 HOLOGRAPHIC OPTICAL NEURAL NETWORKS

We have seen already (Section 12.9.2.2) that optical pattern recognition can be implemented through the use of holographic complex matched filters, although scale and orientation changes of the search pattern require additional preprocessing. Neural networks are of great interest because they are particularly well suited to recognition tasks. A simple model of a neuron is shown in Figure 12.13. The neuron receives inputs in the form of outputs $x_1, x_2, \ldots, x_N$ from N neurons connected to it, these outputs being weighted by $w_1, w_2, \ldots w_N$ respectively. The input to the neuron is $\sum_{i=1}^{N} x_i w_i$, which is a scalar product of vectors $(x_1, x_2, \ldots, x_N)$ and $(w_1, w_2, \ldots w_N)$, and its output is a nonlinear function of the input. By adjusting the weights, one can ensure a particular output from the neuron. This last statement also applies to a highly interconnected *network* of neurons. In other words, we can adjust all the weights so that each of the neurons has a constant output, and the whole network is in one of a number of possible stable states. *These stable states are the patterns that we wish the network to recognize, or remember.*

A neural network can be implemented using holography if the recording material has very high Bragg selectivity and can record a very large number of holographic gratings in the same volume (Figure 12.14). An array of N mutually coherent point sources of light

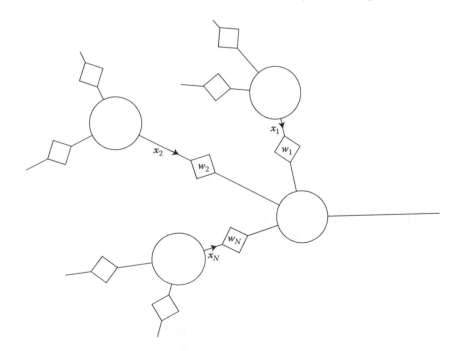

FIGURE 12.13 Neural network.

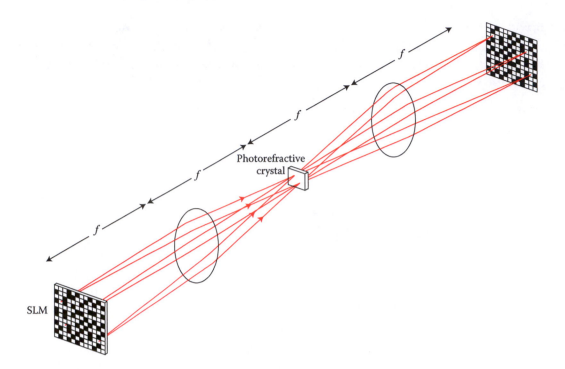

FIGURE 12.14 Holographic neural network.

acting as optical neurons is located in the front focal plane of a lens. In practice, the points are pixels in a SLM. Each of the points emits a spherical light wave, which is converted by the lens into a plane wave, whose direction, or wave vector, depends on the location of the point. Each plane wave interferes with each of the others, in a photorefractive crystal, and $N \times N$ diffraction gratings are recorded. Each grating has a diffraction efficiency, which depends on *exposure time* and on the *intensities* of the two point sources used to record it. *The diffraction efficiencies correspond to the weights in a neural network.*

Subsequently, light from a pixel in the input plane, collimated by the lens and Bragg matched to one of the recorded gratings, is diffracted and focused on an output pixel in the back focal plane of a second lens. The input pixel thus is connected with a weighting to the output pixel, which may be a SEED for example (Section. 12.10). A number of input pixels may be connected to the *same* output pixel if the light from them is traveling in the same direction following diffraction. This is the case *if light from each of the input pixels had originally formed a grating with light from a pixel whose optical image is at the position of the output pixel.* Thus, the combination of an SLM and volume holographic recording material provides an optical analog of a neural network.

A phase conjugate mirror (PCM) may be added to the system (Figure 12.15).

Light from an input pixel is collimated by a lens and diffracted by a grating in the photorefractive crystal, to be focused at an output pixel in the focal plane of a second lens. The *undiffracted* light is phase conjugated to interfere with light *from* the output pixel and record the grating in the crystal [21]. As before, this arrangement connects input and output pixels through the grating, which they have recorded. *However, the light*

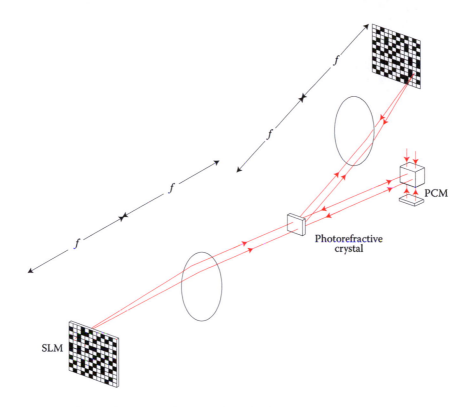

FIGURE 12.15 Holographic neural network with phase conjugator.

amplitudes from the output pixels may be controlled so as to impose any desired weights on grating connections enabling the association of a specific output pattern with a specific input pattern.

12.12 QUANTUM HOLOGRAPHIC DATA STORAGE

In concluding this chapter, it is interesting to consider the question of what is the ultimate data storage capacity that we can achieve. So far, we have only considered HDS systems, in which light is used as the means for storing and retrieving the information. The space needed to store a single data bit is determined by the wavelength. By making use of matter waves, specifically electron waves, whose wavelength is very much shorter than that of light, vastly increased storage capacity is possible [22]. The technique of quantum HDS requires the use of a scanning tunneling microscope (STM).

In an STM, an extremely sharp conducting tip is brought within about 0.5 nm of a surface, which we wish to examine (Figure 12.16). An electron, considered a discrete particle, cannot normally cross the gap but may do so when an electrical bias is applied between tip and surface. This is because the electron is more correctly described by a probability distribution or wave function and may cross the gap by a process known as quantum mechanical tunneling. Data obtained in the form of current as a function of position in the horizontal plane provides a topographical map of the surface as the tip is moved over it in raster scan fashion. If the tip width is of the order of one atom, one can achieve depth

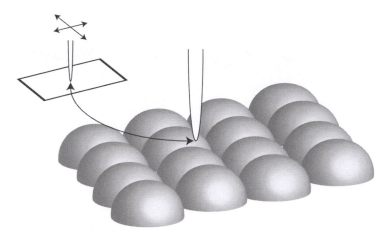

FIGURE 12.16 Scanning tunneling microscopy.

resolution of 0.01 nm and horizontal resolution of 0.1 nm. Scanning tunneling microscopy is a powerful tool for nanoscale research, as we can form images of individual atoms, as well as maneuver them into chosen locations.

Furthermore, we can measure the tunneling current as a function of the bias voltage at any location, which provides data on the local density of energy states. One can exploit this latter technique of *scanning tunneling spectroscopy* for holographic recording and reconstruction.

If the surface is flat pure copper, it is covered by a 2-D gas of conduction electrons, or more precisely, freely propagating electron waves, which have a range or spectrum of electron energies.

A carbon monoxide molecule placed on the surface is surrounded by the electron gas so that the molecule is illuminated by electron waves, which scatter from the molecule and interfere with one another. The local density of states of electron energy is altered accordingly. We now add more CO molecules to form a kind of local molecular corral, which can be probed by the STM (Figure 12.17a). Extending the concept further, CO molecules can be positioned in a pattern that encodes an alphanumeric character in binary form, with binary "1" represented by a high local density of states and binary "0" by a low one.

We now use the STM with an appropriate bias voltage to probe the spatially varying density of states, that is, the hologram, at a corresponding electron wavelength in order to reconstruct the stored character (Figure 12.17b). A remarkable feature of the technique is that, in the same space, one can arrange the CO molecular pattern to store more than one character, each of which is reconstructed at a different bias voltage. Characters are thus reconstructed in a 3-D space, one of the dimensions being energy (Figure 12.17b).

The molecular distribution determines the area in which a 7×5 bit character can be stored. Manoharan and coworkers created CO molecular patterns in an area of 289 nm^2, which they did not minimize [22], but reconstructed characters from an area of 12.5 nm^2 within the molecular corral giving an information density of 2.8 bits nm^{-2}, which is

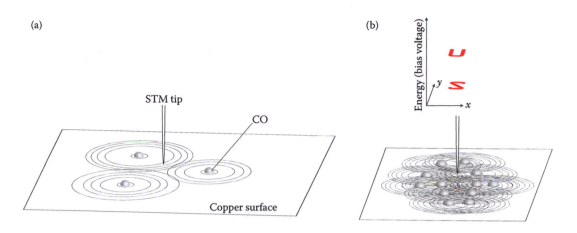

FIGURE 12.17 Quantum holographic data storage. (a) A pattern of electron wave interference formed in a 2-D electron gas by a simple corral of CO molecules. The STM tip is used to probe the pattern with tunneling electrons of a specific energy. (b) CO molecules are placed in computed locations to encode an alphanumeric character. The data is read out by the STM tip, again at a specific tunneling electron energy. The molecular pattern can be computed to encode more than one character, and each is read out at a different energy. (Adapted from C. R. Moon, L. S. Mattos, B. K. Foster, G. Zeltzer, and H. C. Manoharan, Quantum holographic encoding in a two-dimensional electron gas, *Nat. Nanotechnol.*, 4, 167–72, 2009. With permission.)

2.8×10^6 times greater than that possible by conventional optical or holographic means (Section 12.2).

The task of computing the required positions of molecules in order to encode the data is a far from trivial one, and we will not be using quantum HDS systems at home any time soon because of the extremely sophisticated apparatus that is required and the time taken to record and retrieve the data. However, the experiments conclusively demonstrate a local data density far in excess of conventional system capacities.

12.13 SUMMARY

This chapter explains the advantages of HDS over conventional optical disk storage technology. We describe various techniques for high-capacity HDS including angular, peristrophic, polytopic, and shift multiplexing. Special consideration is given to phase-coded reference beam multiplexing because of the additional facility for data processing, which it can provide. We explain the need for exposure scheduling of large numbers of multiplexed holograms in a holographic memory. We discuss the associative recall property of holography and describe data coding and error avoidance techniques in HDS and retrieval. We consider optical data processing with particular reference to techniques based on the optical Fourier transform. We discuss optical logic using holography and holographic schemes for implementation of optical neural networks.

Finally, we look briefly at the technique of quantum HDS.

REFERENCES

1. H. J. Eichler, "Microholographic data storage," Lasers and Electro-Optics Europe CLEO/Europe D. O. I. 10.1109/CLEOE.2005.1568487 (2005).

2. K. Choi, J. W. M. Chon, M. Gu, N. Malic, and R. A. Evans, "Low-distortion holographic data storage media using free-radical ring-opening polymerization," *Adv.Func. Mater.*, 19, 3560–6 (2009).

3. K. Curtis, A. Pu, and D. Psaltis, "Method for holographic storage using peristrophic multiplexing," *Opt. Lett.*, 19, 993–4 (1994).

4. K. Anderson and K. Curtis, "Polytopic multiplexing," *Opt. Lett.*, 29, 1402–4 (2004).

5. D. Psaltis, M. Levene, A. Pu, and G. Barbastathis, "Holographic storage using shift multiplexing," *Opt. Lett.*, 20, 782–4 (1995).

6. G. Barbastathis, G. Levene, and D. Psaltis, "Shift multiplexing with spherical reference waves," *Appl. Opt.*, 35, 2403–17 (1996).

7. K. Contreras, G. Pauliat, C. Arnaud, and G. Roosen, "Application of Lippmann interference photography to data storage," *Journal of the European Optical Society-Rapid Publications*, 3, 08020 (2008).

8. A.M. Darskii and V. Markov, "Shift selectivity of the holograms with a reference speckle wave," *Opt. Spectrosc.*, 65, 392–395 (1988).

9. C. Denz, G. Pauliat, G. Roosen, and T. Tschudi, "Volume hologram multiplexing using a deterministic phase encoding method," *Opt. Commun.*, 85, 171–6 (1991).

10. R. K. R. Yarlagadda and J. E. Hershey, *Hadamard Matrix Analysis and Synthesis with Applications to Communications and Signal/Image Processing*, Kluwer Academic Publishers, Boston, MA (1997).

11. C. Denz, K-O. Muller., T. Heimann, and T. Tschudi, "Volume holographic storage demonstrator based on phase-coded multiplexing," *IEEE J. Sel. Top. Quant. Electron.*, 4, 832–9 (1998).

12. X. Zhang, G. Berger, M. Dietz, and C. Denz, "Unitary matrices for phase-coded holographic memories," *Opt. Lett.*, 31, 1047–9 (2006).

13. K. Curtis and D. Psaltis, "Cross talk in phase-coded holographic memories," *J. Opt. Soc. Am. A.*, 10, 2547–50 (1993).

14. R. John, J. Joseph, and K. Singh, "Phase-image-based content-addressable holographic data storage," *Opt. Commun.*, 232, 99–106 (2004).

15. J-S. Jang and D–H. Shin, "Optical representation of binary data based on both intensity and phase modulation with a twisted-nematic liquid-crystal display for holographic digital data storage," *Opt. Lett.*, 26, 1797–9 (2001).

16. K. M. Johnson, L. Hesselink, and J. W. Goodman, "Holographic reciprocity law failure," *Appl. Opt.*, 23, 218–27 (1984).

17. A. Pu, K. Curtis, and D. Psaltis, "Exposure schedule for multiplexing holograms in photopolymer films," *Opt. Eng.*, 35, 2824–9 (1996).

18. H. Sherif, Characterisation of an Acrylamide-Based Photopolymer for Holographic Data Storage, MPhil thesis, Dublin Institute of Technology (2005).

19. F. H. Mok, G. W. Burr, and D. Psaltis, "System metric for holographic memory systems," *Opt. Lett.*, 21, 896–8 (1996).

20. H. Kuo, Y. K. Lee, Y. Ge, S. Ren, J. E. Roth, T. I. Kamins, D. A. B. Miller, and J. S. Harris, "Strong quantum-confined Stark effect in germanium quantum-well structures on silicon," *Nature*, 437, 1334–6 (2005).

21. D. Psaltis, D. Brady, X-G. Gu, and S. Lin, "Holography in artificial neural networks," *Nature*, 343, 325–30 (1990).

22. C. R. Moon, L. S. Mattos, B. K. Foster, G. Zeltzer, and H. C. Manoharan, "Quantum holographic encoding in a two-dimensional electron gas," *Nat. Nanotechnol.*, 4, 167–72 (2009).

PROBLEMS

1. What should be the approximate thickness of an unslanted reflection phase grating recorded at a wavelength of 532 nm in a photopolymer whose refractive index modulation is 10^{-2}? What is the approximate data storage capacity of a bit format holographic disk of diameter 5 cm in which nonoverlapping gratings of 100% efficiency are recorded using counterpropagating focused beams of wavelength 532 nm. The lenses in the system have focal lengths of 25 cm and the beams are 1.2 mm in diameter. The thickness of the photosensitive layer is 500 μm.

2. Suppose we were to add to the system described in problem 1 the capability of wavelength multiplexing over the range from 550 to 650 nm. What would then be the data storage capacity if the average refractive index is 1.5?

3. Angular multiplexing is used to store data pages as transmission holograms in a 100 μm thick photosensitive layer of refractive index 1.5, with an angle of 60° inside the layer between object and reference beams of wavelength 532 nm. What should be the minimum change in angle of the recording beams when storing each new page of data if each hologram has 5% diffraction efficiency? If the maximum *range* of incident angle of the beams is 10°, how much data could be stored in the same space using a spatial light modulator with 800×600 pixels?

4. Calculate the data capacity of a peristrophic multiplexed data storage system using the parameters of the previous problem. Assume SLM and CCD detector widths of 10 mm and Fourier transform lenses of focal length 20 cm. How does the rotational angle between recordings compare with the angular selectivity of the photopolymer layer and how does the data storage capacity compare with that of the previous problem?

5. A shift multiplexed holographic data storage system utilizes 32 point sources, spaced 1.2 mm apart and Fourier transform lenses of focal length 20 cm. The laser has wavelength 532 nm. Find the number of holograms that can be stored in a recording layer, which is 25 mm in width.

6. A random phase-coded reference beam holographic data storage system could be implemented using two collimated light beams, one passing through an SLM to the hologram plane, the other through a diffuser. The beams are 1 cm in diameter and 532 nm in wavelength and the diffuser and hologram plane are 20 cm apart. What is the diffuser translational shift selectivity of the system? If the diffuser is simply rotated about a normal to its plane between recordings, what are the implications?

7. Using the Walsh–Hadamard algorithm, draw up a table showing the binary phase codes of the reference beams required for the storage of eight data pages. How many orthogonal phase codes are available using an SLM with 800×600 pixels?

8. Show that the 3×3 unitary matrix is

$$\begin{pmatrix} -0.5 + \dfrac{\sqrt{3}}{2}j & -0.5 - \dfrac{\sqrt{3}}{2}j & 1 \\[2ex] -0.5 - \dfrac{\sqrt{3}}{2}j & -0.5 + \dfrac{\sqrt{3}}{2}j & 1 \\[2ex] 1 & 1 & 1 \end{pmatrix}$$

and that the 2×2 unitary matrix is in fact $\begin{pmatrix} 1 & 1 \\ 1 & -1 \end{pmatrix}$.

9. It was stated in Section 12.5.6.2 that binary phase codes, such as the repeating sequence $0, \pi, 0, \pi, \ldots$ are better avoided because they give rise to crosstalk between reconstructed data pages. Explain why this is so.

10. An optical pattern recognition system has Fourier transform lenses of focal length 37.5 cm and a laser of wavelength 633 nm. The object transparency to be searched is positioned symmetrically about the optical axis and has dimensions 24 mm × 36 mm. The highest spatial frequency in it is 20 lines mm^{-1}. What minimum diameter of Fourier lens should we use? What should be the angle of incidence of the reference beam used to record the Fourier holograms of the search patterns?

CHAPTER 13

Digital Holography

13.1 INTRODUCTION

The idea of using a CCD camera to record holograms is obviously a very attractive one since it then becomes possible to avoid all forms of physical or chemical processing of recording materials and many holograms can be recorded per second. The photosensitive elements in the CCD array are regularly spaced in orthogonal directions, with space between them in which no light is detected. Each element of the array can provide a *sample* of the intensity of the light pattern produced by the interference of the object and reference waves. The set of samples is digitized and stored in a computer as a *digital hologram*. We reconstruct the image by *calculating* the result obtained when the digital hologram is illuminated by the reference wave. This last operation is a purely computational one and the reference wave is in fact virtual. The result of the computation is displayed on a computer monitor. This concept of numerical reconstruction leads to some very useful applications. We obviously cannot display a 3-D reconstructed image on the computer monitor, but we can still explore the 3-D space, which was holographically recorded, by calculating the complex amplitude distribution in any chosen plane of observation.

Holographic recording with a discrete CCD array might, on the face of it, appear difficult to implement because such arrays have much lower spatial resolution than the holographic recording materials discussed in Chapter 6. The smallest pixel size in a CCD camera is 6.5 μm, and the distance between neighboring pixels is typically 5 μm so that the maximum resolution is about 100 lines/mm. Even if we could fabricate smaller pixels, each would be illuminated by a smaller number of photons and the shot noise would become more significant. Despite this, there have been significant advances so that digital holography using CCD or CMOS cameras is now a useful technique with applications in metrology and microscopy.

13.2 SPATIAL FREQUENCY BANDWIDTH AND SAMPLING REQUIREMENTS

13.2.1 BANDWIDTH IN DIGITAL FOURIER HOLOGRAPHY

We have seen (Section 5.16) that in order to record the pattern of interference between two mutually coherent, plane wavefronts, we must use a recording material, which is capable of resolving the fringe spacing $\lambda[2\sin(\theta/2)]^{-1}$, where θ is the angle between the wave vectors, and the fringe spacing is typically a fraction of a micrometer for visible light. The use of Gabor, in-line holography (Section 3.4.3) keeps the resolution requirement to an absolute minimum, although we still have to deal with the twin-image problem.

To record a Fourier hologram of a photographic transparency using Gabor's method, we can use a plane wave traveling parallel to the lens axis as reference wave, and the largest spatial frequencies in the Fourier plane are those associated with the periphery of the transparency.

The transparency extends from $y = 0$ to $y = b$ along the y-axis (Figure 13.1), and the highest spatial frequency is obtained using Equation 12.34

$$F(y') \propto \int_y A\delta(y-b)\exp\left(-\frac{jkyy'}{f}\right)dy = A\exp\left(-\frac{jkby'}{f}\right) \tag{13.1}$$

where we have ignored the x dependence. A is the amplitude. The spatial frequency of this wavefront, measured along the y' axis, is

$$B = \frac{b}{\lambda f} \tag{13.2}$$

The spatial frequency bandwidth is therefore $b/\lambda f$. In order to record and reconstruct the expression $A\cos(kby'/f)$, it must be sampled at least *twice* within a single spatial period. By

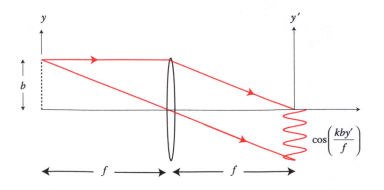

FIGURE 13.1 Spatial bandwidth.

doing so, we can write two equations to solve for the two unknowns, namely the amplitude A and the spatial frequency, that is,

$$A_1 = \cos\left(\frac{kby'}{\lambda f}\right), \quad A_2 = \cos\left[\frac{kb(y' + \Delta y')}{\lambda f}\right]$$

$$\frac{b}{\lambda f} = \frac{\cos^{-1}A_2 - \cos^{-1}A_1}{k\Delta y'} \tag{13.3}$$

with $\Delta y' < \lambda f / b$. The interval between samples must actually be *less* than one spatial period, as samples *exactly* a period apart have $A_1 = A_2$.

Reconstruction of all the spatial frequencies in a range of $b/\lambda f$ requires a spatial sampling rate of at least $2b/\lambda f$. This is the *Nyquist sampling criterion*. The spatial interval between samples is $\lambda f / 2b$.

13.2.2 BANDWIDTH IN DIGITAL FRESNEL HOLOGRAPHY

When recording an ordinary, Fresnel hologram instead of a Fourier hologram, we begin with Equation 2.15,

$$\tilde{A}(x',y') \cong \frac{-je^{jkz}}{\lambda z}\iint_{x,y}\tilde{A}(x,y)\exp\left\{\frac{jk}{2z}\left[(x'-x)^2 + (y'-y)^2\right]\right\}dxdy \tag{13.4}$$

or, in its alternative form, (Equation 2.16),

$$\tilde{A}(x',y') = \frac{-je^{jkz}}{\lambda z}\exp\left[\frac{jk}{2z}(x'^2 + y'^2)\right]$$

$$\iint_{x,y}\left\{\tilde{A}(x,y)\exp\left[\frac{jk}{2z}(x^2 + y^2)\right]\right\}\exp\left[-\frac{jk}{z}(xx' + yy')\right]dxdy \tag{13.5}$$

Equation 13.5 relates the complex amplitude $\tilde{A}(x',y')$ in the x',y' plane to that in the x,y plane. It is the Fourier transform of $\{\tilde{A}(x,y)\exp[(jk/2z)(x^2 + y^2)]\}$, multiplied by the quadratic phase term $\exp[(jk/2z)(x'^2 + y'^2)]$. The Fourier transform of $\{\tilde{A}(x,y)\exp[(jk/2z)(x^2 + y^2)]\}$ has the same spatial frequency bandwidth as before, with the substitution of z for f, determined by the dimensions of the transparency. However, we now have to add the spatial frequencies arising from the term $\exp[(jk/2z)(x'^2 + y'^2)]$. These spatial frequencies are band limited only by the aperture of the hologram in the x',y' plane. If the hologram extends from $y' = 0$ to $y' = h$, the maximum spatial frequency due to the quadratic phase term is $h/2\lambda z$ and the total spatial bandwidth is given by

$$B = \frac{b}{\lambda z} + \frac{h}{2\lambda z} = \frac{2b + h}{2\lambda z} \tag{13.6}$$

The sample rate required is therefore $(2b+h)/\lambda z$. The spatial interval between them is $\lambda z/(2b+h)$, which is rather smaller than that required to record the corresponding Fourier hologram.

By relating these expressions to the physical size of the CCD array and the number of pixels it contains, we can determine the maximum size of object from which a digital hologram can be recorded at distance, z, from the array.

13.3 RECORDING AND NUMERICAL RECONSTRUCTION

The first task is to record the hologram $H(x', y')$, which is the pattern of interference between the object wave $\tilde{O}(x', y')$ and a reference wave $\tilde{R}(x', y')$ in the hologram or CCD, plane x', y'.

$$H(x', y') = \left|\tilde{O}(x', y')\right|^2 + \left|\tilde{R}(x', y')\right|^2 + \tilde{O}(x', y')\tilde{R}^*(x', y') + \tilde{O}^*(x', y')\tilde{R}(x', y') \quad (13.7)$$

We then *multiply* $H(x', y')$ by a reference wave, $\tilde{R}'(x', y')$, which need not be same as that used in the recording, and using the Fresnel integral (Equation 13.5), we calculate the complex amplitude distribution of the reconstructed image, $\tilde{I}(x, y, d)$, in a plane x, y at distance, d, from the hologram to obtain

$$\tilde{I}(x, y, d) = \frac{-j\exp(jkd)}{\lambda d}\exp\left[\frac{jk}{2d}(x^2 + y^2)\right] \quad (13.8)$$

$$\times \iint_{x', y'} H(x', y')\tilde{R}'(x', y')\exp\left[\frac{jk}{2d}(x'^2 + y'^2)\right]\exp\left[-\frac{jk}{d}(xx' + yy')\right]dx'dy'$$

Equation 13.8 can be evaluated for any value of d and for any choice of $\tilde{R}'(x', y')$, including its wavelength. We do not need to satisfy the Bragg condition as it applies to volume holographic recording materials (see Chapter 4), and we can use the conjugate version of the original reference wave so that a real image is reconstructed at the original location of the object.

13.3.1 THE FRESNEL METHOD

The recorded hologram actually consists of a set of discrete samples of the interference pattern between object and reference waves, $\sum_{m=1}^{N}\sum_{n=1}^{N}H(m, n)$, where $N \times N$ is the size of the CCD array with each pixel having area $\Delta x'\Delta y'$. Our task now is to compute the image wavefront by mathematically combining a reconstructing wave with the stored digital hologram and, using the Fresnel integral (Equation 13.5), calculating the complex amplitude in the chosen plane x, y. Each sample $H(m, n)$ is multiplied by an appropriate reference wave term $\tilde{R}'(m, n)$. We calculate $\tilde{I}(x, y)$ at discrete points $p\Delta x, q\Delta y$ in the x, y plane where $p = 1, 2, \ldots,$

N and $q = 1, 2, \ldots, N$ and

$$\tilde{I}(p\Delta x, q\Delta y, d) = \frac{-j}{\lambda d} \exp(jkd) \exp\left(\frac{jk\lambda^2 d}{2}\left[\left(\frac{p\Delta x}{\lambda d}\right)^2 + \left(\frac{q\Delta y}{\lambda d}\right)^2\right]\right)$$

$$\times \sum_{m=1}^{N} \sum_{n=1}^{N} H(m\Delta x', n\Delta y') \tilde{R}'(m\Delta x', n\Delta y')$$

$$\times \exp\left[\frac{jk}{2d}(m^2\Delta x'^2 + n^2\Delta y'^2)\right] \exp\left\{-\frac{jk}{d}\left[(p\Delta x)(m\Delta x') + (q\Delta y)(n\Delta y')\right]\right\}$$

$$(13.9)$$

The size of the hologram is $N\Delta x' \times N\Delta y'$, and from Figure 13.2, the maximum spatial frequencies in the x, y plane are $N\Delta x'/2\lambda d$ and $N\Delta y'/2\lambda d$, which, by the Nyquist criterion, must be sampled at spatial rates of $N\Delta x'/\lambda d$ and $N\Delta y'/\lambda d$. Therefore, the spacings between samples in the image plane are

$$\Delta x = \frac{\lambda d}{N\Delta x'}, \quad \Delta y = \frac{\lambda d}{N\Delta y'} \tag{13.10}$$

The area of a pixel in the image and therefore the resolution in the image depend on the distance, d, between the hologram plane and the plane in which the reconstructed image is calculated, as well as on the area of the CCD array and the wavelength used in the calculation.

Equation 13.9 can be written as

$$\tilde{I}(p, q, d) = \frac{-j}{\lambda d} \exp(jkd) \exp\left(\frac{jk\lambda^2 d}{2}\left[\left(\frac{p}{N\Delta x'}\right)^2 + \left(\frac{q}{N\Delta y'}\right)^2\right]\right) \sum_{m=1}^{N} \sum_{n=1}^{N} H(m, n) \tilde{R}'(m, n)$$

$$\times \exp\left[\frac{jk}{2d}(m^2\Delta x'^2 + n^2\Delta y'^2)\right] \exp\left[-j2\pi\left(\frac{pm}{N} + \frac{qn}{N}\right)\right] \tag{13.11}$$

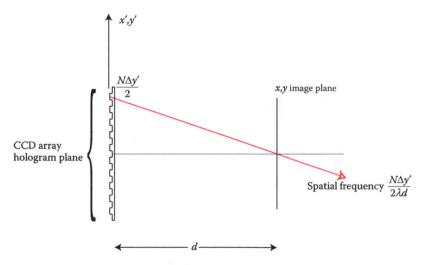

FIGURE 13.2 Digital hologram reconstruction.

We have omitted $\Delta x, \Delta y, \Delta x'$, and $\Delta y'$ where appropriate. Equation 13.11 enables us to calculate the reconstructed image from the recorded, sampled hologram $\Sigma_{m=1}^{N}\Sigma_{n=1}^{N}H(m,n)$ in about 1 s for $N = 1024$, using a personal computer. The result includes the out-of-focus, twin image $\tilde{O}^*(x',y')\tilde{R}(x',y')\tilde{R}'(x',y')$, as well as the zero-order term $\left(|\tilde{O}(x',y')|^2 + |\tilde{R}(x',y')|^2\right)\tilde{R}'(x',y')$. These can be suppressed by methods that we shall shortly discuss. The pure phase term in front of the double summation may be ignored if only the image intensity $|\tilde{I}(p,q)|^2$ is to be calculated. This phase term can also be ignored when one wishes to determine the phase differences between holograms if it does not change between exposures.

A numerical lens $\tilde{P}(m,n)$ may be added, and the reconstructed image appears in the plane that is conjugate to the plane at d. Aberrations of such a lens may also be numerically corrected using an appropriate phase function.

The intensity of the image is calculated from $\text{Re}^2[\tilde{I}(x,y)]+\text{Im}^2[\tilde{I}(x,y)]$. The phase distribution in the image can be calculated from

$$\phi(x,y) = \arctan\left\{\frac{\text{Im}\left[\tilde{I}(x,y)\right]}{\text{Re}\left[\tilde{I}(x,y)\right]}\right\} \tag{13.12}$$

A practical arrangement [1] for the recording and display of a digital hologram is shown in Figure 13.3. The plane reference wave is directed along the camera axis, minimizing the spatial resolution requirement. An example of a result obtained using Equation 13.8 is shown in Figure 13.4.

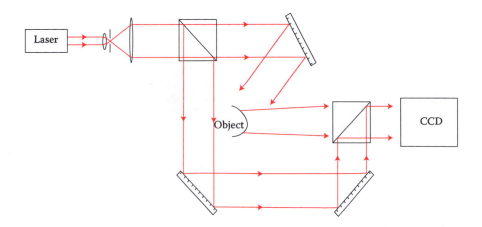

FIGURE 13.3 Digital holography. (Reproduced from Yan Li, Digital Holography and Optical Contouring, PhD Thesis, Liverpool John Moore's University, 2009, by kind permission of the author.)

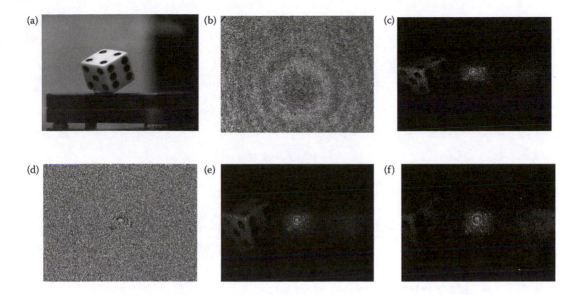

FIGURE 13.4 Digital holographic reconstruction: (a) object; (b) hologram; (c) Fresnel reconstruction from (b) (note the out-of-focus twin image on the right); (d) Fresnel reconstruction of phase image; (e) reconstruction by convolution; and (f) reconstruction by convolution with 5× magnification. (Reproduced from Yan Li, Digital Holography and Optical Contouring, PhD Thesis, Liverpool John Moore's University, 2009, by kind permission of the author.)

13.3.2 THE CONVOLUTION METHOD

The fact that the image resolution depends on the distance, d, between the hologram plane and the reconstruction plane is rather a disadvantage, and the convolution method of reconstruction avoids this difficulty. We can write Equation 13.8 in the form

$$\tilde{I}(x,y,d) = \frac{-j}{\lambda d}\exp(jkd)\iint_{x',y'}\left[H(x',y')\tilde{R}'(x',y')\right]\exp\left[\frac{jk}{2d}(x-x')^2+(y-y')^2\right]dx'dy'$$

(13.13)

Using Equation B.2 (Appendix B), Equation 13.13 becomes

$$\tilde{I}(x,y,d) = \frac{-j\exp(jkd)}{\lambda d}\left[H(x,y)\tilde{R}'(x,y)\right]\otimes\exp\left[\frac{jk}{2d}(x^2+y^2)\right]$$

Applying the convolution theorem (Appendix B)

$$\text{SFT}\left[\tilde{I}(x,y,d)\right] = \text{SFT}\left[H(x',y')\tilde{R}'(x',y')\right]\text{SFT}\left\{\frac{-j\exp(jkd)}{\lambda d}\exp\left[\frac{jk}{2d}(x'^2+y'^2)\right]\right\}$$ (13.14)

The spatial Fourier transform on the right-hand side of Equation 13.14 can be written as

$$\text{SFT}\left\{\frac{-j\exp(jkd)}{\lambda d}\exp\left[\frac{jk}{2d}(x'^2+y'^2)\right]\right\} = \frac{-j\exp(jkd)}{\lambda d}\exp\left[\frac{-jk}{2d}(x^2+y^2)\right]$$

$$\times \iint\exp\left[\frac{jk\left[(x-x')^2+(y-y')^2\right]}{2d}\right]dxdy$$

We use the identity $\int_{-\infty}^{\infty}\exp[jx^2]dx = \sqrt{(\pi/2)}(1+j)$ to obtain

$$\text{SFT}\left\{\frac{-j\exp(jkd)}{\lambda d}\exp\left[\frac{jk}{2d}(x'^2+y'^2)\right]\right\} = \exp\left[\frac{-jk}{2d}(x'^2+y'^2-2d^2)\right]$$

Equation 13.14 now becomes

$$\text{SFT}\left[\tilde{I}(x,y,d)\right] = \exp\left[\frac{-jk}{2d}(x'^2+y'^2-2d^2)\right]\text{SFT}\left[H(x',y')\tilde{R}'(x',y')\right] \quad (13.15)$$

To obtain $\tilde{I}(x,y,d)$, we form the product of $R'(x'y')$ and the recorded hologram $H(x'y')$, calculate its spatial Fourier transform, and multiply by $\exp[(-jk/2d)(x'^2+y'^2-2d^2)]$. We then calculate the inverse spatial Fourier transform of the result. The computation involves one Fourier transform and one inverse transform, both of which can be implemented as fast Fourier transforms.

In the convolution approach, the image pixel size is the same as that of the CCD array, so the resolution in the image is independent of the reconstruction distance, d. We can magnify the image by adding a numerical lens given by the expression $\tilde{L}(x,y) = \exp\left[(jk/2f)(x^2+y^2)\right]$ where f is the focal length (see Equation 2.19) so that Equation 13.15 becomes

$$\text{SFT}\left[\tilde{I}(x,y,d)\right] = \exp\left[\frac{-jk}{2d}(x'^2+y'^2-2d^2)\right]\text{SFT}\left[H(x,y)\tilde{R}(x,y)L(x,y)\right] \quad (13.16)$$

The introduction of a lens of focal length, f, means that the image calculated at distance d is no longer in focus, and we must recalculate it at d' with d and d' related by Equation 1.19, that is, $(1/f) = (1/d) + (1/d')$ and the magnification is (d'/d). As already indicated, the lens aberrations can be corrected by adding an appropriate phase correction term $\tilde{P}(x,y)$, and finally we have

$$\tilde{I}(x,y,d) = \tilde{P}(x,y)\text{SFT}^{-1}\left\{\exp\left[\frac{-jk}{2d}(x'^2+y'^2-2d^2)\right]\text{SFT}\left[H(x,y)\tilde{R}(x,y)L(x,y)\right]\right\} \quad (13.17)$$

The correct form of $\tilde{P}(x,y)$ can be determined by adjusting it until an object in the form of an illuminated pinhole is satisfactorily imaged in reconstruction [2].

13.3.3 THE ANGULAR SPECTRUM METHOD

The Fresnel approximation requires that the distance, d, from the object plane, at which the hologram is recorded, should satisfy the inequality associated with Equation 2.15, that is,

$$d^3 >> \frac{\left[(x'-x)^2 + (y'-y)^2\right]^2}{8\lambda} \tag{13.18}$$

This restriction can be avoided by the angular spectrum method in which the spatial Fourier transform of the recorded hologram is calculated and spatially filtered in order to extract the angular spectrum of the object wave [3]. If we use a plane reference wave at angle θ to the horizontal axis given by $\tilde{R}(x', y') = \exp(jky'\sin\theta)$, Equation 13.7 becomes

$$H(x', y') = 1 + \left|\tilde{O}(x', y')\right|^2 + \tilde{O}(x', y')\exp(-jky'\sin\theta) + \tilde{O}^*(x', y')\exp(jky'\sin\theta) \tag{13.19}$$

Using $-\theta k k_y \sin'$max the convolution theorem (Appendix B),

$$\begin{aligned}
\text{SFT}\left[H(x', y')\right] = H\left(k_{x'}, k_{y'}\right) &= \delta(k_{x'}, k_{y'}) + \tilde{O}(k_{x'}, k_{y'}) \oplus \tilde{O}(k_{x'}, k_{y'}) \\
&+ \tilde{O}(k_{x'}, k_{y'}) \otimes \delta(k_{y'} - k\sin\theta) + \tilde{O}^*(k_{x'}, k_{y'}) \otimes \delta(k_{y'} + k\sin\theta)
\end{aligned} \tag{13.20}$$

The angular spectrum, $H(k_{x'}, k_{y'})$, of $H(x', y')$ is shown in Figure 13.5. The angle θ is chosen so that $k\sin\theta \geq 3k_{y'\text{max}}$, where $k_{y'\text{max}}$ is the highest spatial frequency in the object wave. This is to ensure that the various terms in the angular spectrum $H(k_{x'}, k_{y'})$ do not overlap. A bandpass filter centered on spatial frequency $(0, k\sin\theta)$ is used to isolate $\tilde{O}(k_{x'}, k_{y'})$, which is then spatial frequency shifted to center on $(0, 0)$. The object wave in the hologram plane is obtained from the spatial Fourier transform of $\tilde{O}(k_{x'}, k_{y'})$.

$$\tilde{O}_{z'=0}(x', y') = \iint_{k_x, k_y} \tilde{O}_{z=0}(k_{x'}, k_{y'}) e^{-j(k_{x'}x' + k_{y'}y')} dk_x dk_y \tag{13.21}$$

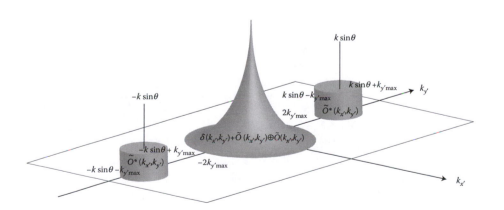

FIGURE 13.5 Angular spectrum of a hologram.

Since the angular spectrum of the object wave consists of a set of plane waves with propagation vectors $k_{x'}, k_{y'}, \sqrt{1-k_{x'}^2-k_{y'}^2}$, the object wave in a plane at $-z'$ is related to that at $z' = 0$ by

$$\tilde{O}_{z'}(x',y') = \tilde{O}_{z'=0}\exp(-jk_{z'}z') = \iint_{kx,ky}\tilde{O}_{z'=0}(k_{x'},k_{y'})e^{-j(k_xx'+k_yy'+k_{z'}z')}dk_xdk_y \qquad (13.22)$$

and we can calculate the complex amplitude of the image without the restriction imposed by the Fresnel approximation.

For comparison, Figure 13.6 shows digital holographic reconstructions obtained from the same hologram by means of the angular spectrum, convolution, and Fresnel methods.

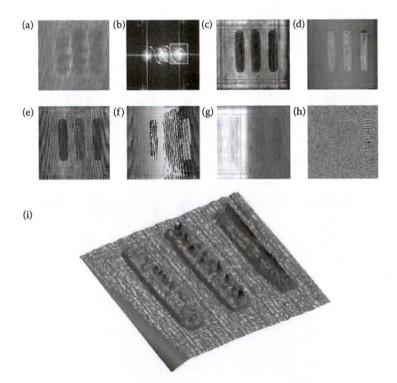

FIGURE 13.6 Holography of a resolution target. The image area is 25×25 μm^2 (452×452 pixels) and the image is at $z = 7$ μm from the hologram: (a) hologram; (b) angular spectrum; (c) amplitude; and (d) phase images by angular spectrum method; (e) amplitude and (f) phase images by convolution method; (g) amplitude and (h) phase images by Fresnel method; (i) 3D grayscale rendering of (d). The individual bars are 2.2 μm wide. (Reproduced from P. Marquet, B. Rappaz, P. J. Magistretti, E. Cuche, Y. Emery, T. Colomb, and C. Depeursinge, Digital holographic microscopy: a noninvasive contrast imaging technique allowing quantitative visualization of living cells with subwavelength axial accuracy *Opt. Lett.*, 30, 468–70, 2005. With permission of Optical Society of America.)

13.4 SUPPRESSION OF THE ZERO-ORDER AND THE TWIN IMAGE

There are three major difficulties with digital holography. We have already referred to one of them, namely the comparatively low spatial resolution of the CCD arrays. Other problems are the twin-image term, $\tilde{O}*(x',y')\tilde{R}'(x',y')$ and the zero-order term, $|\tilde{O}(x',y')|^2 + |\tilde{R}'(x',y')|^2$, in Equation 13.7.

The zero-order and twin-image terms can be suppressed by a number of methods. We have already explained in the discussion of the angular spectrum method of reconstruction (Section 13.3.3) how spatial filtering may be used to isolate the desired image term.

Other methods involve image processing techniques to remove the zero-order term. Still, others employ shutters to block the object and reference beams alternately so that the interference term may be isolated. Phase shifting may be added to facilitate removal of the twin image.

13.4.1 REMOVAL OF ZERO-ORDER TERM BY IMAGE PROCESSING

The simplest method of zero-order suppression is to subtract the average value, of the hologram from $H(m,n)$ so that the hologram is now given by

$$H'(m,n) = H(m,n) - \frac{1}{MN} \sum_{m=1}^{N} \sum_{n=1}^{N} H(m,n) \tag{13.23}$$

The procedure will of course produce some negative values, but these can simply be shifted. Essentially, this is a high-pass spatial frequency filtering operation, which greatly attenuates the effect of the low spatial frequencies associated with $|\tilde{O}(x',y')|^2 + |\tilde{R}'(x',y')|^2$.

Another effective filter is implemented by calculating a local average for $H(m, n)$, which is then subtracted from $H(m, n)$ to give

$$H'(m,n) = H(m,n) - \frac{1}{9}\left[\begin{array}{l} H(m-1,n-1) + H(m-1,n) + H(m-1,n+1) + H(m,n-1) + H(m,n) \\ +H(m,n+1) + H(m+1,n-1) + H(m+1,n) + H(m+1,n+1) \end{array} \right]$$

$$\tag{13.24}$$

Figure 13.7 shows the results obtained using global and local averages to suppress the zero-order term.

13.4.2 SUBTRACTION

The zero-order term can of course be removed by *separately* recording the intensities of the object and reference beams and subtracting them from the hologram. Three CCD arrays of intensity data are obtained using a shutter in the path of the reference beam to record $|\tilde{O}(x',y')|^2$ and another shutter in the path of the object beam to record $|\tilde{R}'(x',y')|^2$, and

FIGURE 13.7 (a) Zero-order suppression by subtracting the mean pixel value from the hologram before numerical reconstruction; (b) using the local average method of Equation 13.21; and (c) zero-order and twin-image suppression by spatial filtering in the Fourier plane. (Reproduced from Yan Li, Digital Holography and Optical Contouring, PhD Thesis, Liverpool John Moore's University, 2009, by kind permission of the author.)

finally both shutters are opened to record $H(x', y')$ and

$$
\begin{aligned}
H'(x', y') &= H(x', y') - \left|\tilde{O}(x', y')\right|^2 - \left|R(x', y')\right|^2 \\
&= \tilde{O}(x', y')R^*(x', y') + \tilde{O}^*(x', y')R(x', y')
\end{aligned}
\tag{13.25}
$$

In a refinement of the method [4], only two frames of data are recorded with a phase change of π in the reference beam between the recordings, which are then subtracted from one another to give $H'(x', y')$ with

$$
H'(x', y')/2 = \left[\tilde{O}(x', y')\tilde{R}^{*\prime}(x', y') + \tilde{O}^*(x', y')\tilde{R}'(x', y')\right]
\tag{13.26}
$$

13.4.3 PHASE SHIFT METHODS

The use of phase shifts in the reference beam between recordings enables removal of *both* the zero-order and the twin image. In the first of these methods, we use the shutter technique to obtain $H'(x', y')$ in Equation 13.27 and then repeat the process with the phase of the reference beam altered by ϕ to obtain $H'_\phi(x', y')$. We then multiply the latter by $\exp(j\phi)$ and form the sum

$$
\begin{aligned}
H'(x', y') + \exp(j\phi)H'_\phi(x', y') &= \tilde{R}\tilde{O}^* + \tilde{R}^*\tilde{O} + \exp(2j\phi)\left[\tilde{R}\tilde{O}^* + \tilde{R}^*\tilde{O}\exp(-2j\phi)\right] \\
&= 2\tilde{R}^*\tilde{O} + \tilde{R}\tilde{O}^*\left[1 + \exp(-2j\phi)\right]
\end{aligned}
\tag{13.27}
$$

If $\phi = \pi/2$, the right-hand side of Equation 13.24 reduces to $2\tilde{R}^*\tilde{O}$.

Finally [4], we can remove the zero-order term and the twin image by using two-phase shifts. We first obtain $|\tilde{O}|^2 + |\tilde{R}|^2$ using the shutter technique and then obtain

$$
\begin{aligned}
H'_{\phi_1}(x', y') - H'(x', y') &= \tilde{O}\tilde{R}^*\left[\exp(-j\phi_1) - 1\right] + \tilde{O}^*\tilde{R}\left[\exp(j\phi_1) - 1\right] \\
H'_{\phi_2}(x', y') - H'(x', y') &= \tilde{O}\tilde{R}^*\left[\exp(-j\phi_2) - 1\right] + \tilde{O}^*\tilde{R}\left[\exp(j\phi_2) - 1\right]
\end{aligned}
\tag{13.28}
$$

$$\frac{H'_{\phi_1}(x',y')-H'(x',y')}{\exp(j\phi_1)-1}=\tilde{O}\tilde{R}*\frac{\exp(-j\phi_1)-1}{\exp(j\phi_1)-1}+\tilde{O}*\tilde{R},$$

$$\frac{H'_{\phi_2}(x',y')-H'(x',y')}{\exp(j\phi_2)-1}=\tilde{O}\tilde{R}*\frac{\exp(-j\phi_2)-1}{\exp(j\phi_2)-1}+\tilde{O}*\tilde{R}$$

Subtraction gives

$$\tilde{O}\tilde{R}*=\left[\frac{H'_{\phi_1}(x',y')-H'(x',y')}{\exp(j\phi_1)-1}-\frac{H'_{\phi_2}(x',y')-H'(x',y')}{\exp(j\phi_2)-1}\right]\left[\frac{\exp(-j\phi_1)-1}{\exp(j\phi_1)-1}-\frac{\exp(-j\phi_2)-1}{\exp(j\phi_2)-1}\right]^{-1}$$

$$(13.29)$$

Phase changes are introduced by a number of methods. All of them effectively require a change in the optical path length of the reference beam. One method is to reflect the reference beam from a mirror whose position is adjusted with subwavelength precision by means of a piezoelectric transducer attached to it. In another method, the reference beam propagates along a single-mode optical fiber coiled around a hollow piezoelectric cylinder, which expands on application of a voltage across the cylinder wall, stretching the fiber. A spatial light modulator (Section 8.8.2.1) may also be used. Alternatively, a single longitudinal mode laser diode can be current modulated to alter its output wavelength.

13.4.4 SUPPRESSION OF ZERO-ORDER AND TWIN IMAGE IN GABOR HOLOGRAPHY

Because the resolution of the CCD camera is so low, the removal of zero-order and twin-image terms is important in Gabor holography. Phase-shift methods [5] can be applied to the problem.

In one method, three frames are recorded with phase shifts of 0, $\pi/2$, and π, and we compute the expression

$$H(x',y')-H_{\pi/2}(x',y')+j\left[H_{\pi/2}(x',y')-H_{\pi}(x',y')\right]$$

$$=\tilde{R}\tilde{O}*+\tilde{O}\tilde{R}*-\tilde{R}\tilde{O}*\exp\left(\frac{j\pi}{2}\right)-\tilde{O}\tilde{R}*\exp\left(\frac{-j\pi}{2}\right)$$

$$+j\left[\tilde{R}\tilde{O}*\exp\left(\frac{j\pi}{2}\right)+\tilde{O}\tilde{R}*\exp\left(\frac{-j\pi}{2}\right)\right]-j\left[\tilde{R}\tilde{O}*\exp(j\pi)+\tilde{O}\tilde{R}*\exp(-j\pi)\right]$$

$$=\tilde{R}\tilde{O}*+\tilde{O}\tilde{R}*-j\tilde{R}\tilde{O}*+j\tilde{O}\tilde{R}*-\tilde{R}\tilde{O}*+\tilde{O}\tilde{R}*+j\tilde{R}\tilde{O}*+j\tilde{O}\tilde{R}*=2\tilde{O}\tilde{R}*(1+j)$$

$$(13.30)$$

$$\tilde{O}(x',y')=\frac{1}{2\tilde{R}*(x',y')}(1-j)\left\{H(x',y')-H_{\pi/2}(x',y')+j\left[H_{\pi/2}(x',y')-H_{\pi}(x',y')\right]\right\}$$

$$(13.31)$$

Alternatively, phase shifts of 0, $\pi/2$, π, and $3\pi/2$ may be used , in which case

$$\tilde{O}(x',y')=\frac{1}{4\tilde{R}*(x',y')}\left\{H(x',y')-H_{\pi}(x',y')+j\left[H_{\pi/2}(x',y')-H_{3\pi/2}(x',y')\right]\right\}$$

$$(13.32)$$

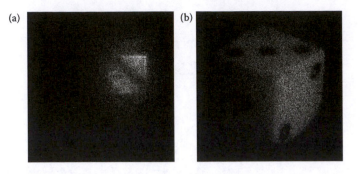

FIGURE 13.8 Reconstruction from (a) in-line hologram and (b) using four-frame algorithm of Equation 13.30. (Reproduced from Yan Li, Digital Holography and Optical Contouring, PhD Thesis, Liverpool John Moore's University, 2009, by kind permission of the author.)

In both cases, the object wave is reconstructed by multiplying the computed expressions in Equations 13.31 and 13.32 by $\tilde{R}(x', y')$. A result obtained using the four-frame algorithm of Equation 13.30 is shown in Figure 13.8.

Another method involves phase shifts of θ_1, θ_2, and θ_3 [6], and $H_F(x', y')$ is given by

$$
\begin{aligned}
H_F(x', y') =& \\
& \left[\left|\tilde{O}(x', y')\right|^2 + \left|R(x', y')\right|^2 + \tilde{O}(x', y')R*(x', y')e^{-j\theta_1} + \tilde{O}*(x', y')R(x', y')e^{j\theta_1} \right] \\
& \times (e^{j\theta_3} - e^{j\theta_2}) + \left[\left|\tilde{O}(x', y')\right|^2 + \left|R(x', y')\right|^2 + \tilde{O}(x', y')R*(x', y')e^{-j\theta_2} + \tilde{O}*(x', y')R(x', y')e^{j\theta_2} \right] \\
& \times (e^{j\theta_1} - e^{j\theta_3}) \left[\left|\tilde{O}(x', y')\right|^2 + \left|R(x', y')\right|^2 + \tilde{O}(x', y')R*(x', y')e^{-j\theta_3} + \tilde{O}*(x', y')R(x', y')e^{j\theta_3} \right] \\
& \times (e^{j\theta_2} - e^{j\theta_1})
\end{aligned}
$$

$$(13.33)$$

The coefficient of $\left|\tilde{O}(x', y')\right|^2 + \left|R(x', y')\right|^2$ is zero, and if $\theta_1 = 0, \theta_2 = 2\pi/3$, and $\theta_3 = 4\pi/3$, the coefficient of $\tilde{O}*(x', y')R(x', y')$ is also zero.

13.5 IMPROVING THE RESOLUTION IN DIGITAL HOLOGRAPHY

There are three ways to improve the holographic performance of a CCD camera. The simplest is to reduce the effective size of the object by forming its image using a diverging lens, that is, one with a negative focal length.

A second method called *aperture synthesis* [7] is used extensively in radio astronomy and radar. The basic principle is that a large hologram may be assembled from a set of holograms from a number of CCD arrays whose locations in the x', y' plane are known very precisely. Alternatively, a single CCD array may be used to record holograms in various precisely known and specified locations in the plane. One could, in principle, record as large a hologram as required.

The third technique enables the Nyquist criterion to be satisfied for higher spatial frequencies than normally possible with a CCD camera. Equation 13.6 shows that the spatial sampling rate, $1/\Delta x$, has to exceed $(2b+h)/\lambda z$. If we assume an object of negligible size, b, then, with $h = N\Delta x$, $z > N\Delta x^2/\lambda$. If this condition is not met, the Nyquist criterion is not satisfied and aliasing occurs. As a result, spatial frequencies lying outside the sample bandwidth are folded into the recorded spatial spectrum. In practice, this means that the two images $\tilde{O}(x',y')$ and $\tilde{O}^*(x',y')$ overlap one another. However, by translating the CCD array through a distance of a fraction of a pixel between recordings [8], the interference pattern can be sampled at spatial intervals smaller than a pixel width. In principle, piezo-electrically controlled translations of 1 μm allow resolution capability comparable with some conventional holographic recording materials although the hologram size essentially remains that of the CCD array.

Despite the limited spatial resolution of the CCD camera, digital holography has a number of important applications. The techniques of holographic interferometry described in Chapter 10, including contouring [8] and vibration analysis [9], can be implemented for the study of small-sized objects, and Figure 13.9 shows some of the excellent results, which have been obtained using digital-pulsed holographic interferometry.

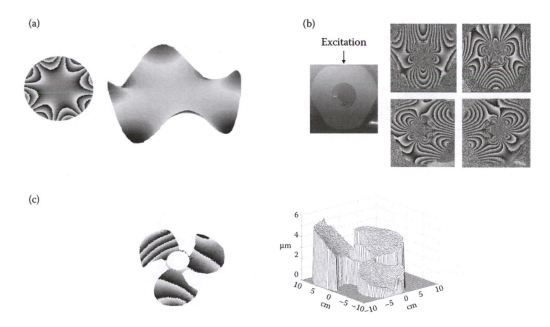

FIGURE 13.9 Pulsed digital holographic interferometry. (a) Plate (diameter 15 cm) vibrating at 4.800 Hz; left, phase map (see Section 10.13); right, displacement map. (b) Tube drawing tool (left) vibrating at 21.5 kHz with phase maps obtained by illumination in four different directions (right). (c) Deformation of a three-blade fan (diameter 21 cm) rotating at 1000 rpm; left, phase map; right, 3-D displacement. (Images by kind permission G. Pedrini, Institut für Technische Optik, Universität Stuttgart.)

13.6 DIGITAL HOLOGRAPHIC MICROSCOPY

Microscopy is used extensively in biology and in medicine to study phase objects, which are largely transparent with only variations in thickness or refractive index to diffract light for imaging purposes. Phase-contrast microscopy, described in Section 1.8.1, is an excellent example of a technique used to solve the problem. Another is dark ground microscopy, in which undiffracted light passing through a phase object is prevented from reaching the image by means of a simple high-pass spatial filter in the form of a small opaque circular disc in the Fourier plane, the disc diameter being approximately that of the resolution limit of the microscope objective. Interference microscopy, in which the object wavefront interferes with a laterally sheared version of itself to produce interference fringes, which are phase contours, is also widely used. All of these techniques are essentially qualitative while digital holographic microscopy (DHM) enables quantitative data to be obtained including measurements on nuclear membranes in cells [10] (Figure 13.10) and neurons [11] (Figure 13.11) with subwavelength accuracy. Object dimensions are obtained from the reconstructed and unwrapped phase image (Equation 13.12). The height difference $h_2 - h_1$ between points in reconstructed images with phases ϕ_2 and ϕ_1 is

$$h_2 - h_1 = \frac{\lambda}{2\pi}\left[\phi_2 - \phi_1\right]$$

(13.34)

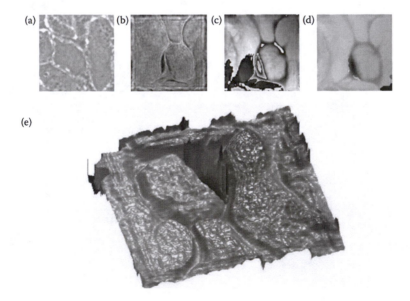

FIGURE 13.10 Holography of confluent SKOV-3 ovarian cancer cells. The image area is 60 × 60 μm² (404 × 404 pixels), and the image is at $z = 10$ μm from the hologram: (a) Zernike phase contrast image; (b) holographic amplitude and (c) phase images; (d) unwrapped phase image; (e) 3-D grayscale rendering of (d). (Reproduced from C. J. Mann, L. Yu, C-M. Lo, and M. K. Kim, High-resolution quantitative phase-contrast microscopy by digital holography, *Opt. Express*, 13, 8693–98, 2005. With permission of Optical Society of America.)

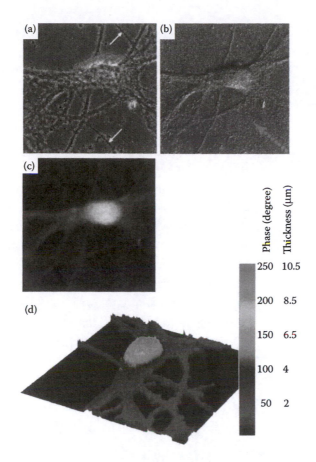

FIGURE 13.11 Images of a living mouse cortical neuron: (a) dark-phase contrast image in culture; (b) differential interference contrast image; (c) raw image; (d) perspective image in gray scale of the phase distribution obtained with DHM. While arrows in (a) indicate artifacts not visible in (b), (a) also shows undesirable halo effects common in phase-contrast images. (Reproduced from P. Marquet, B. Rappaz, P. J. Magistretti, E. Cuche, Y. Emery, T. Colomb, and C. Depeursinge, Digital holographic microscopy: a noninvasive contrast imaging technique allowing quantitative visualization of living cells with subwavelength axial accuracy, *Opt. Lett.*, 30, 468–70, 2005. With permission of Optical Society of America.)

As we know, digital holographic reconstruction allows the reconstructed image to be calculated at any distance from the hologram and displayed as a tomographic or planar section of the reconstructed image space. Although the use of a numerical lens enables us to focus sharply on a plane in image space, with the depth of field determined by the lens, the images of interest may still be obscured by out-of-focus images in other image planes.

An additional problem is that it is assumed in in-line or nearly in-line holography that the light is diffracted just once on its way to the hologram plane. This is known as the first

Born approximation. In practice, the light may be diffracted or scattered many times especially when the object consists of dense biological tissue.

These problems can be avoided by either multiple wavelength or short coherence length digital holography [3], which allows reconstruction only in a plane defined with very high precision, so that very high axial resolution is possible.

13.6.1 MULTIPLE WAVELENGTH METHOD

The multiple wavelength technique is possible because Bragg conditions do not need to be met for digital holographic reconstruction. To see how it works, we record a number of digital holograms using a dye laser stepped through its bandwidth, a hologram being recorded at each step. The object may be an amplitude or phase object. In the former case, laser light is reflected from it; in the latter, light is transmitted through the object. Suppose we record a number of holograms of a phase object at various wavelengths such that the difference in propagation constant is $\Delta k = \Delta(2\pi/\lambda) = -2\pi\Delta\lambda/\lambda^2$. The sum $\tilde{S}(x,y)$ of the reconstructed complex amplitudes at distance, d, from the x,y plane is given by

$$\tilde{S}(x,y,d) = \tilde{O}(x,y,d)\left\{\exp(jkd) + \exp\left[j(k+\Delta k)d\right] + \exp\left[j(k+2\Delta k)d\right] \right. \\ \left. + \cdots + \exp\left[j(k+(N-1)\Delta k)d\right]\right\} \tag{13.35}$$

$$\tilde{S}(x,y,d) = \tilde{O}(x,y,d)\exp(jkd)\left[1 + \exp(j\Delta kd) + \exp(2j\Delta kd) + \cdots + \exp\left\{(N-1)j\Delta kd\right\}\right] \\ = \tilde{O}(x,y,d)\exp(jkd)\frac{1-\exp(jN\Delta kd)}{1-\exp(j\Delta kd)} \tag{13.36}$$

The image intensity is $|\tilde{O}(x,y,d)|^2 \sin^2(N\Delta kd)/\sin^2(\Delta kd)$, which has maximum values at $\Delta kd = n\pi$ with n an integer and of width $2\pi/N$ (see Section 2.2.1.3). This means that the objects in the x,y plane are *repeatedly* imaged in planes separated by distance, d, given by

$$d = \frac{n\pi}{\Delta k} = \frac{n\lambda^2}{2\Delta\lambda} \tag{13.37}$$

and the *depth of field in these image planes* is given by

$$\Delta z = \frac{2\pi}{N\Delta k} = \frac{\lambda^2}{N\Delta\lambda} \tag{13.38}$$

We therefore choose $\Delta\lambda$ small enough to ensure from Equation 13.37 that there is no repetition of images within the region of interest. We choose also a laser bandwidth $N\Delta\lambda$ that provides as small a depth of field as desired.

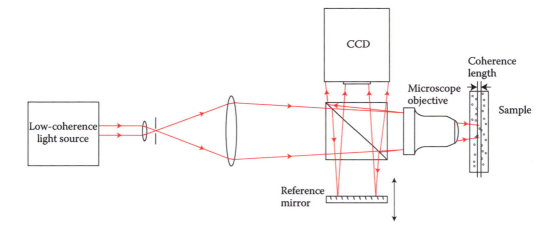

FIGURE 13.12 Optical coherence tomography and DHM. (Reproduced from P. Massatsch, F. Charrière, E. Cuche, P. Marquet, and C. D. Depeursinge, Time-domain optical coherence tomography with digital holographic microscopy, *Appl. Opt.*, 44, 1806–12, 2005. With permission of Optical Society of America.)

13.6.2 OPTICAL COHERENCE TOMOGRAPHY AND DIGITAL HOLOGRAPHIC MICROSCOPY

Another method for obtaining a very small depth of field is *optical coherence tomography* (OCT) [13]. As its name implies, the technique exploits the coherence length of the light source. To see how it works, we recall that interference effects may only be observed if the optical path difference between the interfering light waves is less than the coherence length of the light source. Holography is subject to the same condition, since holograms are recordings of interference patterns. This means that the depth of 3-D image space, which is reconstructed from a hologram, is less than the coherence length, or half the coherence length in the case of a Denisyuk hologram, of the light source used to record it. If we use a light source of very short coherence length, then we can confine the depth of the image reconstruction within a very narrow range. The principle of OCT is illustrated in Figure 13.12. The region of the image space, which is reconstructed, is determined by the position of the plane mirror, which is translated parallel to itself to select different axial regions for recording and reconstruction, the extent of each axial region being determined by the coherence length of the light source. Within each region, numerical focusing enables even greater axial resolution in the reconstructed image.

However, the optical coherence method requires that a hologram be recorded for each image slice to be examined and the need to translate the reference mirror makes for rather slow data acquisition.

13.6.3 OPTICAL SCANNING AND NONSCANNING DIGITAL HOLOGRAPHIC MICROSCOPY

Optical scanning holography can be used to form 3-D images of objects, which emit *incoherent* fluorescent light.

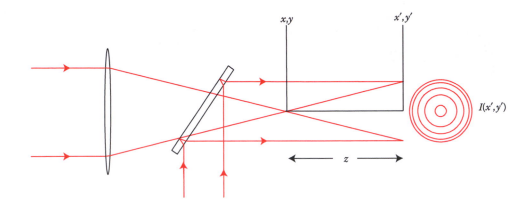

FIGURE 13.13 Fresnel zone interference pattern.

When a collimated beam of light interferes with a collinear, coherent, frequency shifted, focused beam (Figure 13.13), the resulting intensity distribution is given by

$$
I(x',y') = \left| A\exp\left[j(kz - \omega_0 t)\right] - \frac{je^{jkz}}{\lambda z} \iint_{x,y} B\delta(x,y)\exp\left\{\frac{jk}{2z}\left[(x'-x)^2 + (y'-y)^2\right]\right\}dxdy \right.
$$

$$
\left. \times \exp\left[-j(\omega_0 + \Omega)t\right] \right|^2 = \left| A - \frac{j}{\lambda z}B\exp\left[j\left(k\frac{x'^2+y'^2}{2z} - \Omega t\right)\right] \right|^2
$$

$$
= A^2 + \frac{B^2}{\lambda^2 z^2} - \frac{jAB}{\lambda z}\exp\left[j\left(k\frac{x'^2+y'^2}{2z} - \Omega t\right)\right] + \frac{jAB}{\lambda z}\exp\left[-j\left(k\frac{x'^2+y'^2}{2z} - \Omega t\right)\right]
$$

$$
I(x',y') = A^2 + \frac{B^2}{\lambda^2 z^2} + \frac{2AB}{\lambda z}\sin\left[k\frac{x'^2+y'^2}{2z} - \Omega t\right]
$$

$$(13.39)$$

A and B are constants. The frequency shift, Ω, produced by passing the beam through an acoustooptic modulator (Section 8.8.1) is typically much lower than ω_0 so that the wavelength is the same in each beam. Equation 13.39 describes a time-dependent intensity distribution known as a *Fresnel zone pattern* (FZP). It may be moved over the object by means of a scanning mirror rotated about two orthogonal axes in its own plane, as shown in Figure 13.14 [14].

The object is regarded as consisting of a number of points at various locations. Consider a single object point at $x_0, y_0,$ distance z_0 from the focal plane of the lens L1. It is illuminated by the FZP of Equation 13.39, and we obtain an A.C. signal current $i(x',y')$ in the photodetector

$$
i(x,y) \propto \sin\left[k\frac{(x'-x_0)^2 + (y'-y_0)^2}{2z_0} - \Omega t\right] \tag{13.40}
$$

which is multiplied by $\sin(\Omega t)$ giving demodulated signal current,

$$
i_d(x',y') = \frac{1}{2}\left\{\cos\left[k\frac{(x'-x_0)^2 + (y'-y_0)^2}{2z_0}\right] + \cos\left[k\frac{(x'-x_0)^2 + (y'-y_0)^2}{2z_0} - 2\Omega t\right]\right\}
$$

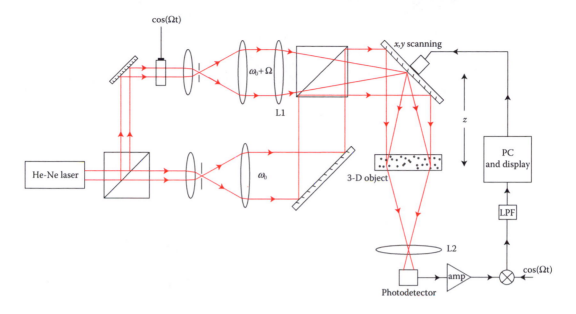

FIGURE 13.14 Optical scanning holography. (Reproduced from T-C. Poon, K. B. Doh, B. W. Schilling, M. H. Wu, K. K. Shinoda, and Y. Suzuki, Three-dimensional microscopy by optical scanning holography, *Optical Engineering*, 34, 1338–44, 1995, by kind permission Society of Photoinstrumentation Engineers.)

which is low-pass filtered to remove the time-dependent term. A bias current is added to give

$$i_d(x', y') = 1 + \cos\left[k\frac{(x'-x_0)^2 + (y'-y_0)^2}{2z_0}\right] \tag{13.41}$$

This signal is an electronic analog of the intensity distribution obtained when a plane wave propagating along the optical axis interferes with the spherical wavefront from the object point at x_0, y_0, z_0. It is recorded and stored along with the positional coordinates of the FZP as an electronic hologram. The significant point to be made here is that this hologram is also obtained if the object point emits fluorescent light when illuminated by the FZP [15]. Numerical reconstruction and focusing may also be implemented.

The technique has been developed further, stimulated by the prospect of holography using not just fluorescent light but incoherent light generally. At the same time, one would like to avoid mechanical scanning because of the time it takes to complete. This has led to the development of Fresnel incoherent correlation holography [6,16] shown schematically in Figure 13.15a.

At the heart of the system is a spatial light modulator, into which is written a diffractive optical element described by the expression

$$R_k(x, y) = 1 + \exp\left[\frac{jk}{2a}(x^2 + y^2) + j\theta_k\right] \tag{13.42}$$

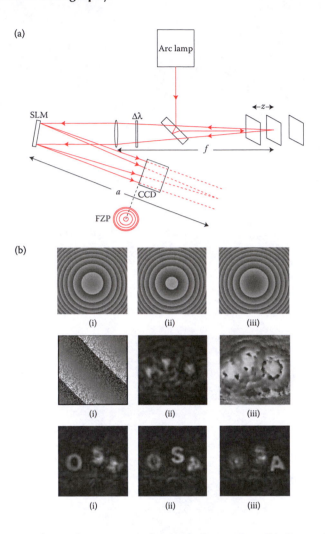

FIGURE 13.15   (a) Fresnel incoherent correlation holography. (b) Fresnel incoherent correlation holography. Top: phase distributions on reflection SLM with (i) $\theta = 0°$, (ii) 120°, (iii) 240°. Middle: (i) enlargement showing half of pixels (randomly chosen) with constant phase (ii) magnitude, and (iii) phase of final hologram. Bottom: Reconstructions at best axial positions of letters (i) O, (ii) S, and (iii) A. (Reproduced from J. Rosen and G. Brooker, Digital spatially incoherent Fresnel holography, *Opt. Lett.*, 32, 912–14, 2007. With permission of Optical Society of America.)

where $k = 1, 2, 3$. The term $j\theta_k$ is needed to deal with the zero-order and twin-image terms. Three holograms are recorded with phase shifts of $\theta = 0°$, 120°, and 240° (Figure 13.15b, top row) in order to apply the algorithm of Equation 13.33 (Section 13.4.4) to remove the zero-order and twin-image terms.

Actually, $R_k(x, y)$ consists of two elements, one a spherical mirror of focal length, a, and the other a plane mirror. Note that although the SLM is reflective in Figure 13.15, it can be transmissive acting simultaneously as a lens of focal length, *a*, and as a transparent plate imposing a spatially uniform phase delay. In either case, *two* wavefronts, one plane and one spherical, propagate from the SLM to the CCD array, which records the FZP of interference between them. Thus, light from every point in object space is recorded as a FZP whose center corresponds to the *x,y* coordinates of the object point and whose fringe spacing in the radial direction is determined by the *z*-coordinate of the object point so that 3-D information is recorded. The SLM thus acts to provide each object wave with its own reference wave.

The mutual coherence between the waves essential to holography is assured by the very short optical path differences between each object wave and its reference wave.

The two phase distributions are both randomly distributed over the SLM, Figure 13.15b, middle row (i), in order to avoid any regular patterns that produce strong diffraction orders with half of the pixels of the SLM assigned to each of the phase distributions.

For *fluorescence* imaging, the object is illuminated using light from an arc lamp reflected at a dichroic filter. The fluorescent light emitted by the object is transmitted by the filter and propagates onwards to the SLM (Figure 13.16) [17].

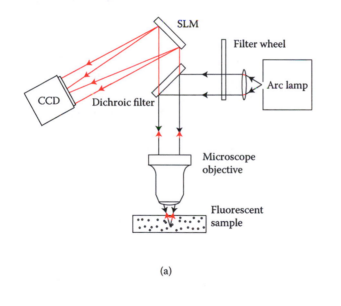

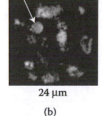

(a)

13 μm

20 μm

24 μm

(b)

FIGURE 13.16 Fluorescence holography: (a) optical arrangement and (b) images reconstructed from a single hologram of fluorescent mixed pollen grains. Arrows indicate features in focus at three different axial locations. Lateral field of view, 140 mm. (Reprinted by permission from Macmillan Publishers Ltd. *Nat. Photonics.* J. Rosen and G. Brooker, Non-scanning motionless fluorescence three-dimensional holographic microscopy, 2, 190–95, copyright 2008.)

Incoherent light and fluorescent light digital holographic microscopy offer great advantages over more conventional microscopies including speed since the whole object space of interest can be captured in a single hologram or, at most, three in order to remove the zero-order and twin-image terms. High numerical apertures can be used in fluorescence studies since the illuminating light plays no part in the holographic recording process so that the transverse spatial resolution rivals that of the best conventional methods.

13.6.4 WAVELENGTH-CODED MICROSCOPY

We can record a set of holographic diffraction gratings using the same wavelength for each of them and use the gratings in an imaging microscopy system with *each of the gratings tuned to a specified wavelength and depth in object space* [18]. A plane reference beam, R, and a plane signal beam, S, both of wavelength λ_1 interfere in a photosensitive layer and an unslanted, transmission grating is recorded (Figure 13.17). The angle between the beams is $2\theta_1$. This grating can be Bragg matched by a beam of wavelength λ_2 incident at an angle θ_2 where $\sin\theta_2 = (\lambda_2 / \lambda_1)\sin\theta_1$. We rotate the layer through an angle $\theta_2 - \theta_1$ and the signal beam through an angle $2(\theta_2 - \theta_1) + \delta$ where δ is an angle, which is small but greater than the angular selectivity. The angle between the beams is now $2\theta_2 + \delta$. We record a second grating again at wavelength λ_1 and then restore the layer to its original orientation. Figure 13.18a shows the wave and grating vector diagram for the first grating recorded at λ_1 and probed at $\lambda_2 > \lambda_1$. Figure 13.18b shows the wave and grating vectors after both gratings have been recorded and the layer restored to its original orientation. The gratings may now *both* be probed at angle θ_2 at *both* wavelengths, and the diffracted beams are at an angle δ apart.

In practice, the recordings are made as shown in Figure 13.19a using a point source, which is moved a distance, Δz, along the axis of the lens L_2 between recordings, by translating the lens L_1 along its axis.

Now, suppose the two gratings are illuminated by two light sources of wavelengths λ_1 and λ_2 at points a small distance, Δz, apart on the optical axis of lens L_1 and near its focus

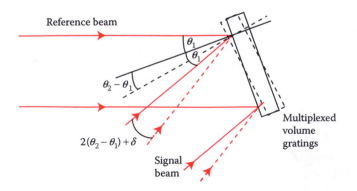

FIGURE 13.17 Principle of wavelength-coded multifocal holography.

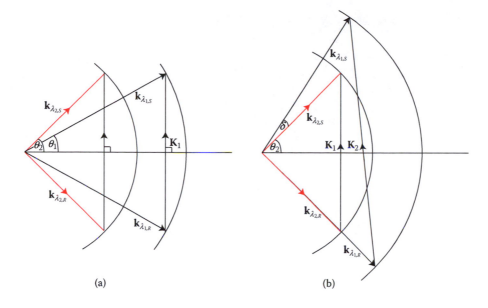

(a)   (b)

FIGURE 13.18 Wavelength-coded multifocal holography; wave and grating vector diagrams. (a) Unslanted grating recorded at λ_1 and probed at λ_2 and (b) both gratings recorded at λ_1 and probed at *either* wavelength, at angle θ_2. (Reproduced from Y. Luo, S. B. Oh, and G. Barbastathis, Wavelength-coded multifocal microscopy, *Opt. Lett.*, 35, 781–83, 2010. With permission of Optical Society of America.)

(Figure 13.19b). The beams are diffracted at angles $2\theta_1$ and $2\theta_1 + \delta$, respectively, relative to the lens axis. They are thus separated by an angle δ and are focused at a distance $f_3\delta$ apart on a CCD array by a lens L_3 of focal length, f_3. Figure 13.19c shows how planar sections of object space, Δz apart on the z-axis, illuminated at different wavelengths may be separated in the x direction when imaged on a CCD array. An example of imagery is shown in Figure 13.19d. The advantages of the method are that longer wavelengths may be used for greater penetration depth within tissue samples, and one may vary the intensity of illumination in each plane independently. Fluorescent specimens may also be imaged [19].

13.6.5 AUTOFOCUSING IN RECONSTRUCTION

We can determine object displacement by measuring the phase shift in the holographically recorded fringe pattern resulting from displacement of a flat part of the object. This operation facilitates updating the distance at which numerical reconstruction is to be carried out to maintain focus on a moving object [20]. Numerical autofocusing for live cell imaging and tracking has also been successfully implemented by four methods [21].

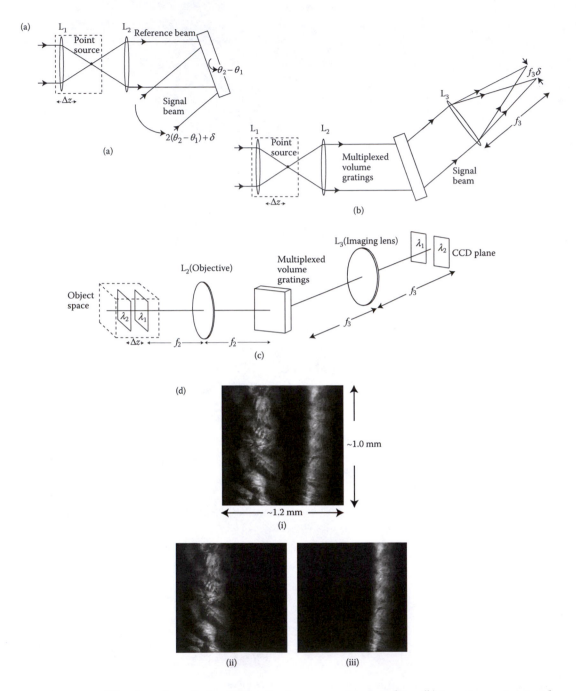

FIGURE 13.19 Wavelength-coded multifocal microscopy: (a) recording; (b) imaging points; and (c) system. (d) (i) Depth-resolved images of onion skin obtained with wavelength-coded multi-focal microscopy using red and blue LED illumination; (ii) blue illumination only; and (iii) red only. (Reproduced from Y. Luo, S. B. Oh, and G. Barbastathis, Wavelength-coded multifocal microscopy, *Opt. Lett.*, 35, 781–83, 2010. With permission of Optical Society of America.)

13.7 OTHER APPLICATIONS OF DIGITAL HOLOGRAPHY

The processing speed and capacity of currently available computing systems enable other interesting applications of digital holography. Among these is correction of optical distortion caused by turbulent atmosphere, which is a problem in long distance imaging. A potential solution involves digital processing of the recorded holograms in order to maximize image acuity, enabling the aberrations due to turbulence to be determined in the form of polynomial phase correction for the removal of image distortion. The technique has been used successfully on a USAF target at a distance of 100 m from the digital holographic recording system [22].

Digital holography has been applied successfully to the problem of combining the outputs from a number of glass optical fibers [23] or fiber amplifiers. The method has features in common with the technique for dispersion compensation described in Section 9.3.2, mainly, that it is based on phase conjugation. A low-power plane wave is sent along the fiber bundle to interfere with a plane wave $\exp(jky\sin\theta)$ propagating at a small angle to the common axis of the fibers (Figure 13.20). The pattern of interference is recorded using a CCD camera. This record constitutes a hologram of the fiber distorted wavefront, $\exp[j\phi(x,y)]$, and is written onto a SLM. The SLM is now illuminated by the conjugate of the reference wave $\exp(-jky\sin\theta)$ to reconstruct the conjugate version, $\exp[-j\phi(x,y)]$, of the distorted wavefront. This wavefront propagates back through the fibers unraveling the wavefront distortion to emerge as a plane wavefront.

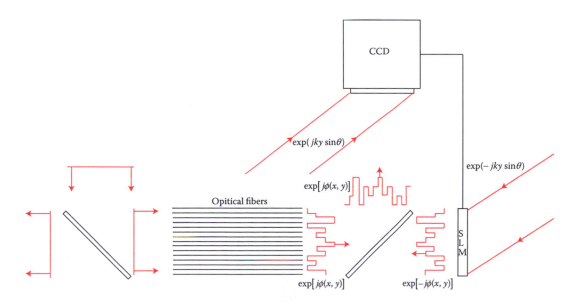

FIGURE 13.20 Coherent combining of outputs from optical fibers. (Reproduced from C. Bellanger, A. Brignon, J. Colineau, and J. P. Huignard, Coherent fiber combining by digital holography, *Opt. Lett.*, 33, 2937–39, 2008. With permission of Optical Society of America.)

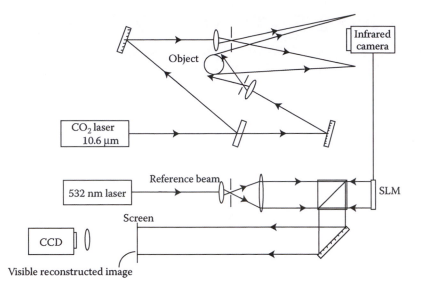

FIGURE 13.21 Long infrared holographic recording, visible light reconstruction. (Reproduced from M. Paturzo, A. Pelagotti, A. Finizio, L. Miccio, M. Locatelli, A. Gertrude, P. Poggi, R. Meucci, and P. Ferraro, Optical reconstruction of digital holograms recorded at 10.6 μm: Route for 3D imaging at long infrared wavelengths, *Opt. Lett.*, 35, 2112–14, 2010. With permission of Optical Society of America.)

Infrared digital holography [24] is possible with infrared cameras using microbolometer or pyroelectric arrays with about 300,000 pixels, each around 25 μm in width and a carbon dioxide (CO_2) laser. The CO_2 laser depends for its operation on population inversion between the quantized vibrational energy levels of the CO_2 molecule and operates at a wavelength of around 10.6 μm, which is well matched to the spectral photosensitivity of the camera, as well as to an atmospheric transmission window. A digital hologram is recorded. Reconstruction in visible light is obtained by rewriting the hologram into a reflective SLM, which is then illuminated by a visible laser (Figure 13.21) to form an image on a screen to be imaged by a CCD camera.

13.8 SUMMARY

In this chapter, we discuss the problems associated with holographic recording using CCD cameras. Despite the limitations imposed by the comparatively poor spatial resolution of CCD arrays, one can successfully exploit digital holography in a wide range of practical applications of holographic interferometry including, static loading, vibration, and surface metrology. We describe the synthetic aperture and pixel shift methods of improving the resolving capability of digital holography. We also consider various methods of numerical reconstruction, as well as techniques for removing the zero-order term and the twin image.

Probably, the most important applications are in microscopy, and much of the rest of this chapter discusses the various techniques in digital holographic microscopy and multiple wavelength and short coherence length holography. We discuss the development of scanning holography for incoherent light and fluorescent light holography, as well as nonscanning, fluorescent light digital holographic microscopy and wavelength-coded holographic microscopy. Finally, we briefly describe applications of digital holography in coherent beam combining and far infrared imaging.

REFERENCES

1. Yan Li, Digital Holography and Optical Contouring, PhD Thesis, Liverpool John Moore's University (2009).
2. A. Stadelmaier and J. H. Massig, "Compensation of lens aberrations in digital holography," *Opt. Lett.*, 25, 1630–2 (2000).
3. L. Yu and L. Cai, "Iterative algorithm with a constraint condition for numerical reconstruction of s three-dimensional object from its hologram," *J. Opt. Soc. Am. A.*, 18, 1033–45 (2001).
4. Y. Takaki, H. Kawai, and H. Ohzu, "Hybrid holographic microscopy free of conjugate and zero-order images," *Appl. Opt.*, 38 4990–6 (1999).
5. I. Yamaguchi and T. Zhang, "Phase-shifting digital holography," *Opt. Lett.*, 22 1268–70 (1997).
6. J. Rosen and G. Brooker, "Digital spatially incoherent Fresnel holography," *Opt. Lett.*, 32 912–4 (2007).
7. D. Claus, "High resolution digital holographic synthetic aperture applied to deformation measurement and extended depth of field method," *Appl. Opt.*, 49, 3187–98 (2010).
8. Y. Li, F. Lilley, D. Burton, and M. Lalor, "Evaluation and benchmarking of a pixel-shifting camera for superresolution lensless digital holography," *Appl. Opt.*, 49, 1643–50 (2010).
9. A. Asundi and V. R. Singh, "Time-averaged in-line digital holographic interferometry for vibration analysis," *Appl. Opt.*, 45, 2391–5 (2006).
10. C. J. Mann, L. Yu, C-M. Lo, and M. K. Kim, "High-resolution quantitative phase-contrast microscopy by digital holography," *Opt. Express*, 13, 8693–8 (2005).
11. P. Marquet, B. Rappaz, P. J. Magistretti, E. Cuche, Y. Emery, T. Colomb, and C. Depeursinge, "Digital holographic microscopy: A noninvasive contrast imaging technique allowing quantitative visualization of living cells with subwavelength axial accuracy," *Opt. Lett.*, 30, 468–70 (2005).
12. M. K. Kim, "Wavelength-scanning digital interference holography for optical section imaging," *Opt. Lett.*, 24, 1693–5 (1999).
13. P. Massatsch, F. Charrière, E. Cuche, P. Marquet, and C. D. Depeursinge, "Time-domain optical coherence tomography with digital holographic microscopy," *Appl. Opt.*, 44, 1806–12 (2005).
14. T-C. Poon, K. B. Doh, B. W. Schilling, M. H. Wu, K. K. Shinoda, and Y. Suzuki "three-dimensional microscopy by optical scanning holography," *Opt. Eng.*, 34, 1338–44 (1995).
15. B. W. Schilling, T-C. Poon, G. Indebetouw, B. Storrie, K. Shinoda, Y. Suzuki, and M. H. Wu, "Three-dimensional holographic fluorescence microscopy," *Opt. Lett.*, 22, 1506–8 (1997).
16. J. Rosen and G. Brooker, "Digital spatially incoherent Fresnel holography," *Opt. Lett.*, 32, 912–4 (2007).
17. J. Rosen and G. Brooker, "Non-scanning motionless fluorescence three-dimensional holographic microscopy," *Nat. Photonics*, 2, 190–5 (2008).
18. Y. Luo, S. B. Oh, and G. Barbastathis, "Wavelength-coded multifocal microscopy," *Opt. Lett.*, 35,781–3 (2010).
19. Y. Luo, P. J. Gelsinger-Austin, J. M. Watson, G. Barbastathis, J. K. Barton, and R. K. Kostuk "Laser-induced fluorescence imaging of subsurface tissue structures with a volume holographic spatial–spectral imaging system," *Opt. Lett.*, 33, 2098–2100 (2008).

20. P. Ferraro, G. Coppola, S. De Nicola, A. Finizio, and G. Pierattini, "Digital holographic microscope with automatic focus tracking by detecting sample displacement in real time," *Opt. Lett.*, 28, 1257–9 (2003).
21. P. Langehanenberg, B. Kemper, D. Dirksen, and G. von Bally, "Autofocusing in digital holographic phase contrast microscopy on pure phase objects for live cell imaging," *Appl. Opt.*, 47, D176–D182 (2008).
22. J. C. Marron, R. L. Kendrick, N. Seldomridge, T. D. Grow, and T. A. Höft, "Atmospheric turbulence correction using digital holographic detection: Experimental results," *Opt. Express*, 17, 11638–51 (2009).
23. C. Bellanger, A. Brignon, J. Colineau, and J. P. Huignard, "Coherent fiber combining by digital holography," *Opt. Lett.*, 33, 2937–9 (2008).
24. M. Paturzo, A. Pelagotti, A. Finizio, L. Miccio, M. Locatelli, A. Gertrude, P. Poggi, R. Meucci, and P. Ferraro, "Optical reconstruction of digital holograms recorded at 10.6 μm: route for 3D imaging at long infrared wavelengths," *Opt. Lett.*, 35, 2112–4 (2010).

PROBLEMS

1. What is the maximum size of transparency whose spatial Fourier transform can be adequately captured by a CCD camera having a square sensor 8 mm × 8 mm with 1000 × 1000 pixels using a lens of focal length 37.5 cm and a laser of wavelength 633 nm?

2. A Fresnel hologram is to be recorded of the transparency of problem 1. What is the minimum distance of the transparency from a hologram 63 mm × 63 mm so that the spatial frequency spectrum of the recorded hologram does not exceed that of the spatial Fourier transform of the transparency?

3. A laser diode whose normal wavelength is 658 nm is used in a digital holographic system. Its output power varies linearly from zero to 35 mW over a current range of 20 to 40 mA. Its output frequency is linearly proportional to the drive current with a frequency change of 2.5 GHz $(mA)^{-1}$. Phase shifting is to be applied in order to suppress the zero-order term by the method of Equation 13.26. What should be the path difference between the object and reference beams in order to introduce a phase change of π between them for a current change of 1 mA? Indicate one possible drawback with this method of phase shifting.

4. Suppose that a laser can be tuned over a range of 100 nm around its midspectral wavelength of 600 nm with a spectral resolution of 0.25 nm. What depth of field can this provide in multiple wavelength digital holography and how does it compare with the depth of field of a microscope objective with numerical aperture of 1.0, used at the same wavelength?

5. In multiple wavelength holography, many holograms are recorded in sequence, the laser wavelength being altered by $\Delta\lambda$ for each new hologram to be recorded and the holographic recordings are added together. Show that the method is equivalent to optical coherence tomography. What are the advantages and disadvantages of each method?

6. A piezoelectric translation device is used to enhance the capability of digital holography. A translation of 15 microns is obtained when a 150 V is applied to the device by means of a 12-bit digital to analog converter. What is the maximum angle between object and reference beams which could be used?

CHAPTER 14

Computer-Generated Holograms

14.1 INTRODUCTION

Besides the fact that they do not require interferometrically stable optical setups for recording, computer-generated holograms (CGH) are attractive for a variety of reasons. There is the intriguing possibility that we can reconstruct any wavefront or combination of wavefronts, even if the corresponding objects do not exist in reality, so long as the wavefronts can be mathematically represented and we can compute the corresponding holograms. Precise computation of the hologram required to reconstruct a very well-defined wavefront facilitates higher standards in optical manufacture. We can now produce complex wavefronts for advanced applications in optical trapping, micromanipulation, and control, for biological investigations, and for assembling complex nanostructured materials.

A number of steps are involved in computer-generated holography. The first is the computation itself, which requires sampling of the wavefont at a sufficiently high spatial frequency to satisfy the Nyquist criterion. In many applications, we compute the discrete spatial Fourier transform, rather than the wavefront itself.

In the second step, we add the complex amplitude values of the object wave, or the coefficients of its discrete spatial Fourier transform to those of a plane reference wave, and obtain the squared modulus of the sum, for each point in the hologram plane.

We must then find a way of physically presenting the computed hologram data to a light beam to construct the desired image wavefront. One method is to use a laser, inkjet, or bubble jet device to print a large-scale version of the hologram, which we then photographically reduce onto a photographic transparency. Sometimes, we use two transparencies: one is constant thickness but spatially varying amplitude transmissivity, and the other is constant transmissivity and variable thickness or refractive index, to represent the phase. We may also use lithographic techniques as they provide sufficiently high spatial resolution to enable direct writing of the hologram. We can also write data into a spatial light modulator (SLM), which allows sequencing of holograms for dynamical light wave reconstruction.

In fact, one needs only to calculate the term corresponding to the product of the conjugate reference wave and the object wave. A bias may be added to the result to ensure positive values everywhere for the printing device. A significant advantage is that the zero-order term does not appear in the reconstruction.

Generally, we will assume that Fourier transform holograms are to be computed. This means we compute the spatial Fourier transform of the object wave. We show in the next section how we can *physically* represent the transform, illuminate it with an appropriate reference wave, and Fourier transform the result, to obtain an image.

14.2 METHODS OF REPRESENTATION

14.2.1 BINARY DETOUR—PHASE METHOD

Binary detour phase [1,2] is a method of representing a sampled, or discrete, spatial Fourier transform of an object wave. The hologram plane is divided up into an array of opaque cells, all with the same dimensions. Within each of them, we specify a transparent, rectangular area, whose dimensions correspond to the amplitude of the Fourier component at the cell location and whose position in the cell corresponds to the phase of the Fourier component. The hologram is binary in the sense that the amplitude transmittance at every point is either one or zero. Such holograms are well suited to fabrication by binary printing devices, which deposit, or do not deposit, ink at any given location and can readily produce printed squares or rectangles with precise dimensions. Ink deposited in the form of rectangular patches gives rise to higher orders in reconstruction since the diffraction pattern of a rectangular aperture has the form of a sin c function. This problem can be alleviated by the use of circular printed pixels of varying size [3].

We consider first a single rectangular aperture of area $w_x \times w_y$ centered at x_0, y_0 in an otherwise opaque, flat screen in the x,y plane. Its complex amplitude transmissivity is

$$\tilde{t}(x, y) = \text{rect}\left(\frac{x - x_0}{w_x}\right)\text{rect}\left(\frac{y - y_0}{w_y}\right) \tag{14.1}$$

It is illuminated by a plane wave $\exp(jky\sin\theta)$ so that the transmitted complex amplitude is

$$\tilde{A}(x, y) = \exp(jky\sin\theta)\text{rect}\left(\frac{x - x_0}{w_x}\right)\text{rect}\left(\frac{y - y_0}{w_y}\right) \tag{14.2}$$

The spatial Fourier transform of $\tilde{A}(x, y)$ is, from Equation 2.22,

$$\tilde{A}(x', y') = \frac{1}{\lambda f}\int_{-(w_x/2)}^{(w_x/2)}\int_{-(w_y/2)}^{(w_y/2)} \exp(jky\sin\theta)\text{rect}\left(\frac{x - x_0}{w_x}\right)\text{rect}\left(\frac{y - y_0}{w_y}\right)\exp\left[-\frac{jk}{f}(xx' + yy')\right]dxdy$$

$$= \frac{1}{\lambda f}\int_{-(w_x/2)+x_0}^{(w_x/2)+x_0}\int_{-(w_y/2)+y_0}^{(w_y/2)+y_0} \exp[jk(Y + y_0)\sin\theta]\text{rect}\left(\frac{X}{w_x}\right)\text{rect}\left(\frac{Y}{w_y}\right)$$

$$\times \exp\left\{-\frac{jk}{f}[(X + x_0)x' + (Y + y_0)y']\right\}dXdY$$

where f is the focal length of the Fourier transform lens, and $X = x - x_0$ and $Y = y - y_0$.

$$= \frac{w_x w_y}{\lambda f} \exp\left(\frac{-jkx_0 x'}{f}\right) \exp\left[\frac{-jky_0(y' - f\sin\theta)}{f}\right]$$

$$\times \int_{-(1/2)}^{(1/2)} \int_{-(1/2)}^{(1/2)} \mathrm{rect}(X')\mathrm{rect}(Y') \exp\left\{-\frac{jk}{f}[X'w_x x' + Y'w_y(y' - f\sin\theta)]\right\} dX' dY'$$

with $X' = X/w_x$, $Y' = Y/w_y$

$$\tilde{A}(x', y') = \frac{w_x w_y}{\lambda f} \mathrm{sinc}\left(\frac{kw_x x'}{2f}\right) \mathrm{sinc}\left[\frac{kw_y(y' - f\sin\theta)}{2f}\right]$$

$$\times \exp\left\{\frac{-jk[x_0 x' + y_0(y' - f\sin\theta)]}{f}\right\} \tag{14.3}$$

If $w_y \ll \lambda/\sin\theta$, the aperture is much narrower than one spatial cycle of the illuminating plane wave in the x,y plane, in other words the angle θ is small and we can neglect the shift $f\sin\theta$ in the term $\mathrm{sinc}[kw_y(y' - f\sin\theta)/2f]$. Then, letting both of the sin c terms approximate to unity, we have

$$\tilde{A}(x', y') = \frac{w_x w_y}{\lambda f} \exp(-jky_0 \sin\theta) \exp\left[\frac{-jk(x_0 x' + y_0 y')}{f}\right] \tag{14.4}$$

We now divide the opaque screen in the x,y plane into regularly spaced cells, all with the same area $\Delta x \Delta y$, one of them centered at the origin, each containing an aperture with its midpoint located at $(n\Delta x, m\Delta y + \delta y_{n,m})$ where $n = 0, 1, 2, \ldots, N_x-1$ and $m = 0, 1, 2, \ldots, M_y-1$ and $N_x \times M_y$ is the number of apertures and cells (Figure 14.1). The term $\delta y_{n,m}$ is introduced so that the aperture in cell n,m centered at $(n\Delta x, m\Delta y)$ is displaced from the midpoint, $m\Delta y$, of the cell for phase coding purposes. The area of each aperture $w_{x,n} w_{y,m}$ depends on the cell n,m in which it is located and determines the amplitude of light transmitted by the hologram at that location. Note that only $w_{x,n}$ actually varies, while the vertical dimension, $w_{y,m}$, of the aperture does not. This is to ensure that the condition $w_{y,m} \ll \lambda/\sin\theta$ can always be met.

We rewrite Equation 14.4, to include all of the cells, in discrete form as

$$\tilde{A}(x', y') = \frac{1}{\lambda f} \sum_{n=0}^{N_x-1} \sum_{m=0}^{M_y-1} w_{x,n} w_{y,m} \exp\left[-j2\pi(m\Delta y + \delta y_{n,m})\frac{\sin\theta}{\lambda}\right]$$

$$\times \exp\left[\frac{-jk(x'n\Delta x + y'm\Delta y + y'\delta y_{n,m})}{f}\right] \tag{14.5}$$

Setting $\Delta y = \lambda/\sin\theta$, we have

$$\tilde{A}(x', y') = \frac{1}{\lambda f} \sum_{n=0}^{N_x-1} \sum_{m=0}^{M_y-1} w_{x,n} w_{y,m} \exp\left[\frac{-j2\pi\delta y_{n,m}}{\Delta y}\right] \exp\left[\frac{-jk(x'n\Delta x + y'm\Delta y + y'\delta y_{n,m})}{f}\right] \tag{14.6}$$

where $x_0 = n\Delta x$, $y_0 = m\Delta y + \delta_{n,m}$.

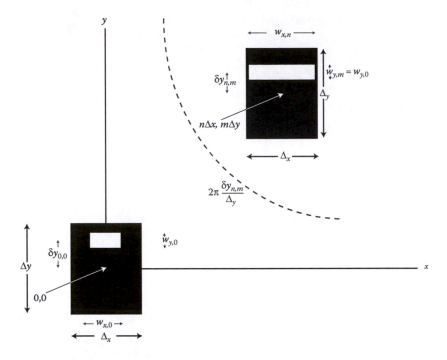

FIGURE 14.1 Binary detour phase holography. In the Fourier hologram, transparent rectangles, with area $w_{x,n} \times w_{y,0}$ representing amplitude, are located in regularly spaced, contiguous opaque cells with their centers at $m\Delta x, n\Delta y$. The offset, $\delta y_{n,m}$, of a rectangle from the midpoint of its cell represents the phase, $2\pi\delta y_{n,m}/\lambda$, at $m\Delta x, n\Delta y$.

Setting

$$\exp\left\{\frac{-jky'\,\delta y_{n,m}}{f}\right\} \approx 1$$

$$\tilde{A}(x',y') = \frac{1}{\lambda f}\sum_{n=0}^{N_x-1}\sum_{m=0}^{M_y-1} w_{x,n}w_{y,m}\exp\left(-j2\pi\frac{\delta y_{n,m}}{\Delta y}\right)\exp\left[\frac{-jk(x'\,n\Delta x + y'\,m\Delta y)}{f}\right] \quad (14.7)$$

We let the expression

$$\tilde{A}(n\Delta x, m\Delta y) = w_{x,n}w_{y,m}\exp\left(-j2\pi\frac{\delta y_{n,m}}{\Delta y}\right) \quad (14.8)$$

be the computed Fourier transform of the complex amplitude distribution in the original object wave. The amplitude of the Fourier component at $n\Delta x, m\Delta y$ is represented by $w_{x,n}w_{y,m}$ and its phase by $-2\pi\delta y_{n,m}/\Delta y$. Then $\tilde{A}(x',y')$ in Equation 14.7 is the Fourier transform of $\tilde{A}(n\Delta x, m\Delta y)$, that is, the image, which appears in the back focal plane of the lens, when the Fourier hologram is illuminated by the original reference wave $\exp(jky\sin\theta)$. When an ink printer is used, each element is printed in reverse contrast and a reduced scale, photographic negative has the desired complex amplitude transmissivity. There is always a minimum increment associated with the movement of the printer head. This means that both the area which it prints and its offset are always quantized, and this produces some noise in the reconstructed image.

To satisfy the Nyquist criterion, we must calculate the spatial Fourier transform at spatial intervals of, at most, $\Delta x = \lambda f / 2a$ and $\Delta y = \lambda f / 2b$ where f is the focal length of the lens, which forms the spatial Fourier transform of the object whose dimensions parallel to the x and y axes, respectively, are a and b (see Section 13.2.1).

An example of a binary detour phase CGH is shown in Figure 14.2, along with the image reconstructed from it.

A variation of the method [4] is based on the principle that the amplitude and phase of any complex quantity (i.e., a Fourier component) can be represented by two orthogonal component vectors: one directed along either the positive (phase = 0°) or the negative (phase = 180°) real axis and the other along the positive (phase = 90°) or negative (phase = 270°) imaginary axis (Figure 14.3a). Each cell is therefore divided into four subcells, two of which are opaque, each of the others having transmittance proportional to the magnitude of the vector that it represents.

In this way, each cell specifies completely the amplitude and phase of the Fourier component at its location in the hologram plane.

A simplification is possible using just three subcells with phases of 0°, 120°, and 240° [5] (Figure 14.3b).

The physical representation of Fourier transform holograms can be problematical because of the very large variation in amplitude of the various spatial frequency components, the amplitudes of low frequency being very large compared with the amplitudes of the high-frequency components so that a recording material with limited

(a) Binary detour phase hologram

(b) Reconstructed image

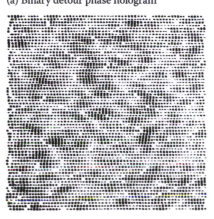

FIGURE 14.2 Binary detour phase hologram (a) and reconstructed image (b). (Reproduced with permission from B. R. Brown and A. W. Lohmann, Computer generated binary holograms, *IBM J. Res. Dev.*, 13, 160–67, © 1969 IEEE.)

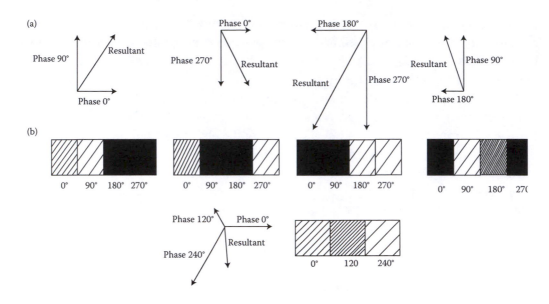

FIGURE 14.3 Use of subcells: (a) Lee's detour phase method (after W. H. Lee, Sampled Fourier transform hologram generated by computer, *Appl. Opt.*, 9, 639–43, 1970) and (b) Burckhardt's method (after C. B. Burkhardt, A simplification of Lee's method of generating holograms by computer, *Appl. Opt.*, 9, 1949, 1970).

dynamic range may be unable to reproduce the computed hologram with a high degree of fidelity.

The problem is basically that encountered when we attempt to make a hologram of a largely transparent object such as a glass vase, or of an object, which reflects light into the hologram. Light transmitted or reflected by the object in the direction of the hologram is reconstructed as glare in the image. It is much better to illuminate such objects through a *diffuser* in a direction in which cases the object beam to miss the recording medium altogether if neither diffuser nor object is present. The diffuser has the effect of adding a random phase to the object wave, and of course, it gives rise to speckle in the image, although this is usually acceptable if the speckle size is kept small (see Section 7.4.2), and the image is often of much higher quality with a wider range of viewing angles.

We can implement the equivalent of a diffuser by adding a random phase to the complex amplitude of each sampled point on the object before calculating the spatial Fourier transform. The result is that the spatial Fourier spectrum is much more uniform in amplitude although speckle again appears in the reconstructed image.

14.2.2 THE KINOFORM

We need not calculate the amplitudes of the Fourier components if they are all assumed to have the same value of 1. The assumption of unit amplitude everywhere removes the need to vary the area of the transparent rectangles in the opaque cells as in the binary detour phase method. Instead, we calculate only the phase at each point in the hologram plane, representing it by an appropriate gray level, which in turn determines the intensity of illumination of a photosensitive material, such as silver halide. Following processing with carefully controlled bleaching, the intensity variations, representing phases, are recorded as thickness or refractive index variations, or both. The result is a device called a kinoform [6], which modulates the phase of a beam of light passing through it producing the image as before in the back focal plane of a lens. Alternatively, we may use a self-processing photopolymer to obtain a similar result. In either case, it is very important to ensure that the phase retardation imposed by the device covers just the range 0 to 2π. Kinoforms are essentially pure phase holograms so have very high diffraction efficiency. An example is shown in Figure 14.4 with the image reconstructed from it.

14.2.3 REFERENCELESS OFF-AXIS COMPUTED HOLOGRAM (ROACH)

The methods discussed so far have all followed holographic principles, in that a reference wave is assumed to be present. This is so that the hologram can be realized in the form of a transparency with a real and positive amplitude transmittance. However, if we separate the amplitudes and phases of the Fourier components and code amplitude as transmittance of one layer in a photosensitive medium and phase as the thickness or the refractive of another, we can dispense with the computation of the interference of the reference wave and the Fourier transform of the object wave. The method can be implemented in color reversal film normally used for photographic projection transparencies. The film consists of layers, each of which is photosensitive in a different region of the spectrum of visible light. We start by using the computed values of the amplitudes of the Fourier transform components to modulate the intensity of the electron beam in a cathode ray tube (CRT) display, which is photographed through a red filter. The phases of the Fourier components

(a) Gray-level representation of phase

(b) Image recosntructed from phase hologram

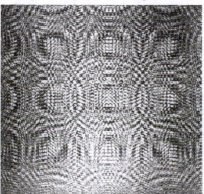

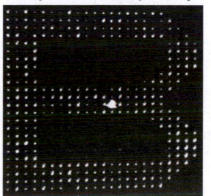

FIGURE 14.4 Kinoform: (a) gray-level representation of phase and (b) image reconstructed from phase hologram. (Reproduced with permission from L. B. Lesem, P. M. Hirsch, and J. A. Jordan, The kinoform: A new wavefront reconstruction device, *IBM J. Res. Dev.*, 13, 150–55, © 1969 IEEE.)

are now also used to modulate the display, which is photographed through a blue–green filter. After processing, the transparency is illuminated by red light. The transmissivity of the red sensitive layer is spatially modulated and encodes the amplitude, while the blue and green layers are transparent to red light but have been spatially modulated in thickness, thus encoding the phase.

We may substitute spatial scanning red and green lasers for the CRT and color filters. Amplitude-only and phase-only SLM may also be used in place of color reversal film.

Kinoforms and ROACHs have a distinct advantage since they can encode a Fourier component in a single resolution element of the recording material while binary detour phase holograms may use up a significant number of resolution elements. However, phase error in the latter is much easier to avoid.

14.3 THREE-DIMENSIONAL OBJECTS

In order to compute a hologram of a three-dimensional object, we can divide the object into a set of two-dimensional thin slices normal to the optical axis. The computation is demanding: first, we need to consider what exactly would be seen when looking back toward the object from any particular point on the hologram plane in order to identify distant objects, which are obscured by proximate ones and ensure that they remain so in the reconstructed image. Second, we need to consider the large amount of computation involved, consisting of a Fourier transform calculation for each of the slices. Third, we need to calculate a hologram of sufficiently large aperture size so that the reconstructed image will exhibit acceptably wide parallax.

Procedures have been developed to keep the computational workload within manageable proportions. One method sacrifices vertical parallax, thus requiring only that we calculate a one-dimensional Fourier transform for each horizontal line in the hologram plane, rather like the Benton hologram of Section 8.4.

Another approach involves calculating narrow, vertical strip holograms of the object wave as seen from viewpoints which lie on a horizontal line. The holograms are then stitched together to form a spatially multiplexed hologram similar to the spatially multiplexed hologram described in Section 8.6. Since the real image from the spatially multiplexed hologram is two dimensional, we can make an image plane copy, which can be illuminated with white light to obtain an achromatic image (see Section 8.4.1).

Many researchers have used analytical descriptions of objects, first breaking them down into simple shapes, such as straight lines, polygons, and ellipses, which have appropriate dimensions and orientations in space and whose transforms are not so onerous to compute, for example, using angular spectrum methods [7]. Developments include coping with solid objects [8,9] and variations in shade and surface texture [9]. The introduction of an analytical formula for the frequency distribution of a diffusely shaded triangle further allows rendering to be performed directly in spatial frequency space [10], thus avoiding Fourier transformation for each triangle. A single transform is required when the angular spectra of all the triangles have been obtained.

Object waves have been analytically described using an algorithm largely based on the methods of computer graphics [11], which readily facilitate the inclusion of shading, texture, and hidden objects.

In parallel with developments in the description of the object wave, access to vastly increased computing power and speed at ever decreasing cost, as well as graphics processing units and the use of look-up tables for storing interference fringe pattern data [12], have helped to advance computer-generated holography. Computer-generated holography for video display has also advanced considerably, and we discussed some of these developments in Section 8.8.

A special purpose computer has been constructed [13] for the generation of holograms, each consisting of one million points, displayed once per second on a liquid crystal device.

14.4 OPTICAL TESTING

High-performance optical components including prisms, lenses, optical flats, and mirrors are interferometrically tested. Typically, a plane wave transmitted or reflected by the component interferes with a reference wave in a Twyman-Green interferometer, as shown in Figure 14.5. In the prism testing arrangement of Figure 14.5a, the auxiliary plane mirror on the right is oriented so that the plane wave incident on it is directed back through the prism to be superimposed on the reference plane wave. In the case of a lens, the auxiliary component is a convex mirror whose center of curvature coincides with the lens focus (Figure 14.5b). A plane wave transmitted through the lens and reflected from the convex mirror has its curvature negated on the return trip to become again superimposed on the reference plane wave. The arrangement in Figure 14.5b can also be used to test a convex mirror.

The interference fringes, which are produced, are indicative of aberration due to the errors in the surface profile, such as lack of flatness of a plane mirror or prism surface, or incorrect local curvature on a lens or curved mirror surface. The errors may be corrected by direct polishing of the optical surfaces until an acceptably small number of fringes are seen in the output plane of the interferometer.

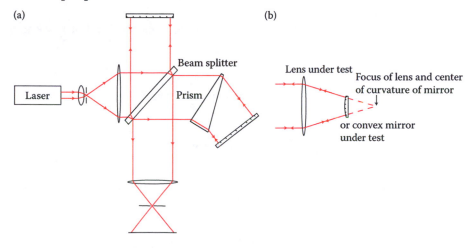

FIGURE 14.5 Twyman-Green interferometer for testing (a) prism and (b) lens.

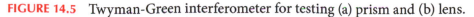

14.4.1 OPTICAL TESTING USING COMPUTER-GENERATED HOLOGRAMS

Optical components are often aspherical in order to minimize the optical aberrations of purely spherical surfaces. This means that interferometric comparison with a spherical or plane reference wavefront may result in a fringe pattern, which can be difficult to interpret. A solution is to use a CGH [14] illuminated by a plane wave. The reconstructed wavefront closely resembles the wavefront, which the test component is required to produce when it is similarly illuminated. Superposition of the wavefronts from the hologram and the test component results in a comparatively simple fringe pattern, indicative of any error in the component.

14.4.2 COMPUTER-GENERATED INTERFEROGRAMS

The CGH may be required to produce a wavefront whose phase varies continuously over many integer multiples of 2π. In a binary detour phase computer-generated hologram, we require the phase to be continuous across the boundaries between cells. If, for example, the value of the phase in consecutive cells has been rising, to reach a value of 1.95π and the value in the next cell is 0.05π, we must unwrap the latter by adding 2π to it (see Section 10.13 and Figure 10.15). Unwrapped values of phase in adjacent cells, which are respectively close to 2π and 0, mean that the corresponding apertures are close to the boundary between the cells and may partially overlap in the hard copy version of the CGH, resulting in an error in the phase. Such errors in phase may occur quite frequently if the unwrapped phase is to vary over many multiples of 2π.

An alternative to binary detour phase CGHs for optical testing purposes, is possible since there are no amplitude variations required in the reconstructed waveform [15].

Consider a simple example of a CGH, which reconstructs a divergent spherical wavefront coaxial with the z-axis, when illuminated by a plane wave, $\exp(jky\sin\theta)$, as shown in Figure 14.6a.

The phase difference between the phases of the spherical wavefront at x, y and at 0,0 is given by

$$\phi(x,y) = \frac{2\pi}{\lambda}\left[\left(f^2 + x^2 + y^2\right)^{\frac{1}{2}} - f\right] \approx \frac{\pi\left(x^2 + y^2\right)}{\lambda f} \tag{14.9}$$

and the intensity transmissivity of the hologram is given by

$$\left|\exp\left[\frac{j\pi\left(x^2 + y^2\right)}{\lambda f}\right] + \exp\left(jky\sin\theta\right)\right|^2 = 2 + 2\cos\left[\frac{\pi\left(x^2 + y^2\right)}{\lambda f} - ky\sin\theta\right] \tag{14.10}$$

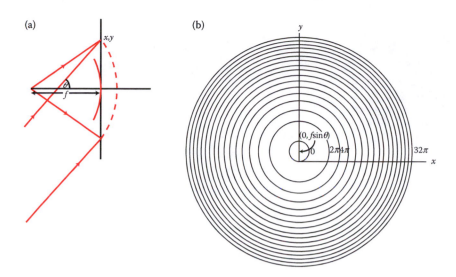

FIGURE 14.6 Computer-generated interferogram. (a) The plane reference wave exp(*jkysinθ*) interferes with the spherical wavefront exp[*jπ* (*x²* + *y²*)/*λf*]. (b) The skeletalized fringe pattern from (a) is photoreduced to form the final computer-generated interferogram.

Production of a hard copy of Equation 14.10 requires a plotting device, which can provide a continuous range of gray-level output to cope with the continuous range of intensity transmission. However, we only have to produce a wavefront with variable phase. Variation in amplitude is not required, so the CGH can be represented by a plotted pattern with variable spacing but without variation in contrast. We can simply plot contours of constant phase, which are skeletal versions of the fringes described by Equation 14.10, by which we mean that each fringe is represented by a line joining its points of maximum intensity. The contours are conveniently represented by the equation

$$\frac{\pi\left(x^2 + y^2\right)}{\lambda f} - ky\sin\theta = 2n\pi$$

(14.11)

with *n* an integer and are circles with centers at 0, $f\sin\theta$, and radii, r_n, given by

$$r_n = \sqrt{2n\lambda f + f^2 \sin\theta^2}$$

(14.12)

We choose values of *λ* and *f*, which are compatible with the plotter resolution (Figure 14.6b) and then make a reduced scale photographic copy. Such phase-only CGHs for optical testing are known as computer-generated interferograms.

In Figure 14.7, a computer-generated interferogram is used to test an aspherical mirror. The wavefront reflected from the test mirror interferes with a plane wave reflected from the flat reference mirror. The resulting interference fringes are superimposed on the computer-generated interferogram. The latter is a skeletal version of the computed pattern of interference between the plane reference wavefront and the wavefront that would be reflected from

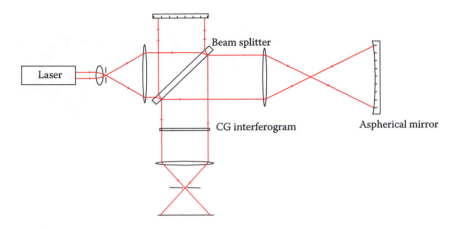

FIGURE 14.7 Optical testing using a computer-generated interferogram.

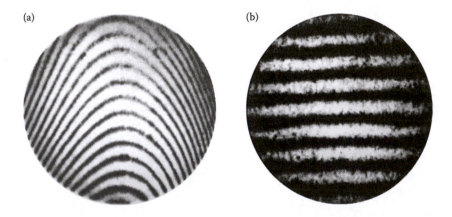

FIGURE 14.8 Typical fringe pattern (a) without CGH and (b) with CGH. (Reproduced from J. C. Wyant and P. V. Bennett, Using computer-generated holograms to test aspheric wavefronts, *Appl. Opt.*, 11, 2833–39, 1972. With permission of Optical Society of America.)

the test component if it met its specification. The resulting moiré pattern is used to assess the aberrations of the test component.

Typical patterns are shown in Figure 14.8.

It is important to isolate the wavefronts of interest, namely that from the component being tested and the wavefront obtained by illuminating the hologram with the plane reference wave. We must therefore ensure that the plane wave vector is at a greater angle to the axis than that of any wave vector from the test component, and a small aperture is placed in the focal plane of the final lens at the bottom of Figure 14.7.

The hologram must also be designed so that its diffracted spatial frequency spectra do not overlap. Suppose the maximum number of fringes across the hologram width is f. Recall from Section 13.2 (Figure 13.5) that the hologram is then required to have a carrier frequency of $3f$ so that the maximum spatial frequency in its fringe pattern is $4f$. If the plotter has a spatial resolution of Δd, the error in fringe position is $\Delta d/2$, and the wavefront error expressed as a fraction of a wavelength is

$$\frac{\Delta d/2}{(1/4f)} \tag{14.13}$$

So to ensure a maximum wavefront error of $\lambda/10$, the number of fringes in the hologram width must not exceed $1/(20\Delta d)$.

Lateral displacement between the wavefronts, or shear, must also be avoided. This requires that to maintain a wavefront error of less than $\lambda/10$, the errors in hologram width, W, and in positioning must both be less than $W/(10f)$.

The computational load involved in generating holograms for testing highly aspherical components has led to a compromise approach to testing. An approximately correct test wavefront can be produced by a conventional, so-called null optical component, whose errors are corrected by a hologram for which the computation is a less demanding task. This approach has advantage that the CGH design can then be extended by ray tracing methods to include compensation for the aberrations introduced by the other components of the interferometer. Ray tracing software packages such as OSLO© and ZEMAX© include this facility for the design of CGHs.

14.5 OPTICAL TRAPS AND COMPUTER-GENERATED HOLOGRAPHIC OPTICAL TWEEZERS

Nanoscience and nanoengineering have become almost ubiquitous, if largely unnoticed, in our daily lives, with a rapidly growing range of applications in physics, chemistry, and materials and life sciences. Among the most important research tools of nanoscience are optical traps and tweezers.

In the following sections, we focus on CGHs as tools for trapping and manipulating micrometer and submicrometer-sized objects.

The basic concept here is the use of the electromagnetic forces accompanying a beam of light such as a laser, which can be tightly focused, to trap a small object. Controlled movement of the beam then enables us to manipulate the object. We can also transport and rotate objects. Furthermore, using arrays of light beams, we may control many objects simultaneously.

14.5.1 OPTICAL TRAPPING

There are two forces at work on an object of wavelength or smaller size, in a focused light beam (Figure 14.9). In region close to the focus, the object becomes electrically polarized, and its resulting dipole moment is acted on by the gradient of the electric field so that the object is drawn closer to the focus of the laser (Figure 14.9a) [16]. Objects near the focus but displaced along the beam axis in the direction opposite to propagation are pushed by radiation pressure toward the focus. Once there, the balance of the gradient forces maintains the object in position.

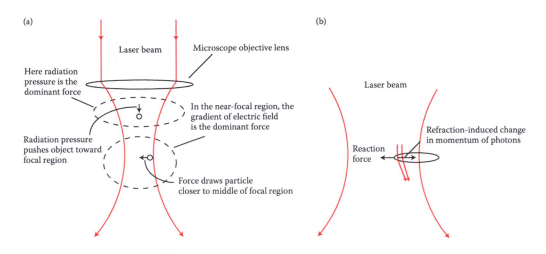

FIGURE 14.9 Optical trapping: (a) object of wavelength size and (b) larger sized object.

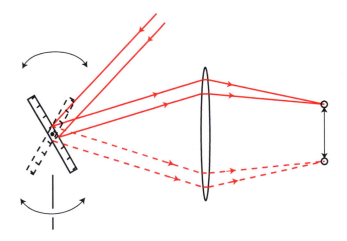

FIGURE 14.10 Scanning trap.

Larger objects refract the light so that it changes direction (Figure 14.9b). The accompanying change in direction of the momentum vector of the photons constitutes a force, in reaction to which the object is pulled toward the focus. Such forces are exerted on large optical components such as lenses, prisms, and mirrors, but the effect in these cases is negligible.

Optical traps using beams with just a few mW of power can hold objects in place, whose dimensions range from a few nm to about 1 μm. Once the object has been trapped, the laser beam can be used to perform a tweezing action. We can move the object in a plane by means of a mirror rotated about orthogonal axes in its plane (Figure 14.10). Thus, a single trapping beam can be made to scan rapidly, trapping an object briefly, transporting it to a desired location, and then moving on to deal with another object [17].

14.5.2 HOLOGRAPHIC OPTICAL TWEEZERS

Arrays of optical traps are an important advance on single-beam traps, providing the facility to trap and manipulate a large number of objects at the same time. We can implement such an array in the form of a reflection or transmission CGH (Figure 14.11), serving as a multiple holographic optical trap (HOT), which diffracts a single input beam, forming a set of output beams, each propagating in a different direction [16].

The combination of lenses L_1 and L_2 forms an image of the hologram at the entrance pupil of the microscope objective O_1. Trapped objects are illuminated from above through the objective O_2, and the objective O_1 forms images of them on the faceplate of a charge-coupled device (CCD) camera. The dichroic filter prevents backscattered trapping light from reaching the camera. The CGH is a phase-only device so that it may be implemented as a kinoform (Section 14.2.2) with a randomly chosen phase assigned to each of the traps to produce arrays of beams with quite uniform intensity. When a significant number of traps is produced in this way, they are usually only a little more intense than the zero-order beam that must therefore be prevented from reaching the objects by means of an opaque stop.

Obviously much greater versatility is possible using holographic optical tweezers, which can be rapidly reconfigured. Such devices may be implemented in computer-controlled spatial light modulators (SLM) allowing refresh rates at up to 100 Hz. The development of fast algorithms for the computation of dynamic HOTs using SLMs is the focus of much research effort and arrays of traps numbering 20×20 have been implemented [18].

Although the use of SLMs means that ghost traps appear, due to diffraction by the regular pixellated structure of such devices, these are typically weaker than the designed traps. The traps may vary in intensity quite significantly, but this is not normally a serious

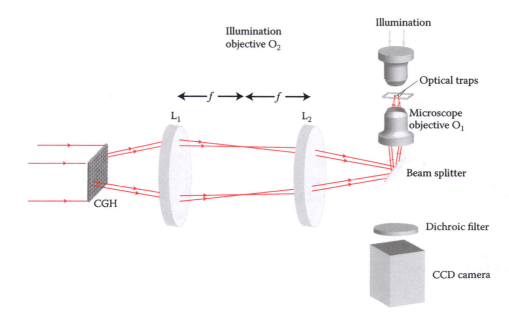

FIGURE 14.11 Holographic optical trap.

problem [19]. SLMs have limited spatial resolution and are used for trapping within a single plane at a time rather than for tweezing in three dimensions.

By altering the directions of some of the trapping beams, one can transport selected objects within a plane. Similarly, an object may also be shifted between traps so that it follows a predetermined path.

Interfaces enabling keystroke-based manual trapping and computer mouse- or joystick-controlled transport of objects have been developed [20], although the need for human interaction may be avoided when feature recognition techniques are brought to bear [21] so that manipulation may then be automated. Spatial bandpass filtering, digitizing, and thresholding of an image enables the centers of spherical objects to be identified to an accuracy of about 100 nm. Then, the moments of the intensity distribution around the centers are calculated, facilitating location of objects with even greater precision, about 10 nm. These techniques can also be used for distinguishing objects, which have distinctly different diameters.

Having located all the objects in a set, suppose we wish to arrange some of them into a specified pattern. First, the hologram needed to trap the required number of objects at their identified locations is calculated and applied before the objects have had time to diffuse away from those locations. The time τ_D taken by a particle to diffuse a distance, D, from its initial location is estimated by an ensemble-averaged solution to the Langevin equation as

$$\tau_D = \frac{3\pi\eta D^3}{4k_B T}$$

(14.14)

where k_B is Boltzmann's constant, T the temperature, and η the viscosity. The example given by Chapin et al. [21] is that of water ($\eta = 0.001$ Pa s) dispersed silica particles with diameter D about 1 μm. As this is larger than the laser wavelength, the traps are also about one diameter in width, and τ_D is about 0.6 s, so this is the time available to calculate the hologram and trap the particles.

The next stage is to plan the paths that the particles will follow in order to take up their new positions. Destinations and shortest paths to them are determined. Each trap must then be moved to a new location, typically no more than a particle diameter away to ensure that the particle does not escape. If the rate at which a trap is implemented at consecutive locations is f, then the speed is fD. Although the refresh rate may be increased as SLM technology improves, the speed of translation of the particles is limited by competition between the optical trap forces and the viscous drag force. Collisions are avoided by exchanging destinations if the colliding particles are identical. If they are not and it is important that each particle arrives at its originally intended destination, then avoidance is achieved by moving the objects in orthogonal but opposite directions in the same way that polite pedestrians attempt to pass by one another in opposite directions on a crowded street. If the problem remains unresolved after two such side steps, a random step is applied to each particle. In Figure 14.12, examples of object sorting array formation and repatterning are shown.

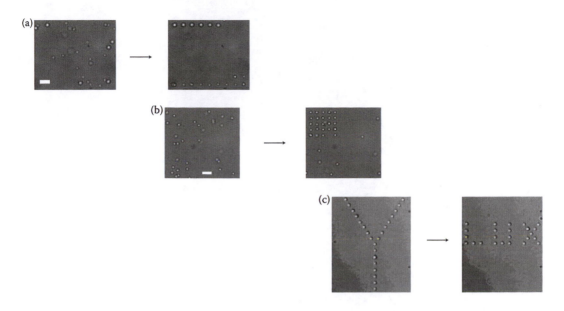

FIGURE 14.12 Automated (a) sorting of silica particles by size (b) formation of a 5 × 5 grid (scale bars represent 5 mm) and (c) repatterning taking a time of 3 s. (Reproduced from S. C. Chapin, V. Germain, and E. R. Dufresne, Automated trapping, assembly, and sorting with holographic optical tweezers, *Opt. Exp.*, 14, 13095–100, 2006. With permission of Optical Society of America.)

14.5.3 OTHER FORMS OF OPTICAL TRAPS

Our discussion so far has been confined to HOTs and tweezers, which essentially consist of multiple collimated light beams derived from a static or dynamic CGH and brought to a tight focus by a microscope objective.

However, the fact that an arbitrary phase distribution may be applied to a light beam by holographic means opens the way to creating exotic states of light. Among these are Bessel and helical modes of light.

14.5.3.1 Bessel Mode

Recall Equation 1.9 for an electromagnetic wave

$$\left(\nabla^2 - \frac{1}{c^2} \frac{\partial^2}{\partial t^2} \right) E(r,t) = 0 \tag{14.15}$$

A solution of Equation 14.15 is

$$E(x, y, z, t) = \exp[j(\beta z - \omega t)] \int_0^{2\pi} \tilde{A}(\phi) \exp[j\alpha(x \cos \phi + y \sin \phi)] d\phi \tag{14.16}$$

where $\beta^2 + \alpha^2 = (\omega/c)^2 = (2\pi/\lambda)^2$. If β is real, *the intensity does not alter with the distance z* [22]. Suppose that $\tilde{A}(\phi)$ is independent of ϕ, meaning that the beam is cylindrically symmetrical about the z-axis. Equation 14.16 becomes

$$E(x,y,z,t) \propto \exp[j(\beta z - \omega t)]\frac{1}{2\pi}\int_0^{2\pi} \exp[j\alpha(x\cos\phi + y\sin\phi)]d\phi \tag{14.17}$$

Letting $x = r\cos\theta$, $y = r\sin\theta$

$$E(x,y,z,t) \propto \exp[j(\beta z - \omega t)]\frac{1}{2\pi}\int_0^{2\pi} \exp[j\alpha r\cos(\phi - \theta)]d\phi \tag{14.18}$$

$$E(x,y,z,t) \propto \exp\left[j(\beta z - \omega t)\right]\frac{1}{2\pi}\int_{-\theta}^{2\pi-\theta} \exp\left[j\alpha r\cos\phi'\right]d\phi' \text{ where } \phi' = \phi - \theta$$

$$\int_{-\theta}^{2\pi-\theta} \exp(j\alpha r\cos\phi')d\phi' = \int_{-\theta}^0 \exp(j\alpha r\cos\phi')d\phi'$$

$$+ \int_0^{2\pi} \exp(j\alpha r\cos\phi')d\phi' - \int_{2\pi-\theta}^{2\pi} \exp(j\alpha r\cos\phi')d\phi'$$

which, if we increment ϕ' by 2π in the first integral on the right-hand side,

$$= \int_{2\pi-\theta}^{2\pi} \exp\left[j\alpha r\cos(2\pi + \phi')\right]d\phi' + \int_0^{2\pi} \exp(j\alpha r\cos\phi')d\phi' - \int_{2\pi-\theta}^{2\pi} \exp(j\alpha r\cos\phi')d\phi',$$

$$E(x,y,z,t) \propto \exp\left[j(\beta z - \omega t)\right]\frac{1}{2\pi}\int_0^{2\pi} \exp(j\alpha r\cos\phi')d\phi' = \exp\left[j(\beta z - \omega t)\right]J_0(\alpha r)$$

$J_0(\alpha r)$ is a zero-order Bessel function of the first kind and the beam is called a Bessel beam.

$$I(x,y,z) \propto J_0^2(\alpha r) \tag{14.19}$$

which, being independent of z, has a constant intensity profile. When $\alpha = 0$, the wave is plane, but for $0 < \alpha < \omega/c$, the intensity of the beam decreases with distance, r from the axis, first reaching zero at $\alpha r = 2.405$, so that if we set α at its maximum value of $\omega/c = 2\pi/\lambda$, then the central part of the beam has a minimum diameter of approximately 0.75λ. The intensity profile of the beam has the form shown in Figure 14.13a. The spatial period of the rings, *measured in the radial direction,* in the x, y plane tends toward a constant value with increasing distance. Putting it in another way, all of the wave vectors of the beam lie on the surface of a cone of half angle

$$\theta = \arcsin\left(\frac{\alpha\lambda}{2\pi}\right) \tag{14.20}$$

and are directed from the cone apex to its base as shown in Figure 14.13b.

This means that the spatial Fourier transform of a Bessel beam consists of a circle. Reversing the process, a Bessel beam may be formed using an annular aperture located in the front focal plane of a lens of radius R as shown in Figure 14.13c and illuminated by a plane wave. The central lobe of the beam extends beyond the lens by a distance z_c, which, from the geometry of Figure 14.13c and Equation 14.20, is given by

$$z_c = \frac{R}{\tan\theta} = R\sqrt{\left(\frac{2\pi}{\alpha\lambda}\right)^2 - 1} \tag{14.21}$$

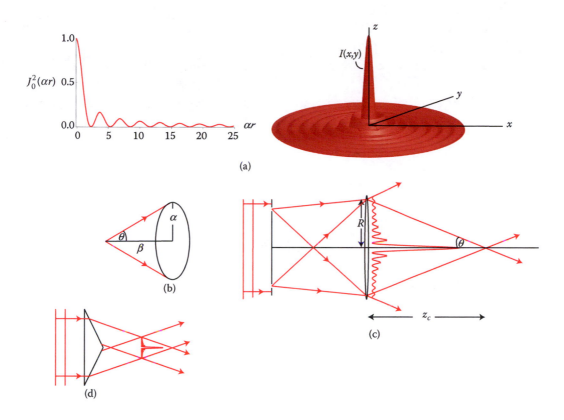

FIGURE 14.13 Zero-order Bessel beam: (a) intensity distribution in a Bessel beam; (b) angular spectrum; (c) Bessel beam from an annular aperture in the front focal plane of a lens; and (d) Bessel beam from a bulk optical axicon.

Suppose we aim for a central spot diameter of 100 μm. Then, α has a value of 4.8×10^4 m^{-1}, and at wavelength 532 nm and a lens diameter of 10 mm, we have $z_c = 80$ mm. Thus, we see that the central lobe of the Bessel beam can extend for a large distance without changing its width. The beam acts as an extended optical trap able to hold microscopic objects in a line on the beam axis. The price we have to pay is that the total power in each of the rings is about the same so that if there are 20 rings, the central spot only has about 5% of the total power.

As an alternative to an annular aperture, which is very wasteful of light, from Equation 14.17, if a plane wave has imposed on it a conical phase function given by

$$\phi(r) = \gamma r \tag{14.22}$$

where r is the distance from the beam axis and γ a constant, the result is a light wave of the form described by Equation 14.18. Such a device, called an axicon, takes the form of a cone-shaped bulk optical component or SLM-based, static, or dynamic CGH.

Although the *total* power in each ring of a zero-order Bessel beam is the same, the amplitude decreases with reciprocal distance from the axis, and this means that objects tend to move from outer rings toward inner ones, the tendency being greater for less massive objects, which can thus be sorted from more massive ones. The process is facilitated by the random Brownian motion of objects, which are more likely to end up in a stronger than in a weaker trap. Even if it is partially obstructed or distorted, the beam is reconstructed when it has propagated over a characteristic distance thus enabling the formation of multiple traps [23].

Higher order Bessel beams may be produced using appropriate CGH. An example is a first-order beam (Figure 14.14) in which the first maximum intensity occurs at $\alpha r = 1.75$ and objects may be trapped within the first ring or in the dark region enclosed by it.

14.5.3.2 Helical Modes

The use of phase functions enables us to produce even more exotic optical traps. Consider the phase function

$$\phi(\theta) = l\theta \tag{14.23}$$

imposed on a longitudinal mode of a laser, where θ is the azimuthal angle around the beam axis, and l is an integer number. Figure 14.15 shows an example with $l = 6$, the phase being represented in gray scale. In the region of the beam axis, the intensity is zero due to destructive interference between wavefront elements, which spatially overlap and are out of phase with one another. Overlap, and therefore interference between out of phase wavefront elements, decreases with distance from the axis. The result is that the intensity profile has an annular form. The region of zero intensity extends further from the axis as l increases, since overlap between out-of-phase wavefront elements is increased accordingly. The annular radius is proportional to l.

The beam acts as an annular trap for small objects, while larger ones whose diameter is less than that of the annulus can be trapped within it. Thus, objects can be selectively trapped by size, by altering the value of l. A "dark" trap of this type is very useful for objects that reflect or absorb light or that can be damaged by light.

Furthermore, the wavefront, defined in Section 1.1 as a surface on which the phase is everywhere the same, has a helical shape, and each photon in it has angular momentum $l\hbar$ where $\hbar$ is Planck's constant divided by 2π. This angular momentum imparts a torque on an object trapped in the annular region so that it describes a circular path within it. The beam is effectively an optical vortex and a tool, which can move objects trapped within and uniformly distributed around the annulus in a circular orbit. Interference of an optical vortex with a plane wave can also be used as an optical rotator of trapped objects.

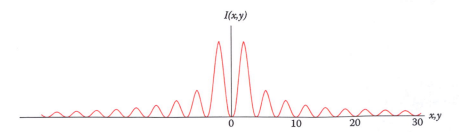

FIGURE 14.14 Intensity profile of first-order Bessel beam.

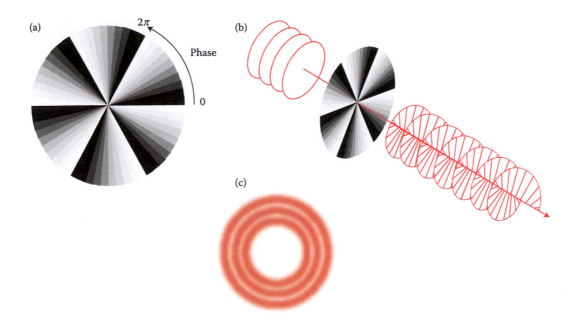

FIGURE 14.15 (a) A SLM phase function $\phi(\theta) = l\theta$, where θ is azimuthal angle around the beam axis, (b) converts a longitudinal mode laser beam into a helical one and (c) with an annular intensity distribution imposing angular momentum $l\hbar$ on a particle trapped within the beam. Here, $l = 6$.

14.5.4 APPLICATIONS OF HOLOGRAPHIC OPTICAL TWEEZERS

Applications of these tools range from the assembly of new materials and mechanical devices from nanosized building blocks to the use of subnanoNewton forces for biological studies at the cellular and subcellular levels. A pair of optical traps can be used to draw apart objects, which are bound together by elastic forces, or to push them together, in order to study their mechanical properties. The forces at work in optical traps and the torques in helical beams have been determined [24] and can be greater than 100 pN and controlled with a precision of 100 aN.

Permanent three-dimensional nanostructures and components may be assembled by combining translation and rotation of building blocks [25] followed by in situ photopolymerization (Figures 14.16 and 14.17).

Optical microrotors and machines have also been demonstrated (Figure 14.18). Similarly, gearwheels may be rotated by suitably shaped objects trapped within the annulus of an optical vortex of appropriate radius [26].

Microfluidic applications [27] include optically assembled and optically controlled pumps and valves (Figure 14.19) enabling flow control and purification at the molecular level.

The biological applications of HOTs and tweezers include sorting of cells. The use of such tools on biological tissue at the cellular and subcellular level has led to better understanding of the changes in mechanical properties, which accompany disease [28]. Optical traps may also be able to provide insights into biocompatibility at the cellular level and lead

Top view   Side view   Structure formed

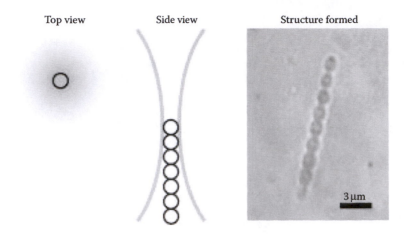

FIGURE 14.16 Multiple 0.64-mm silica particles oriented in an optical trap and photopolymerized by the trapping beam. (Reproduced with permission from A. Terray, J. Oakey, and D. W. M. Marr, *Appl. Phys. Lett.*, 81, 1555, 2002. Copyright 2002 by the American Physical Society.)

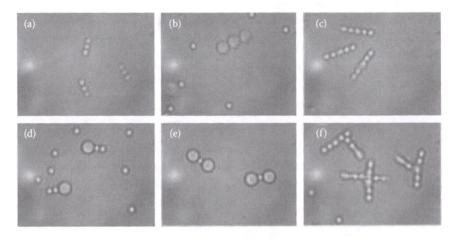

FIGURE 14.17 Homogeneous (a–c, f) and heterogeneous (d, e) colloidal structures assembled as in Figure 14.16. (Reproduced with permission from A. Terray, J. Oakey, and D. W. M. Marr, *Appl. Phys. Lett.*, 81, 1555, 2002. Copyright 2002 by the American Physical Society.)

to much more successful and reliable transplant and artificial implant surgery, as well as more effective and better targeted drug treatment.

In summary, automated real-time design and updating of CGH for precise shaping and directing of optical wavefronts is a rapidly maturing skill. It can provide the tools for a range of scientific and engineering disciplines to be able to operate on the microscopic scale. We are likely to see further significant progress in the integration of such tools for automated sorting, separation, directed motion, and in situ assembly.

(a)

(b)

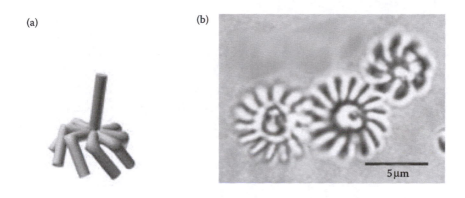

FIGURE 14.18 Micromachine: (a) propeller model and (b) propeller (top right), held and rotated by laser tweezers, drives two engaged cogwheels. The components are fabricated by in situ assembly and photopolymerization. (Reproduced with permission from P. Galajda and P. Ormos, Complex micromachines produced and driven by light, *Appl. Phys. Lett.*, 78, 249, 2001, Copyright 2001, American Institute of Physics.)

(a) (b) (c)

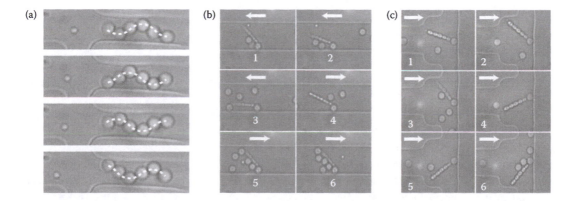

FIGURE 14.19 Micropump and microvalves. (a) Peristaltic micropump made from 3-mm colloidal silica. Optical tweezers induce sinusoidal motion of the pump causing right to left movement of the 1.5-μm tracer particle. (b) Colloidal flap valve. Arrows indicate the direction of fluid flow. In images 1–3, a flow of 2 nl/hour pushes the flap valve against the channel wall, allowing particulates in the flow to pass. In images 4–6, the flow is reversed, swinging the valve across the channel, restricting the flow of the 3 μm colloids but not the smaller 1.5-μm tracer particles. The valve was optically assembled and polymerized in a deeper region of the microfluidic network and optically maneuvered into position. (c) Three-way colloidal valve. In images 1–3, the valve directs 3-μm particles to the lower channel. In images 4–6, the optically switched valve directs particles into the upper channel. In such systems, the valve is often actuated by the laser used for its assembly. (Reproduced from Alex Terray, John Oakey, and David. M. W. Marr, Microfluidic control using colloidal devices, *Science*, 296, 1841–44, 2002. Reprinted with permission of AAAS.)

14.6 SUMMARY

In this chapter, we explain the rationale for computer-generated holography and describe some methods for the realization of CGHs with emphasis on how to encode the phase.

In discussing applications, we concentrate first on computer-generated holograms and computer-generated interferograms for testing high-specification optical components.

We then turn to techniques for optical trapping of microscopic and nanoscopic objects and show how these techniques may be implemented using computer-generated holography. We describe some of the simpler methods for generating unconventional wavefronts for micromanipulation including rotation. We conclude with a brief discussion of the applications and potential of holographic optical tweezers.

REFERENCES

1. B. R. Brown and A. W. Lohmann, "Complex spatial filtering with binary masks," *Appl. Opt.*, 5, 967–9 (1966).
2. B. R. Brown and A. W. Lohmann, "Computer generated binary holograms," *IBM J. Res. Dev.*, 13, 160–7 (1969).
3. E. Buckley and T. Wilkinson, "Laser-printer computer-generated holograms exhibiting suppressed higher orders in the replay field," *Opt. Lett.*, 31, 1397–8 (2006).
4. W. H. Lee, "Sampled Fourier transform hologram generated by computer," *Appl. Opt.*, 9, 639–43 (1970).
5. C. B. Burckhardt, "A simplification of Lee's method of generating holograms by computer," *Appl. Opt.*, 9, 1949 (1970).
6. L. B. Lesem, P. M. Hirsch, and J. A. Jordan, "The kinoform: a new wavefront reconstruction device," *IBM J. Res. Dev.*, 13,150–5 (1969).
7. D. Leseberg and C. Frère, "Computer-generated holograms of 3-D objects composed of tilted planar segments," *Appl. Opt.*, 27, 3020–4 (1988).
8. K. Matsushima and A. Kondoh, "Wave optical algorithm for creating digitally synthetic holograms of three-dimensional surface objects," Practical Holography XVII and Holographic Materials IX, T. H. Jeong and S. H. Stevenson, (eds.) Proc. SPIE 5005, 190 (2003).
9. K. Matsushima, "Computer-generated holograms for three-dimensional surface objects with shade and texture," *Appl. Opt.*, 44, 4607–14 (2005).
10. L. Ahrenberg, P. Benzie, M. Magnor, and J. Watson, "Computer generated holograms from three dimensional meshes using an analytic light transport model," *Appl. Opt.*, 47, 1567–74 (2008).
11. M. Janda, I. Hanák, and O. Levent, "Hologram synthesis for photorealistic reconstruction," *J. Opt. Soc. Am. A.*, 25, 3083–96 (2008).
12. S.-C. Kim and E.-S. Kim, "Fast computation of hologram patterns of a 3D object using run-length encoding and novel look-up table methods," *Appl. Opt.*, 48, 1030–41 (2009).
13. Y. Ichihashi, H. Nakayama, T. Ito, N. Masuda, T. Shimobaba, A. Shiraki, and T. Sugie, "HORN-6 special-purpose clustered computing system for electroholography," *Opt. Exp.*, 17, 13895–903 (2009).
14. J. C. Wyant and P. V. Bennett, "Using computer-generated holograms to test aspheric wavefronts," *Appl. Opt.*, 11, 2833–9 (1972).
15. W. H. Lee, "Binary synthetic holograms," *Appl. Opt.*, 13 1677–82 (1974).
16. D. G. Grier, "A revolution in optical manipulation," *Nature*, 424, 810–6 (2003).

17. K. Sasaki, M. Koshioka, H. Misawa, N. Kitamura, and H. Masuhara, "Pattern formation and flow control of fine particles by laser-scanning micromanipulation," *Opt. Lett.*, 16, 1463–5 (1991).

18. J. E. Curtis, B. A. Koss, and D. G. Grier, "Dynamic holographic optical tweezers," *Opt. Commun.*, 207, 169–75 (2002).

19. D. G. Grier and Y. Roichman, "Holographic optical trapping," *Appl. Opt.*, 45, 880–7 (2006).

20. J. Leach, K. Wulff, G. Sinclair, P. Jordan, J. Courtial, L. Thomson, G. Gibson, et al., "Interactive approach to optical tweezers control," *Appl. Opt.*, 45, 897–903 (2005).

21. S. C. Chapin, V. Germain, and E. R. Dufresne, "Automated trapping, assembly, and sorting with holographic optical tweezers," *Opt. Exp.*, 14, 13095–100 (2006).

22. J. Durnin, "Exact solutions for nondiffracting beams. I. The scalar theory," *J. Opt. Soc. Am. A.*, 4, 651–4 (1987).

23. V. Garcés-Chávez, D. McGloin, H. Melville, W. Sibbett, and K. Dholakia, "Simultaneous micromanipulation in multiple planes using a self-reconstructing light beam," *Nature*, 419, 145–7 (2002).

24. B. Sun, Y. Roichman, and D. G. Grier, "Theory of holographic optical trapping," *Opt. Exp.*, 16, 15765–76 (2008).

25. A.Terray, J. Oakey, and D. W. M. Marr, "Fabrication of linear colloidal structures for microfluidic applications," *Appl. Phys. Lett.*, 81, 1555–7 (2002).

26. P. Galajda and P. Ormos, "Complex machines produced and driven by light," *Appl. Phys. Lett*, 78, 249–51 (2001).

27. A.Terray, J. Oakey, and D. W. M. Marr, "Microfluidic control using colloidal devices," *Science*, 296, 1841–4 (2002).

28. T. W. Remmerbach, F. Wottawah, J. Dietrich, B. Lincoln, C. Wittekind, and J. Guck, "Oral cancer diagnosis by mechanical phenotyping," *Cancer Res.*, 69, 1728–32 (2009). See also the review by K. Ramser, and D. Hanstorp, "Optical manipulation for single-cell studies," *J. Biophoton.*, 3, 187–206 (2010).

PROBLEMS

1. Estimate the communication bandwidth required for a broadcast videoholography system. The dynamic display is 50 cm wide and 30 cm high and the angle between object and reference beams of wavelength 550 nm in the hologram recording system is 60°. The hologram refresh rate is 30 holograms per second.

2. A square, binary detour phase Fourier hologram is made up of square cells of area w^2. The amplitude of each Fourier component is proportional to the area of a clear rectangular aperture with dimensions $w_{x,n} w_y$ in each cell, where $w_{x,n}$ ranges from 0 to w and $w_y = 0.4_w$, the remainder of the cell being opaque. The object is a single point of light located on the z-axis. Find an expression for the diffraction efficiency of the hologram if the reference wave is plane and incident normally at the hologram.

3. Estimate the mechanical force exerted by a 1 mW beam of laser light of wavelength 532 nm on a flat, disc-shaped particle. Assume that the particle completely reflects the light and is 100 nm in diameter. The light is a collimated beam 1 mm in diameter and is focused normally on the disc using a 20× microscope objective. (Photons have momentum h/λ, where h is Planck's constant).

4. A transparent hemispherical particle of radius 5 μm and refractive index 1.5 is illuminated near its edge by a 1 mW beam of light of wavelength 532 nm, 1 mm in diameter, focused on the particle's flat surface by a lens with a focal length of 20 mm, the light being normally incident. Find the direction and approximate magnitude of the force on the particle.

CHAPTER 15

Holography and the Behavior of Light

15.1 INTRODUCTION

We saw in Section 13.6.1 how we can restrict the depth of field in digital holography by using a broadband light source. Such a source has a correspondingly short coherence length, and therefore, the range of object size, or depth, which may be recorded in a hologram, is thereby restricted. Short coherence length can thus be a distinct advantage in holographic applications where good depth resolution is required.

Similar considerations apply if the light source is pulsed with the physical length of the pulse replacing the coherence length. In other words, the temporal duration of the pulse replaces the average duration of a wavetrain.

Suppose a single laser pulse is used to record a hologram as shown in Figure 15.1. The pulse is split into two by a beam splitter to form object and reference beams, which traverse different routes in space to arrive finally at the hologram plane. Clearly, they must arrive together in time, or more correctly, their times of arrival should not differ by more than the pulse duration, otherwise they do not interfere at all. If the pulse duration is extremely short, say 1 ps, then its physical length is just 0.3 mm and only that portion of the object, whose extent is the projection of the pulse length onto the object surface, can be holographically recorded.

Using this basic principle, we can apply holography to the study of light itself, a subject known as light-in-flight (LIF). Holography enables us to realize physically many of the mental pictures we have of optical phenomena such as reflection, refraction, diffraction, and focusing of light beams, and which we normally represent in diagrammatic form.

15.2 THEORY OF LIGHT-IN-FLIGHT HOLOGRAPHY

In the holographic setup [1] shown in Figure 15.2, a pulse of laser light is split so that part of it goes on to illuminate a diffusing screen at an angle of incidence close to 90°. The light is

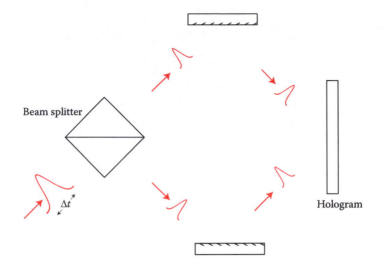

FIGURE 15.1 A hologram of a light pulse can only be recorded if both object and reference pulses arrive at the hologram plane within time interval Δt of one another.

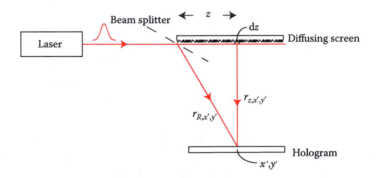

FIGURE 15.2 Holographic recording of a laser pulse trajectory. (Reproduced from Y. N. Denisyuk and D. I. Staselko, Light-in-flight recording: high-speed holographic motion pictures of ultrafast phenomena: Comment, *Appl. Opt.*, 31, 1682–84, 1992. With permission of Optical Society of America.)

then scattered toward the hologram plane. The purpose of the diffusing screen is to ensure that the hologram receives light, which has been scattered from points that are located at progressively increasing distances from the beam splitter. The other part of the pulse is reflected at the beam splitter, directly toward the hologram. The light scattered by element, dz, of the diffuser, at distance z from the origin at the beam splitter, provides a contribution, $d\tilde{O}$, to the object wave at a point x', y' on the hologram plane, given by

$$d\tilde{O} = dz \exp[jg_O(k_0, x', y')] \int_0^\infty E(k) \exp\{j[k(z + r_{z,x',y'} + f(z)) - \omega t + \phi(k)]\}dk \quad (15.1)$$

where $E(k)$ and $\phi(k)$ are the spectral and phase distributions of the light in the pulse. The phase factor $g_O(k_0, x', y')$ can be placed outside the integral sign since the phase change arising from propagation from $z = 0$ to $z = z$ and on to x', y' does not vary much in a small area close to x', y', assuming that the spectral bandwidth of the light is fairly narrow. The term $f(z)$ is a geometrical path factor that varies randomly due to the surface profile of the screen and contributes a random phase shift while $r_{z,x',y'}$ is the distance between the element at z and x', y'.

The reference wave at x', y' is given by

$$\tilde{R} = \exp[jg_R(k_0, x', y')] \int_0^\infty E(k') \exp\{j[k'r_{R,x',y'} - \omega t + \phi(k')]\}dk' \tag{15.2}$$

The recorded intensity is $\int_{-\infty}^\infty (d\tilde{O} + \tilde{R})(d\tilde{O} + \tilde{R})^* dt$, and we assume that the amplitude transmissivity at the point x', y' of the hologram is proportional to this last expression. Illuminating the hologram by purely monochromatic light, we obtain

$$\int_{-\infty}^\infty R^*(d\tilde{O})\tilde{R}dt = dz \exp\{j[g_O(k_0, x', y') - \omega t]\} \int_0^\infty E^2(k) \exp\{[jk(z + r_{z,x',y'} + f(z))]\}dk \tag{15.3}$$

as the exponentials involving the time, produce $\delta(k - k')$.

On looking through the hologram at x', y', we observe the square of the term on the right-hand side of Equation 15.3. The distance $z + r_{z,x',y'}$ in the exponential term is strictly related to *time* through the velocity of light so that, as we change viewpoint, we observe the pulse at different times in its flight. In other words, *as we scan our viewpoint across the hologram, we observe the recorded trajectory of the pulse.* Starting from the extreme left of the hologram, we observe the reconstruction of the earliest arriving part of the laser light pulse.

A somewhat modified version of the experiment using an Argon ion laser with a short coherence length [2] is shown in Figure 15.3a. Here, a flat white-painted surface is obliquely illuminated by the object beam. The arrangement is such that the path of the object beam from the pinhole of the spatial filter to the midpoint of the object surface and thence to the midpoint of the hologram is equal to that of the reference beam via the two mirrors, to the same point on the hologram. Optical paths ending at points left of the hologram midpoint are shorter, whereas those ending at points to the right are longer so that, again, the hologram is a recording of the history of the arriving light. The width of the photosensitive layer determines the time interval, over which one can record the trajectory of the object beam. Thus, for example, the *maximum* time taken by a light pulse to traverse a photosensitive layer 10 cm in width is 0.33 ns if the angle of incidence is 90°. At smaller angles of incidence, the time is shorter.

15.3 REFLECTION AND OTHER PHENOMENA

In Figure 15.3b, a small plane mirror is mounted vertically on the surface to reflect the central part of the wavefront. Note that the part of the wavefront outside the perimeter of the object surface is not recorded. The result is seen in Figure 15.3c.

Abramson has also demonstrated holographic recording of light focused by a cylindrical lens using the apparatus of Figure 15.3 (see Figure 15.4a) [3], as well as the optical delay imposed on a light pulse passing normally through a rectangular block of polymethylmethacrylate (PMMA) (Figure 15.4b and 15.4c) [4].

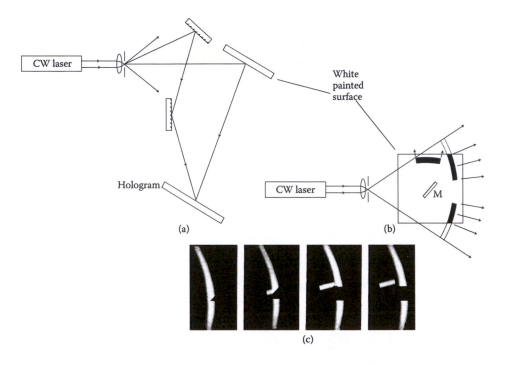

FIGURE 15.3 Holographic recording of a reflected wavefront: (a) optical arrangement; (b) reflection from a mirror; and (c) reconstructions from the hologram. The first picture on the left shows the wavefront arriving at the mirror, M, in (b). The second shows the beginning of reflection, followed by almost complete separation of reflected and incident light leaving a hole in the middle of the incident wavefront. (Reproduced from N. Abramson, Light-in-flight recording by holography, *Opt. Lett.*, 3, 121–23, 1978. With permission of Optical Society of America.)

Other LIF experiments [5] enable recording and reconstruction of the refracted trajectory of a laser pulse through a rectangular block of glass, as well as total reflection within the glass and diffraction by a grating. In Figure 15.5, the object pulse is expanded, collimated, and then directed to illuminate a transmitting diffuser at an oblique angle so that its time of arrival at the diffuser varies with the distance along the diffuser from the left side. Likewise, the time of arrival at the hologram plane varies with distance, z, from the left. In this way the refracted trajectory of a light pulse through a rectangular glass block, is demonstrated.

LIF phenomena may also be demonstrated using electronic speckle pattern interferometry (ESPI) (see Section 10.14.2). The basic arrangement is that of an out-of-plane sensitive ESPI system [6] with a variable optical delay line to ensure that the optical path difference in the interferometer is less than the coherence length of the laser source and enable one to alter the region of object space for which this is the case. One can implement image subtraction to obtain the interference term by introducing a phase shift of π in every alternate image frame and adding images on the camera faceplate. Although the image quality of the reconstructed optical events (a spherical wavefront impinging on a flat surface and a spherical wavefront traversing a small glass plate) is not very high, the system has the

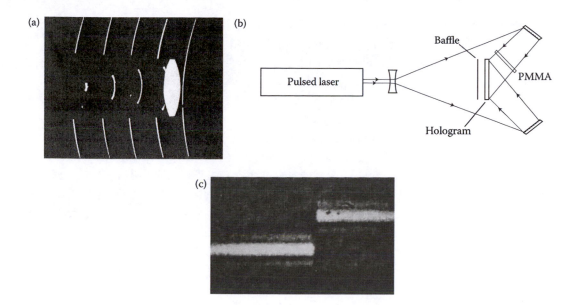

FIGURE 15.4 Focusing and delay. (a) Focusing: a series of reconstructions from a single LIF hologram. A 1ps laser pulse, 3 mm in width, approaches a lens (highlighted) from the right and is progressively focused. (b) Recording: an optical delay introduced by a plexiglass plate (c) holographic reconstruction before (left) and after (right) insertion of plexiglass plate in (b). (Reproduced from N. Abramson, Light-in-flight recording: High speed motion pictures of ultrafast phenomena, *Appl. Opt.*, 22, 215–31, 1983. With permission of Optical Society of America.)

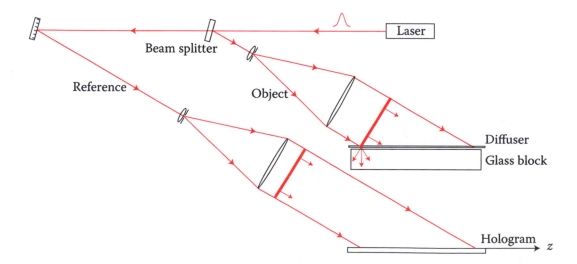

FIGURE 15.5 Refraction through a rectangular glass block. (Reproduced from T. Kubota and Y. Awatsuji, Observation of light propagation by holography with a picosecond pulsed laser, *Opt. Lett.*, 27, 815–17, 2002. With permission of Optical Society of America.)

advantages of all ESPI systems, namely the elimination of hologram recording and processing and the display of the results at television frame rates.

One can also use digital holography with numerical reconstruction (See Chapter 13) to display LIF phenomena [7].

There are nonholographic methods of recording high-speed optical phenomena. One method exploits the fact that for certain combinations of laser wavelength and fluorescent dye, the energy required to cause fluorescence in a dye is twice the energy of the photons in the laser pulse. Fluorescence therefore only occurs when two pulses overlap in a solution of the dye, with spatial extent approximately twice that of each pulse. We can record the resulting spatial distribution of fluorescence intensity using a high-speed, rather costly, streak camera [8]. The speed of response of the recording system is limited by the electronic bandwidth of the camera, whereas no such limitation applies to the holographic method.

15.4 EXTENDING THE RECORD

As explained in Section 15.2, the duration, T, of the recorded trajectory of the light is determined by the width of the hologram because the reference beam pulse incident at angle α as shown in Figure 15.6a traverses the hologram of width L in time, T, given by

$$T = L\sin\alpha/c \tag{15.4}$$

One could use a wider hologram in order to extend the duration. Alternatively, the beam can be reflected back and forth across the hologram aperture by means of mirrors placed on either side of it [1]. Another method is to reduce the velocity of the light by allowing it to propagate in a medium of high refractive index.

Pettersson et al. [9,10] used a diffraction grating to extend the duration. Suppose a light pulse illuminates a grating over a region of width W as shown in Figure 15.6b. The path

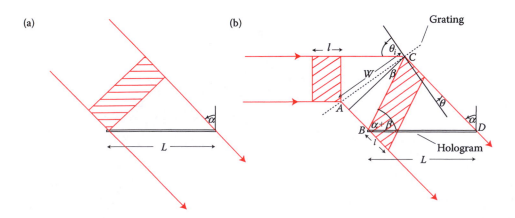

FIGURE 15.6 Extending hologram transit time of a pulse: (a) normal transit and (b) using a diffraction grating. (Reproduced from S.-G. Pettersson, H. Bergstrom, and N. Abramson, Light-in-flight recording. 6: Experiment with view-time expansion using a skew reference wave, *Appl. Opt.*, 28, 766–70, 1989. With permission of Optical Society of America.)

difference between light rays *arriving* at the extremities of the illuminated area is $W \sin \theta_i$, where θ_i is the angle of incidence at the grating. The path difference between rays *leaving* from the extremities is $W \sin \theta$, with θ the diffraction angle. The total path difference is

$$W(\sin \theta_i - \sin \theta) = AB \tag{15.5}$$

The pulse has effectively been stretched in length l to $l + AB$, along its direction of propagation. Suppose it is now incident at angle α at the plane of the recording layer. The transit time is T, given by

$$T = \frac{CD}{c} = \frac{L\sin(\alpha + \beta)}{c \cos \beta} \tag{15.6}$$

which increases with the angle β. If β is zero, the transit time is again given by Equation 15.4.

A different approach is used in digital holography because the angle α must be kept very small so that the low-resolution CCD array can satisfactorily cope with the pattern of interference between object and reference beams [7]. A set of parallel-sided transparent plates with differing thicknesses is placed in the path of the reference beam as shown in Figure 15.7a. Each plate of thickness, t, and refractive index, n, introduces a delay τ in the time of arrival of the reference beam, given by

$$\tau = (n-1)t / c \tag{15.7}$$

For example, a PMMA plate with a refractive index of 1.5 and thickness 6 mm produces a delay of 10 ps. This method does not *extend* the total duration of the recorded history of the pulse

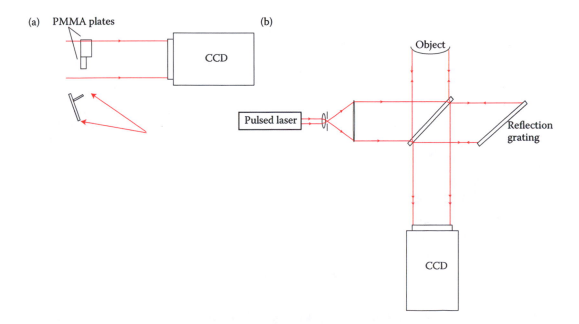

FIGURE 15.7 Delayed reference wave in digital light-in-flight holography: (a) pile of plates method and (b) reflection grating.

trajectory but allows one to reconstruct *parts* of the trajectory, interspersed by gaps. A disadvantage of the method is that diffraction occurs at the edges of the plates, and the resulting interference effects are recorded by the CCD array. Thus, some parts of the recorded holograms cannot be used in computing the reconstruction. This drawback can be overcome using a reference beam, which is incident on a reflection grating in an optical setup similar to a Michelson interferometer. The grating may be of the conventional type, known as a Littrow grating [11], or it may be a holographically recorded reflection grating, and it produces a continuous delay in the time of arrival of the reference pulse across the CCD array (Figure 15.7b).

15.5 APPLICATIONS OF LIGHT-IN-FLIGHT HOLOGRAPHY

15.5.1 CONTOURING

One of the great advantages of LIF holography using a pulsed laser is that mechanical and thermal instabilities present no difficulty, so one application of the technique is the contouring of a rapidly rotating fan using a 13 ps pulse from a frequency doubled NdYag laser, the contour resolution being about 2 mm [4]. The contour observed through a point x', y' of the hologram is a line joining the points on the object, for which the object and reference beams arrive at x', y' within a pulse duration of one another.

This method of contouring leads to a method of shape comparison [12]. Suppose we use as object illuminating light a temporal sequence of pulses having a constant delay between consecutive ones. The longest delayed pulse traverses an optical path to the point on the object surface nearest to the hologram plane, whereas the shortest delayed pulse traverse paths to points on the object surface, which are progressively more distant from the hologram plane. This should result in all pulses reflected by the object arriving at the hologram plane at the same time, or more precisely, within one pulse duration of each other. If the object is then replaced by another one nominally having the same detailed surface profile and oriented in space with its principal axes coinciding with those of the first object, then the duration of the holographically reconstructed light is a measure of the similarity of the two surface shapes.

An extension of the method, for application to shape recognition, involves *computing* a hologram of the light trajectory that would be obtained from the master object [12]. Nilsson and Carlsson [13] describe a digital method of shape measurement.

The ESPI technique combined with LIF enables comparison of a surface shape with a flat reference surface [14] using an optical setup similar to a Michelson interferometer.

We recall Equation 10.46, which describes the resulting image intensity when two images are subtracted from one another, following a change in phase, $\psi(x, y)$,

$$\Delta I(x,y) = I(x,y) - I'(x,y) = 4\sqrt{I_O I_R} \sin\left(\phi + \frac{\psi}{2}\right)\sin\left(\frac{\psi}{2}\right) \quad (15.8)$$

and note that $\Delta I(x, y)$ cycles through maximum and minimum values as the reference surface is translated out of plane. The resulting image $\Delta I(x, y)$ is computed for each incremented position of the reference surface, as it is translated through equal small fractions of

(a)

(b)

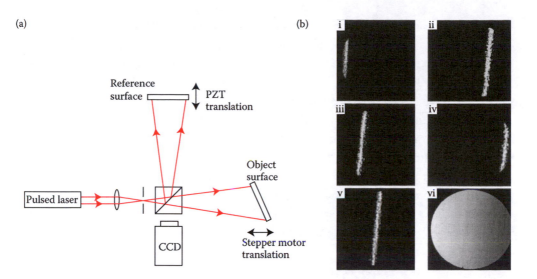

FIGURE 15.8 Shape comparison by LIF speckle interferometry: (a) optical arrangement and (b) reconstructions (i–v) of contours of object surface represented by different intersections of object and reference surfaces as the latter is translated and (vi) final gray-level image of object surface. (Reproduced from A. Wei and T. E. Carlsson, Direct object-shape comparison by light-in-flight speckle interferometry, *Opt. Lett.*, 22, 1538–60, 1997. With permission of Optical Society of America.)

a wavelength by means of a piezoelectric transducer. This procedure enables one to obtain maximum intensity and minimum intensity images. Subtraction of these from each other produces a single contour line joining all the points on the object surface, for which the optical path difference in the interferometer is less than the length of a laser pulse. The object surface is then translated by a stepper motor to a new position to obtain a new contour.

An interesting contouring method combines LIF and rainbow holography (Section 8.4) enabling the contours to be color coded and displayed together in a single reconstructed image [15]. The method makes use of the fact that at each point across the width of the hologram, we observe the reconstructed light pulse at a specific instant in time. One may therefore regard the hologram as a series of vertical slit holograms, with each slit representing a small time interval although no physical slit is present. The hologram is illuminated by a conjugate reference wave so that a real image is reconstructed of the pulse trajectory and a second, image plane hologram (see Section 8.4.1) is recorded in the conventional manner. Finally, one reconstructs a real orthoscopic image from this hologram using a white light source.

15.5.2 PARTICLE VELOCIMETRY

We have seen in Section 9.2 that holography provides a useful method for studying small dispersed particles and their velocities. One can view the reconstructed images through a microscope whose limited depth of field effectively allows for examination of the particles within a narrow range of depth. Nonetheless, out-of-focus images inevitably accompany those that are in focus.

One way of avoiding this problem is to use a thin sheet of pulsed light to illuminate the object space so that one records a hologram only of the particles in the plane illuminated by the sheet [16]. One can introduce multiple reference beams at different angles of incidence at the hologram plane, each of them carefully matched in its path length range across the hologram to the path length range associated with one of the illuminated sections of object space. However, the method is optically complex and sampling in planar sections, which one must select at the outset, is somewhat restrictive.

The LIF technique does allow one to ensure that the object space, which is reconstructed from a particular section of the hologram, is depth limited by the duration of the light pulse used in recording the hologram [17]. Optimal depth limitation is achieved by illuminating the object space in the same direction in which the reconstructed, backscattered light is observed through the hologram (Figure 15.9).

15.5.3 TESTING OF OPTICAL FIBERS

Optical fibers can be tested by examining the distorted shape of a light pulse after it has propagated through a length of the fiber [18]. The pulse then illuminates a white screen at a very oblique angle so that light is scattered toward the hologram plane, which is also illuminated obliquely by collimated reference light (Figure 15.10). Removal of the fiber allows one to record a hologram of the trajectory of the undistorted pulse for comparison.

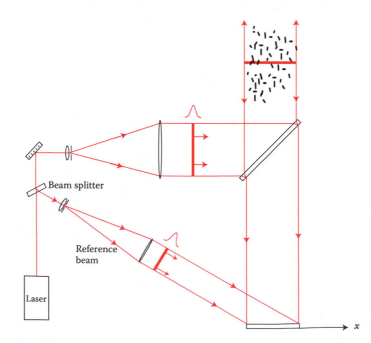

FIGURE 15.9 Particle imaging by LIF. Planes parallel to and at various distances from the recorded hologram are observed through it at different points.

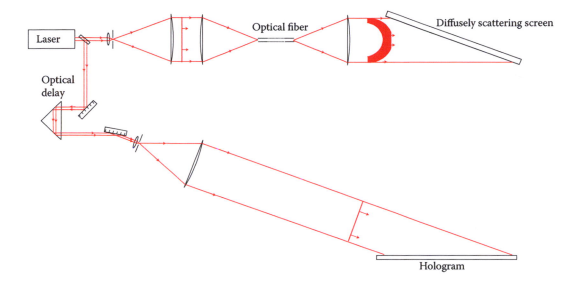

FIGURE 15.10 Testing optical fiber by LIF. (Reproduced from N. Abramson, Optical fiber testing using light-in-flight recording by holography, *Appl. Opt.*, 26, 4657–59, 1987. With permission of Optical Society of America.)

The laser pulse shape is altered and its duration increased significantly at the edges, due to the longer path taken by light whose wave vector is slightly off axis and which undergoes multiple total internal reflections while on-axis light takes the shortest route.

15.5.4 IMAGING THROUGH SCATTERING MEDIA

The possibility of obtaining images of objects, which are surrounded by optically scattering media, attracts considerable interest. The effects of the scattering medium surrounding the object may be comparatively slight as in the case of atmospheric turbulence or very light mist but an extreme case, of perhaps greatest interest, is the effect of surrounding tissue on images of internal organs. X-rays have been used extensively for many years, but one would like to avoid the use of ionizing radiation on living tissue.

The phase conjugation techniques described in Section 9.3 could in principle be used to extract images by unraveling the wavefront distortions imposed on an object wave [19], but a better solution to the problem may be a technique known as first-arriving light. If we send a short laser pulse through the scattering medium, the pulse becomes stretched out in time as it scatters at the inhomogeneities. We can think of the light as traversing a zigzag path, continuing in this manner after diffraction at the object. A small amount of light, which reaches the object without scattering and is diffracted so as to reach the hologram plane, again without scattering, will have a least time trajectory and will arrive there first (Figure 15.11). Any detection system, which discriminates in favor of first-arriving light, could be used to form an image of the object. LIF holography provides a method of doing this since we can adjust the delay of the reference beam and its angle of incidence at the hologram plane so that the time interval covered by the width of the hologram

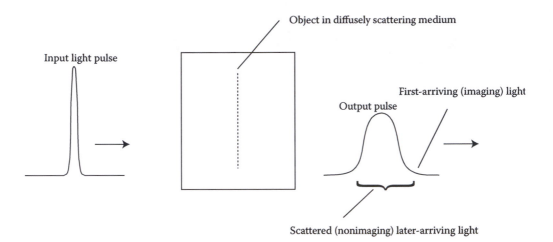

FIGURE 15.11 First-arriving light (transillumination).

encompasses the interval of arrival time of first-arriving object beam light. This first-arriving light will inevitably be accompanied by some light, which is scattered toward the image plane even before it can reach the object. However, this situation is avoided in transillumination experiments [20] in which the light source and the hologram plane are placed on either side of the diffusing medium as shown in Figure 15.11, in what is effectively a Gabor holography arrangement.

An interesting approach to LIF holography of first-arriving light involves the use of a laser capable of being tuned over a wide range of wavelengths [21]. If the hologram recording time is the same as, or an integer multiple of the time it takes to tune the laser over its bandwidth, the result is the same as if the hologram were recorded using a broadband light source and therefore one with a short coherence length. For LIF purposes, this is equivalent to a short-pulse laser. Later-arriving light is also recorded, but not holographically, and produces a non-imaging background. If we record a *digital* hologram using a CCD camera, the time interval for LIF, represented by the width of the CCD array, is extremely short. This means that we need to take care that the reference beam is delayed by just the right amount so that its arrival at the hologram plane coincides with first-arriving object light. For this purpose, we record a set of digital holograms, one at each of the discrete wavelengths emitted by the laser source. We then store all these holograms in computer memory and add the reconstructions together. We adjust the phase of the reconstructing reference light, before each addition operation to compensate for the difference in the time of arrival of the reference and that of the first-arriving light. Only the latter is likely to produce an image structure, and the remainder of the object light gives rise only to random speckle patterns. The procedure is computationally intensive as we, or rather the image processor, must look to see if any image-like features, such as edges or points, have emerged after each addition operation. When this happens, the phase (that is the time delay) is adjusted around the current value to sharpen this image. An example of the result obtained is shown in Figure 15.12a). Of course, the search may involve more than one image.

(a)

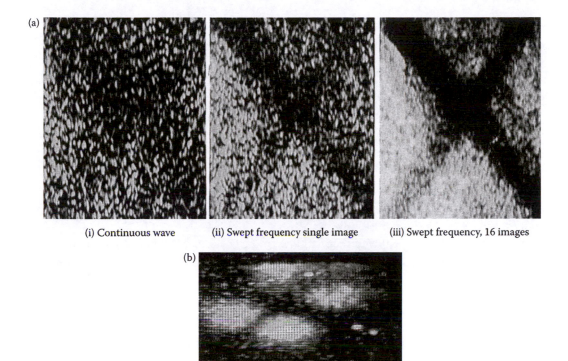

(i) Continuous wave   (ii) Swept frequency single image   (iii) Swept frequency, 16 images

(b)

FIGURE 15.12 Digital holography with first-arriving light. (a) 1-mm and 1.5-mm diameter wires imaged through a layer of chicken meat using the swept frequency technique: (i) continuous wave, (ii) swept frequency single image, and (iii) swept frequency, 16 images. (Reproduced from E. Leith, H. Chen, Y. Chen, D. Dilworth, J. Lopez, J. Rudd, and P.-C. Sun Imaging through scattering media with holography, *J. Opt. Soc. Am. A.*, 9, 1148–53, 1992. With permission of Optical Society of America.) (b) Pair of crossed wires (2 mm and 1.5 mm diameter.) imaged through a living human hand. To avoid effect of hand motion, exposure time was 1 ms. Averaging over 25 images at intervals exceeding coherence time means speckle is uncorrelated. (Reproduced from H. Chen, M. Shih, E. Arons, E. Leith, L. Lopez, D. Dilworth, and P. C. Sun, Electronic holographic imaging through living tissue, *Appl. Opt.*, 33, 3630–32, 1994. With permission of Optical Society of America.)

The LIF technique has also been used successfully in imaging through living human tissue [22]. An exposure time is chosen, which is too short (5–20 ms) for natural motion of living tissue to cause excessive smearing of recorded interference fringes, but long enough for adequate exposure on a cooled CCD array with up to 18 bits of gray-level resolution, enabling even very low-contrast interference patterns to be recorded. Laser speckle is decorrelated if there is a sufficiently long delay *between exposures* so that the addition of reconstructed images increases the signal-to-noise ratio. A low-pass spatial filter helps to eliminate high spatial frequency scattered light from the recordings and matches the spatial frequency bandwidth of the light to that of the CCD array. An image is shown in Figure 15.12b).

15.6 SUMMARY

In this chapter, we consider the uniquely holographic technique of LIF, highlighting the fact that for most practical purposes, we may use either a pulsed- or a continuous-wave laser. This is because the temporal resolution of LIF is determined equally well by the physical length of a laser pulse or the coherence length of the light from a continuous-wave laser. Emphasizing that a hologram is a historical record of the trajectory of a light beam, we discuss the basic theory of LIF and explain its use in demonstrating fundamental optical phenomena such as propagation, reflection, refraction, total internal reflection, and diffraction of light.

There follow brief descriptions of alternative methods of implementing LIF recording such as ESPI and digital holography. We describe methods for extending the duration of the holographic record including recording with CCD cameras with their small arrays.

We then go on to consider some applications of LIF, such as surface contouring, particle imaging and the testing of optical fibers.

Finally, we discuss the use of LIF for the holographic recording of first-arriving light because of its potential for holographic recording in diffusing media, including living human tissue.

REFERENCES

1. D. I. Staselko and Y. N. Denisyuk, "Holographic registration of temporal coherence of a wave train of a pulse radiation source," *Opt. Spectrosc.*, 26, 413 (1969). See also Y. N. Denisyuk and D. I. Staselko, "Light-in-flight recording: high-speed holographic motion pictures of ultrafast phenomena: comment," *Appl. Opt.*, 31, 1682–4 (1992).
2. N. Abramson, "Light-in-flight recording by holography," *Opt. Lett.*, 3, 121–3 (1978).
3. N. Abramson, "Light-in-flight recording: High speed motion pictures of ultrafast phenomena," *Appl. Opt.*, 22, 215–31 (1983).
4. N. Abramson, "Single pulse light-in-flight recording by holography," *Appl. Opt.*, 26, 1834–41 (1989).
5. T. Kubota and Y. Awatsuji, "Observation of light propagation by holography with a picosecond pulsed laser," *Opt. Lett.*, 27, 815–7 (2002).
6. H. Rabal, J. Pomarica, and R. Arizaga, "Light-in-flight digital holography display," *Appl. Opt.*, 33, 4358–60 (1994).
7. J. Pomarico, U. Schnars, H.-J. Hartmann, and W. Juptner, "Digital recording and numerical reconstruction of holograms: A new method for displaying light-in-flight," *Appl. Opt.*, 34, 8095–9 (1995).
8. J. A. Giordmaine, P. M. Rentzepis, S. L. Shapiro, and K. W. Wecht, "Two-photon excitation of fluorescence by picosecond light pulses," *Appl. Phys. Lett.*, 11, 216–8 (1967).
9. S.-G. Pettersson, H. Bergstrom, and N. Abramson, "Light-in-flight recording. 5: Theory of slowing down the faster-than-light motion of the light shutter," *Appl. Opt.*, 28, 759–65 (1989).
10. S.-G. Pettersson, H. Bergstrom, and N. Abramson, "Light-in-flight recording. 6: Experiment with view-time expansion using a skew reference wave," *Appl. Opt.*, 28, 766–70 (1989).
11. N. Abramson, "Time reconstruction in light-in-flight recording by holography," *Appl. Opt.*, 30, 1242–52 (1991).
12. D. Mendlovic and N. Avishay, "Three-dimensional shape recognition using computer-generated holograms and temporal light-in-flight technique," *Appl. Opt.*, 34, 6621–5 (1996).

13. B. Nilsson and T. E. Carlsson, "Direct three-dimensional shape measurement by digital light-in-flight holography," *Appl. Opt.,* 37, 7954–9 (1998).

14. A. Wei and T. E. Carlsson, "Direct object-shape comparison by light-in-flight speckle interferometry," *Opt. Lett.,* 22, 1538–60 (1997).

15. G. Molesini and F. Quercloli, "White light-in-flight holography," *Appl. Opt.,* 24, 3406–14 (1985). See also G. Molesini and F. Quercloli, "Pseudocolor contouring based in light-in-flight holography," *J. Opt. Soc. Am. A,* 3, 738–42 (1986).

16. H. Hinrichs and K. D. Hinsch, *Developments in Laser Techniques and Applications to Fluid Mechanics,* R. J. Adrian, D. F. G. Durao, F. Durst, M. V. Heitor, M. Maeda, and J. Whitelaw, (eds.), Springer-Verlag, Berlin, 408–22 (1966).

17. H. Hinrichs, K. D. Hinsch, J. Kickstein, and M. Bohmer, "Light-in-flight holography for visualization and velocimetry in three-dimensional flows," *Opt. Lett.,* 22, 818–30 (1997).

18. N. Abramson "Optical fiber testing using light-in-flight recording by holography," *Appl. Opt.,* 26, 4657–9 (1987).

19. H. J. Gerritsen, "Holography and four-wave mixing to see through the skin," *Analog Optical Processing and Computing,"* H. J. Caulfield (ed.), *Proc. SPIE* 519, 128–31 (1989).

20. K. G. Spears, J. Serafin, N. H. Abramson, X. Zhu, and H. Bjelkhagen, "Chrono-coherent imaging for medicine," *IEEE Transactions on Biomedical Engineering,* 36, 1210–21 (1989).

21. E. Leith, H. Chen, Y. Chen, D. Dilworth, J. Lopez, J. Rudd, and P.-C. Sun, "Imaging through scattering media with holography," *J. Opt. Soc. Am. A.,* 9, 1148–53 (1992).

22. H. Chen, M. Shih, E. Arons, E. Leith, L. Lopez, D. Dilworth, and P. C. Sun, "Electronic holographic imaging through living tissue," *Appl. Opt.,* 33, 3630–2 (1994).

PROBLEMS

1. Light-in-flight is used for measuring the radius of curvature of spherical convex surfaces by means of the arrangement shown in Figure 15.7. Find an expression for minimum radius of curvature, which can be measured, if the reflection grating is at an angle of 45° to the light incident on it and the optical path difference is zero on the axis of symmetry.

2. Estimate the thickness of the PMMA block in Figure 15. 4b by taking measurements from the photograph in Figure 15.4c. The refractive index of PMMA is approximately 1.5.

3. Equation 15.5 can be rearranged to tell us the speed at which a laser pulse crosses the hologram. It exceeds the speed of light. Is this reasonable?

4. The arrangement of Figure 15.1 can be used simply to measure the duration of a laser pulse. Using the Uncertainty Principle in the form

$$\Delta E \Delta t \approx \hbar$$

where $\hbar = h/2\pi$ and ΔE is the uncertainty in pulse energy, find an expression for the coherence length of the laser and show that it is approximately equal to the physical length of the pulse.

CHAPTER 16

Polarization Holography

16.1 INTRODUCTION

In our discussion up to this point, we have assumed that the interfering light beams used to record a hologram have the same polarization state, usually plane, or linear, with the polarization planes parallel to one another. This is essential in order to obtain maximum contrast in the interference pattern to be recorded and hence to obtain maximum diffraction efficiency of the hologram. When the polarizations are not linear and parallel, some of the dynamic range of the recording medium is used up in recording a spatially uniform light intensity. However, if we use a recording medium that has sensitivity to the polarization state of the recording beams, we can record a *polarization hologram*. Among the very interesting consequences are the possibility of obtaining 100% diffraction efficiency, even in thin layers of recording material, and achromatic response. Besides holographic displays and data storage, other interesting applications are possible including the analysis of polarized light, bifocal lenses, and electrically switchable optical devices.

16.2 DESCRIPTION OF POLARIZED LIGHT

We discussed linearly, circularly, and elliptically polarized light in Sections 2.9.1 and 2.9.2. For convenience, we will briefly review each of these polarization states.

Any pure polarization state is fully described by two orthogonally linearly polarized light waves, which are phase shifted with respect to one another, with their resultant given by

$$\mathbf{E}(z,t) = \hat{\mathbf{x}}E_{0x}\cos(kz - \omega t) + \hat{\mathbf{y}}E_{0y}\cos(kz - \omega t + \phi) \tag{16.1}$$

Equation 16.1 describes a plane wave propagating along the z-axis. It has orthogonal components of amplitude E_{0x} parallel to the x-axis and E_{0y} parallel to the y-axis with a phase difference ϕ between them. If $\phi = 2n\pi$, the two components are in phase and the resultant is linearly polarized, with electric field at $\tan^{-1}(E_{0y}/E_{0x})$ to the x-axis, and given by

$$\mathbf{E} = (\hat{\mathbf{x}}E_{0x} + \hat{\mathbf{y}}E_{0y})\cos(kz - \omega t) \tag{16.2}$$

If $\phi = 2n\pi - \pi/2$, we have

$$\mathbf{E}(z,t) = \hat{\mathbf{x}}E_{0x}\cos(kz - \omega t) + \hat{\mathbf{y}}E_{0y}\cos\left(kz - \omega t - \frac{\pi}{2}\right) \tag{16.3}$$

which means that the resultant electric field vector rotates *clockwise*, with time, around the z-axis as we look in the $-z$ direction toward the source of the light. The wave is *right circularly* polarized if $E_{0x} = E_{0y}$ and *right elliptically* polarized if $E_{0x} \neq E_{0y}$ (Figure 2.19a).

If $\phi = 2n\pi + \pi/2$, we have

$$\mathbf{E}(z,t) = \hat{\mathbf{x}}E_{0x}\cos(kz - \omega t) + \hat{\mathbf{y}}E_{0y}\cos\left(kz - \omega t + \frac{\pi}{2}\right) \tag{16.4}$$

and the resultant electric field vector rotates *anticlockwise* around the z-axis, with time, and is *left circularly* polarized if $E_{0x} = E_{0y}$ and *left elliptically* polarized if $E_{0x} \neq E_{0y}$ (Figure 2.19b).

Rewriting Equation 16.1 as

$$\mathbf{E}(z,t) = \hat{\mathbf{x}}E_{0x}\cos(kz - \omega t) + \hat{\mathbf{y}}E_{0y}\cos(kz - \omega t + \phi) = \mathbf{E}_x + \mathbf{E}_y \tag{16.5}$$

$$\frac{\mathbf{E}_y}{\hat{\mathbf{y}}E_{0y}} = \cos(kz - \omega t)\cos\phi - \sin(kz - \omega t)\sin\phi = \frac{\mathbf{E}_x}{\hat{\mathbf{x}}E_{0x}}\cos\phi - \sqrt{1 - \left(\frac{\mathbf{E}_x}{\hat{\mathbf{x}}E_{0x}}\right)^2}\sin\phi \tag{16.6}$$

$$\left(\frac{\mathbf{E}_y}{\hat{\mathbf{y}}E_{0y}} - \frac{\mathbf{E}_x}{\hat{\mathbf{x}}E_{0x}}\cos\phi\right)^2 = \left[1 - \left(\frac{\mathbf{E}_x}{\hat{\mathbf{x}}E_{0x}}\right)^2\right]\sin^2\phi \tag{16.7}$$

$$\left(\frac{\mathbf{E}_y}{\hat{\mathbf{y}}E_{0y}}\right)^2 - 2\frac{\mathbf{E}_y\mathbf{E}_x}{\hat{\mathbf{y}}E_{0y}\hat{\mathbf{x}}E_{0x}}\cos\phi + \left(\frac{\mathbf{E}_x}{\hat{\mathbf{x}}E_{0x}}\right)^2 = \sin^2\phi \tag{16.8}$$

that is,

$$\left(\frac{E_y}{E_{0y}}\right)^2 + \left(\frac{E_x}{E_{0x}}\right)^2 - 2\frac{E_yE_x}{E_{0y}E_{0x}}\cos\phi = \sin^2\phi \tag{16.9}$$

Equation 16.9 describes an ellipse (Figure 16.1) whose major axis is at an angle ψ to the x-axis (Figure 16.1).

To find ψ, we write Equation 16.9 in matrix form

$$\begin{pmatrix} E_x & E_y \end{pmatrix} \begin{pmatrix} \dfrac{1}{E_{0x}^2} & \dfrac{-\cos\phi}{E_{0x}E_{0y}} \\[2ex] \dfrac{-\cos\phi}{E_{0x}E_{0y}} & \dfrac{1}{E_{0y}^2} \end{pmatrix} \begin{pmatrix} E_x \\ E_y \end{pmatrix} = \sin^2\phi \tag{16.10}$$

and rotate the axes to obtain a simpler form

$$\begin{pmatrix} E_x' & E_y' \end{pmatrix} \begin{pmatrix} \cos\psi & \sin\psi \\ -\sin\psi & \cos\psi \end{pmatrix} \begin{pmatrix} \dfrac{1}{E_{0x}^2} & \dfrac{-\cos\phi}{E_{0x}E_{0y}} \\[2ex] \dfrac{-\cos\phi}{E_{0x}E_{0y}} & \dfrac{1}{E_{0y}^2} \end{pmatrix} \begin{pmatrix} \cos\psi & -\sin\psi \\ \sin\psi & \cos\psi \end{pmatrix} \begin{pmatrix} E_x' \\ E_y' \end{pmatrix} = \sin^2\phi \tag{16.11}$$

$$= \begin{pmatrix} E_x' & E_y' \end{pmatrix} \begin{pmatrix} \cos\psi & \sin\psi \\ -\sin\psi & \cos\psi \end{pmatrix} \begin{pmatrix} \dfrac{1}{E_{0x}^2} & \dfrac{-\cos\phi}{E_{0x}E_{0y}} \\[2ex] \dfrac{-\cos\phi}{E_{0x}E_{0y}} & \dfrac{1}{E_{0y}^2} \end{pmatrix} \begin{pmatrix} \cos\psi E_x' - \sin\psi E_y' \\ \sin\psi E_x' + \cos\psi E_y' \end{pmatrix}$$

$$= \begin{pmatrix} E_x' & E_y' \end{pmatrix} \begin{pmatrix} \cos\psi & \sin\psi \\ -\sin\psi & \cos\psi \end{pmatrix} \begin{pmatrix} \dfrac{1}{E_{0x}^2}\left(\cos\psi E_x' - \sin\psi E_y'\right) - \dfrac{\cos\phi}{E_{0x}E_{0y}}\left(\sin\psi E_x' + \cos\psi E_y'\right) \\[3ex] -\dfrac{\cos\phi}{E_{0x}E_{0y}}\left(\cos\psi E_x' - \sin\psi E_y'\right) + \dfrac{1}{E_{0y}^2}\left(\sin\psi E_x' + \cos\psi E_y'\right) \end{pmatrix}$$

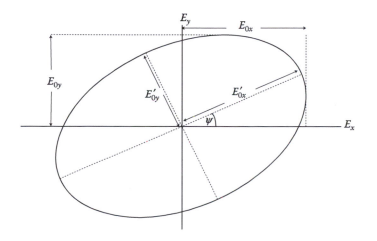

FIGURE 16.1 Elliptical polarization (Equation 16.9).

$$
\begin{aligned}
=\left(E'_x\ E'_y\right) &\left(\begin{bmatrix} \dfrac{1}{E_{0x}^2}\left(\cos^2\psi E'_x-\sin\psi\cos\psi E'_y\right)-\dfrac{\cos\phi}{E_{0x}E_{0y}}\left(\sin\psi\cos\psi E'_x+\cos^2\psi E'_y\right) \\[2mm] -\dfrac{\cos\phi}{E_{0x}E_{0y}}\left(\sin\psi\cos\psi E'_x-\sin^2\psi E'_y\right)+\dfrac{1}{E_{0y}^2}\left(\sin^2\psi E'_x+\sin\psi\cos\psi E'_y\right) \\[4mm] \dfrac{1}{E_{0x}^2}\left(-\sin\psi\cos\psi E'_x+\sin^2\psi E'_y\right)+\dfrac{\cos\phi}{E_{0x}E_{0y}}\left(\sin^2\psi E'_x+\sin\psi\cos\psi E'_y\right) \\[2mm] -\dfrac{\cos\phi}{E_{0x}E_{0y}}\left(\cos^2\psi E'_x-\sin\psi\cos\psi E'_y\right)+\dfrac{1}{E_{0y}^2}\left(\sin\psi\cos\psi E'_x+\cos^2\psi E'_y\right) \end{bmatrix}\right)
\end{aligned}
$$

$$
\begin{aligned}
=&\ \frac{1}{E_{0x}^2}\left(\cos^2\psi E_x'^2-\sin\psi\cos\psi E'_xE'_y\right)-\frac{\cos\phi}{E_{0x}E_{0y}}\left(\sin\psi\cos\psi E_x'^2+\cos^2\psi E'_xE'_y\right) \\[2mm]
&-\frac{\cos\phi}{E_{0x}E_{0y}}\left(\sin\psi\cos\psi E_x'^2-\sin^2\psi E'_xE'_y\right)+\frac{1}{E_{0y}^2}\left(\sin^2\psi E_x'^2+\sin\psi\cos\psi E'_xE'_y\right) \\[2mm]
&+\frac{1}{E_{0x}^2}\left(-\sin\psi\cos\psi E'_xS'_y+\sin^2\psi E_y'^2\right)+\frac{\cos\phi}{E_{0x}E_{0y}}\left(\sin^2\psi E'_xE'_y+\sin\psi\cos\psi E_y'^2\right) \\[2mm]
&-\frac{\cos\phi}{E_{0x}E_{0y}}\left(\cos^2\psi E'_xE'_y-\sin\psi\cos\psi E_y'^2\right)+\frac{1}{E_{0y}^2}\left(\sin\psi\cos\psi E'_xE'_y+\cos^2\psi E_y'^2\right)
\end{aligned}
$$

Setting the coefficient of $E'_xE'_y$ to zero, we obtain

$$
\psi=\frac{1}{2}\tan^{-1}\left(\frac{2E_{0x}E_{0y}\cos\phi}{E_{0x}^2-E_{0y}^2}\right) \tag{16.12}
$$

and Equation 16.9 becomes

$$
E_x'^2\left(\frac{\cos^2\psi}{E_{0x}^2}-\frac{\sin(2\psi)\cos\phi}{E_{0x}E_{0y}}+\frac{\sin^2\psi}{E_{0y}^2}\right)+E_y'^2\left(\frac{\sin^2\psi}{E_{0x}^2}+\frac{\sin(2\psi)\cos\phi}{E_{0x}E_{0y}}+\frac{\cos^2\psi}{E_{0y}^2}\right)=\sin^2\phi \tag{16.13}
$$

which is written as

$$
\frac{E_x'^2}{E_{0x}'^2}+\frac{E_y'^2}{E_{0y}'^2}=1 \tag{16.14}
$$

and the ellipticity is

$$
\frac{E'_{0y}}{E'_{0x}}=\left(\frac{E_{0x}^2\sin^2\psi+E_{0y}^2\cos^2\psi-2\sin\psi\cos\psi\cos\phi}{E_{0x}^2\cos^2\psi+E_{0y}^2\sin^2\psi+2\sin\psi\cos\psi\cos\phi}\right)^{\frac{1}{2}} \tag{16.15}
$$

If $\phi=\pi/2$, $\psi=0$, we obtain ellipticity E_{0y}/E_{0x} as we would expect.

16.3 JONES VECTORS AND MATRIX NOTATION

It is convenient to use complex notation, thus

$$\tilde{E}(z,t) = \hat{x}E_{0x}\exp\left[j(kz - \omega t)\right] + \hat{y}E_{0y}\exp\left[j(kz - \omega t + \phi)\right] \tag{16.16}$$

and then we write the expression for $\tilde{E}(z,t)$ as a *column vector*, that is

$$\tilde{E}(z,t) = \begin{pmatrix} E_{0x} \\ E_{0y}\exp(j\phi) \end{pmatrix}\exp\left[j(kz - \omega t)\right] \tag{16.17}$$

whose intensity, $\tilde{E}(z,t)\tilde{E}^*(z,t)/2 = (E_{0x}^2 + E_{0v}^2)/2$, we set to unity.

If the light is linearly polarized at angle θ to the x-axis, the column vector is $\begin{pmatrix} \cos\theta \\ \sin\theta \end{pmatrix}$.

Various polarization states can be specified by the column vectors $\begin{pmatrix} 1 \\ 0 \end{pmatrix}$ for horizontally plane polarized light, $\begin{pmatrix} 0 \\ 1 \end{pmatrix}$ for vertically plane polarized light, $(1/\sqrt{2})\begin{pmatrix} 1 \\ j \end{pmatrix}$ for right circularly polarized light, $(1/\sqrt{2})\begin{pmatrix} 1 \\ j \end{pmatrix}$ for left circularly polarized light, $\begin{pmatrix} a \\ -jb \end{pmatrix}$ for right elliptically polarized light, and $\begin{pmatrix} a \\ jb \end{pmatrix}$ for left elliptically polarized light.

These vectors are known as *Jones vectors*. They describe pure polarization states, which are characteristic of purely monochromatic light such as that produced by a laser source operating in a single longitudinal mode.

Consider a beam of collimated, monochromatic light incident normally on a parallel-sided plate made from transparent birefringent material with thickness t, whose optic axis coincides with the vertical y-axis. The light is linearly polarized at angle θ to the horizontal, x-axis, so its Jones vector is $\begin{pmatrix} \cos\theta \\ \sin\theta \end{pmatrix}$.

The plate can be described using a 2×2 *Jones matrix*

$$M = \begin{pmatrix} \exp\left(\dfrac{j2\pi n_o t}{\lambda}\right) & 0 \\ 0 & \exp\left(\dfrac{j2\pi n_e t}{\lambda}\right) \end{pmatrix} \tag{16.18}$$

Without loss of generality, we can divide M by the factor $\exp(j2\pi n_o t/\lambda)$ and

$$M = \begin{pmatrix} 1 & 0 \\ 0 & \exp\left(\dfrac{j2\pi(n_e - n_o)t}{\lambda}\right) \end{pmatrix} \tag{16.19}$$

The phase difference introduced between the component of linear polarization parallel to the y-axis and that parallel to the x-axis is

$$\phi = \frac{2\pi}{\lambda}(n_e - n_o)t \tag{16.20}$$

Thus, for example, a half wave plate has $\phi = \pi$ and is designated by the matrix $\begin{pmatrix} 1 & 0 \\ 0 & -1 \end{pmatrix}$ and the light emerging from the plate has Jones vector

$$\begin{pmatrix} 1 & 0 \\ 0 & -1 \end{pmatrix}\begin{pmatrix} \cos\theta \\ \sin\theta \end{pmatrix} = \begin{pmatrix} \cos\theta \\ -\sin\theta \end{pmatrix} \tag{16.21}$$

that is, a wave, which is linearly polarized at $-\theta$ to the x-axis. If θ is 45°, the plane of polarization is rotated by 90° as we saw in Section 8.3.1.2.

Similarly, a *half wave plate* designated $\begin{pmatrix} 1 & 0 \\ 0 & -1 \end{pmatrix}$ converts right circularly polarized light $1/\sqrt{2}\begin{pmatrix} 1 \\ -j \end{pmatrix}$ into left circularly polarized light and vice versa, since

$$\begin{pmatrix} 1 & 0 \\ 0 & -1 \end{pmatrix}\frac{1}{\sqrt{2}}\begin{pmatrix} 1 \\ -j \end{pmatrix} = \frac{1}{\sqrt{2}}\begin{pmatrix} 1 \\ j \end{pmatrix} \tag{16.22}$$

A *quarter wave plate*, designated $\begin{pmatrix} 1 & 0 \\ 0 & j \end{pmatrix}$, converts linearly polarized light with its plane at 45° to the x-axis into left circularly polarized light, since

$$\begin{pmatrix} 1 & 0 \\ 0 & j \end{pmatrix}\begin{pmatrix} \frac{1}{\sqrt{2}} \\ \frac{1}{\sqrt{2}} \end{pmatrix} = \frac{1}{\sqrt{2}}\begin{pmatrix} 1 \\ j \end{pmatrix} \tag{16.23}$$

Note that the quarter wave plate in Equation 16.23 has its fast axis in the horizontal x direction, since it is the y component of the output field, which is retarded in phase by $\pi/2$ relative to the x component.

A homogeneous circular polarizer (HCP) consists of two quarter wave plates with orthogonal fast axes placed on either side of a linear polarizer, whose transmission axis is at 45° to the fast axes of the waveplates. Thus, its form is

$$\text{HCP} = \begin{pmatrix} j & 0 \\ 0 & 1 \end{pmatrix}\begin{pmatrix} 1 & 0 \\ 0 & 1 \end{pmatrix}\begin{pmatrix} 1 & 0 \\ 0 & j \end{pmatrix} \tag{16.24}$$

and right circularly polarized passes through it. However, left circularly polarized light cannot pass through it.

$$\begin{pmatrix} j & 0 \\ 0 & 1 \end{pmatrix}\begin{pmatrix} 1 & 1 \\ 1 & 1 \end{pmatrix}\begin{pmatrix} 1 & 0 \\ 0 & j \end{pmatrix}\frac{1}{\sqrt{2}}\begin{pmatrix} 1 \\ j \end{pmatrix} = \begin{pmatrix} 0 \\ 0 \end{pmatrix} \tag{16.25}$$

In-plane rotation of the waveplates by 90° allows only left circularly polarized light to pass through and blocks right circularly polarized light.

16.4 STOKES PARAMETERS

Jones vectors are very useful when dealing with pure polarization states, and the light emitted by a single atom in going from a higher to a lower energy state is in a fairly pure state of elliptical polarization, the ellipticity being zero if the light is linearly polarized or 1.0 if circular. However, as the emission is of finite duration, the electric field varies in amplitude (Section 2.8.2), and therefore, the orientation of the ellipse is not fixed. Thus, the polarization state changes with time, although this happens slowly in the case of monochromatic or nearly monochromatic light. The light from any practical source consists of electromagnetic waves from many different atoms in transition between different higher and lower energy states, and the polarization state of the light is not at all pure but a mixture of many different states.

In order to deal with this situation, we use an experimental procedure, which enables us to characterize the polarization properties of any light source, using the *Stokes parameters* named after their originator, George Strokes, and derived from intensity measurements, which we can obtain by placing three polarizing devices in turn between the source and a photodetector. The first device is a neutral density filter, which transmits exactly half the light intensity resulting in an intensity reading I_0. The second is a linear polarizer with horizontal transmission axis, and we obtain intensity I_1. The polarizer is then rotated so that its transmission axis is at 45° to the horizontal and it now transmits intensity I_2. The third device is a homogeneous right circular polarizer, which transmits intensity I_3. The Stokes parameters are defined as

$$\begin{aligned}
S_0 &= 2I_0 \\
S_1 &= 2I_1 - 2I_0 \\
S_2 &= 2I_2 - 2I_0 \\
S_3 &= 2I_3 - 2I_0
\end{aligned} \tag{16.26}$$

The first, S_0, is proportional to the total intensity of the light, and the sign of S_1 tells us whether the light tends to align with the x-axis ($S_1 > 0$) or with the y-axis ($S_1 < 0$). If $S_1 = 0$, neither direction is preferred, and the light could be elliptical at with major axis $\pm$ 45° or circular or completely unpolarized. S_2 indicates the tendency to align at 45° ($S_2 > 0$) or $-45°$ ($S_2 < 0$) or neither ($S_2 = 0$), and S_3 indicates a tendency to right circular polarization ($S_3 > 0$) or left circular polarization ($S_3 < 0$).

If the light is nearly monochromatic, we can rewrite Equation 16.9 in terms of measurable quantities, namely intensities. We replace the electric fields by their *time averages*, then multiply by $4E_{0x}^2 E_{0y}^2$

$$4E_{0y}{}^2 \langle E_x{}^2 \rangle + 4E_{0x}{}^2 \langle E_y{}^2 \rangle - 8E_{0x}E_{0y} \langle E_y E_x \rangle \cos\phi = 4E_{0x}{}^2 E_{0y}{}^2 \sin^2\phi$$

Noting that

$$\langle E_x^2(t) \rangle = E_{0x}^2 / 2, \quad \langle E_y^2(t) \rangle = E_0^2 / 2, \quad 2\langle E_x(t)E_y(t) \rangle = E_{0x}E_{0y}\cos\phi$$

we obtain

$$4E_{0x}^2 E_{0y}^2 - 4E_{0x}^2 E_{0y}^2 \cos \phi = 4E_{0x}^2 E_{0y}^2 \sin^2 \phi$$

$$\left(E_{0x}^2 + E_{0y}^2\right)^2 - \left(E_{0x}^2 - E_{0y}^2\right)^2 - \left(2E_{0x}E_{0y} \cos \phi\right)^2 = \left(2E_{0x}E_{0y} \sin \phi\right)^2 \qquad (16.27)$$

The bracketed terms in Equation 16.27 are again the Stokes parameters S_0, S_1, S_2, and S_3 with

$$\begin{aligned}
S_0 &= E_{0x}^2 + E_{0y}^2 \\
S_1 &= E_{0x}^2 - E_{0y}^2 \\
S_2 &= 2E_{0x}E_{0y} \cos \phi \\
S_3 &= 2E_{0x}E_{0y} \sin \phi
\end{aligned} \qquad (16.28)$$

16.5 PHOTOINDUCED ANISOTROPY

Polarization holography is possible in a material, which becomes anisotropic (see Section 2.9.4) when illuminated by polarized light. We assume that the material is normally isotropic, meaning that the oscillators in it are randomly distributed and randomly oriented. Elliptically polarized light with its major axis at angle ψ to the horizontal axis, as shown in Figure 16.1, causes the material to become anisotropic. We restrict our discussion to linearly photoanisotropic materials, in which only the refractive index depends on direction, although all materials used in polarization holography absorb light anisotropically, that is, they are dichroic to some extent. Furthermore, many materials exhibit photoinduced circular anisotropy, the refractive index then being different for left and right circularly polarized light.

The photoinduced refractive index is $n_\parallel$ in the direction of the electric field and $n_\perp$ in the direction normal to field. The photoinduced birefringence is

$$\Delta n = n_\parallel - n_\perp = \Delta n_\parallel - \Delta n_\perp \qquad (16.29)$$

and the average photoinduced change in refractive index is

$$\Delta n_{av} = \frac{\Delta n_\parallel + \Delta n_\perp}{2} \qquad (16.30)$$

where $\Delta n_\parallel$ and $\Delta n_\perp$ are the *changes* in refractive index due to the illumination given by

$$\Delta n_\parallel = k_\parallel E_{0x}^2 + k_\perp E_{0y}^2 \quad \text{and} \quad \Delta n_\perp = k_\perp E_{0x}^2 + k_\parallel E_{0y}^2 \qquad (16.31)$$

Here $k_\parallel$ and $k_\perp$ are constants, which determine the changes in refractive index parallel and perpendicular to the field direction, respectively.

The Jones matrix describing the anisotropic *changes* in the material is

$$n_{E_{0x},E_{0y}} = \begin{pmatrix} \Delta n_{\parallel} & 0 \\ 0 & \Delta n_{\perp} \end{pmatrix} = \begin{pmatrix} \Delta n_{av} + \Delta n/2 & 0 \\ 0 & \Delta n_{av} - \Delta n/2 \end{pmatrix}$$

(16.32)

From Equation 16.31,

$$n_{E_{0x},E_{0y}} = \begin{pmatrix} \dfrac{(k_{\parallel}+k_{\perp})}{2}\left(E_{0x}^2 + E_{0y}^2\right) + \dfrac{(k_{\parallel}-k_{\perp})}{2}\left(E_{0x}^2 - E_{0y}^2\right) & 0 \\ 0 & \dfrac{(k_{\parallel}+k_{\perp})}{2}\left(E_{0x}^2 + E_{0y}^2\right) - \dfrac{(k_{\parallel}-k_{\perp})}{2}\left(E_{0x}^2 - E_{0y}^2\right) \end{pmatrix}$$

(16.33)

The matrix $n_{E_{0x},E_{0y}}$ is referenced to the axes of the ellipse, so we rotate it by $-\psi$ to reference it to the x,y coordinate system.

$$n_{x,y} = \begin{pmatrix} \cos\psi & -\sin\psi \\ \sin\psi & \cos\psi \end{pmatrix} n_{E_{0x},E_{0y}} \begin{pmatrix} \cos\psi & \sin\psi \\ -\sin\psi & \cos\psi \end{pmatrix}$$

(16.34)

$$n_{x,y} = \begin{pmatrix} \dfrac{(k_{\parallel}+k_{\perp})}{2}\left(E_{0x}^2 + E_{0y}^2\right) + \dfrac{(k_{\parallel}-k_{\perp})}{2}\left(E_{0x}^2 - E_{0y}^2\right)\cos(2\psi) & \dfrac{(k_{\parallel}-k_{\perp})}{2}\left(E_{0x}^2 - E_{0y}^2\right)\sin(2\psi) \\ \dfrac{(k_{\parallel}-k_{\perp})}{2}\left(E_{0x}^2 - E_{0y}^2\right)\sin(2\psi) & \dfrac{(k_{\parallel}+k_{\perp})}{2}\left(E_{0x}^2 + E_{0y}^2\right) - \dfrac{(k_{\parallel}-k_{\perp})}{2}\left(E_{0x}^2 - E_{0y}^2\right)\cos(2\psi) \end{pmatrix}$$

(16.35)

The Stokes parameters S_0, S_1, and S_2 are

$$S_0 = \left(E_{0x}\cos\psi\right)^2 + \left(E_{0y}\sin\psi\right)^2 = E_{0x}^2 + E_{0y}^2$$

$$S_1 = \left(E_{0x}\cos\psi - E_{0y}\sin\psi\right)^2 - \left(E_{0y}\cos\psi - E_{0x}\sin\psi\right)^2 = \left(E_{0x}^2 - E_{0y}^2\right)\cos(2\psi)$$

$$S_2 = \left[E_{0x}\cos(\psi-45^\circ) - E_{0y}\sin(\psi-45^\circ)\right]^2 - \left[E_{0y}\cos(\psi-45^\circ) - E_{0x}\sin(\psi-45^\circ)\right]^2$$

$$= \left(E_{0x}^2 - E_{0y}^2\right)\sin(2\psi)$$

(16.36)

Combining Equations 16.34 and 16.35

$$n_{x,y} = \begin{pmatrix} \dfrac{(k_{\parallel}+k_{\perp})}{2}S_0 + \dfrac{(k_{\parallel}-k_{\perp})}{2}S_1 & \dfrac{(k_{\parallel}-k_{\perp})}{2}S_2 \\ \dfrac{(k_{\parallel}-k_{\perp})}{2}S_2 & \dfrac{(k_{\parallel}+k_{\perp})}{2}S_0 - \dfrac{(k_{\parallel}-k_{\perp})}{2}S_1 \end{pmatrix}$$

(16.37)

We can use Equation 16.37 for different polarization holographic recording and reconstruction configurations. A comprehensive and detailed study of different configurations has been given by Nikolova and Ramanujam [1]. We will consider some examples, following their treatment.

16.6 TRANSMISSION POLARIZATION HOLOGRAPHY

16.6.1 LINEARLY POLARIZED RECORDING WAVES

A photoanisotropic layer in the ξ,η plane is illuminated by mutually coherent plane, linearly polarized waves, O and R, of wavelength λ, incident at angles $-\theta$ and θ to the normal, and their electric fields being, respectively, parallel and orthogonal to the η axis (Figure 16.2). *They cannot therefore interfere and a conventional holographic diffraction grating cannot be recorded.* A conventional recording material such as those discussed in Chapter 6 would simply undergo a spatially uniform change in transmittance or refractive index, or both.

It makes matters simpler if we reference the waves to x,y axes, which are rotated +45° with respect to the η axis so that the Jones vectors of the waves are

$$O = \frac{1}{2}\begin{pmatrix}1\\1\end{pmatrix}\exp(jk\xi\sin\theta); \quad R = \frac{1}{2}\begin{pmatrix}1\\-1\end{pmatrix}\exp(-jk\xi\sin\theta) \tag{16.38}$$

with $k = 2\pi/\lambda$, and the interference between the two waves has Jones vector.

$$O+R = \frac{1}{2}\begin{pmatrix}\exp[jk\xi\sin\theta]\\\exp[jk\xi\sin\theta]\end{pmatrix} + \frac{1}{2}\begin{pmatrix}\exp[-jk\xi\sin\theta]\\-\exp[-jk\xi\sin\theta]\end{pmatrix} = \begin{pmatrix}\cos[k\xi\sin\theta]\\j\sin[k\xi\sin\theta]\end{pmatrix} \tag{16.39}$$

Equation 16.39 *describes a wave, whose polarization state is spatially modulated along the ξ direction.* Figure 16.3 shows the polarization state for different values of the term $k\xi\sin\theta$.

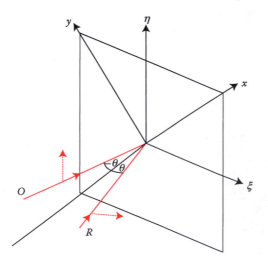

FIGURE 16.2 Polarization holographic recording with orthogonally plane polarized light waves (dashed arrows indicate polarization directions).

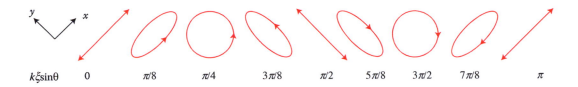

$k\xi\sin\theta$ 0 $\pi/8$ $\pi/4$ $3\pi/8$ $\pi/2$ $5\pi/8$ $3\pi/2$ $7\pi/8$ π

FIGURE 16.3 Resultant of two orthogonal plane polarized light waves with spatially varying phase difference.

As a result of illumination by $O + R$, the layer develops a spatially modulated birefringence.

The Stokes parameters of $O + R$ obtained from Equations 16.28 and 16.38 are

$$S_0 = 1; \quad S_1 = \cos(2k\xi\sin\theta); \quad S_2 = 0 \tag{16.40}$$

Inserting these values in Equation 16.37, we obtain

$$n_{x,y} = \begin{pmatrix} \dfrac{\left(k_\parallel + k_\perp\right)}{2} + \dfrac{\left(k_\parallel - k_\perp\right)}{2}\cos(2k\xi\sin\theta) & 0 \\ 0 & \dfrac{\left(k_\parallel + k_\perp\right)}{2} - \dfrac{\left(k_\parallel - k_\perp\right)}{2}\cos(2k\xi\sin\theta) \end{pmatrix} \tag{16.41}$$

The transmissivity of the exposed layer is given by

$$T_{x,y} = \exp\left[jkt\left(n_0 + \frac{\left(k_\parallel + k_\perp\right)}{2}\right)\right] + \begin{pmatrix} \exp\left[j\phi\cos(2\delta)\right] & 0 \\ 0 & \exp\left[-j\phi\cos(2\delta)\right] \end{pmatrix} \tag{16.42}$$

where t is the layer thickness, $\phi = kt[(k_\parallel - k_\perp)/2]$ and $\delta = k\xi\cos\theta$.

Ignoring the term $\exp[jkt(n_0 + (k_\parallel + k_\perp)/2)]$, we expand Equation 16.42 as a series [2]

$$T_{x,y} = 2\begin{pmatrix} \dfrac{1}{2}B_0(\phi) + jB_1(\phi)\cos(2\delta) \\ -B_2(\phi)\cos(4\delta) + \cdots & 0 \\ 0 & \dfrac{1}{2}B_0(\phi) + jB_1(-\phi)\cos(2\delta) \\ & -B_2(-\phi)\cos(4\delta) + \cdots \end{pmatrix} \tag{16.43}$$

We now illuminate the recording with a plane wave, R, linearly polarized at angle α to the x-axis and incident at angle θ to the normal. We obtain output beams

$$\text{zero order}: \quad B_0(\phi)\begin{pmatrix} \cos\alpha \\ \sin\alpha \end{pmatrix}\exp(-j\delta) \tag{16.44}$$

$$+1\text{ order}: \quad jB_1(\phi)\begin{pmatrix} \cos\alpha \\ -\sin\alpha \end{pmatrix}\exp(j\delta); \quad -1\text{ order}: \quad jB_1(\phi)\begin{pmatrix} \cos\alpha \\ -\sin\alpha \end{pmatrix}\exp(-3j\delta) \tag{16.45}$$

$$+2 \text{ order}: \quad -B_2(\phi)\begin{pmatrix}\cos\alpha\\\sin\alpha\end{pmatrix}\exp(3\delta); \quad -2 \text{ order}: \quad -B_2(\phi)\begin{pmatrix}\cos\alpha\\\sin\alpha\end{pmatrix}\exp(-5j\delta) \quad (16.46)$$

The recorded polarization grating acts as a half wave plate, since the first-order diffracted beams are linearly polarized with their polarization plane rotated by -2α with respect to the x-axis. Note that the illuminating light does not have to be incident at angle θ. Changing the angle of incidence simply changes the propagation directions of the output beams.

The second-order beams have the same polarization as the illuminating light.

If we use right circularly polarized light $1/\sqrt{2}\begin{pmatrix}1\\-j\end{pmatrix}$ to illuminate the grating, again at incident angle θ the output +1 and −1 order waves are given by

$$2\begin{pmatrix}jB_1(\phi)\cos(2\delta) & 0\\0 & jB_1(-\phi)\cos(2\delta)\end{pmatrix}\frac{1}{\sqrt{2}}\begin{pmatrix}1\\-j\end{pmatrix}\exp(-j\delta)$$

$$=\sqrt{2}\begin{pmatrix}jB_1(\phi)\exp(j\delta)+jB_1(\phi)\exp(-3j\delta)\\B_1(-\phi)\exp(j\delta)+B_1(-\phi)\exp(-3j\delta)\end{pmatrix}=2\sqrt{2}j\begin{pmatrix}1\\j\end{pmatrix}\left[\exp(j\delta)+\exp(-3j\delta)\right] \quad (16.47)$$

which are both left circularly polarized.

The second-order diffracted beams are given by

$$-\sqrt{2}\begin{pmatrix}B_2(\phi)\cos(4\delta) & 0\\0 & B_2(-\phi)\cos(4\delta)\end{pmatrix}\begin{pmatrix}1\\-j\end{pmatrix}\exp(-j\delta)$$

$$=-\sqrt{2}J_2(\phi)\left[\begin{pmatrix}1\\-j\end{pmatrix}\exp(3\delta)+\begin{pmatrix}1\\-j\end{pmatrix}\exp(-5\delta)\right] \quad (16.48)$$

whose polarization states are that of the illumination.

16.6.2 CIRCULARLY POLARIZED RECORDING WAVES

If the recording waves are right and left circularly polarized given by

$$O=\begin{pmatrix}1\\-j\end{pmatrix}\exp(j\delta); \quad R=\begin{pmatrix}1\\j\end{pmatrix}\exp(-j\delta) \quad (16.49)$$

The resultant

$$O+R=\begin{pmatrix}\cos\delta\\\sin\delta\end{pmatrix}; \quad S_0=1; \quad S_1=\cos(2\delta); \quad S_2=\sin(2\delta) \quad (16.50)$$

is shown in Figure 16.4 for different values of δ.

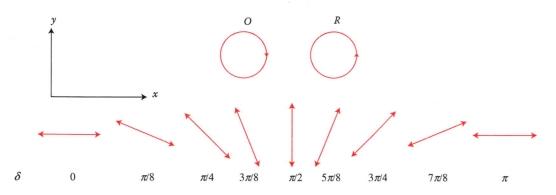

FIGURE 16.4 Resultant of two orthogonal circularly polarized light waves with spatially varying phase difference.

To simplify matters, we use the Jones matrix (Equation 16. 32) at $\delta = 0$,

$$n_{x,y} = \begin{pmatrix} \Delta n_x & 0 \\ 0 & \Delta n_y \end{pmatrix} = \begin{pmatrix} \Delta n_{av} + \dfrac{\Delta n}{2} & 0 \\ 0 & \Delta n_{av} - \dfrac{\Delta n}{2} \end{pmatrix} \qquad (16.51)$$

and the transmissivity after recording is

$$T_{x,y} = \exp\left(jkn_0 t\right)\exp\left(jkn_{av}t\right)\begin{pmatrix} \exp\dfrac{jkt\left[\Delta n_x - \Delta n_y\right]}{2} & 0 \\ 0 & \exp\dfrac{-jkt\left[\Delta n_x - \Delta n_y\right]}{2} \end{pmatrix} \qquad (16.52)$$

Then, for other orientations of $O + R$, we rotate $T_{x,y}$ through angle δ.

Ignoring terms outside the matrix and letting $\phi = jkt[\Delta n_x - \Delta n_y]/2$, we have

$$T_{x,y,\delta} = \begin{pmatrix} \cos\delta & \sin\delta \\ -\sin\delta & \cos\delta \end{pmatrix}\begin{pmatrix} \exp(j\phi) & 0 \\ 0 & \exp(-j\phi) \end{pmatrix}\begin{pmatrix} \cos\delta & -\sin\delta \\ \sin\delta & \cos\delta \end{pmatrix}$$

$$= \begin{pmatrix} \cos\phi + j\sin\phi\cos(2\delta) & -j\sin\phi\sin(2\delta) \\ -j\sin\phi\sin(2\delta) & \cos\phi - j\sin\phi\cos(2\delta) \end{pmatrix}$$

$$= \begin{pmatrix} \cos\phi + j\sin\phi\dfrac{\exp(2j\delta) + \exp(-2j\delta)}{2} & -j\sin\phi\dfrac{\exp(2j\delta) - \exp(-2j\delta)}{2j} \\ -j\sin\phi\dfrac{\exp(2j\delta) - \exp(-2j\delta)}{2j} & \cos\phi - j\sin\phi\dfrac{\exp(2j\delta) + \exp(-2j\delta)}{2} \end{pmatrix}$$

$$T_{x,y,\delta} = \begin{pmatrix} \cos\phi & 0 \\ 0 & \cos\phi \end{pmatrix} + \frac{j\sin\phi}{2}\left[\begin{pmatrix} 1 & j \\ j & -1 \end{pmatrix}\exp(2j\delta) + \begin{pmatrix} 1 & -j \\ -j & -1 \end{pmatrix}\exp(-2j\delta)\right] \quad (16.53)$$

If we illuminate this grating with a wave linearly polarized at angle α to the x-axis, we obtain $\cos\phi\begin{pmatrix}\cos\alpha\\\sin\alpha\end{pmatrix}$ for the zero order. The +1 order and −1 order diffracted waves are, respectively

$$\frac{j\sin\phi}{2}\begin{pmatrix} 1 \\ j \end{pmatrix}\exp(j\alpha)\exp(2j\delta)$$

$$\text{and } \frac{j\sin\phi}{2}\begin{pmatrix} 1 \\ -j \end{pmatrix}\exp(-j\alpha)\exp(-2j\delta) \quad (16.54)$$

If $\phi = \pi/2$, *all* the light is diffracted equally into the first orders so that the diffraction efficiency is 50%.

If we illuminate the grating using *right elliptically* polarized light, $\begin{pmatrix} a \\ -jb \end{pmatrix}$, we obtain

$$\frac{j\sin\phi}{2}\left[\begin{pmatrix} 1 & j \\ j & -1 \end{pmatrix}\begin{pmatrix} a \\ -jb \end{pmatrix}\exp(2j\delta) + \begin{pmatrix} 1 & -j \\ -j & -1 \end{pmatrix}\begin{pmatrix} a \\ -jb \end{pmatrix}\exp(-2j\delta)\right]$$

$$= \frac{j\sin\phi}{2}\left[\begin{pmatrix} a+b \\ j(a+b) \end{pmatrix}\exp(2j\delta) + \begin{pmatrix} a-b \\ -j(a-b) \end{pmatrix}\exp(-2j\delta)\right] \quad (16.55)$$

Equation 16.55 shows that if the illuminating light is *right circularly* polarized, it is diffracted only into the +1 order with *left* circular polarization and the diffraction efficiency is 100% if $\phi = \pi/2$. If on the other hand, the illuminating beam is *left circularly* polarized, we obtain only *right circularly* polarized −1 order diffracted light.

16.6.2.1 Polarization Diffraction Grating Stokesmeter

We can apply the results obtained in Section 16.6.2 in the measurement of the Stokes parameters. Suppose we illuminate the grating described by Equation 16.54 with a beam having arbitrary polarization

$$E = \begin{pmatrix} E_x \\ E_y \exp(j\psi) \end{pmatrix} \quad (16.56)$$

The zero-order output beam has intensity $|\cos\phi|^2$ with no change in polarization, and the diffracted beams are given by

$$E_{+1} = \frac{j\sin\phi}{2}\begin{pmatrix} 1 & j \\ j & -1 \end{pmatrix}\begin{pmatrix} E_x \\ E_y \exp(j\psi) \end{pmatrix}\exp(2j\delta)$$

$$= \frac{j\sin\phi}{2}\left[E_x + jE_y \exp(j\psi)\right]\begin{pmatrix} 1 \\ j \end{pmatrix}\exp(2j\delta) \quad (16.57)$$

$$E_{-1} = \frac{j\sin\phi}{2}\begin{pmatrix} 1 & -j \\ -j & -1 \end{pmatrix}\begin{pmatrix} E_x \\ E_y\exp(j\psi) \end{pmatrix}\exp(-2j\delta)$$

$$= \frac{j\sin\phi}{2}\left[E_x - jE_y\exp(j\psi)\right]\begin{pmatrix} 1 \\ -j \end{pmatrix}\exp(-2j\delta) \tag{16.58}$$

so that, regardless of the polarization state of the illuminating beam, the output diffracted beams are, respectively, left and right circularly polarized.

The intensities I_{+1} and I_{-1} of the diffracted beams are related to the Stokes parameters S_0 and S_3 of the illuminating light

$$I_{+1} = c\left[E_x^2 + E_y^2 + 2E_xE_y\sin\psi\right] = c(S_0 + S_3)$$

$$I_{-1} = c\left[E_x + E_y^2 - 2E_xE_y\sin\psi\right] = c(S_0 - S_3) \tag{16.59}$$

with c a constant.

We can use the polarization grating of Equation 16.53 to measure all four Stokes parameters [3]. Broadband light is incident on the polarization diffraction grating. The zero-order light passes straight through to a thin conventional diffraction grating where it is diffracted into two beams: one passing through a horizontal linear polarizer to a linear detector array, producing an intensity measurement $I_{1,\lambda}$ with total measurement efficiency $\eta_{1,\lambda}$; the other through a linear polarizer at 45° to the horizontal plane to a second array to obtain $I_{2,\lambda}$ with efficiency $\eta_{2,\lambda}$. The beams diffracted by the polarization grating are incident on third and fourth linear detector arrays giving intensity measurements $I_{3,\lambda}$ and $I_{4,\lambda}$ with efficiency $\eta_{3,\lambda}$ and $\eta_{4,\lambda}$, respectively. Each efficiency $\eta_{1,\lambda}$, $\eta_{2,\lambda}$, $\eta_{3,\lambda}$, and $\eta_{4,\lambda}$ includes the diffraction efficiency of the relevant grating and the transmissivity of the linear polarizer (in the cases of $\eta_{1,\lambda}$ and $\eta_{2,\lambda}$), as well as the photodetection efficiency of the array.

We now have

$$I_{1,\lambda} = \eta_{1,\lambda}E_{x,\lambda}^2; \quad I_{2,\lambda} = \eta_{2,\lambda}\left(E_{x,\lambda}^2 + E_{y,\lambda}^2 + 2E_{x,\lambda}E_{y,\lambda}\cos\psi_\lambda\right);$$

$$I_{3,\lambda} = \eta_{3,\lambda}E_{\text{lcp},\lambda}^2; \quad I_{4,\lambda} = \eta_{4,\lambda}E_{\text{rcp},\lambda}^2 \tag{16.60}$$

where the subscripts *rcp* and *lcp* refer to right and left circular polarizations.

The efficiencies $\eta_{1,\lambda}$, $\eta_{2,\lambda}$, and $\eta_{3,\lambda}$, $\eta_{4,\lambda}$ are obtained by corresponding measurements of a beam of horizontal linearly polarized light of known spectral intensity distribution, in which case, since $E_{y,\lambda} = 0$, Equation 16.60 becomes

$$I_{1,\lambda} = \eta_{1,\lambda}E_{x,\lambda}^2; \quad I_{2,\lambda} = \eta_{2,\lambda}E_{x,\lambda}^2; \quad I_{3,\lambda} = \eta_{3,\lambda}E_{\text{lcp},\lambda}^2; \quad I_{4,\lambda} = \eta_{4,\lambda}E_{\text{rcp},\lambda}^2 \tag{16.61}$$

We finally have

$$S_0 = I_{3,\lambda}\eta_{3,\lambda}^{-1} + I_{4,\lambda}\eta_{4,\lambda}^{-1}; \quad S_1 = 2I_{1,\lambda}\eta_{1,\lambda}^{-1} - \left(I_{3,\lambda}\eta_{3,\lambda}^{-1} + I_{4,\lambda}\eta_{4,\lambda}^{-1}\right)$$

$$S_2 = I_{2,\lambda}\eta_{2,\lambda}^{-1} - \left(I_{3,\lambda}\eta_{3,\lambda}^{-1} + I_{4,\lambda}\eta_{4,\lambda}^{-1}\right); \quad S_3 = I_{3,\lambda}\eta_{3,\lambda}^{-1} - I_{4,\lambda}\eta_{4,\lambda}^{-1} \tag{16.62}$$

16.6.3 RECORDING WAVES WITH PARALLEL LINEAR POLARIZATIONS

In ordinary holographic recording, we try to use object and reference waves, which are in the same state of polarization, and only the intensity of the light is spatially modulated.

If the recording material is photoanisostropic, we obtain a polarization diffraction grating, as well as the normal one.

Assuming that

$$O = \frac{1}{\sqrt{2}}\begin{pmatrix} 1 \\ 0 \end{pmatrix}\exp(j\delta); \quad R = \frac{1}{\sqrt{2}}\begin{pmatrix} 1 \\ 0 \end{pmatrix}\exp(-j\delta); \quad O + R = \sqrt{2}\begin{pmatrix} \cos\delta \\ 0 \end{pmatrix} \tag{16.63}$$

$$S_0 = 2\cos^2\delta = S_1; \quad S_2 = 0 \tag{16.64}$$

From Equation 16.37,

$$n_{x,y} = \begin{pmatrix} k_{\parallel}\left[1 + \cos(2\delta)\right] & 0 \\ 0 & k_{\perp}\left[1 + \cos(2\delta)\right] \end{pmatrix} \tag{16.65}$$

The transmissivity is given by

$$T = \begin{pmatrix} \exp\left\{jk_{\parallel}t\left[1 + \cos(2\delta)\right]\right\} & 0 \\ 0 & \exp\left\{jk_{\perp}t\left[1 + \cos(2\delta)\right]\right\} \end{pmatrix} \tag{16.66}$$

$$T = \begin{pmatrix} \exp\left(jk_{\parallel}t\right) & 0 \\ 0 & \exp\left(jk_{\perp}t\right) \end{pmatrix}\begin{pmatrix} \exp\left[jk_{\parallel}t\cos(2\delta)\right] & 0 \\ 0 & \exp\left[jk_{\perp}t\cos(2\delta)\right] \end{pmatrix} \tag{16.67}$$

We assume a small polarization modulation so that

$$T = j\begin{pmatrix} \exp\left(jk_{\parallel}t\right) & 0 \\ 0 & \exp\left(jk_{\perp}t\right) \end{pmatrix}\begin{pmatrix} k_{\parallel}t\cos(2\delta) & 0 \\ 0 & k_{\perp}t\cos(2\delta) \end{pmatrix} \tag{16.68}$$

If the grating is illuminated with a linearly polarized wave $\begin{pmatrix} \cos\alpha \\ \sin\alpha \end{pmatrix}$, we obtain diffracted light $jt\cos(2\delta)\begin{pmatrix} \exp(jk_{\parallel}t)k_{\parallel}\cos\alpha \\ \exp(jk_{\perp}t)k_{\perp}\sin\alpha \end{pmatrix}$. Setting $\alpha = 0$ enables us to measure $k_{\parallel}$, while setting $\alpha = \pi/2$ enables $k_{\perp}$ to be measured, and we can then find the photoanisotropy of the recording material.

16.7 REFLECTION POLARIZATION HOLOGRAPHIC GRATINGS

We assume counterpropagating beams incident along the normal to a recording layer, which has linear photoanisotropy so that the recorded grating is unslanted. The problem is that the photoanisotropy is a periodic function of depth in the layer. We therefore divide each complete photoanisotropic period into four equally thick sublayers, and we assume no change in photoanisotropy with depth in a sublayer. We will only summarize the results here. Two important recording configurations are considered. These are

I. Orthogonal linear polarizations

$$O = \frac{1}{2}\begin{pmatrix} 1 \\ -1 \end{pmatrix} \exp(j\delta); \quad R = \frac{1}{2}\begin{pmatrix} 1 \\ 1 \end{pmatrix} \exp(-j\delta);$$

$$O + R = \begin{pmatrix} \cos\delta \\ -j\sin\delta \end{pmatrix} \text{ as shown in Figure 16.3} \tag{16.69}$$

$$S_0 = 1; \quad S_1 = \cos(2\delta); \quad S_2 = 0 \tag{16.70}$$

$$n_{x,y} = \begin{pmatrix} \dfrac{(k_\parallel + k_\perp)}{2} + \dfrac{(k_\parallel - k_\perp)}{2}\cos(2\delta) & 0 \\[2ex] 0 & \dfrac{(k_\parallel + k_\perp)}{2} - \dfrac{(k_\parallel - k_\perp)}{2}\cos(2\delta) \end{pmatrix} \tag{16.71}$$

The recorded grating behaves as a *half wave plate* for *diffracted* light. Its axes are at 45° to the polarization planes of the recording waves. The diffraction efficiency is independent of the polarization state of the illuminating light.

II. Orthogonal circular polarizations

$$O = \frac{1}{2}\begin{pmatrix} 1 \\ -j \end{pmatrix} \exp(j\delta); \quad R = \frac{1}{2}\begin{pmatrix} 1 \\ j \end{pmatrix} \exp(-j\delta); \quad O + R = \begin{pmatrix} \cos\delta \\ \sin\delta \end{pmatrix} \tag{16.72}$$

Note that *both beams are right circularly polarized* (looking toward their sources), but since they are counterpropagating, the Jones vectors are those for right and circularly polarized light.

$$S_0 = 1; \quad S_1 = \cos(2\delta); \quad S_2 = \sin(2\delta) \tag{16.73}$$

$$n_{x,y} = \begin{pmatrix} \dfrac{(k_\parallel + k_\perp)}{2} + \dfrac{(k_\parallel - k_\perp)}{2}\cos(2\delta) & \dfrac{(k_\parallel - k_\perp)}{2}\sin(2\delta) \\[2ex] \dfrac{(k_\parallel - k_\perp)}{2}\sin(2\delta) & \dfrac{(k_\parallel + k_\perp)}{2} - \dfrac{(k_\parallel - k_\perp)}{2}\cos(2\delta) \end{pmatrix} \tag{16.74}$$

If the recorded grating is illuminated by *left elliptically polarized light in the same direction as the recording* beam *R*, the diffracted light is *right circularly polarized* with intensity depending on the ellipticity of the illuminating light. Maximum intensity of diffracted light is obtained when the illuminating light is left circularly polarized and zero intensity when it is right circularly polarized.

The properties of reflection polarization gratings are thus very similar to those of transmission polarization gratings.

16.8 PHOTOANISOTROPIC RECORDING MATERIALS FOR POLARIZATION HOLOGRAPHY

A number of materials have been used successfully in polarization holography and its applications. We have described photodichroic materials in Section 6.6.1, but here we will focus our attention on photoanisotropic materials because they are rather more useful and have been extensively studied [1].

In the context of polarization holography, the important thing about many of these materials is that they exhibit the phenomenon of *cis–trans isomerism* (from the Latin *cis*, meaning on this side and *trans*, meaning on the other, or opposite side). By this, we mean that the molecular units comprising the material alter their basic shapes in response to light. An example is azobenzene (Figure 16.5). The *trans* form of this molecule consists of two benzene rings on opposite sides of a nitrogen–nitrogen double bond while the *cis* form has the rings on the same side. The molecule is essentially planar in both states and isotropic, that is, long and thin. The *trans* state can absorb linearly polarized light in the spectral range from 450 nm to 550 nm and change into the higher energy *cis* state. The absorption spectra of the two states overlap so that the process is readily reversed and the cycle can continue. Upon each return to the *trans* state, the molecular dipole moment is randomly reoriented in space until its dipole moment is directed at 90° to the plane of polarization of the light, and the process stops since now the light is no longer absorbed. The result is that the azobenzene is anisotropic in regions where it has been illuminated. The induced photoanisotropy does not last as the molecules eventually become randomly oriented again, and to avoid this, the azobenzene may be attached as a side chain to a long polymer molecule to form an azopolymer.

16.8.1 SURFACE RELIEF

As in many other materials such as silver halide, photopolymers, and photoresist, photoanisotropic materials also develop surface relief in conformity with the illumination pattern in holographic recording. For some applications, such as data storage, surface relief should be suppressed because any higher spatial harmonics can diffract into pixels other than those actually intended. We can suppress surface relief by using azopolymers with

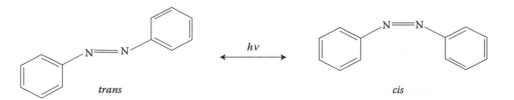

FIGURE 16.5 *Trans* and *cis* isomers of azobenzene.

longer polymer chains, which results in self-assembly at the surface in an ordered fashion [4]. It is also found that recording beams linearly polarized normal to the grating vector [5] do not produce any significant surface relief, whereas beams linearly polarized parallel to the grating vector, do. In these recordings, a single beam of laser light was passed through an amplitude mask, so the photoanisotropic recording layer was illuminated by light whose intensity was spatially modulated at a frequency of about 40 lines mm^{-1}, which could be the cause of the surface relief modulation. However, using two orthogonally plane polarized light beams and therefore no spatial variation of light intensity, a surface relief grating was obtained. The grating profile gradually changes during recording from one matching the spatial modulation of the photoinduced anisotropy to twice this frequency [6].

We can distinguish quantitatively between surface relief and volume gratings in polarization holography. We start by adapting the result for a phase grating (Section 3.2.2 and see also Figure 11.2), whose amplitude diffraction efficiency η is given by

$$\eta = B_1(\phi_0) \tag{16.75}$$

where ϕ_0 is now taken to be the maximum phase modulation, imposed by the surface grating and is given by (Figure 16.6)

$$\phi_0 = \frac{4\pi}{\lambda}\Delta n t_0 \tag{16.76}$$

with t_0 the amplitude of the surface height modulation, and Δn the difference between the refractive indices of the layer and the surrounding medium (Figure 16.6).

The matrix describing the effect of the surface relief is

$$T_{sfc} = \begin{pmatrix} \exp\left[j4\pi\Delta n t_0 \cos(\delta+\delta_0)\right] & 0 \\ 0 & \exp\left[j4\pi\Delta n t_0 \cos(\delta+\delta_0)\right] \end{pmatrix} \tag{16.77}$$

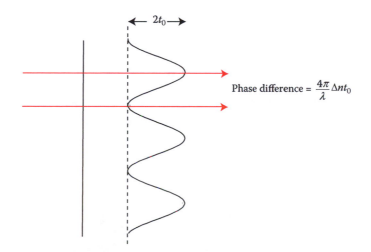

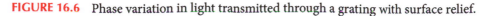

FIGURE 16.6 Phase variation in light transmitted through a grating with surface relief.

where δ_0 is a possible phase difference between the surface grating and the polarization grating. We multiply T_{sfc} by the matrix, which describes the polarization grating in the layer, for example, that in Equation 16.42 or 16.52 although many other polarization states of the recording beams are possible, to obtain the total matrix describing the transmissivity of the combination of surface and polarization gratings. Finally, as before, we multiply the total transmissivity matrix by the Jones vector for the illuminating beam used in reconstruction in order to obtain the amplitudes and intensities of the output diffracted beams. Measurements of the intensities of the various polarization components enable one to determine the relative strengths of the two gratings and of the phase shift δ_0.

16.9 APPLICATIONS OF POLARIZATION HOLOGRAPHY

We have described (in Section 16.6.2.1) a Stokesmeter based on a polarization holographic grating, recorded with orthogonal circularly polarized light beams. We now look at some other applications. The unique characteristics of polarization holography should be emphasized. One is that very high diffraction efficiencies can be obtained even in very thin layers of recording material. Another is that the hologram may be used at a wavelength other than that at which it was recorded, although the diffraction efficiency changes with wavelength, since it depends on the induced birefringence. This means, for example, that polarization holographic optical elements do not exhibit chromatic effects. A third characteristic is that a surface relief hologram may be recorded at twice the spatial frequency of the polarization grating in an azobenzene polymer layer [6], allowing for the possibility of recording holographic optical elements with dual functions, the obvious example being a bifocal lens [7].

16.9.1 HOLOGRAPHIC DISPLAY

Although large-format polarization holographic recording materials are not readily feasible, Fourier holograms of two-dimensional photographic transparencies can be recorded, since the dimensions of the holograms can then be rather small (~1 mm²) and three-dimensional images of small objects can also be recorded on photoanisotropic films of a few cm² in area using either transmission or Denisyuk, reflection holography. To ensure recording of a true reflection polarization hologram, one uses circularly polarized light, some of which passes through the photoanisotropic layer and a quarter wave plate, converting it to linearly polarized light to illuminate the object. The light reflected from the object assumes the same circular polarization state as the reference light on passing through the quarter wave plate, but, since it is propagating in the opposite direction to the reference wave, the two waves are now in orthogonal circularly polarized states (Section 16.7.2). Polarization rainbow holograms (Section 8.4.2) may also be recorded.

16.9.2 POLARIZATION HOLOGRAPHIC DATA STORAGE

Polarization holographic data storage has been implemented [8] using orthogonal circularly polarized light beams, one of which is transmitted through a spatial light intensity modulator in which data pages are written in the form of bit maps. There are significant advantages in such a system. The holographic memory may be probed at a different wavelength from that used in recording, so the data may be read nondestructively. Problems associated with shrinkage of the recording material (Section 12.4) simply do not arise. The Fourier optical design of the system means that the hologram can be as small as 0.048 mm^2 in area with a bit storage density of $1\mu\text{m}^{-2}$.

The majority of multiplexing techniques discussed in Chapter 12 for volume holographic data storage are not possible here, since the recording layer is very thin but phase-coded multiplexing could be implemented.

16.9.3 MULTIPLEXING AND LOGIC

We have seen (Section 16.6.2) that we can record a polarization hologram using orthogonal circularly polarized beams. We can record multiplexed holograms [9]: first of a transparency with corresponding object wave $O_{1\text{lcp}}$, using a reference wave $R_{1\text{rcp}}$ and second of a transparency with corresponding object wave $O_{2\text{rcp}}$, using a reference wave $R_{1\text{lcp}}$. Note that we have exchanged the polarizations of the object and reference beams for the second recording. Suppose we now illuminate the hologram using a beam of light which has passed through a half wave plate followed by a quarter wave plate, allowing complete control of the ellipticity of the reconstructing light and therefore the intensity of the reconstructed image (Equation 16.55). When the polarization of the reconstructing light matches that of the reference beam used in the first recording, O_1 is reconstructed. When the polarization of the reconstructing light matches that of the reference beam used in the second recording, O_2 is reconstructed. For any other polarization of the reconstructing light, both O_1 and O_2 are reconstructed, their relative intensities determined by the ellipticity of the reconstructing light.

16.9.4 ELECTRICALLY SWITCHABLE DEVICES

A polarization grating induces alignment of liquid crystal molecules, which are brought into contact with it, and electrically switchable optical devices may be fabricated in this way [10]. Glass plates are coated with indium tin oxide transparent electrodes and one of them is spin coated with a layer of azodye-doped polyimide between 20 and 30 nm in thickness. Orthogonal circularly polarized beams are then used to record polarization gratings with diffraction efficiency of only about 10^{-6} because the recording layer is so thin. A cell is assembled from the plates and filled with liquid crystals, which become aligned according to the pattern of Figure 16.4. Because of the birefringence of the liquid crystals, the diffraction efficiency of the device increases by a factor of 10^4 and can be controlled by applying an electric field. As the field increases, the spatial extent of the liquid crystal alignment penetrates deeper into the liquid crystal layer from the polarization grating plane, and the diffraction grating increases up to 14%. Provenzano et al. [11] obtained diffraction efficiency close to 100% in a similar device having alignment layers in the form of polarization gratings on both plates, which may be rotated in-plane by 90° relative to one another to form a two-dimensional grating [12].

16.10 SUMMARY

In this chapter, we consider polarization holography which has a number of unique characteristics and useful applications. We begin by describing the various polarization states that light may have and use Jones vectors to define each state. The Jones matrix describing a polarizing optical component can then be used to see how the polarization state of a beam of light is altered upon transmission through the component. We derive the four Stokes parameters because they fully describe the polarization of any light beam including one that is not in a pure polarization state and are important in determining the characteristics of different polarization holographic gratings.

We next consider photoanisotropy, the phenomenon exploited most commonly in polarization holography, and discuss some examples of how a spatially varying photoinduced anisotropy can lead to holographic grating formation even in the absence of a spatially varying light intensity.

Finally, we look at some applications of polarization holography including those that are unique to it.

REFERENCES

1. L. Nikolova and P. S. Ramanujam, *Polarization Holography*, Cambridge University Press, Cambridge (2009).
2. I. S. Gradshteyn and I. M. Ruzhik, *Table of Integrals, Series, and Products*, 6th edition, A. Jeffrey and D. Zwillinger (eds.), Academic Press, San Diego (2000).
3. T. Todorov, L. Nikolova, G. Stoilov, and B. Hristov, "Spectral Stokesmeter 1: Implementation of the device," *Appl. Opt.*, 46, 6662–8 (2007).
4. F. X. You, M. Y. Paik, M. Hackel, L. Kador, D. Kropp, H. W. Schmidt, and C. K. Ober, "Control and suppression of surface relief gratings in liquid-crystalline perfluoroalkyl-azobenzene polymers," *Adv. Funct. Mater.*, 16, 1577–81 (2006).
5. N. C. R. Holme, L. Nikolova, S. Hvilsted, P. H. Rasmussen, R. H. Berg, and P. S. Ramanujam, "Optically induced surface-relief phenomena in azobenzene polymers," *Appl. Phys. Lett.*, 66, 1166–8 (1999).
6. I. Naydenova, L. Nikolova, T. Todorav, N. C. R. Holme, P. S. Ramanujam, and S. Hvilsted, "Diffraction from polarization holographic gratings with surface relief in side-chain polymers," *J. Opt. Soc. Am. B*, 15, 1257–65 (1998).
7. G. Martinez-Ponce, T. Petrova, N. Tomova, V. Dragostinova, T. Todorov, and L. Nikolova, "Bifocal polarization holographic lens," *Opt. Lett.*, 29, 1001–3 (2004).
8. N. C. R. Holme, S. Hvilsted, E. Lörincz, A. Matharu, L. Nedelchev, L. Nikolova, and P. S. Ramanujam, "Azobenzene polyesters for polarization holographic storage: Part I: Materials and characterization," *Handbook of Organic Electronics and Photonics, Vol. 2 Photonic Materials and Devices*, H. S. Nalwa (ed.), 183–211, American Scientific Publishers, New York (2007). See also E. Lörincz, P. Koppa, G. Erdei, A. Süto, F. Ujhelyi, P. Várhegyi, P. S. Ramanujam, and P. I. Richter, "Azobenzene polyesters for polarization holographic storage: Part II. Technology and system," *Handbook of Organic Electronics and Photonics*, H. S. Nalwa (ed.), 213–234, American Scientific Publishers, New York (2007).
9. T. Todorov, L. Nikolova, K. Stoyanova, and N. Tomova, "Polarization holography. 3. Some applications of polarization recording," *Appl. Opt.*, 24, 785–8 (1985).

10. L. M. Blinov, G. Cipparone, A. Mazzulla, C. Provenzano, S. P. Palto, M. I. Barnik, A. B. Arbuzov, et al., "Electric field controlled polarization grating based on a hybrid structure "photosensitive polymer-liquid crystal," *Appl. Phys. Lett.*, 87, 061105 (2005). See also L. M. Blinov, G. Cipparone, A. Mazzulla, C. Provenzano, S. P. Palto, M. I. Barnik, A. B. Arbuzov, et al., "A nematic liquid crystal as an amplifying replica of a holographic polarization grating," *Mol. Cryst. Liq. Crist.*, 449, 147–60 (2006).

11. C. Provenzano, P. Pagliusi, and G. Cipparrone, "Highly efficient liquid crystal based diffraction grating induced by polarization hologram at the aligning surfaces," *Appl. Phys. Lett.*, 89, 121105 (2006).

12. C. Provenzano, P. Pagliusi, and G. Cipparrone, "Electrically tunable two-dimensional liquid crystal gratings induced by polarization holography," *Opt. Express,* 15, 5872–8 (2007).

PROBLEMS

1. Find the eigenvalues of the matrix of Equation 16.10 and hence obtain Equation 16.14.

2. Show that the addition of Jones vectors for right and left circularly polarized light produces horizontally plane polarized light and demonstrate this by superposition of the two waves.

3. Show that a homogeneous circular polarizer consisting of two quarter wave plates on either side of a linear polarizer and described by Equation 16.24 transmits right circularly polarized light. Prove Equation 16.25 to show that it does not transmit left circularly polarized light.

4. Show that linearly polarized light, which is transmitted through a half wave plate followed by a quarter wave plate, becomes elliptically polarized.

5. A half wave plate introduces a phase difference of π between light that is plane polarized parallel to its fast axis and light that is plane polarized in the orthogonal direction. Calculate the minimum thickness of such a half wave plate for light of wavelength 633 nm made from quartz whose birefringence is 0.009. The device could have any thickness, which produces a phase difference which is an integer multiple of 2π plus π. If its thickness is 170 μm, how would you get it to work as a half wave plate? What if it were 180 μm thick?

6. Suppose that in Equation 16.41, the induced birefringence is small and the grating is illuminated again by R, what precisely would be obtained in reconstruction and what would be the diffraction efficiency?

7. Show that when a polarization holographic grating described by Equation 16.53 is illuminated by left circularly polarized light, only right circularly polarized light is obtained if $\phi = \pi/2$.

8. Explain *physically* how the result obtained in the previous problem occurs.

CHAPTER 17

Holographic Sensors and Indicators

17.1 INTRODUCTION

In Chapter 12 in which we discussed holographic data storage systems and techniques, it was pointed out that the storage medium must not undergo any change during or after recording, which might lead to a change in the Bragg angle, with consequent difficulty in retrieving the data.

For data storage applications, we must avoid changes in refractive index modulation and average refractive index, recorded fringe spacing, and hologram thickness, and considerable research effort has been expended in the development of very robust, low-shrinkage recording materials. However, for materials that do exhibit changes in response to external stimuli, applications in sensing are possible.

We now consider how we can exploit dimensional and other changes in a hologram, enabling it to perform a sensing action. The formation of a hologram can also be a sensing action. Thus a wide range of holographic sensors is possible, as we shall see.

17.2 BASIC PRINCIPLES

There are several holographic sensing principles. The first is the change in diffraction efficiency due to a change in the refractive index modulation of the hologram (Figure 17.1a), which arises as a result of a physical or chemical interaction between one or more of the chemical constituents of the hologram and the analyte to be sensed or detected. The second is a change in the spacing of the recorded fringe pattern, which changes the angular position of the Bragg peak in a transmission hologram (Figure 17.1b) or the spectral position of the Bragg peak in the case of a reflection hologram (Figure 17.1c). A change in thickness of a *slanted* grating also leads to a change in direction of the diffracted light (Figure 17.1d). Yet another possibility is the *formation* of a hologram as a consequence of the presence of an analyte, which we will consider in Section 17.7.

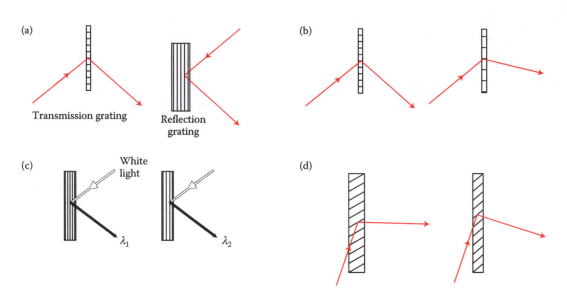

FIGURE 17.1 Holographic sensing: (a) change in refractive index modulation—change in diffraction efficiency; (b) change in recorded fringe spacing—change in direction of diffracted light (transmission grating); (c) shrinkage/swelling-induced change in recorded fringe spacing—change in wavelength of diffracted light (reflection grating illuminated by white light); and (d) shrinkage in a slanted grating.

We can also exploit change in average refractive index or changes in the physical dimensions of the hologram due to external influences.

17.3 THEORY

17.3.1 VOLUME-PHASE TRANSMISSION HOLOGRAPHIC GRATINGS

To see how the diffraction efficiency of a volume-phase grating changes as a result of change in any one of a number of parameters, including refractive index modulation, thickness, wavelength and angle of incidence of the illuminating light, we use Equation 4.33.

$$\eta = \frac{\sin^2 \left(\xi^2 + \upsilon^2\right)^{\frac{1}{2}}}{\left(1 + \xi^2 / \upsilon^2\right)} \tag{17.1}$$

If we assume the grating is illuminated on-Bragg $\xi = 0$ and

$$\eta = \sin^2 \left(\frac{\pi n_1 T}{\lambda_{\text{air}} \cos \theta}\right) \tag{17.2}$$

Differentiating gives

$$\Delta \eta = 2 \eta \arcsin \sqrt{\eta} \cot \left(\arcsin \sqrt{\eta}\right) \left(\frac{\Delta n_1}{n_1} + \frac{\Delta T}{T}\right) \tag{17.3}$$

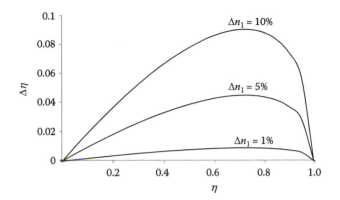

FIGURE 17.2 Change in diffraction efficiency versus initial value for a volume-phase transmission grating.

where we have assumed that the illuminating wavelength and angle do not change. We see that there is a linear relationship between the change in diffraction efficiency and change in thickness. Figure 17.2 shows the change in diffraction efficiency plotted against the original value for different initial index modulations, assuming no change in thickness.

The optimal initial diffraction efficiency is about 0.7 at which a change of 10% in refractive index modulation causes the diffraction efficiency to change by about 9%.

17.3.2 VOLUME-PHASE REFLECTION HOLOGRAPHIC GRATINGS

From Equation 4.49 assuming on-Bragg illumination,

$$\eta = \tanh^2\left(\frac{\pi n_1 T}{\lambda_{air}\cos\psi}\right) \tag{17.4}$$

Differentiating, we obtain

$$\Delta\eta = 2\eta\sqrt{\eta}(1-\eta)\frac{\pi\Delta n_1 T}{\lambda_{air}\cos\psi} \tag{17.5}$$

where again we have assumed no change in grating thickness, illumination wavelength, or angle.

Figure 17.3 shows the change in diffraction efficiency plotted against the original value for different changes in refractive index modulation in a reflection grating 30 μm in thickness with initial index modulation of 0.005. The wavelength is 500 nm, and Bragg angle is 30°.

Clearly, changes in refractive index modulation in reflection-phase gratings can also give rise to substantial changes in diffraction efficiency.

Alternatively, a change in recorded fringe spacing, d, in a reflection grating illuminated by white light can give rise to a change in the spectral position, λ_{air}, of the Bragg peak as

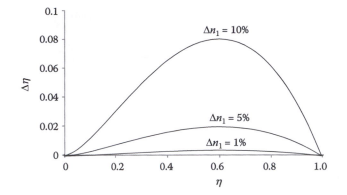

FIGURE 17.3 Change in diffraction efficiency versus initial value for a volume-phase reflection grating.

Equation 4.15 shows

$$2n_0 d \sin\theta = \lambda_{air} \tag{17.6}$$

Similarly, the spectral position of the Bragg peak depends linearly on the average refractive index, n_0. Thus, changes in either the recorded fringe spacing or the average refractive index, or both, result in a change in color of diffracted light when a reflection grating or hologram is illuminated using white light.

In a slanted transmission grating, the recorded fringe planes rotate when the grating undergoes shrinkage, resulting in change in the angular position of the Bragg peak (Section 7.6.2), and this effect may also be used for sensing purposes.

17.4 PRACTICAL SENSORS AND INDICATORS

Although transmission-phase gratings can be utilized for sensing purposes, they require monochromatic light for their illumination, as well as a photodetector and associated electronic circuitry, whether to monitor a change in signal level as diffraction efficiency changes or a change in direction of the diffracted light by position-sensitive photodetection.

On the other hand, reflection holograms as indicators are the focus of research because change in color is easily observed by eye under white light illumination conditions while accurate measurement of the location of the spectral peak in the diffracted light enables precisely calibrated sensing.

17.5 SENSORS BASED ON SILVER HALIDE AND RELATED MATERIALS

Holographic sensors for protease, pH, water, alcohol, glucose, lactate, spore germination, and humidity have all been developed.

To detect the protease, trypsin, for the diagnosis of pancreatic disease [1], a reflection-phase grating is recorded in silver halide in a gelatin matrix. The action of trypsin on peptide bonds weakens the gelatin, enabling it to swell so that the Bragg peak is shifted toward longer wavelength. The weakening of the gelatin also results in a reduced diffraction efficiency.

Another sensor, this time for the detection of water in organic solvents, relies on swelling due to the absorption of water by the gelatin in a silver halide reflection grating [2]. Similarly the presence of alcohol may be indicated by the swelling, which it causes in cross-linked poly(hydroxyethyl methacrylate) film [3] again causing a Bragg shift in a reflection holographic grating.

A color changing pH indicator [4] relies on molecular groups that can be ionized; these groups being attached to a hydrogel in the photosensitive layer. The swelling is due to the repulsive forces between groups that become ionized when the pH is altered.

Glucose sensing has been implemented using phenylboronic acid derivatives in a biocompatible hydrogel in the photosensitive layer. These derivatives can bind to glucose causing the hydrogel to swell, again changing the Bragg peak to longer wavelengths. Phenylboronic acids are themselves nonselective with respect to glucose and lactate, but the selectivity of the sensor for glucose over lactate can be improved by appropriate design of the hydrogel [5] and vice versa [6].

Spore germination involves the release of divalent calcium ions, which can bind to a methacrylated analogue of nitrilotriacetic acid monomer in the matrix material of a holographic grating [7]. The bound calcium shields repulsive forces between ionized monomer units in the matrix backbone causing the matrix to contract so that the Bragg peak is shifted to shorter wavelength.

It is well known that silver halide materials are sensitive to humidity, and indeed, we can use this sensitivity for control of color in display holograms [8], although care is need to avoid high humidity as excessive moisture can completely adversely affect the recorded fringe pattern.

This is even more important in the case of holograms recorded in dichromated gelatin, which should be completely sealed immediately after processing and drying, as any moisture entering the matrix will greatly reduce the refractive index modulation (Section 6.3). Gelatin has been used in a humidity sensor in the form of a wavelength-dependent holographic filter with fiber-optic delivery and collection of the illuminating and diffracted light [9], but the results obtained were not entirely satisfactory because of accompanying temperature dependence, hysteresis effects, as well as variable sensitivity.

17.6 PHOTOPOLYMER-BASED HOLOGRAPHIC SENSORS AND INDICATORS

All of the sensors discussed so far involve the use of silver halide or other materials, which require chemical processing. Photopolymer recording materials (Section 6.6.4) are an alternative offering some advantages, being self-processing in some cases. Dopants may also be incorporated in the photopolymer layer, usually in the form of nanosized zeolite particles, which, besides their chemical properties, have unique physical structures. In addition, the spatial distribution of these nanoparticles may be altered in conformity with the recorded fringe pattern, a process called holographic photopatterning. We may take advantage of the characteristics of zeolites and of the photopatterning effect in a number of ways applicable to sensing.

17.6.1 HUMIDITY SENSING

A holographic humidity sensor must be sensitive to moisture, but not damaged by it so that it does not deteriorate over time. Continuous humidity monitoring requires a reversible sensor. Irreversible humidity indicators may be required to verify that goods received have not been exposed to different humidity levels if such exposure would compromise quality. Irreversible humidity indicators can be used for assurance of product integrity and for authentication if the exposure to different humidity is treated as evidence of possible tampering or substitution.

A hologram, which changes color, when exposed to high humidity level through simply breathing on it, may also be used as an authentication device.

A humidity sensor with a fast response time of 350 ms has been recorded in a polymer-dispersed liquid crystal material [10].

A color changing, reversible, humidity sensitive hologram has been developed [11,12] using an acrylamide-based photopolymer similar to that described in Section 6.6.4. Again, no wet processing is required following the recording stage. The main reason for the color change is a change in the thickness of the polymer film when it is exposed to a changing humidity level [12], which implies a change in recorded fringe spacing, d.

Differentiating Equation 17.6, we obtain

$$\frac{\Delta\lambda_{air}}{\lambda_{air}} = \frac{\Delta d}{d} + \frac{\Delta n_0}{n_0} + \cot\theta\Delta\theta \tag{17.7}$$

where $\Delta\lambda_{air}$, Δn_0, Δd, and $\Delta\theta$ are the changes in wavelength, average refractive index, grating period, and Bragg angle, respectively.

Equation 17.7 shows that to obtain a large change in the spectral position of the Bragg peak, we should record the grating at the longest available wavelength, in a material with low-average refractive index, n_0, which suggests a porous material. We should also record at high spatial frequency, that is, small d, although this is not compatible with long wavelength recording.

Assuming no change in average refractive index, or Bragg angle, the change in Bragg peak position is proportional to the change in recorded fringe spacing. Thus, a 20% change in thickness of the grating will typically result in change in wavelength of about 100 nm, of the diffracted light, in white light illumination.

In order to calibrate a grating for humidity sensing purposes, it is enclosed in a transparent walled chamber in which the relative humidity can be precisely controlled. The grating is illuminated by collimated white light at the Bragg angle and the diffracted light is spectrally analyzed. Figure 17.4 shows a typical result.

A typical example of a color-changing hologram is shown in Figure 17.5, with the effect of breathing shown in Figure 17.6.

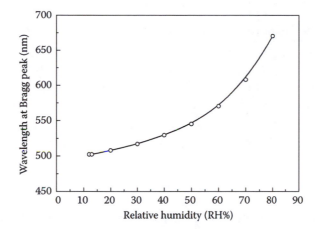

FIGURE 17.4 Humidity response of a 30-mm-thick holographic grating recorded at 29% RH.

| 10% | 30% | 50% | 70% | 80% |

FIGURE 17.5 **(See color insert.)** Hologram color at different levels of relative humidity (RH%). (Courtesy of I. Naydenova and R. Jallapuram.)

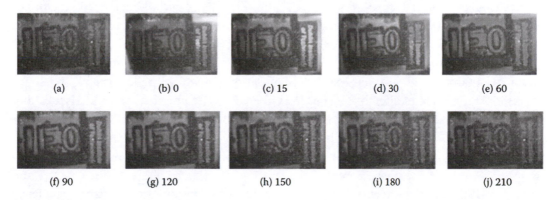

| (a) | (b) 0 | (c) 15 | (d) 30 | (e) 60 |

| (f) 90 | (g) 120 | (h) 150 | (i) 180 | (j) 210 |

FIGURE 17.6 **(See color insert.)** Color change of a hologram induced by breathing (a) before, (b)–(j) time (sec) afterwards. (Courtesy of I. Naydenova and R. Jallapuram.)

Color change can be seen, whether the relative humidity changes or not, by simply altering the direction either of the illumination or of the observation. For this reason, in order to ensure that the color change is solely due to a change in the grating thickness, the angle of illumination must be fixed or the grating must be thick enough so that only a narrow spectral range of color is seen.

Some additional features of these humidity sensors are worth noting. Response is reversible. Sensitivity to temperature is low with very little variation in the spectral position of the Bragg peak for temperatures up to 50° C at lower humidity. At higher RH (~45%), the peak position can vary by up to 20 nm within the range 25°C–50°C. Heating of the grating causes the calibration curve to shift toward the shorter wavelength end of the spectrum. One can also shift the calibration curve of the grating by changing the RH value at which it is recorded. Finally, the thickness of the photopolymer layer has a strong influence on the speed of response, with thinner layers responding more rapidly than thicker ones, as expected.

17.6.2 NANOPARTICLE-DOPED PHOTOPOLYMERS

As we saw in Section 6.6.4.2, photopolymerization during holographic recording is accompanied by diffusion of monomer from regions of low illumination (dark fringes) into higher illumination (bright fringes) driven by the concentration gradient, which arises as monomer is depleted by polymerization. Similarly short-chain polymer molecules diffuse in the opposite direction although longer chain ones cannot easily do so. Consideration of these processes leads to the idea that nanoparticles dispersed in a photopolymer layer might undergo spatial redistribution during holographic recording, and this is found to be the case (Figure 17.7), with electron probe microanalysis showing that nanoparticles are expelled from bright into dark fringe regions [13]. This has been confirmed by measuring the phase difference between the illuminating interference fringe pattern and the recorded refractive index pattern [14]. One of the significant benefits of redistribution is that if the refractive index of the nanoparticles differs from that of the host photopolymer, we can obtain an increase in refractive index modulation and thus of diffraction efficiency. Solid nanoparticles such as TiO_2 [15,16], SiO_2 [17], and ZrO_2 [18] have all been used. The nanoparticles should also help to reduce shrinkage for holographic data storage applications.

Zeolite nanoparticles (Figure 17.8) widely used as sieves for purification [19] may also be added to photopolymers, helping to reduce shrinkage but, perhaps more importantly, acting as analyte traps. The sensing action here is one in which the local refractive index in regions of higher zeolite concentration is altered by the presence of an analyte, either physically trapped within the zeolite (Figure 17.9) or bound to it by intermolecular forces.

A simple example is that of an irreversible humidity sensor [20] using the aluminophosphate zeolite ALPO-18 whose initially uniform spatial distribution changes during holographic recording of a phase transmission grating. Heating to about 120°C expels water trapped within the pore structure of the ALPO-18. We then note the diffraction efficiency at low relative humidity of 15%. Exposure of the grating to relative humidity of 60% causes the pores to refill with moisture (refractive index, 1.33), reducing the refractive index modulation and the diffraction efficiency. Finally, if we restore the humidity to 15% we find that diffraction efficiency of the grating does not return to its original value of 70% in contrast to the behavior of an undoped grating, whose efficiency does reach its original value.

Another example is the sensing of the organic solvent, toluene, whose refractive index is 1.50, and which can alter the average refractive index of a reflection grating recorded in an undoped acrylate polymer [21], thus altering the spectral position of the Bragg peak.

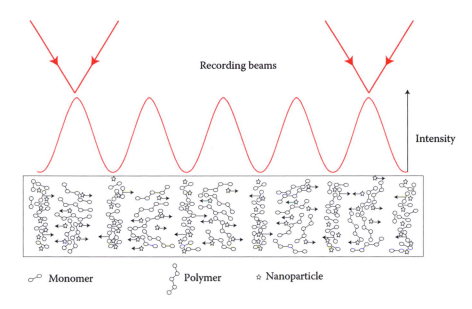

FIGURE 17.7 Photopatterning of nanoparticles.

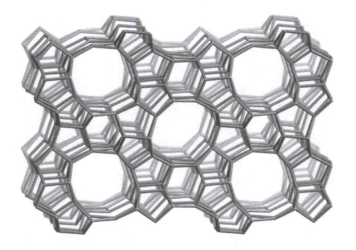

FIGURE 17.8 Structure of a typical zeolite. In this example (MFI) the particle size ranges from 10 to 70 nm and pore dimensions are typically 0.53 × 0.56 nm. (Reproduced by kind permission of Dr. Christian Baerlocher.)

Change in average refractive index should have no effect on the diffraction efficiency of a transmission grating recorded in an undoped photopolymer although there is evidence of dimensional change that does alter the efficiency [22]. However, addition of nanoparticulate zeolite Beta and its redistribution during holographic recording of a grating results in reduction in diffraction efficiency by more than 20% when exposed to toluene (19 ppm) compared with about 10% in a pure photopolymer grating. It is believed that adsorption of toluene in the zeolite-rich regions (Figure 17.9), which have been produced by photopatterning during holographic recording, raises the refractive index, resulting in a reduction in refractive index modulation and hence in diffraction efficiency.

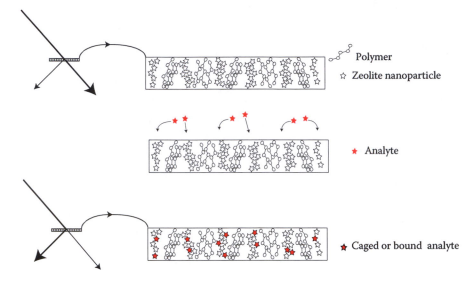

FIGURE 17.9 Sensing action of photopatterned nanozeolite.

17.7 SENSING BY HOLOGRAM FORMATION

The idea that the formation of a hologram could itself be a sensing action is based on the composition of certain photopolymer systems used in holography.

Many photopolymerizable materials for holographic recording using visible light consist of a dye photosensitizer, a free radical generator, or cosensitizer and are usually prepared as dry films, when a binder such as polyvinyl alcohol is included, or as liquid films sandwiched between transparent cover plates if the binder is omitted. Light of appropriate wavelength triggers a photopolymerization reaction leading ultimately to a change in the film's refractive index as explained in Section 6.6.4.1. The role of the dye photosensitizer is to absorb visible light and transfer the absorbed energy to the appropriate cosensitizer so that polymerization can begin. (In the case of illumination by UV light, we can dispense with the dye as UV light has sufficient energy to cause direct conversion of monomers into free radicals and polymerization proceeds directly). The correct choice and concentration of dye are important in order to ensure optimum sensitivity to the wavelength of laser light used in the recording process and to obtain efficient polymerization. The dye is normally included in the photopolymer composition at the preparation stage to ensure its uniform distribution throughout the photopolymer so that the holographic response will be spatially uniform.

However, assuming that the sensitizer is chemically stable, it can be added at a later stage and a hologram can still be recorded [23], always providing that the unsensitized dry layer is sufficiently permeable to facilitate diffusion of externally deposited sensitizer into the bulk photopolymer.

As shown in Figure 17.10, an unsensitized photopolymer solution is prepared by mixing all the components, except for the dye sensitizer. We deposit some of the solution on a glass substrate and leave it to dry. Dye sensitizer in dry or liquid form is then deposited on the dry film, which becomes locally sensitized as the dye diffuses into it. The film is now illuminated by an interference pattern produced by two coherent laser beams to record a holographic diffraction grating, but only in the bulk photopolymer directly underneath the deposited dye although it is to be expected that some lateral diffusion will also occur. The grating is a transmission or a reflection one depending on the illumination geometry. The familiar rainbow effect produced by a transmission diffraction grating is readily observed in white light. In the case of a reflection grating, a single color is observed in white light illumination. In either case, examination in white light reveals the presence and the exact location of the dye. Dyes such as Erythrosine B, Eosin Y, and Eosin-5 isothiocyanate in concentrations as low as 10^{-8} M can be visually detected by recording holographic transmission diffraction gratings.

Obviously, a holographic recording setup is not actually required. We may simply copy an existing transmission or reflection diffraction grating into the dye deposition region by placing it in close proximity to the film and illuminating with a single beam of light.

This simple process, known as *dye deposition holography* (Figures 17.10 and 17.11), can be utilized in a detection scheme, in which a dye-labeled chemical or biochemical analyte may be used to photoactivate an otherwise unsensitized photopolymer film. The presence of the analyte, as well as its location, is revealed by diffraction of light as a result of the formation of a holographic grating. The scheme offers the potential for detection of dye-labeled analytes by providing a visual, easily interpreted signal. Dye deposition holography involves an amplification process in the form of a polymerization chain reaction, triggered

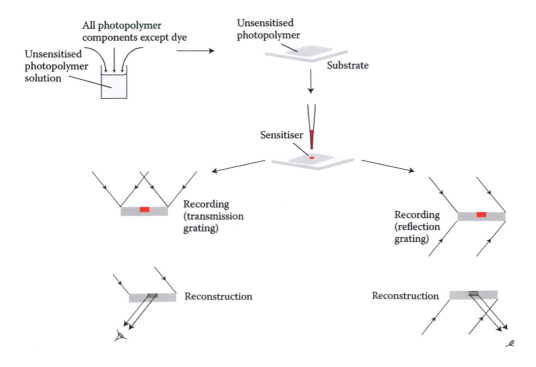

FIGURE 17.10 Dye deposition holography.

by a single excited dye molecule and leading to polymerization of thousands of monomer molecules. In addition, diffraction by the recorded grating produces an optical signal in a direction in which no light was previously detected so that the signal-to-noise ratio is intrinsically very high.

In current commercially available systems, a dye-labeled analyte is detected by the fluorescent light which the dye emits when it is illuminated by laser light. This is a well-established method, but fluorescent emission is isotropic, necessitating careful design of the optical system to ensure that as much of the fluorescence as possible reaches the detector [24].

We have seen that dye deposition holography works well when the dye is simply deposited on the unsensitized film. In order to verify that photopolymerization and grating formation are possible when the dye is attached to an analyte, we can carry out a simple demonstration experiment.

Dye labels for biological molecules usually have high fluorescence yield and are not suitable as sensitizers for photopolymerization, for which high triplet yield and long-lived triplet states are required (Section 6.6.4.2). Fortunately, a number of suitable dyes are available,

FIGURE 17.11 **(See color insert.)** Examples of dye deposition holography: (a) Left: photopolymer sample that has been sensitized but not exposed to light. The sensitized areas are not visible to the eye as the dye concentration is extremely low. Right: Following exposure to an interference pattern, transmission holographic gratings are recorded in the sensitized areas and are visible in transmitted white light. (b) In these examples, dye is deposited using a soft brush. (Courtesy of I. Naydenova.)

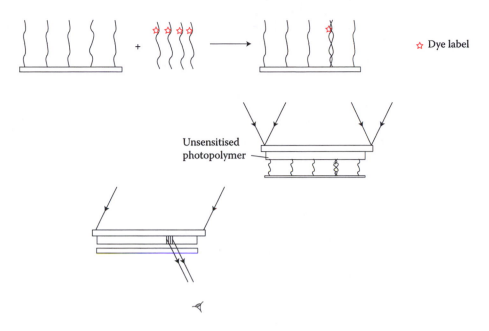

☆ Dye label

Unsensitised photopolymer

FIGURE 17.12 DNA hybridization triggers formation of a transmission grating by dye deposition holography.

such as Eosin-5-isothiocyanate. This dye was used at a concentration of 10 mM as a label for 17-mer single stranded DNA. Solution volumes of 0.5 µl containing dye-labeled DNA in concentrations from 10^{-3} to 10^{-9} M were deposited on a microarray substrate so that the amount of DNA at each location varied between 0.5 nmols and 0.5 fmols. A separately prepared unsensitized dry photopolymer layer was peeled from its substrate and placed on top of the immobilizeddye-labeled DNA (Figure 17.12). Holographic transmission diffraction gratings were recorded in the photopolymer layer in just those areas in contact with the dye-labeled DNA. With no attempt at optimization of the dye label, photopolymer composition or recording conditions, 50 fmol of DNA was visually detected.

The use of liquid instead of dry unsensitized layers could further improve the sensitivity of the detection method, as the diffusion depth of the sensitizer would not be a limiting factor. The diffraction efficiency of the recorded hologram depends on both the refractive index modulation and the thickness of the hologram. For a given diffraction efficiency, a hologram with an increased thickness facilitates the detection of a lower refractive index modulation. The light-induced refractive index modulation resulting from a small number of sensitizing molecules is expected to be proportional to their number so that greater effective hologram thickness would provide increased sensitivity. The addition of photodetection systems with high directional sensitivity can further enhance sensitivity of the method.

Finally, we note that if we use just one beam to illuminate the film, the local change in the photopolymer refractive index will be spatially homogeneous. One can control the dye deposition so that upon illumination the refractive index change is patterned to produce photonic structures that confine, focus, or direct the light propagating through them [23]. This process is known as *dye deposition lithography*. However, lateral diffusion of dye and resolution of the deposition system, set a limit to the smallest size of feature which we can create in such structures.

17.8 SUMMARY

In this chapter, we consider holographic sensing. We survey various applications in which a color change takes place in a reflection hologram or holographic grating. In most cases, color change is due to a change in physical dimensions, which occurs as a result of a closely specified and controlled interaction between the analyte and the hologram support medium. A number of biological and chemical sensors have been developed.

We then consider the potential of zeolite-doped photopolymers for sensing applications based on the chemical or physical affinity of zeolite nanoparticles for specific analytes.

Finally, we discuss the principle and application of dye deposition holography as a method of sensing an analyte, which has been separately labeled with a photosensitizing dye.

REFERENCES

1. R. B. Millington, A. G. Mayes, and C. R. Lowe, "A holographic sensor for proteases," *Anal. Chem.*, 67, 4229–33 (1995).
2. J. Blyth, R. B. Millington, A. G. Maye, E. R. Frears, and C. R Lowe, "Holographic sensor for water in solvents," *Anal. Chem.*, 68, 1089–94 (1996).
3. A. G. Mayes, J. Blyth, M. Kyro1ainen-Reay, R. B. Millington, and C. R. Lowe, "A holographic alcohol sensor," *Anal. Chem.* 71, 3390–6 (1999).
4. A. J. Marshall, J. Blyth, C. A. B. Davidson, and C. R. Lowe, "pH-sensitive holographic sensors," *Anal. Chem.*, 75, 4423–31 (2003).
5. S. Kabilan, A. J. Marshall, F. K. Sartain, M.-C. Lee, A. Hussain, X. Yang, J. Blyth, et al., "Holographic glucose sensors," *Biosensors Bioelectronics*, 20, 1602–10 (2005).
6. F. K. Sartain, X. Yang, and C. R. Lowe, "Holographic lactate sensor," *Anal. Chem.*, 78 5664–70 (2006).
7. D. Bhatta, G. Christie, B. Madrigal-Gonzalez, J. Blyth, and C. R. Lowe, "Holographic sensors for the detection of bacterial spores," *Biosens. Bioelectron.*, 23, 520–7 (2007).
8. D. R. Wuest and R. S. Lakes, "Color control in reflection holograms by humidity," *Appl. Opt.*, 30, 2363–7 (1991).
9. R. C. Spooncer, F. A. Al-Ramadhan and B. E. Jones, "A humidity sensor using a wavelength-dependent holographic filter with fibre optic links," *Int. J. Optoelectron.*, 7 449–52 (1992).
10. J. Shi, V. K. S. Hsiao, and T. J. Huang, "Nanoporous polymeric transmission gratings for high-speed humidity sensing," *Nanotechnology*, 18, 465501 (2007).
11. I. Naydenova, R. Jallapuram, V. Toal, and S. Martin, "A visual indication of environmental humidity using a colour changing hologram recorded in a self-developing photopolymer," *Appl. Phys. Lett.*, 92, 031109 (2008).
12. I. Naydenova, J. Raghavendra, V. Toal, and S. Martin, "Characterisation of the humidity and temperature responses of a reflection hologram recorded in an acrylamide-based photopolymer," *Sensor Actuat B: Chem*, 139, 35–8 (2009).
13. Y. Tomita, K. Chikama, Y. Nohara, N. Suzuki, K. Furushima, and Y. Endoh, "Two-dimensional imaging of atomic distribution morphology created by holographically induced mass transfer of monomer molecules and nanoparticles in a silica-nanoparticle-dispersed photopolymer film," *Opt. Lett.*, 31, 1402–4 (2006).

14. N. Suzuki and Y. Tomita, "Real-time phase-shift measurement during formation of a volume holographic grating in nanoparticle-dispersed photopolymers," *Appl. Phys. Lett.*, 88, 011105 (2006).

15. N. Suzuki, Y. Tomita, and T. Kojima, "Holographic recording in TiO$_2$ nanoparticle-dispersed methacrylate photopolymer films," *Appl. Phys. Lett.*, 81, 4121–3 (2002).

16. C. Sanchez, M. Escuti, C. Heesh, C. Bastiaansen, D. Broer, J. Loos, and R. Nussbaumer, "TiO$_2$ nanoparticle-photopolymer composites for volume holographic recording," *Adv. Funct. Mat.*, 15, 1623–9 (2005).

17. N. Suzuki and Y. Tomita, "Silica-nanoparticle-dispersed methacrylate photopolymers with net diffraction efficiency near 100%," *Appl. Opt.*, 43, 2125–9 (2004).

18. N. Suzuki, Y. Tomita, K. Ohmori, M. Hidaka, and K. Chikama, "Highly transparent ZrO$_2$ nanoparticle-dispersed acrylate photopolymers for volume holographic recording," *Opt. Express*, 14, 12712–9 (2006).

19. *Ordered Porous Solids: Recent Advances and Prospects*, V. Valtchev, S. Mintova, and M. Tsapatsis (eds.), Elsevier (2008).

20. E. Leite, Tz. Babeva, E.-P. Ng, V. Toal, S. Minatova, and I. Naydenova, "Optical properties of photopolymer layers doped with aluminophosphate nanocrystals," *J. Phys. Chem. C*, 2010, 114, 16767–75 (2010).

21. V. Hsiao, W. Kirkey, F. Chen, A. Cartwright, P. Prasad, and T. Bunning, "Organic solvent vapor detection using holographic photopolymer reflection gratings," *Adv. Mater.*, 17, 2211–4 (2005).

22. E. Leite, I. Naydenova, S. Mintova, L. Leclerq, and V. Toal, "Photopolymerisable nanocomposites for holographic recording and sensor application," *Appl. Opt.*, 49, 3652–60 (2010).

23. I. Naydenova, S. Martin, and V. Toal, "Photopolymerisation: beyond the standard approach to sensitization," *J. Eur. Opt. Soc.: Rapid Publications*, 4, 09042 (2009).

24. P. Macko and M. Whelan, "Aberration-free lithography setup for fabrication of holographic diffractive elements," *Opt. Lett.*, 34, 3006–9 (2009).

PROBLEMS

1. An unslanted reflection holographic diffraction grating is to be used as a color-changing strain gauge, the strain being applied in the plane of the hologram in the direction of its long axis. The grating is recorded at a wavelength of 500 nm in a layer of material, which is 5 cm long, 100 μm thick, and 1 cm in width and whose Young's modulus of elasticity is 10^8 Pa. Find the change in Bragg peak wavelength per Newton of applied force assuming the ratio of vertical to horizontal strain is 0.3.

2. If the grating in problem 1 is placed in a pressure chamber and the pressure altered by 5×10^3 N m^{-2}, what is the change in Bragg peak wavelength?

3. Find an expression for the change in slant angle of a volume transmission grating as a function of pressure applied normally to its surface. A holographic transmission grating is recorded in a photopolymer using coherent light beams whose angles of incidence are 49° and 0°. The value of Young's modulus for the photopolymer is 10^8. If the grating is subjected to a pressure increase of 500 atm, what is the change in the Bragg angle?

4. Analysis of a transmission holographic grating recorded in a layer of zeolite-doped photopolymer shows that the zeolite particles have been redistributed with a spatial modulation of 15%. Suppose the zeolite particles, which are initially empty, have a selective affinity for toluene, whose refractive index is 1.5, what would be the percentage change in diffraction efficiency if the grating were immersed in toluene and then allowed to dry.

5. A photosensitizing dye is deposited on a photopolymer layer using an inkjet printer whose spatial resolution is 250 l mm^{-1}. The printed pattern is then exposed to ordinary light. Find the shortest focal length of lens of diameter 1 cm, which could be fabricated in this way.

The Fresnel–Kirchoff Integral

Our task here is to find the intensity distribution in a plane beyond the region where diffraction has occurred. To do this, we start with a monochromatic wave

$$\tilde{\mathbf{E}}(x, y, z)\exp(-j\omega t),$$

which, when substituted in the wave equation (Equation 1.9)

$$\nabla^2\tilde{\mathbf{E}}\exp(-j\omega t) = \frac{1}{c^2}\frac{\partial^2\tilde{\mathbf{E}}\exp(-j\omega t)}{\partial t^2}$$

results in Helmholtz's equation

$$\nabla^2\tilde{\mathbf{E}} + k^2\tilde{\mathbf{E}} = 0 \tag{A.1}$$

The function $\exp(jkr')/r'$ also satisfies Helmholtz's equation since

$$\nabla^2\left(\frac{e^{jkr'}}{r'}\right) + k^2\left(\frac{e^{jkr'}}{r'}\right) = 0 \tag{A.2}$$

Hence

$$\iiint_V\left\{\tilde{\mathbf{E}}\nabla^2\left(\frac{e^{jkr'}}{r'}\right) - \left(\frac{e^{jkr'}}{r'}\right)\nabla^2\tilde{\mathbf{E}}\right\}dV = \iiint_V\left\{-k^2\left(\frac{e^{jkr'}}{r'}\right)\tilde{\mathbf{E}} + \left(\frac{e^{jkr'}}{r'}\right)k^2\tilde{\mathbf{E}}\right\}dV = 0$$

By Green's theorem

$$\iiint_V\left\{\tilde{\mathbf{E}}\nabla^2\left(\frac{e^{jkr'}}{r'}\right) - \left(\frac{e^{jkr'}}{r'}\right)\nabla^2\tilde{\mathbf{E}}\right\}dV = \iint_S\left\{\tilde{\mathbf{E}}\nabla\left(\frac{e^{jkr'}}{r'}\right) - \frac{e^{jkr'}}{r'}\nabla\tilde{\mathbf{E}}\right\}\cdot d\mathbf{S}$$

so $$\iint_S\left\{\tilde{\mathbf{E}}\nabla\left(\frac{e^{jkr'}}{r'}\right) - \frac{e^{jkr'}}{r'}\nabla\tilde{\mathbf{E}}\right\}\cdot d\mathbf{S} = 0 \tag{A.3}$$

The calculation of the left-hand integral over a surface, S_1, enclosing a point P from which r' is measured, involves the difficulty that, as $r' \rightarrow 0$, some terms become infinite. We therefore exclude P by enclosing it in a second, spherical, surface S_A (Figure A.1).

On the surface S_A, the gradient is directed away from P while the unit vector normal is towards P and

$$\nabla\left(\frac{e^{jkr'}}{r'}\right) = -\left\{-\frac{1}{r'^2} + \frac{jk}{r'}\right\}e^{jkr'}\mathbf{n}$$

$$\iint_{S_2}\left\{\tilde{\mathbf{E}}\nabla\left(\frac{e^{jkr'}}{r'}\right) - \frac{e^{jkr'}}{r'}\nabla\tilde{\mathbf{E}}\right\}\cdot d\mathbf{S} = \iint_{S_2}\left\{\frac{\tilde{\mathbf{E}}e^{jkr'}}{r'^2} - \frac{\tilde{\mathbf{E}}jke^{jkr'}}{r'} - \frac{e^{jkr'}}{r'}\nabla\tilde{\mathbf{E}}\right\}r'^2 d\Omega$$

with $d\Omega$ the solid angle subtended by the surface element $d\mathbf{S}$ at P.

As the radius of S_A goes to zero, the second and third terms in the integrand become zero and $e^{jkr'} = 1$. Thus, the surface integral becomes $4\pi\tilde{\mathbf{E}}_P$ where $\tilde{\mathbf{E}}_P$ is $\tilde{\mathbf{E}}$ at the point P.

So, on the surface S_1

$$\iint_{S_1}\left\{\tilde{\mathbf{E}}\nabla\left(\frac{e^{jkr'}}{r'}\right) - \frac{e^{jkr'}}{r'}\nabla\tilde{\mathbf{E}}\right\}\cdot d\mathbf{S} = 4\pi\tilde{\mathbf{E}}_P$$

$$\tilde{\mathbf{E}}_P = \frac{1}{4\pi}\iint_{S_1}\left\{\tilde{\mathbf{E}}\nabla\left(\frac{e^{jkr'}}{r'}\right) - \frac{e^{jkr'}}{r'}\nabla\tilde{\mathbf{E}}\right\}\cdot d\mathbf{S} \qquad (A.4)$$

This equation is Kirchoff's integral theorem and it allows us to calculate the amplitude $\tilde{\mathbf{E}}_P$ at a point P, if we know $\tilde{\mathbf{E}}$ and its gradient everywhere on a surface enclosing P.

For example, if a point source emits spherical wavefronts of complex amplitude

$$\tilde{\mathbf{E}}(r,t) = \frac{E_0}{r}e^{j(kr - \omega t)}$$

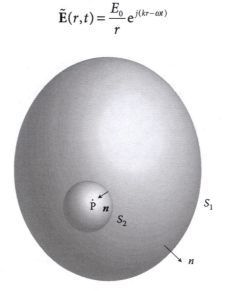

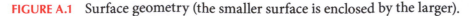

FIGURE A.1 Surface geometry (the smaller surface is enclosed by the larger).

then, omitting time dependence for now,

$$\tilde{\mathbf{E}}_{\mathrm{P}} = \frac{1}{4\pi}\iint_S \left\{ \frac{E_0}{r}e^{jkr}\frac{\partial\left(\frac{e^{jkr'}}{r'}\right)}{\partial r'} \right\}(-\hat{r}')\cdot d\mathbf{S} - \frac{1}{4\pi}\iint_S \left\{ \frac{e^{jkr'}}{r'}\frac{\partial\left(\frac{E_0}{r}e^{jkr}\right)}{\partial r} \right\}(-\hat{r})\cdot d\mathbf{S}$$

$$= \frac{E_0}{4\pi}\iint_S \left\{ -\frac{e^{jkr}}{r}e^{jkr'}\left(\frac{jk}{r'} - \frac{1}{r'^2}\right)\hat{r}'\cdot d\mathbf{S} + \frac{e^{jkr'}}{r'}e^{jkr}\left(\frac{jk}{r} - \frac{1}{r^2}\right)\hat{r}\cdot d\mathbf{S} \right\} \qquad (A.5)$$

The terms in this equation involving $1/r'^2$ and $1/r^2$ are neglected because λ is extremely small compared with either r' or r so that finally

$$\tilde{\mathbf{E}}_{\mathrm{P}} = -\frac{E_0 j}{2\lambda}\iint_S \frac{e^{jk(r+r')}}{rr'}(\hat{\mathbf{r}}' - \hat{\mathbf{r}})\cdot d\mathbf{S} \qquad (A.6)$$

This is the Fresnel–Kirchoff diffraction formula and it enables us to calculate the complex amplitude of the light at a point, P, if we know the amplitude everywhere on a surface S that encloses the point.

Again, to avoid the possibility that some terms may become infinite, we surround the point P by a sphere (Figure A.2) whose radius will be reduced to zero and $\hat{\mathbf{r}}\cdot d\mathbf{S}$ becomes $\cos\pi$ while $\hat{\mathbf{r}}'\cdot d\mathbf{S}$ is $\cos\theta$.

$$\tilde{\mathbf{E}}_{\mathrm{P}} = \frac{E_0 e^{-j\left(\omega t+\frac{\pi}{2}\right)}}{2\lambda}\iint_S \frac{\cos\theta+1}{rr'}e^{j[k(r+r')]}d\mathbf{S} \qquad (A.7)$$

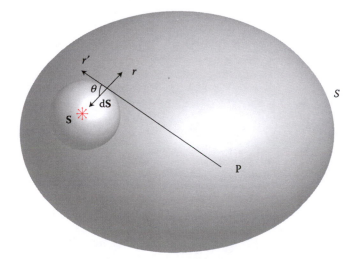

FIGURE A.2 Effect at point P of a light source at S (the smaller surface enclosed by the larger).

We have restored the time dependence. Equation A.7 can be used to find the optical disturbance at a point P by summing (integrating) all of the contributions made to that disturbance from a surface, S, enclosing P.

Equation A.7 is a very important mathematical statement of Huyghens' principle, which is that every point (or vanishingly small element dS) on a wavefront is a source of secondary spherical wave fronts and the subsequent form of the original wavefront at any later time is the surface that is tangential to the secondary wavefronts at that time. The term $(\cos\theta +1)/2$ is called the obliquity factor and it ensures that the secondary wavefronts do not produce a wave traveling back towards the source since $(\cos\theta +1)/2 = 0$ when $\theta = \pi$.

We now assume that light from a source illuminates a plane x,y where diffraction takes place and we want to determine the disturbance at some point in a plane $x',\ y'$ beyond the diffraction plane x,y. The x,y plane is taken to be infinite in extent. We assume that there can be no contribution to the disturbance at P from points immediately to the right of opaque parts of the x,y plane. We also set up a hemispherical surface S, which completely encloses the point P (see Figure A.3) and which is of such large radius that there is no contribution from it to the disturbance at P. This means that only those regions of the x,y plane, which do not obscure the light wavefront from the source, can contribute to the disturbance at P.

We now assume the source is sufficiently far away from the $x,\ y$ plane so that r is effectively constant over $x,\ y$.

Similarly, the $x,\ y$ and $x',\ y'$ planes are assumed sufficiently far apart so that r' in the denominator of the integrand in Equation A.7 can be taken as a constant (Figure A.3b). This assumption does not allow us to regard $e^{jkr'}$ as a constant because $k = 2\pi/\lambda$ and λ is

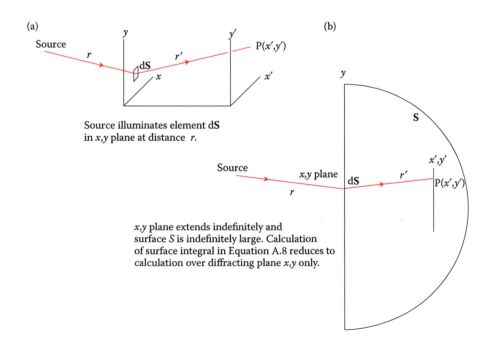

(a)

Source illuminates element d**S** in x,y plane at distance r.

(b)

x,y plane extends indefinitely and surface S is indefinitely large. Calculation of surface integral in Equation A.8 reduces to calculation over diffracting plane x,y only.

FIGURE A.3 Equivalence of integral over surface, S, and integral over diffracting (x,y) plane.

very small and even small variations in r' mean that kr' varies significantly. Finally, the term θ is also assumed to be very small so we have

$$\tilde{\mathbf{E}}_P = \frac{E_0 e^{j\left(kr - \omega t - \frac{\pi}{2}\right)}}{\lambda r r'} \iint_S e^{jkr'} d\mathbf{S} \tag{A.8}$$

All of these assumptions are collectively known as the Fraunhofer diffraction conditions.

The Convolution Theorem

The 2-D Fourier transforms of functions $f(x, y)$ and $g(x, y)$ are, respectively,

$$F(x', y') = \int_{-\infty}^{\infty} \int_{-\infty}^{\infty} f(x, y) \exp\left[-\frac{jk}{f}(xx' + yy')\right] dx dy$$

and

$$G(x'y') = \int_{-\infty}^{\infty} \int_{-\infty}^{\infty} g(x, y) \exp\left[-\frac{jk}{f}(xx' + yy')\right] dx dy$$

(B.1)

The convolution $h(x, y)$ of $f(x, y)$ and $g(x, y)$ is given by

$$h(x, y) = f(x, y) \otimes g(x, y) = \int_{-\infty}^{\infty} \int_{-\infty}^{\infty} f(x'', y''')g(x - x'', y - y'') dx'' dy''$$

(B.2)

x'', y'' being dummy variables.

The Fourier transform of $h(x, y)$ is

$$H(x', y') = \int_{-\infty}^{\infty} \int_{-\infty}^{\infty} h(x, y) \exp\left[-\frac{jk}{f}(xx' + yy')\right] dx dy$$

(B.3)

$$H(x', y') = \int_{-\infty}^{\infty} \int_{-\infty}^{\infty} \int_{-\infty}^{\infty} \int_{-\infty}^{\infty} f(x'', y'')g(x - x'', y - y'') \exp\left[-\frac{jk}{f}(xx' + yy')\right] dx dy dx'' dy''$$

(B.4)

Letting $X = x - x''$, $Y = y - y''$

$$H(x', y') =$$

$$\int_{-\infty}^{\infty} \int_{-\infty}^{\infty} \int_{-\infty}^{\infty} \int_{-\infty}^{\infty} f(x'', y'')g(X, Y) \exp\left[-\frac{jk}{f}(Xx' + Yy')\right] \exp\left[-\frac{jk}{f}(x''x' + y''y')\right] dX dY dx'' dy''$$

(B.5)

$$H(x', y') = \left[\int_{-\infty}^{\infty} \int_{-\infty}^{\infty} f(x'', y'') \exp\left[-jk(x''x' + y''y')\right] dx'' dy''\right] G(x', y') = F(x', y')G(x', y')$$

(B.6)

Thus, the Fourier transform of the convolution of two functions is the product of their transforms. This is known as the convolution theorem.

The corollary is that the convolution of two functions is given by the Fourier transform of the product of their transforms.

The cross correlation of $f(x, y)$ and $g(x, y)$ is written as $f(x,y) \oplus g(x,y)$ and given by

$$f(x,y) \oplus g(x,y) = \int_{-\infty}^{\infty} \int_{-\infty}^{\infty} f^*(x'',y'')g(x+x'',y+y'')dx''dy'' \tag{B.7}$$

and, from Equation B.2

$$f(x,y) \oplus g(x,y) = f(x,y) \otimes g(-x,-y) \tag{B.8}$$

and the *autocorrelation* of $f(x,y)$ is

$$f(x,y) \oplus f(x,y) = \int_{-\infty}^{\infty} \int_{-\infty}^{\infty} f^*(x'',y'')f(x''+x,y''+y)dx''dy'' = f^*(x,y) \otimes f(-x,-y) \tag{B.9}$$

Applying the correlation theorem, we obtain

$$FT[f(x,y) \oplus f(x,y)] = |F(x'y')|^2 \tag{B.10}$$

which is the autocorrelation, or Weiner–Kinchin theorem.

INDEX